Environmental Science and Engineering

Series Editors

Ulrich Förstner, Buchholz, Germany

Wim H. Rulkens, Department of Environmental Technology, Wageningen, The Netherlands

The ultimate goal of this series is to contribute to the protection of our environment, which calls for both profound research and the ongoing development of solutions and measurements by experts in the field. Accordingly, the series promotes not only a deeper understanding of environmental processes and the evaluation of management strategies, but also design and technology aimed at improving environmental quality. Books focusing on the former are published in the subseries Environmental Science, those focusing on the latter in the subseries Environmental Engineering.

Nitish Kumar
Editor

Heavy Metal Toxicity

Human Health Impact and Mitigation Strategies

Springer

Editor
Nitish Kumar
Department of Biotechnology
Central University of South Bihar
Gaya, Bihar, India

ISSN 1863-5520 ISSN 1863-5539 (electronic)
Environmental Science and Engineering
ISBN 978-3-031-56641-7 ISBN 978-3-031-56642-4 (eBook)
https://doi.org/10.1007/978-3-031-56642-4

This Springer imprint is published by the registered company Springer Nature Switzerland AG
The registered company address is: Gewerbestrasse 11, 6330 Cham, Switzerland

Preface

Heavy metals, in general trace elements, are one of the major environmental problems. Nowadays, increasing environmental and global public health concerns related with environmental contamination by heavy metals are well known. Moreover, human exposure has risen dramatically because of an exponential increase of their use in several activities such as agricultural, industrial, technological, and urban applications. They are present in soils, water, and atmosphere and they are a serious risk for the food chain.

Approximately 10 million contaminated sites have been reported globally, occupying approximately 49.42 million acres of land, of which >50% is contaminated with toxic heavy metals. This situation warrants immediate attention to limit the introduction of heavy metals into soil systems and to remove the prevailing heavy metals from polluted soils. However, the majority of existing heavy-metal-removal technologies are expensive, inefficient, or generate secondary pollutants. Therefore, it is of great importance to develop cheaper, environmentally friendly, and sustainable approaches (including the development of new immobilizing agents) to manage and rehabilitate heavy-metal-contaminated soils.

In the light of the aforementioned facts, this book sheds light on this global environmental issue, and proposes solutions to contamination through multi-disciplinary approaches and case studies from different parts of the world. It addresses sustainable heavy metal contamination remediation strategies using the potential applications of recent biological technology such as biotechnology, bioremediation, phytoremediation, biochar, absorbent, genetic engineering, and nanotechnology approaches. This book is of interest to researchers, teachers, environmental scientists, environmental engineers, environmentalists, and policy makers. Also, the book serves as additional reading material for undergraduate and graduate students of environmental microbiology, biotechnology, eco-toxicology, environmental remediation, waste management, and environmental sciences as well as the general audience.

Gaya, Bihar, India | Nitish Kumar

Acknowledgments

First and foremost, I would like to praise and thank God, the Almighty, who has granted countless blessing, knowledge, and opportunity to accomplish the book project work. Thanks to all the authors of the various chapters for their contributions. It had been a bit of a long process from the initial outlines to developing the full chapters and then revising them in the light of reviewer's comments. I sincerely acknowledge the author's willingness to go through this process. I also acknowledge the work and knowledge of the members of our review panels, many of which had to be done at short notice. Thanks to all the people at Springer Nature, especially Dr. Christian Witschel and Mr. Chandra Sekaran Arjunan with whom I corresponded for their advice and facilitation in the production of this book. I am grateful to my family members Mrs. Kiran (Wife), Miss Kartika Sharma and Laavanya Sharma (Daughters), and parents for their incredible and selfless support all the time.

Gaya, Bihar, India Nitish Kumar

Contents

Chapter 1
Heavy Metal Toxicity and Their Effects on Environment

Varun Dhiman

Abstract To better understand the effects of heavy metal toxicity on the local flora and fauna and how these metals affect their health and the ecosystem, this article will analyze relevant literature. The use of heavy metals to pollute the environment has grown significantly in the last several decades. Since heavy metals are now more common in nature and can bioaccumulate and biomagnify in the appropriate food chain due to increased urbanization and industrial growth, exposure to these metals appears to be normal in the modern era. Many heavy metals that interact with the biotic ecosystem, like mercury, lead, cadmium, chromium, copper, zinc, and so forth, can cause major health problems, oxidative stress, and molecular abnormalities like DNA damage, lipid peroxidation, protein modification, cellular damage, apoptosis, and interference with electrolyte channels. Free radicals are produced by heavy metals and have a variety of negative effects on an organism's ability to perform its normal metabolic activities. We have outlined the harmful consequences of heavy metals on the environment in this article, which eventually have an impact on human health.

Keywords Environment · Environmental pollutants · Health impacts · Heavy metals · Toxicity

1.1 Introduction

A multitude of anthropogenic and natural activities have released loads of hazardous substances into the environment. Toxic contaminants are eventually introduced into various environmental domains by a few variables, including fast urbanization, industrial development at environmental expense, and technological breakthroughs. One of the most harmful environmental pollutants, heavy metals have toxic effects that prevent ecosystems from normal functioning (Dhiman and Pant 2022). From

V. Dhiman (✉)
Department of Environmental Science, Government College, Una 174303, Himachal Pradesh, India
e-mail: varundhiman79@gmail.com

N. Kumar (ed.), *Heavy Metal Toxicity*, Environmental Science and Engineering,
https://doi.org/10.1007/978-3-031-56642-4_1

a biochemical and biophysical perspective, metals are significant because of their critical roles in maintaining an organism's metabolism. However, the functions of heavy metals are incredibly diverse, and they can be harmful even at extremely low exposure levels. These metals have a specific weight of greater than 5 g/cm^3 and are extremely dense (Agarwal and Darbar 2023). The hazardous effects on the exposed flora and fauna of several environmental spheres have intensified in recent years as a result of increased exposure to these metals. It is recognized that heavy metals such as mercury, lead, cadmium, arsenic, aluminum, copper, nickel, and zinc can be found in wastewater, deteriorated soil, and the atmosphere. Their presence in the environment has detrimental effects on living things, including humans; promotes immune system diseases, mental dysfunction, aberrant hormonal changes, and birth deformities. These metals are present in trace amounts in a variety of goods, including cosmetics, pharmaceuticals, and pesticides used in agriculture (Arshad et al. 2020). Recent studies have demonstrated that heavy metals exhibit toxicity not just as an individual toxicant but also as part of a variety of toxic complexes, which have more damaging effects on the environment (Anyanwu et al. 2018). It is impossible to completely avoid exposure to heavy metals because they can enter the human body by various pathways besides typical ones, such as drinking water, food, drinks, and ambient exposure (Witkowska et al. 2021). On interactions within different environmental spheres, these metals undergo numerous transformations (Fig. 1.1).

The risk of exposure to heavy metals can be reduced, though, by adopting a healthy lifestyle and making investments in eco-friendly technology. However, different metal chelators are used for heavy metal chelation (Rong and Han 2009). They are target metal-oriented and remove heavy metal toxicity by forming numerous complexes (Table 1.1).

Because of their extreme persistence in the environment, heavy metals are unable to decompose and metabolize into other intermediary compounds (Dhiman et al. 2020). These metals move cyclically across different ecosystems. Humans are more susceptible to the toxicity of heavy metals as their gastrointestinal tract absorbs these

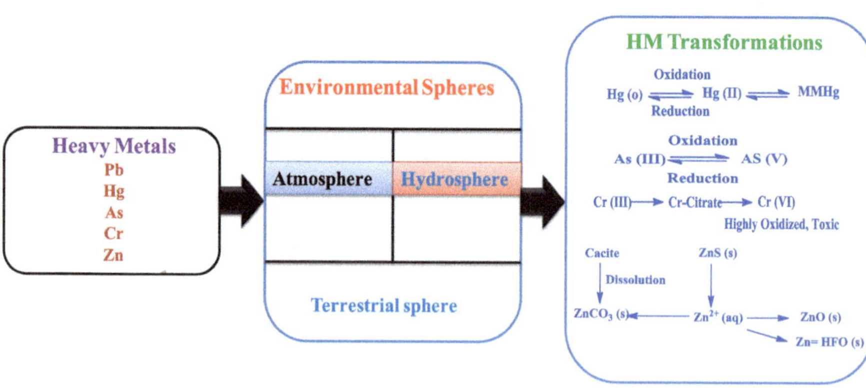

Fig. 1.1 Heavy metals transformations in different environmental spheres

Table 1.1 List of heavy metal chelators with their other details (Adopted from Rong and Han 2009)

Chelating agent	Common name	IUPAC name	Chemical structure	Molecular weight (g/mol)	Coordination groups	Activation metabolism	Target metal
Succimer (DMSA)	Chemet, Dim-sa	(2*R*,3*S*)-2,3-disulfanylbutanedioic acid	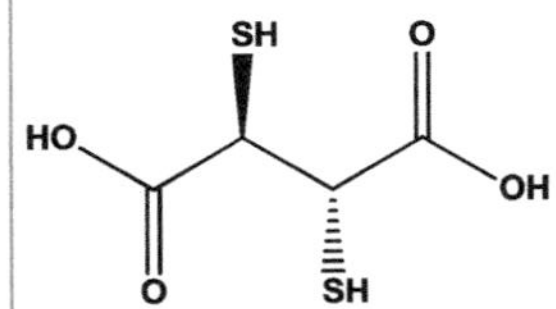	182.21	Carboxylic group, thiol group	Excretion via urine >90% as DMSA—cysteine disulfide conjugates	Lead, mercury, arsenic
Deferox-amine (DFOA)	Desferal	*N*′-{5-[Acetyl(hydroxy)amino]pentyl}-*N*-[5-({4-[(5-aminopentyl)(hydroxy)amino]-4-oxobutanoyl}amino)pentyl]-*N*-hydroxysuccinamide	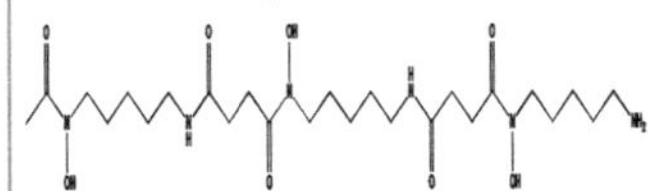	560.7	Bidendate hydroxamic groups	Excreted in the feces via the bile	Iron, aluminium
Editic acid (EDTA)	Versene, cheelox, calcium disodium versenate	2,2′,2′′,2′′′-(Ethane-1,2-diyldinitrilo)tetraacetic acid	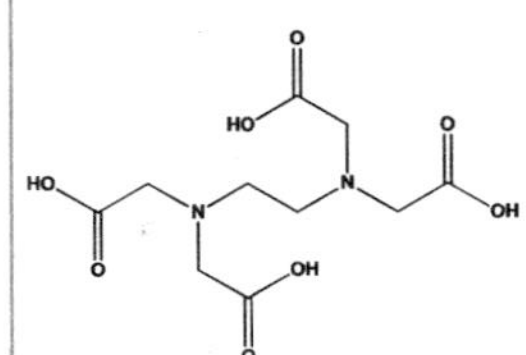	292.24	Amino and carboxylate	Not metabolized. Excreted unchanged, generally coordinated with a different divalent cation	Lead, iron

(continued)

Table 1.1 (continued)

Chelating agent	Common name	IUPAC name	Chemical structure	Molecular weight (g/mol)	Coordination groups	Activation metabolism	Target metal
Deferasirox	Oleptiss, asunra, exjade	4-(3,5-Bis(2-hydroxyphenyl)-1H-1,2,4-triazol-1-yl)benzoic acid	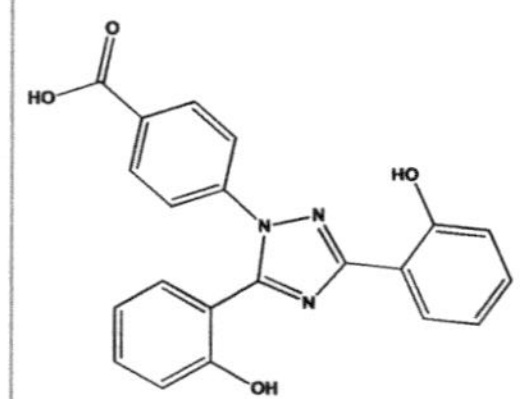	373.36	Triazolyl nitrogen, phenolic oxygen atoms	*Excreted largely* via *bile and feces* (total ≥90%)	Iron
DMPS	Unithiol, unitiol, dimaval	2,3-Bis(sulfanyl)propane-1-sulfonic acid	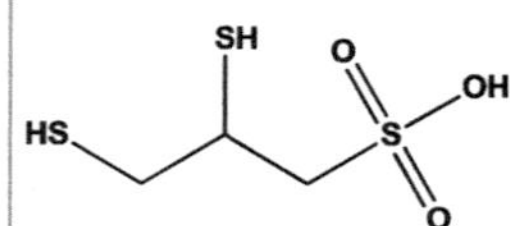	188.27	Oxygen and sulfhydry	84% of IV dose excreted through urine	Mercury
BAL (British anti-Lewisite)	Panobal, dithioglycerol, sulfactin	2,3-bis(sulfanyl)propan-1-ol	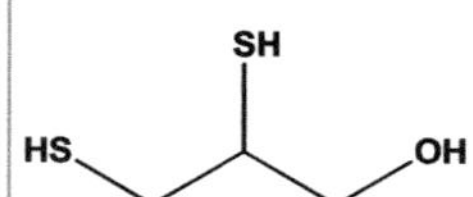	124.22	Sulfhydryl and hydroxyl	Excreted unchanged in urine	Mercury, lead, gold arsenic
Penicillamine	3-Mercapto-L-valine, L-penicillamine	(2*R*)-2-amino-3-methyl-3-sulfanylbutanoic acid	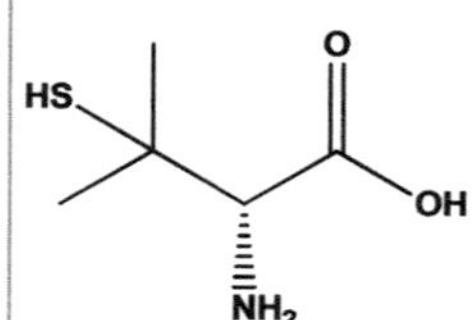	149.21	Thiol, amine and carboxylic acid	Rarely excreted unchanged; excreted mainly as disulfides	Copper, zinc, lead, mercury

metals from contaminated food and these metals circulate into the other vital organs of the body with the help of the circulatory system.

1.2 Occurrence and Influencing Factors

The presence of heavy metals in several environmental spheres such as soil, water, and air is a significant cause for concern. The terrestrial sphere, for instance, is a vital component for all living things and is severely contaminated by volcanic eruption, pedo-genetic rock weathering, and—more importantly—anthropogenic activities such as metal smelting, mineral mining, electroplating, coal burning, effluent discharge, dry deposition, oil well drilling, incineration, steel manufacturing, and vehicle emissions. Because of its innate ability to absorb metals, soil serves as an effective sink for heavy metal adsorption (Dhiman et al. 2022). Moreover, one apparent major source of heavy metal contamination appears to be the atmospheric dry deposition caused by vehicle pollution (Ali et al. 2019). One of the world's most populous countries, China, is particularly concerned about heavy metal contamination of its soil. Due to its extensive exploration of new mineral resource locations, mining is a major industry in the area. The MEP study, 2014 states that because of the region's high mining activity, 33.4% of mining sites had heavy metal concentrations over the threshold limit (Zhong et al. 2020). Similarly, a variety of heavy metals are present in drinking water (Singh et al. 2022). Among the most common heavy metals added into water systems are arsenic, cadmium, lead, and mercury. One of the well-known examples of heavy metal concentrations that are known to be present in wastewater is that of dissolved and particulate forms, including Cu, Cr, Ni, Pb, Mn, Fe, Cd, and Zn. Additional research validates the existence of heavy metals in the atmosphere. Because of its greater amount of metal-to-particulate-matter binding potential, atmospheric particulate matter is a cause for concern in this situation. Due to fine particulate matter's larger surface area, heavy metals have been shown to adsorb on its surface (Popoola et al. 2018). Through the process of dry deposition, these metals have infiltrated the terrestrial environment and have since bio-accumulated in both plants and human tissue (He et al. 2023). According to studies, there are several avenues for heavy metals to migrate and change in the environment. These pathways are greatly influenced by things like pH levels and other variables. Below is a discussion of a few of these influential factors.

1.2.1 pH Conditions

Various research works validate the significance of ambient pH levels in preserving the leaching and bioavailability of heavy metals. The mobility and pH dependency of heavy metals released from metallurgical slag are evaluated in a study by Król et al. (2020). They discovered that the pH conditions had a significant impact on heavy metal leaching. The leaching and mobility of heavy metals are significantly impacted by acidic pH conditions, according to research that conducts batch leaching experiments and sequential extraction. Analyzing the behavior of Cd, Ni, and Cu released from river sediment observed that the alkali water solution facilitates the leaching of these metals into the river environment (Zhang et al. 2018).

1.2.2 Mean Annual Precipitation (MAP)

The discharge of heavy metals and their stability in the environment are likewise related to the MAP. The investigations attest to precipitation's part in the environmental release of metals. Unconfined compressive strength (UCS) and toxicity characteristic leaching procedures (TCLP) are two methods used in studies to determine its function. It has been noted that acidic rain precipitation influences heavy metals' tendency to leach and encourages their mobility in the appropriate environment (Wang et al. 2021).

1.2.3 Temperature

Many research investigations have demonstrated the impact of temperature fluctuations on the environmental mobilization of heavy metals. The impact of temperature on the leaching capacity of heavy metals is validated by these investigations. The cumulative leaching and lead leaching concentration were observed to rise as the temperature rose, according to a recent study by Wang et al. (2022) on the influence of temperature on the leaching behavior of lead. Semi-dynamic leaching test is used in this particular study as a means of confirmation (Wang et al. 2022).

1.2.4 Curing Agent

Understanding the role of the curing agent is also beneficial in comprehending the mechanism of leaching and bioavailability of heavy metals. Research indicates that as the concentration of curing agents increases, the leaching concentration of heavy metals such as lead, cadmium, and copper decreases. This is because the curing agent forms a chemical connection with the leached metals, facilitating additional adsorption in an alkaline environment (Li et al. 2021).

1.3 Toxicokinetics and Toxicodynamics

Both the toxicokinetics and toxicodynamics of heavy metals are crucial components that aid in our comprehension of kinetics as a whole and the associated impacts of a certain heavy metal or toxicant exposed to living things. Understanding the toxicokinetics of a chemical requires an understanding of its absorption, distribution, metabolism, and excretion routes inside the organism's body (Shanmugam et al. 2023). Furthermore, with the aid of toxicodynamics research, it is crucial to comprehend the generated consequences, cellular toxicity, and diverse processes by which any toxicant might modify the body's regular metabolism (Srivastava et al. 2022). Various heavy metals and other toxicants interact with the body of the organism at the cellular level, resulting in a variety of dysfunctions and abnormalities. In toxicodynamics, chemical proximity and reactivity are the two key variables (Dellafiora et al. 2018), whereas in toxicokinetics, the physiochemical nature and biological makeup play a larger role in determining the outcome (Locher 2020). Lead, for example, has strong binding properties to the carboxyl and sulfhydryl groups of many structural and functional proteins in the body when discussing its toxicokinetics (Goering 1993). By acting antagonistically, it will replenish the calcium and further disrupt the body's calcium pathway (Rădulescu and Lundgren 2019). Its thermodynamic analysis demonstrates that it modifies the structure and alters the integrity of intercellular junctions and biochemical alterations in renal proximal tubule cells (Navarro-Moreno et al. 2009). It is easily absorbed in the skin, GI tract, and pulmonary system (Wani et al. 2016).

1.4 Toxicity of Heavy Metals and Their Environmental Effects

1.4.1 Lead

On ecosystems, wildlife, and human health, lead exposure can have several negative effects. The usage of lead-based products (such as leaded gasoline and lead-based paints) and mining operations are the common ways that humans might release the harmful heavy metal lead into the environment. Lead can build up in the soil, particularly in the vicinity of industrial zones, former building sites, and highways. Plants and soil microbes may suffer as a result of this contamination. Lead-contaminated soil can damage terrestrial ecosystems overall, impede plant growth, and interfere with nutrient intake (Collin et al. 2022). It can come from polluted soil and leach into both surface and groundwater. It can be a serious hazard to aquatic life in aquatic habitats. Lead can have a domino impact on the food chain by preventing aquatic plants, insects, and other species from growing and reproducing (Kumar et al. 2020). Lead exposure can occur in birds, animals, and other species through contaminated food and water sources. Animals with lead poisoning may develop many medical conditions, including weaker immune systems, difficulties reproducing, and brain impairment. The health effects of lead exposure are seriously concerning. Various health concerns, particularly in children, such as behavioral disorders, learning challenges, and developmental delays, might result from it (Flora et al. 2006). Adult exposure to lead can result in high blood pressure, renal damage, and problems in conceiving (Hu et al. 2007). Contaminated water, soil, and dust are major causes of exposure since lead may enter the human body through ingestion, inhalation, and skin contact. Burning leaded gasoline, incinerating garbage and other industrial activities can all emit lead into the atmosphere. Even though many nations have phased out leaded gasoline, lead pollution in metropolitan areas has persisted due to past use (World Health Organization 2023). Lead airborne particle inhalation can cause respiratory troubles as well as other health problems (Yasmeen and Hafeez 2023). Consumer goods may also be contaminated with lead, particularly antique jewelry, paints, and toys (Gul et al. 2023). Individuals who come into contact with these goods may be at risk for health problems. Numerous studies have demonstrated that lead exposure, particularly in young children, can cause behavioral issues, lowered IQ, and impaired cognitive development (Ruebner et al. 2019). The "Rochester Lead-in-Dust Study," which focused on the link between lead exposure and lowered IQ in children, is one of the noteworthy studies. Figure 1.2 clearly shows the numerous toxic effects of some major heavy metals.

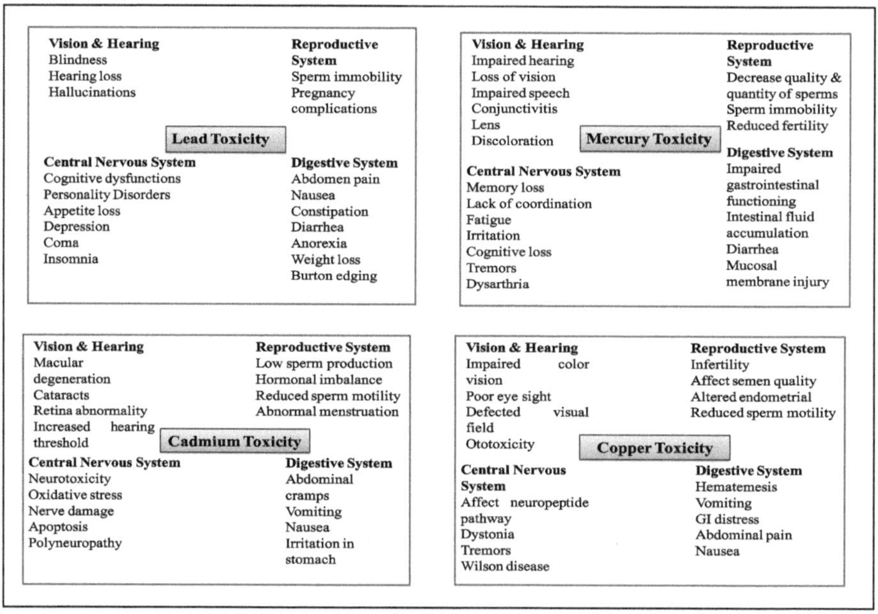

Fig. 1.2 Numerous toxic effects of some major heavy metals

1.4.2 Mercury

Mercury is a very hazardous and persistent environmental contaminant that can harm both humans and ecosystems in a variety of ways (Driscoll et al. 2013). In aquatic ecosystems, mercury can accumulate in living things and move upwards in the food chain. This is particularly dangerous when it comes to the highly hazardous organic form of mercury, methylmercury (Chiang et al. 2021). Larger fish at the top of the food chain can bioaccumulate extremely high levels of methylmercury, and small aquatic species like plankton and fish can absorb and concentrate this toxin from the water. Aquatic ecosystems may be disrupted by elevated mercury levels in bodies of water. The health and survival of fish, invertebrates, and other aquatic life can be negatively impacted by mercury's effects on the development and reproduction of aquatic animals (de Almeida Rodrigues et al. 2019). Eating mercury-contaminated fish and shellfish, particularly methylmercury, can be extremely harmful to a person's health. Mercury exposure can cause neurological and developmental issues, especially in developing fetuses and young infants (Dack et al. 2022). For this reason, several fish species are not advised to be consumed, especially by youngsters and pregnant women.

1.4.3 Cadmium

The negative consequences of being exposed to the heavy metal cadmium are referred to as cadmium poisoning. Although it is a naturally occurring metal in the crust of the Earth, human activities including mining, industrial processes, and the use of items containing cadmium also release this element into the environment (Suhani et al. 2021). Even at low concentrations, prolonged exposure to cadmium can have detrimental effects on health (Rahimzadeh et al. 2017). Through a variety of channels, including agricultural runoff, industrial discharges, and atmospheric deposition, cadmium can infiltrate aquatic bodies. It can build up in aquatic life and sediments. Water with high cadmium concentrations can harm aquatic life and disturb ecosystems (Ray 1984). Additionally, cadmium can pollute soil due to industrial operations, sewage sludge application, and application of phosphate fertilizers (Krishnamurti et al. 2005). Cadmium is persistent in the environment due to its great affinity for soil particles (Dutta et al. 2020). Plants can absorb it and ingest it into the food chain. Due to its lengthy half-life in the environment, cadmium can linger in an area and cause harm for a long time. Because of its tenacity, it is very difficult to control (Genchi et al. 2020).

1.4.4 Arsenic

Because exposure to this naturally occurring element might pose health concerns, environmental arsenic poisoning is a serious concern (Kapp 2018). There are many different types of arsenic, but the most harmful ones are inorganic arsenic compounds (Mandal and Suzuki 2002). Naturally occurring in the Earth's crust, arsenic can be released into the environment through rock weathering and volcanic activity, among other natural processes (Masuda 2018). The usage of insecticides containing arsenic and other human activities like mining and smelting can also lead to environmental arsenic contamination (Garelick et al. 2008). There are several ways that arsenic can enter the human body, but the most common one is through contaminated food and water. In some work environments, breathing in dust or fumes containing arsenic may also be dangerous. Both people and animals can have major health consequences from arsenic exposure. Chronic exposure to drinking water or food contaminated with low concentrations of arsenic can cause a variety of health difficulties, such as skin lesions, malignancies (especially skin, lung, and bladder cancers), cardiovascular disorders, and developmental abnormalities in children (Villaescusa and Bollinger 2008). Table 1.2 thoroughly describes different sources, their coordination complexes, the pollution index, health and environmental impacts of major heavy metals.

1.4.5 Aluminum

The third most prevalent metal in the Earth's crust, aluminum is a naturally occurring element. Although aluminum is not intrinsically hazardous to the environment, it can have some unfavorable consequences if it is present in large amounts in some forms or situations. Aluminum can contaminate water and soil, particularly in acidic environments (Sorenson et al. 1974). Aluminum ions become more soluble in acidic conditions and can be harmful to aquatic life, including fish and invertebrates, by obstructing their ability to breathe and absorb nutrients (Gensemer and Playle 1999). The soil can get contaminated with aluminum in places where there is a high concentration of the metal, such as bauxite mining sites (Pradhan et al. 2023). Local ecosystems may be impacted by the contamination's effects on plant growth and nutrient availability for plants. Some plants are susceptible to excessive soil levels of aluminum. Elevated levels of aluminum in plants can impede their ability to form roots and absorb nutrients, which can result in lower agricultural yields and poorer vegetative health overall (Zhao and Shen 2018).

Table 1.2 Different sources, their coordination complexes, pollution index, health and environmental impacts of major heavy metals

Heavy metals	Permissible limit in water	Pollution index	HM sources	Metal co-ordination complexes	Environmental/health impacts	References
Lead	0.01–0.015 mg L^{-1}	7.95	• Lead-based paints • Lead dust • Metal smelting • Polluted drinking • Water and air, • Folk medicines, • Toys and jewelry, • Lead-glazed ceramics	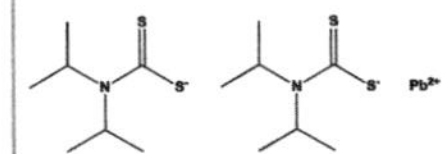 lead(II) N,N-diisopropyldithiocarbamate 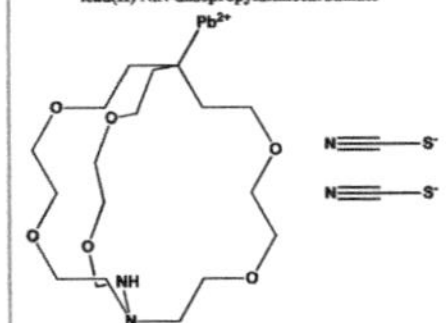 (4,7,13,16,21,24-hexaoxa-1,10-diazabicyclo[8.8.8]hexacosane) lead(II) thiocyanate 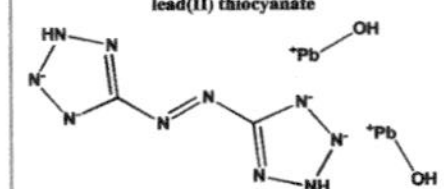 bis[hydroxolead(II)] 5,5'-azotetrazolediide	• Biodiversity loss • Altered reproduction of plants and animals • Renal dysfunction • Anemia • Hypertension • Neuronal damage • Hearing and speech impairment	Gidlow (2004)

(continued)

Table 1.2 (continued)

Heavy metals	Permissible limit in water	Pollution index	HM sources	Metal co-ordination complexes	Environmental/health impacts	References
Mercury	0.001–0.002 mg L^{-1}	–	• Mining • dry deposition • Electronic devices • Paint • Coal • Metal ores	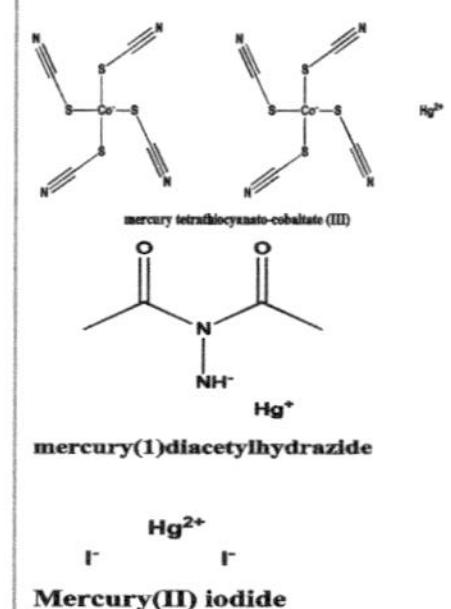	• Weaken immunity • Altered enzymatic functioning • Distorted coordination and senses • Altered nervous, digestive system • Behavioral disorders	Rice et al. (2014)
Chromium	0.05–0.1 mg L^{-1}	2.24	• Ferro chromate refractory material • Chromium steel • Oil well drilling • Metal plating • Combustion • Electroplating	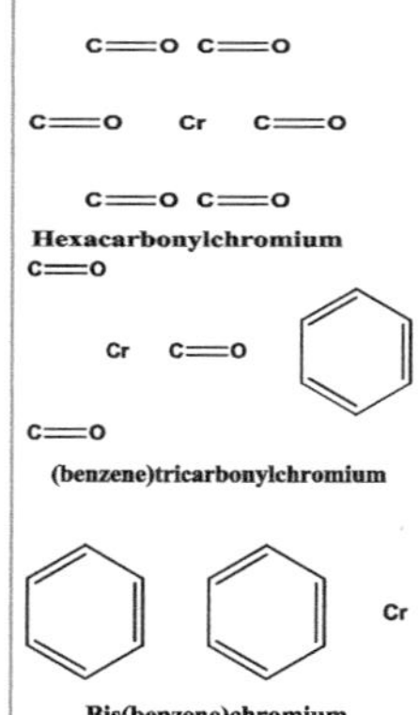	• Affect plant metabolic activities • Hampering plant growth • DNA damage • Lethality of liver and kidney • Inhibition of enzymatic functioning • Oxidative stress • Reduced grain quality	Guertin (2004)

(continued)

Table 1.2 (continued)

Heavy metals	Permissible limit in water	Pollution index	HM sources	Metal co-ordination complexes	Environmental/health impacts	References
Cadmium	0.005 mg L^{-1}	3.81	• Smelters • Nickel–cadmium batteries • Fossil fuel burning • Incineration	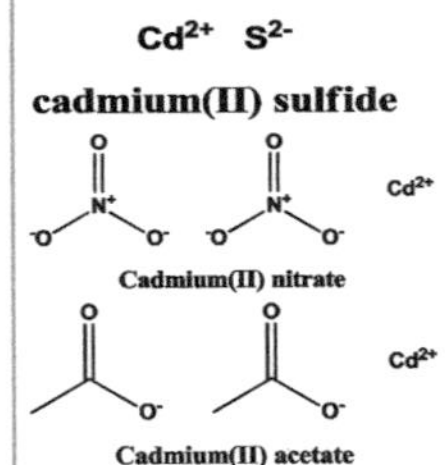	• Carcinogenic activity • Oxidative stress causes liver and kidney diseases • In vivo and in vitro epigenetic changes in cells • Peripheral artery disease • Neurodegenerative diseases	Friberg (2018)
Copper	0.05–1.5 mg L^{-1}	0.03	• Oil refineries • Pesticides • Chemical industries • Coal burning • Combustion By-products • Metal pipes	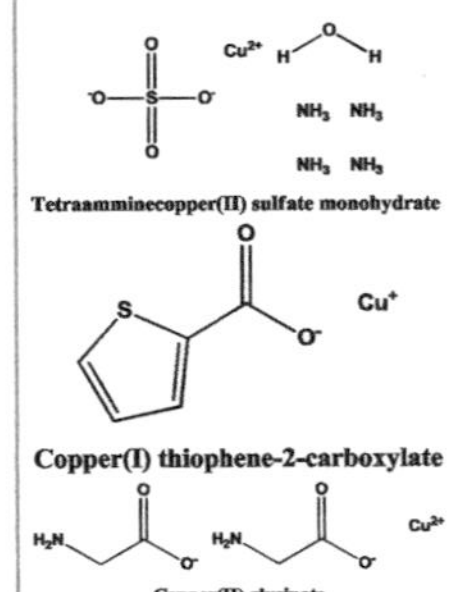	• Disrupt metabolic processes of marine organisms • Inhibition of nutrient mineralization • Its exposure causes nervous system damage, kidney, irritation of the respiratory tract	Flemming and Trevors (1989)

(continued)

Table 1.2 (continued)

Heavy metals	Permissible limit in water	Pollution index	HM sources	Metal co-ordination complexes	Environmental/health impacts	References
Zinc	5 ppm	0.04	• Steel manufacturing • Coal burning • Mining activities • Metal smelting process • Emissions from wastewater treatment plants	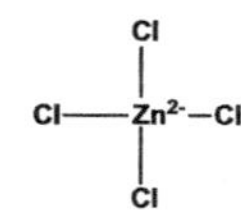 **Tetrachlorozincate** 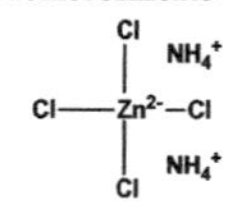 **ammonium tetrachlorozincate** 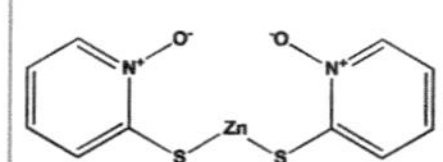 **bis(2-pyridylthio)zinc 1,1'-dioxide**	• High concentration reduces microbial diversity • Phytotoxic to plants • Metabolism dysregulation • Disturbed homeostasis of biological systems • Causes health concerns like chest pain, breathing distress, gastrointestinal problems, epigastric pain, etc.	Li et al. (2019)
Arsenic	0.01–0.05 mg L^{-1}	3.35	• Ores extraction • Rock sediments • Groundwater • Mining • Ore smelting	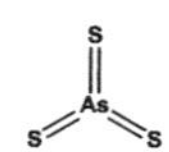 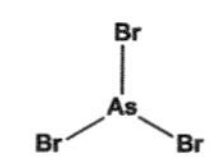 **Arsenic trisulfide** **Tribromoarsane** 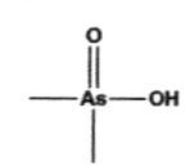 **Dimethylarsinic acid**	• Induces neurological disorders • Cause numerous types of cancers • Arsenic poisoning induces anorexia, reflex loss, weakness, skin lesions, etc.	Hughes (2002)

(continued)

Table 1.2 (continued)

Heavy metals	Permissible limit in water	Pollution index	HM sources	Metal co-ordination complexes	Environmental/health impacts	References
Aluminum	0.2 mg L^{-1}	–	• Cooking utensils • Packaging material • Coal-fired power plants • Smelters	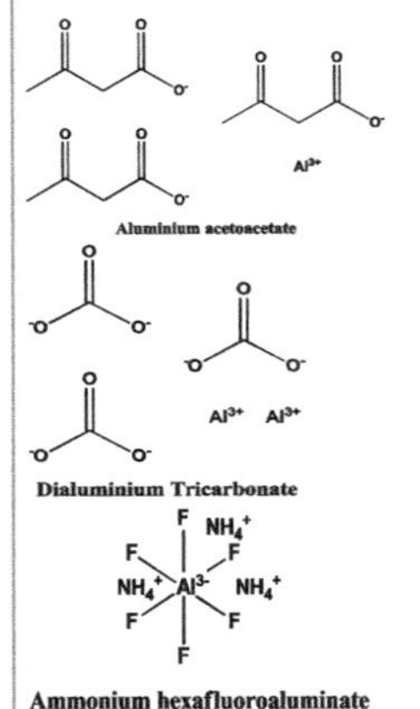	• Disturbed osmoregulation in animals and humans • Act as a neurotoxin • Its excess exposure is linked to Alzheimer's disease, epilepsy, and autism • Premature bone mineralization • Impaired renal activity	Becaria et al. (2002)
Nickel	0.1 mg L^{-1}	1.37	• Coal and oil-burning • Incineration • Metal manufacturing waste • Coal fly ash	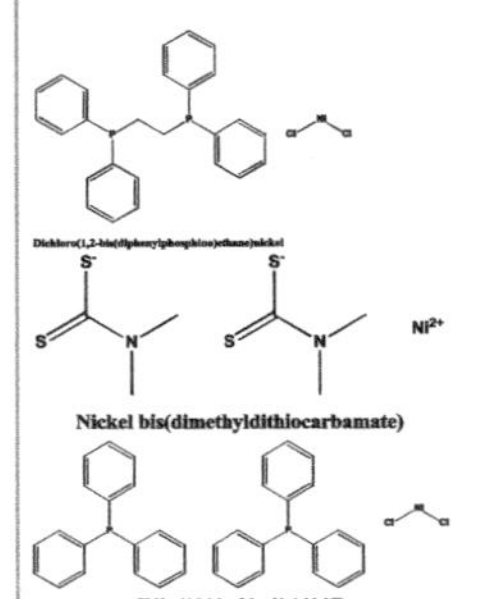	• Long-term exposure causes cardiovascular and kidney disease, lung fibrosis, nasal cancer, etc. • Oxidative stress, mitochondrial dysfunction at the molecular level • Chloroplast structure distorted due to Nickel toxicity	Denkhaus and Salnikow (2002)

(continued)

Table 1.2 (continued)

Heavy metals	Permissible limit in water	Pollution index	HM sources	Metal co-ordination complexes	Environmental/health impacts	References
Iron	0.5 mg L^{-1}	5.55	• Soil and weathered rocks • Heavy fuel oil • Hard coal • Well underground water	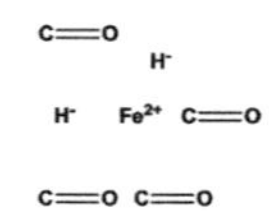 Iron tetracarbonyl dihydride 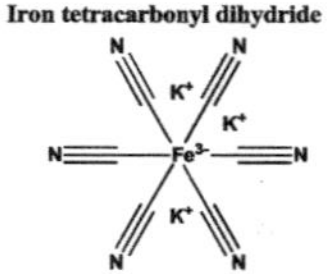 Potassium hexacyanoferrate(III) 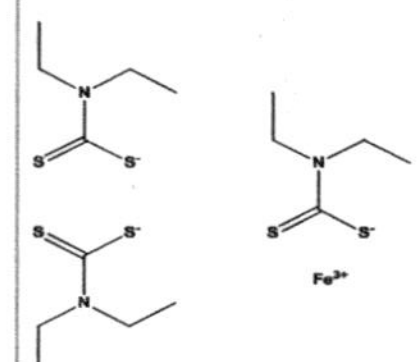 Iron tris(diethyldithiocarbamate)	• Higher concentration causes metabolic, physiological, and morphological alterations in plants • Damage plant cell membrane, affect yield, and reduce growth of plants • Abdominal pain and nausea, vomiting in humans due to iron poisoning • The generation of free radicals induces stress, resulting in apoptosis	Papanikolaou and Pantopoulos (2005)

(continued)

Table 1.2 (continued)

Heavy metals	Permissible limit in water	Pollution index	HM sources	Metal co-ordination complexes	Environmental/health impacts	References
Manganese	0.05 mg L^{-1}	2.17	• Foundries • Fossil fuel combustion • Sewage sludge • Municipal waste • Mineral processing • Rocks	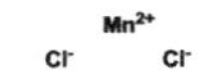 Manganese(II) chloride 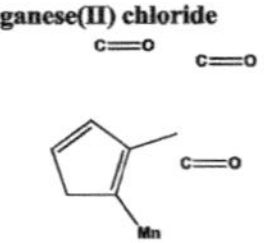 2-Methylcyclopentadienyl manganese tricarbonyl 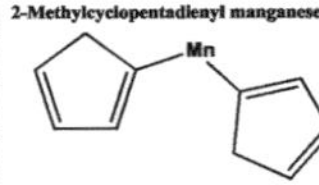 Bis(cyclopentadienyl)manganese	• Its exposure inhibits the dopaminergic neuronal enzyme activity • Promote infant mortality rate • Induces hepatic and cardio-toxicity • Induce psychiatric abnormalities • Induces cytotoxic effects	O'Neal and Zheng (2015)

(continued)

Table 1.2 (continued)

Heavy metals	Permissible limit in water	Pollution index	HM sources	Metal co-ordination complexes	Environmental/health impacts	References
Cobalt	1–2 ppb	–	• Nickel refining process • Hard coal-fired power plants • Processing of copper-containing ores	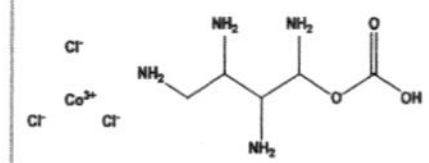 carbonatbis(ethylenediamine)cobalt(III) chloride 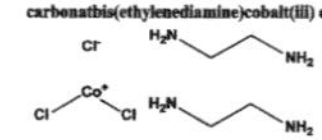 Cis-Dichlorobis(ethylenediamine)cobalt(III) chloride 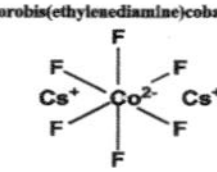 **Caesium hexafluorocobaltate(IV)**	• Reduces antioxidant enzymatic activities • Enhance oxidative stress • Lung hemorrhage, respiratory irritation on inhalation • Liver disorders, pneumonia, fibrosis-like diseases • Sublethal effects on the growth of certain plants like maize	Leyssens et al. (2017)
Beryllium	0.004 mg L^{-1}	–	• Coal and oil-burning • Volcanic dust • Industrial manufacturing • Incineration of municipal solid waste	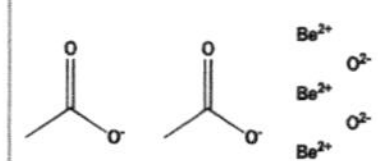 Beryllium oxide acetate 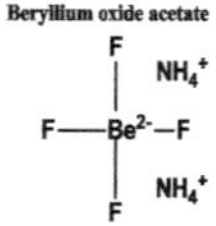 **Ammonium tetrafluoroberyllate** 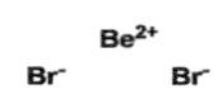 **Beryllium bromide**	• Inhalation leads to carcinogenicity in humans • Retarded plant growth and yield • Occupational exposure causes acute beryllium pneumonitis, Granulomatous lung disease in humans	Cooper and Harrison (2009)

1.5 Conclusion

The toxicity and environmental impacts of many heavy metals, such as lead, mercury, chromium, cadmium, copper, zinc, arsenic, aluminum, nickel, iron, manganese, cobalt, and beryllium, were reviewed in this article with particular attention to human health. Even at ppb levels, it has been noted that prolonged exposure to them will cause major health risks as well as several environmental hazards. The only ways to prevent their exposure are to live a green lifestyle and use green technologies. Additionally, heavy metals biomonitoring appears to be a very effective way to reduce occupational exposure. Because heavy metals are highly persistent, they can easily travel in different environmental spheres and eventually interact with living things, imposing adverse effects on them in the form of altered physiological and biological activities. This article discusses various environmental and health concerns of living flora and fauna along with the most recent heavy metal reference studies. As such, it could be a valuable resource for both the general public and the scientific community. It has been determined that the only ways to effectively address the toxicity of heavy metals are to employ environmentally friendly methods and create affordable biomonitoring and remediation protocols.

Acknowledgements The author is grateful to the Govt. College, Una, Himachal Pradesh for providing necessary facilities for this work.

Disclosure Statement No potential conflict of interest was reported by the author.

Data Availability Statement Data sharing is applicable to this article.

References

Agarwal S, Darbar S (2023) Microorganism assisted synthesized metal and metal oxide nanoparticles for removal of heavy metal ions from the wastewater effluents. In: Singh P, Kumar V, Bakshi M, Hussain CM, Sillanpää M (eds) Environmental applications of microbial nanotechnology. Elsevier, pp 127–148. https://doi.org/10.1016/B978-0-323-91744-5.00017-5

Ali H, Khan E, Ilahi I (2019) Environmental chemistry and ecotoxicology of hazardous heavy metals: environmental persistence, toxicity, and bioaccumulation. J Chem 6730305:14. https://doi.org/10.1155/2019/6730305

Anyanwu BO, Ezejiofor AN, Igweze ZN, Orisakwe OE (2018) Heavy metal mixture exposure and effects in developing nations: an update. Toxics 6. https://doi.org/10.3390/toxics6040065

Arshad H, Mehmood MZ, Shah MH, Abbasi AM (2020) Evaluation of heavy metals in cosmetic products and their health risk assessment. Saudi Pharm J 28:779–790. https://doi.org/10.1016/j.jsps.2020.05.006

Becaria A, Campbell A, Bondy SC (2002) Aluminum as a toxicant. Toxicol Ind Health (SAGE Publications Ltd STM) 18:309–320. https://doi.org/10.1191/0748233702th157oa

Chiang G, Kidd KA, Díaz-Jaramillo M, Espejo W, Bahamonde P, O'Driscoll NJ, Munkittrick KR (2021) Methylmercury biomagnification in coastal aquatic food webs from western Patagonia and western Antarctic Peninsula. Chemosphere 262:128360. https://doi.org/10.1016/j.chemosphere.2020.128360

Collin S, Baskar A, Geevarghese DM, Ali MNVS, Bahubali P, Choudhary R, Lvov V et al (2022) Bioaccumulation of lead (Pb) and its effects in plants: a review. J Hazard Mater Lett 3:100064. https://doi.org/10.1016/j.hazl.2022.100064

Cooper RG, Harrison AP (2009) The uses and adverse effects of beryllium on health. Indian J Occup Environ Med (Wolters Kluwer–Medknow Publications) 13:65

Dack K, Fell M, Taylor CM, Havdahl A, Lewis SJ (2022) Prenatal mercury exposure and neurodevelopment up to the age of 5 years: a systematic review. Int J Environ Res Public Health 19. https://doi.org/10.3390/ijerph19041976

de Almeida Rodrigues P, Ferrari RG, dos Santos LN, Conte Junior CA (2019) Mercury in aquatic fauna contamination: a systematic review on its dynamics and potential health risks. J Environ Sci 84:205–218. https://doi.org/10.1016/j.jes.2019.02.018

Dellafiora L, Dall'Asta C, Galaverna G (2018) Toxicodynamics of mycotoxins in the framework of food risk assessment—an in silico perspective. Toxins 10. https://doi.org/10.3390/toxins10020052

Denkhaus E, Salnikow K (2002) Nickel essentiality, toxicity, and carcinogenicity. Crit Rev Oncol Hematol 42:35–56. https://doi.org/10.1016/S1040-8428(01)00214-1

Dhiman V, Pant D (2022) π-π interaction: defining the role and relevance in environmental detoxification of heavy metals from soil. In: Shit PK, Adhikary PP, Bhunia GS, Sengupta D (eds) Soil health and environmental sustainability: application of geospatial technology. Springer International Publishing, Cham, pp 659–672. https://doi.org/10.1007/978-3-031-09270-1_29.

Dhiman V, Pant D, Kumari S, Kumar S (2020) Recent advances in novel remediation processes towards heavy metals removal from wastewaters. In: Advances in environmental pollution management: wastewater impacts and treatment technologies. Agro Environ Media, pp 77–99

Dhiman V, Giri A, Pant D (2022) Pollutants bioremediation using biosurfactants: a novel approach for improving soil health. In: Shit PK, Adhikary PP, Bhunia GS, Sengupta D (eds) Soil health and environmental sustainability: application of geospatial technology. Springer International Publishing, Cham, pp 489–502. https://doi.org/10.1007/978-3-031-09270-1_21.

Driscoll CT, Mason RP, Chan HM, Jacob DJ, Pirrone N (2013) Mercury as a global pollutant: sources, pathways, and effects. Environ Sci Technol (American Chemical Society) 47:4967–4983. https://doi.org/10.1021/es305071v

Dutta A, Patra A, Jatav HS, Jatav SS, Singh SK, Sathyanarayana E, Verma S, Singh P (2020) Toxicity of cadmium in soil-plant-human continuum and its bioremediation techniques. In: Larramendy ML, Soloneski S (eds) Soil contamination, Ch. 4. IntechOpen, Rijeka. https://doi.org/10.5772/intechopen.94307

Flemming CA, Trevors JT (1989) Copper toxicity and chemistry in the environment: a review. Water Air Soil Pollut 44:143–158. https://doi.org/10.1007/BF00228784

Flora SJS, Flora G, Saxena G (2006) Environmental occurrence, health effects and management of lead poisoning. In: Casas JS, Sordo J (eds) Lead. Elsevier Science B.V., Amsterdam, pp 158–228. https://doi.org/10.1016/B978-044452945-9/50004-X.

Friberg L (2018) Cadmium in the environment. CRC Press

Garelick H, Jones H, Dybowska A, Valsami-Jones E (2008) Arsenic pollution sources. In: Reviews of environmental contamination Volume 197: International perspectives on arsenic pollution and remediation. Springer New York, New York, NY, pp 17–60. https://doi.org/10.1007/978-0-387-79284-2_2

Genchi G, Sinicropi MS, Lauria G, Carocci A, Catalano A (2020) The effects of cadmium toxicity. Int J Environ Res Public Health 17. MDPI: 3782. 32466586. PubMed. https://doi.org/10.3390/ijerph17113782

Gensemer RW, Playle RC (1999) The bioavailability and toxicity of aluminum in aquatic environments. Crit Rev Environ Sci Technol (Taylor & Francis) 29:315–450. https://doi.org/10.1080/10643389991259245

Gidlow DA (2004) Lead toxicity. Occup Med 54:76–81. https://doi.org/10.1093/occmed/kqh019

Goering PL (1993) Lead-protein interactions as a basis for lead toxicity. Neurotoxicology 14:45–60

Guertin J (2004) Toxicity and health effects of chromium (all oxidation states). In: Chromium (VI) handbook. CRC Press Boca Raton, FL, pp 215–234

Gul A, Gul D-e-S, Mohiuddin S (2023) Metals as toxicants in event-based expedited production of children's jewellery. Environ Sci Pollut Res 30:73964–73973. https://doi.org/10.1007/s11356-023-27620-y

He L, Wang S, Liu M, Chen Z, Xu J, Dong Y (2023) Transport and transformation of atmospheric metals in ecosystems: a review. J Hazard Mater Adv 9:100218. https://doi.org/10.1016/j.hazadv.2022.100218

Hu H, Shih R, Rothenberg S, Schwartz BS (2007) The epidemiology of lead toxicity in adults: measuring dose and consideration of other methodologic issues. Environ Health Perspect (National Institute of Environmental Health Sciences) 115:455–462

Hughes MF (2002) Arsenic toxicity and potential mechanisms of action. Toxicol Lett 133:1–16. https://doi.org/10.1016/S0378-4274(02)00084-X

Kapp RW (2018) Arsenic toxicology☆. In: Reference module in biomedical sciences. Elsevier. https://doi.org/10.1016/B978-0-12-801238-3.62151-7

Krishnamurti GSR, McArthur DFE, Wang MK, Kozak LM, Huang PM (2005) Biogeochemistry of soil cadmium and the impact on terrestrial food chain contamination. In: Huang PM, Gobran GR (eds) Biogeochemistry of trace elements in the rhizosphere. Elsevier, Amsterdam, 197–257. https://doi.org/10.1016/B978-044451997-9/50009-X.

Król A, Mizerna K, Bożym M (2020) An assessment of pH-dependent release and mobility of heavy metals from metallurgical slag. J Hazard Mater 384:121502. https://doi.org/10.1016/j.jhazmat.2019.121502

Kumar A, Kumar A, Cabral-Pinto MMS, Chaturvedi AK, Shabnam AA, Subrahmanyam G, Mondal R et al (2020) Lead toxicity: health hazards, influence on food chain, and sustainable remediation approaches. Int J Environ Res Public Health 17. https://doi.org/10.3390/ijerph17072179

Leyssens L, Vinck B, Van Der Straeten C, Wuyts F, Maes L (2017) Cobalt toxicity in humans—a review of the potential sources and systemic health effects. Toxicology 387:43–56. https://doi.org/10.1016/j.tox.2017.05.015

Li XF, Wang PF, Feng CL, Liu DQ, Chen JK, Wu FC (2019) Acute toxicity and hazardous concentrations of zinc to native freshwater organisms under different pH values in China. Bull Environ Contam Toxicol 103:120–126. https://doi.org/10.1007/s00128-018-2441-2

Li X, Yang R, Li H, Yi H, Jing H (2021) Experimental study on solidification and stabilization of heavy-metal-contaminated soil using cementitious materials. Materials 14. https://doi.org/10.3390/ma14174999

Locher M (2020) Characterization of physicochemical parameters in toxicology. In: Reichl F-X, Schwenk M (eds) Regulatory toxicology. Springer Berlin Heidelberg, Berlin, Heidelberg, pp 1–8. https://doi.org/10.1007/978-3-642-36206-4_30-2

Mandal BK, Suzuki KT (2002) Arsenic round the world: a review. Talanta 58:201–235. https://doi.org/10.1016/S0039-9140(02)00268-0

Masuda H (2018) Arsenic cycling in the Earth's crust and hydrosphere: interaction between naturally occurring arsenic and human activities. Prog Earth Planet Sci 5:68. https://doi.org/10.1186/s40645-018-0224-3

Navarro-Moreno LG, Quintanar-Escorza MA, González S, Mondragón R, Cerbón-Solorzáno J, Valdés J, Calderón-Salinas JV (2009) Effects of lead intoxication on intercellular junctions and biochemical alterations of the renal proximal tubule cells. Toxicol in Vitro 23:1298–1304. https://doi.org/10.1016/j.tiv.2009.07.020

O'Neal SL, Zheng W (2015) Manganese toxicity upon overexposure: a decade in review. Curr Environ Health Rep 2:315–328. https://doi.org/10.1007/s40572-015-0056-x

Papanikolaou G, Pantopoulos K (2005) Iron metabolism and toxicity. Toxicol Appl Pharmacol 202:199–211. https://doi.org/10.1016/j.taap.2004.06.021

Popoola LT, Adebanjo SA, Adeoye BK (2018) Assessment of atmospheric particulate matter and heavy metals: a critical review. Int J Environ Sci Technol 15:935–948. https://doi.org/10.1007/s13762-017-1454-4

Pradhan G, Tripathy B, Ram DK, Digal AK, Das AP (2023) Bauxite mining waste pollution and its sustainable management through bioremediation. Geomicrobiol J (Taylor & Francis) 1–10. https://doi.org/10.1080/01490451.2023.2235353

Rădulescu A, Lundgren S (2019) A pharmacokinetic model of lead absorption and calcium competitive dynamics. Sci Rep 9:14225. https://doi.org/10.1038/s41598-019-50654-7

Rahimzadeh MR, Rahimzadeh MR, Kazemi S, Moghadamnia A (2017) Cadmium toxicity and treatment: an update. Caspian J Intern Med 8:135 (Babol University of Medical Sciences)

Ray S (1984) Bioaccumulation of cadmium in marine organisms. Experientia 40:14–23. https://doi.org/10.1007/BF01959097

Rice KM, Walker Jr EM, Wu M, Gillette C, Blough ER (2014) Environmental mercury and its toxic effects. J Prevent Med Public Health 47:74 (Korean Society for Preventive Medicine)

Rong G, Han D (2009) Chelation therapy for mercury poisoning. Biochem Insights 2. https://doi.org/10.4137/BCI.S3036

Ruebner RL, Hooper SR, Parrish C, Furth SL, Fadrowski JJ (2019) Environmental lead exposure is associated with neurocognitive dysfunction in children with chronic kidney disease. Pediatr Nephrol 34:2371–2379. https://doi.org/10.1007/s00467-019-04306-7

Shanmugam PST, Sampath T, Jagadeeswaran I, Bhalerao VP, Thamizharasan S, Krithaksha V, Saha J (2023) Toxicokinetics. In: Shanmugam PST, Sampath T, Jagadeeswaran I (eds) Biocompatibility protocols for medical devices and materials. Academic Press, pp 175–186. https://doi.org/10.1016/B978-0-323-91952-4.00012-X

Singh A, Sharma A, Verma RK, Chopade RL, Pandit PP, Nagar V, Aseri V et al (2022) Heavy metal contamination of water and their toxic effect on living organisms. In: Dorta DJ, de Oliveira DP (eds) The toxicity of environmental pollutants, Ch. 2. IntechOpen, Rijeka. https://doi.org/10.5772/intechopen.105075

Sorenson John RJ, Campbell Irene R, Tepper Lloyd B, Lingg Robert D (1974) Aluminum in the environment and human health. Environ Health Perspect 8:3–95. https://doi.org/10.1289/ehp.7483

Srivastava AK, Kumar D, Singh D, Singh RK (2022) Toxicokinetics and toxicodynamics of xenobiotics in cancer development. In: Srivastava AK, Kumar D, Singh D, Singh RK (eds) Xenobiotics in chemical carcinogenesis. Academic Press, pp 61–82. https://doi.org/10.1016/B978-0-323-90560-2.00009-1.

Suhani I, Sahab S, Srivastava V, Singh RP (2021) Impact of cadmium pollution on food safety and human health. Curr Opin Toxicol 27:1–7. https://doi.org/10.1016/j.cotox.2021.04.004

Villaescusa I, Bollinger J-C (2008) Arsenic in drinking water: sources, occurrence and health effects (a review). Rev Environ Sci Bio/technol 7:307–323. https://doi.org/10.1007/s11157-008-9138-7

Wang P, Liu X, Zeng G, Ma J, Xia F (2022) Effects of temperature on the leaching behavior of Pb from cement stabilization/solidification-treated contaminated soil. Separations 9. https://doi.org/10.3390/separations9120402

Wang Y, Tang J, Li Z (2021) Strength and leaching behavior of heavy metal contaminated sludge solidified/stabilized by compound binders. Environ Res 197:111053. https://doi.org/10.1016/j.envres.2021.111053

Wani AL, Ara A, Usmani JA (2016) Lead toxicity: a review. Interdiscip Toxicol 8:55–64. https://doi.org/10.1515/intox-2015-0009

Witkowska D, Słowik J, Chilicka K (2021) Heavy metals and human health: possible exposure pathways and the competition for protein binding sites. Molecules 26. https://doi.org/10.3390/molecules26196060

World Health Organization (2023) Preventing disease through healthy environments: exposure to lead: a major public health concern. World Health Organization

Yasmeen R, Hafeez F (2023) Effect of particulate emissions on the respiratory system of workers belonging to different industries: an overview. Arch Respir Res 2

Zhang Y, Zhang H, Zhang Z, Liu C, Sun C, Zhang W, Marhaba T (2018) pH effect on heavy metal release from a polluted sediment. Edited by George Z. Kyzas. J Chem (Hindawi) 7597640. https://doi.org/10.1155/2018/7597640.

Zhao XQ, Shen RF (2018) Aluminum–nitrogen interactions in the soil–plant system. Front Plant Sci (Frontiers Media SA) 9:807
Zhong X, Chen Z, Li Y, Ding K, Liu W, Ye, Liu, Yuan Y et al (2020) Factors influencing heavy metal availability and risk assessment of soils at typical metal mines in Eastern China. J Hazard Mater 400:123289. https://doi.org/10.1016/j.jhazmat.2020.123289

Chapter 2
Heavy Metals Dynamics in the Soil-Plant Continuum

Senad Murtić

Abstract Soil is a biologically active and physically unique and dynamic medium that has developed mainly as a result of pedogenic processes through the weathering of rocks and sediments. It consists of mineral and organic constituents, water, air and living organisms and serves among others as a medium for plant growth. Unfortunately, in the past few decades, the increase in anthropogenic activities due to rapid industrialization and modern agricultural practices has drastically increased soil pollution with heavy metals, thereby reducing soil fertility. The most commonly used indicator for soil heavy metal pollution is heavy metal bioavailability. Bioavailability of heavy metals in soils and thus their toxicity depends on numerous factors including heavy metal speciation, soil chemical and physical properties, physicochemical and biological processes that control heavy metal transformation from one form to another, as well as the soil-plant-microbe interactions. In this light, a better understanding of the patterns and driving factors of plant-soil-microbe interactions as well as the physicochemical processes that influence the behavior of heavy metals in the soils such as complexation, cation exchange, specific adsorption/desorption reactions and precipitation can help to clarify the processes that determine the fate of heavy metals in soil-plant continuum. Consequently, this knowledge would greatly contribute to the remediation of heavy metal polluted soils.

Keywords Bioavailability · Heavy metals · Soil-plant interactions · Remediation

2.1 Introduction

Any economic development, especially those related to rapid industrialization and increased use of agricultural soils, inevitably leads to an increase in the standard of living of the population, but at the same time leads to soil pollution and degradation. Soil pollution with undesirable substances, among which heavy metals occupy a

S. Murtić (✉)
Department of Plant Physiology, Faculty of Agriculture and Food Science, University of Sarajevo, Sarajevo, Bosnia and Herzegovina
e-mail: murticsenad@hotmail.com

N. Kumar (ed.), *Heavy Metal Toxicity*, Environmental Science and Engineering,
https://doi.org/10.1007/978-3-031-56642-4_2

leading position, negatively affects food safety by both reducing soil's ability to yield food and by causing crops produced from polluted soils to be unsafe for animal and human consumption. The importance and seriousness of this issue is also reflected in the fact that soil pollution by heavy metals adversely affects the quality of water resources, habitats and biodiversity, causing irreparable damage to entire ecosystems. In this regard, there is urgency in controlling the heavy metal soil pollution in order to minimize further damage (Chen et al. 2015).

Heavy metals are generally defined as elements which possess a specific density of more than 4 g/cm^3 and are able to exert adverse effects on the environment and living organisms (Järup 2003). Cadmium (Cd), lead (Pb), chromium (Cr), mercury (Hg), and arsenic (As) are among the most harmful heavy metals in the environment. They are considered as systemic toxicants because of their adverse effects on human health as well as plants and animals, even at very low concentration (Tchounwou et al. 2012). Some heavy metals such as manganese (Mn), copper (Cu) and zinc (Zn) are essential nutrients for plant growth; however, they can also be toxic at concentrations above the values required for their function (Angulo-Bejarano et al. 2021).

The presence of heavy metals in the soil can be both of geogenic and anthropogenic origin. Geogenic sources include weathering of heavy metal-bearing rocks and sediments, while anthropogenic sources include various industrial and agricultural activities such as mining, smelting, release of industrial waste and intensive use of mineral fertilizers and pesticides (Alsafran et al. 2023).

The most commonly used indicator for soil heavy metal pollution is their total content. However it should be noted that heavy metal bioavailability rather than total heavy metal content determines the adverse effects of heavy metals. Namely, the high total heavy metal content in the soil does not necessarily mean its negative effect on the plant, if they are not available to the plant (Sulaiman and Hamzah 2018). In addition, seasonal and spatial variations of heavy metals in the soil cannot be ignored in assessing the degree of heavy metal soil pollution. In this sense, the level of heavy metals in soils is expected to be higher in the dry season than in the wet season due to leaching of the heavy metal ions down the soil profile by rainfall (Zhai et al. 2022).

The availability of heavy metals in soils and consequently their toxicity is also closely related to heavy metal speciation i.e. the chemical form in which they occur in the soil. Heavy metals in soil can be found as free ions in soil solution, as structural components of the lattices of soil minerals as well as insoluble precipitates with other soil components (Qayyum et al. 2016). However, changes in environmental conditions can greatly affect the speciation of heavy metals across the solid and liquid phases, and thus their mobility and bioavailability (Caporale and Violante 2016). Furthermore, the bioavailability and thus toxicity of heavy metals in the soil strongly depends on soil properties which include, among others, soil pH, organic matter content, redox conditions and cation exchange capacity i.e. soil's ability to retain positively charged ions (Borch et al. 2010).

Considering all these facts, it is obvious that assessment of total heavy metal contents alone in soils cannot provide sufficient information about the degree of heavy metal soil pollution. In this light, using data of both total and available heavy metal forms and identifying the relationships between soil properties and the bioavailability of heavy metals can help to understand the impact of heavy metals on the plant growth and development as well as the dynamics of heavy metals in the soil-plant continuum (Kandziora-Ciupa et al. 2021).

It is worth mentioning that the heavy metal soil pollution affects not only the plant growth and development, but also the plant community composition. Namely, high levels of heavy metals in soils can drastically change plant community structure, indicating that the plant species composition and distribution are closely related to the degree of soil contamination with heavy metals (Woch et al. 2016). Many studies in this regard have shown that various plant species have various abilities to cope with heavy metals. Some plants may thrive well on soils polluted by heavy metals; however, some plants do not have that ability, which mainly depends on plant genotype and soil-plant-microbe interactions (Mapaure et al. 2011; Goh et al. 2013).

Soil-plant interactions refer to the biological, chemical, and physical effects exerted by soil on plant growth and development, and reciprocal effects of plants on soil physical and chemical properties as well as on soil microbial activity (Bararzadeh Ledari et al. 2020). Unfortunately, the soil plant interactions as well as heavy metals dynamics in the soil-plant continuum has not been studied enough, although it is well-known that plant-soil interactions are crucial for the growth and development of any plant species in both polluted and healthy soils. However, such studies are necessary within a context of the maintenance of soil health and the reduction of heavy metal soil pollution.

This study aimed to investigate the impact of heavy metals on the plant growth and development as well as the dynamics of heavy metals in the soil-plant continuum. This study also provides insight into the role of soil-plant interactions in plant response to heavy metal stress.

2.2 The Dynamics of Heavy Metals in the Soil-Plant Interactions

Plants are often considered as passive elements of the environment although they play a key role in shaping the environment. Among all multicellular organisms, only plants are able to create sugar and oxygen from carbon dioxide, water and sunlight through photosynthesis. Humans and all other living organisms, with a few exceptions such as some microorganisms, completely depend on plants as the primary source of both food and oxygen. Plants also greatly influence soil chemical and physical properties through various mechanisms, including root penetration, water extraction and root exudation. In addition, plants prevent soil erosion and provide organic matter that forms the basis for all microbial activity in the soil (Baluška and Mancuso 2020).

On the other hand, soils provide physical support, oxygen, water and nutrients to plants. Soils also are habitats for beneficial soil microbes that feed the plant by converting nutrients in the soil. Soil microbes can also improve significantly plant performance under stress conditions (Jacoby et al. 2017).

Although many aspects of soil are beneficial to plants, it can also restrict plant growth, notably through limiting root penetration into deeper soil layers. Besides, soil is a place for storing heavy metals and other toxic substances which can negatively affect plant growth and development (Alengebawy et al. 2021). It is worth mentioning that the interaction between plants and soil traits not only occur in natural environments, but also in agricultural ecosystems, significantly influencing food production, especially crop growth and productivity (Ehrmann and Ritz 2014). However, the relationships among soil properties and plants in both polluted and healthy soils have not been sufficiently explored. Better understanding of soil-plant interactions is required to improve sustainability of agricultural production systems especially in areas polluted by heavy metals.

Soil-plant interactions under the high content of heavy metals in the soil represent an ecological complexity that still needs to be elucidated. Namely, variability in soil properties, especially differences in soil pH and organic matter content, preclude the ability to definitively predict the behavior and fate of heavy metal ions in soil-plant interactions (Hou et al. 2019). This already complex situation is further complicated by differences in heavy metal uptake efficiencies among plant species. Nonetheless, a complete understanding of the factors that influence the behavior of heavy metals in the soils including the distribution of heavy metals between solid phase and soil solution can help to clarify the processes that determine the fate of heavy metals in soil-plant interactions.

Heavy metals exist in soils in several forms characterized by different availability (Chu et al. 2022). Available form of any heavy metal refers to the sum of the soluble amount in the liquid phase i.e. free ions and the amount retained by the solid phase that can be transferred to the liquid phase i.e. exchangeable form (Ram and Verloo 1985). Heavy metals in soil can also be found in several forms that are largely unavailable to plants such as heavy metals occluded in the lattices of soil minerals or insoluble precipitates with other soil components (Aydinalp and Marinova 2003).

The convert of heavy metals from unavailable forms to available or vice versa primarily depends on the physicochemical processes that control heavy metal transformation from one form to another such as complexation, adsorption–desorption processes, cation exchange, and precipitation (Manikandan et al. 2023).

Distribution and forms of heavy metals in soils can also be modified by biological processes i.e. root exudation and soil microbial activity. For example, the hydrogen ions secreted by plant roots can displace heavy metal cations adsorbed on soil particles, while some root exudates, particularly organic acids can form strong bonds with heavy metal ions, thus affecting their availability. In addition, root exudates provide various carbon and energy sources for soil microbes, which in turn reduce the harmful effects of heavy metals, notably through their immobilization in soils (Jiang et al. 2023).

Key processes influencing the distribution of heavy metals between the solid phase and soil solution and thus their bioavailability are summarized in Fig. 2.1.

Most scientists agree that adsorption–desorption reactions are the most important processes affecting the bioavailability of heavy metals in soils (Caporale and Violante 2016; Yu et al. 2023). These reactions are strongly influenced by the nature of soil

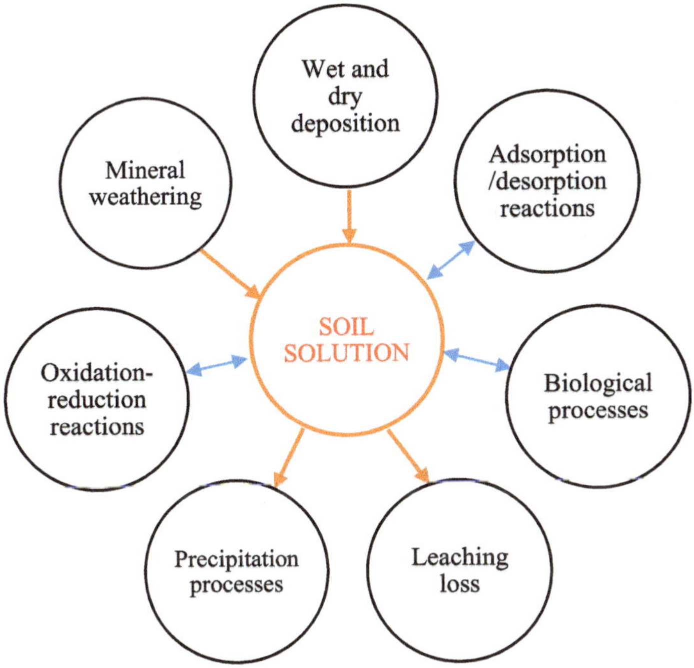

Fig. 2.1 Key processes affecting heavy metal distribution between the solid phase and soil solution

components, pH value and the composition of the soil solution. Soil components differ greatly in their ion-exchange capacities and their adsorption site energies. A common observation is that soil heavy metal adsorption increases with decreasing particle size (Madzhieva et al. 2014). This can be explained by the fact that smaller soil particles such as clay particles have a larger surface area which provides more sites for adsorption of heavy metal ions. Soil pH also strongly influences the adsorption–desorption reactions in the soils. In general, the ability of soil constituents to adsorb heavy metals decreases with increasing soil acidity due to competition between hydrogen ions and heavy metal ions for available charged sites on the adsorbent surface.

Other chemical processes that greatly affect the heavy metal bioavailability in the soils are oxidation–reduction (redox) reactions (Violante et al. 2010). Namely, most heavy metals can exist in soil in more than one oxidation state and, in this regard, the oxidation–reduction reactions play a crucial role in controlling their bioavailability. For instance, As is more bioavailable in their lower oxidation state i.e. reduced form (As-III), whereas Cr is more bioavailable and thus toxic in their oxidized form (Cr-VI). However, redox reactions are not uniform even in the same soil as they are continually influenced by soil chemical properties and soil microbial activity (Szabó et al. 2022).

Heavy metal precipitation is process that also strongly influences the behavior and fate of heavy metals in the soils (Biswas et al. 2018). It is a dynamic process that involves the conversion of soluble heavy metal ions into insoluble forms that will than precipitate. This process in the soil environment is influenced mainly by pH and heavy metal chemical speciation. It is well known that heavy metals at high pH tend to form insoluble salts such as heavy metal phosphates and carbonates, while at low pH they can be found mainly as free ionic species and are therefore more mobile and bioavailable. pH soil is also the key parameter affecting the migration of heavy metals from soil to groundwater. This process is known as leaching (Sun et al. 2022).

Soil organic matter is another soil chemical parameter that have a strong influence on the partition of heavy metals between the solid phase and soil solution and thus their bioavailability (Shi et al. 2018). Organic matter particles carry a large number of negative charges due to the presence of hydroxyl, carboxyl and phenolic groups on their surface and therefore they have the ability to attract and retain positively charged heavy metal ions. In this point of view, soil organic matter helps to minimize heavy metal mobility in the soil and thus their availability to plants. On the other hand, degradation processes of organic matter can lead to release of heavy metal ions into the soil solution, thus increasing their bioavailability (Aigberua 2018).

Bioavailability of heavy metals in heavy metal polluted soils also depends on soil physical properties including soil texture, aggregate stability and density (Olaniran et al. 2013). These properties have a great effect on the processes that affect the behaviour of heavy metals in the soils, especially on the adsorption–desorption reactions. In this regard, clay soils have a higher heavy metal adsorption capacities than sandy soils mainly due to their large surface area and permanent or variable charges, thus reducing the bioavailability of heavy metals in soils (Angelaki et al. 2022).

Through root exudates, plants can also change soil properties and consequently the bioavailability of heavy metals in the soil (Yan et al. 2020). Under heavy metal stress conditions, plant roots release large amounts of different metabolites in the soil rhizosphere, some of which can reduce the bioavailability of toxic heavy metals either by changing pH or by forming heavy metal-metabolite complexes. Root exudates also provide various energy sources to microbial communities which in turn support plant growth and health. In addition, plant-associated microbes can alleviate heavy metal toxicity in soil by immobilizing or converting heavy metals to less toxic forms (Tiwari and Lata 2018). These points increasingly lead to the conclusion that the relationships between plants and their associated microbial communities play a key role not only in regulating heavy metal bioavailability in the soils but also in establishing healthy soil-plant interactions.

Interestingly, heavy metals can also affect soil-plant interactions through its impact on the composition and activity of the soil microbial community. The effects of these metals on soil microbial community depends on a number of factors, including, among others, pH, soil temperature, soil texture, organic matter content and heavy metal chemical speciation (Chu 2018). Furthermore, the change in bioavailability of one heavy metal can change the bioavailability of another in the soil, which significantly influences, directly or indirectly, plant-soil interactions.

The dynamics of heavy metals in plant-soil interactions is also strongly influenced by climatic conditions, including, in particular precipitation and temperature. For instance, increase in atmospheric temperature even a small amount can drastically change the transpiration rate of plants, which in turn, can affect the heavy metal uptake by plant (Dodgen et al. 2015). In addition, an increase in atmospheric temperature accelerates the decomposition rate of organic matter that strongly influences, among others, the mobility and potential availability of heavy metals (Rodrigues et al. 2023).

Furthermore, it is worth mentioning that plant genotype and variability in soil physicochemical properties as well as their interaction trigger variations in plant traits that can regulate the dynamics of heavy metals in soil-plant interactions with the aim to enhance plant tolerance to heavy metal toxicity (Emamverdian et al. 2015). In this regard, plants have evolved different heavy metal avoidance or/and tolerance mechanisms to ensure their survival in heavy metal polluted environment (Viehweger 2014). Avoidance mechanisms are based on the ability of plants to limit the uptake of heavy metals into the root system or to restrict their movement from the roots to the aerial parts of parts. Some of the most important and widely used mechanisms of heavy metal avoidance are heavy metal immobilization through mycorrhizal association, embedding the heavy metals in the root cell walls and heavy metal complexation via root exudates (Sladkovska et al. 2022). On the other hand, heavy metal tolerance mechanisms are based on the ability of plants to detoxify heavy metals within their body by various processes including heavy metal sequestration, chelation of heavy metals in the cytosol and compartmentation of heavy metals in the vacuole. Some mechanisms, usually described as ubiquitous, can reduce the negative effects of various heavy metals, and some mechanisms are specific only to certain heavy metals (Narayanan and Ma 2023a, b).

In general, the dynamics of heavy metals in the soil-plant interactions are complex since it depends not only on plant species, microbial soil composition and soil properties but also on specific combinations of numerous environmental factors. Therefore, it is very difficult to predict behavior of heavy metals in soil-plant interactions (Guala et al. 2010). Nevertheless, a better knowledge of plant-soil-microbe interactions can help in developing a correct agronomic, breeding or biotechnological approaches with the aim of improving plant response to stress caused by heavy metals. These approaches must take into consideration site-specific biotic and abiotic interactions between plants and their associated microbial communities (Agrahari et al. 2020).

2.3 Plant-Soil Feedbacks in Heavy Metal Polluted Environments

Plant-soil feedbacks refer to the ability of plants to modify biotic and abiotic soil conditions that in turn affect the plant's ability to grow in the same soil in the future (Beals et al. 2020). The key process involved in the plant-soil feedbacks is activity of microbial community in the soil. Namely, plants have complex influences on soil microbial communities and consequently on soil chemical properties, and this, in turn, can affect plant fitness as well as genetically based plant traits. Although it is known that composition of soil microbial community plays a key role in the resulting plant-soil feedbacks, the mechanisms underlying these feedbacks are still unclear (Dastogeer et al. 2022).

In general, plant-associated microbial communities that shape plant-soil feedbacks can be classified into three groups: microbial pathogens, microbial symbionts and microbial decomposers. They are the main biotic determinants of the direction and the strength of plant-soil feedbacks (Zhu et al. 2022).

Plant-soil feedbacks can vary from positive to negative depending on whether they promote or suppress the growth of subsequently grown plants (Forero et al. 2021). In this regard, positive plant-soil feedbacks occur when plants grow better on the soils where these plants were previously grown ('self' or 'home' soils) than on the non-native soils ('other' or 'away' soils), while negative plant-soil feedbacks occur when plants grow better on 'other' soils than on the 'home' soils (Kulmatiski 2019).

The reasons for the negative plant-soil feedbacks can be, among others, nutrient depletion in soil, changes in soil microbial community, autotoxicity i.e. excretion of toxic substances that inhibit the growth and development of the same plant species as well as accumulation of soil pathogens (Luo et al. 2019). In agriculture, the negative effects of plant-soil feedbacks can be avoided by implementing a sustainable crop rotation. This practice not only improves soil fertility, but also reduces pest and disease pressure and nutrient loss.

On the other hand, a positive plant-soil feedbacks occurs when plants change soil properties in a way that facilitates their growth in the same soil in the future. Interactions between root exudates, soil microbiota and soil components play a key role in achieving this. Many studies in this regard have shown that the plant-soil feedbacks have a great influence on the soil properties and consequently on the plants that grow subsequently in that soil (Ma et al. 2017; De Long et al. 2023). However, plant-soil feedbacks have mostly been investigated in isolation, under controlled conditions, and therefore cannot be generalized to conditions in natural environments where plants are exposed to a variety of abiotic and biotic stressors. Plant-soil feedbacks under stress conditions are more complex and therefore less predictable mainly due to stress-related metabolic changes in plants and microorganisms.

Root exudates play a crucial role in plant interactions with soil properties and rhizosphere microbial communities under stress conditions (Pan et al. 2022). In this regard, several studies have shown that the total amount of root exudates increases with increasing stress level or stress duration (Chai and Schachtman 2022; Chen et al. 2022). It is furthermore worth mentioning that the stressful conditions not only change the amount of root exudates, but also their composition. The amount and composition of root exudates also depends on the plant genotype, plant developmental stage as well as the properties of the soil in which plants grow. Stress events also strongly affect the microbial metabolism as well as composition of rhizosphere microbial communities (Olanrewaju et al. 2019).

Understanding the relationship between the stress-induced changes in plant and microbial metabolism and the resulting changes in plant-soil feedbacks is therefore essential for developing new strategies to improve not only plant performance under stress conditions, but also soil quality and bioremediation techniques for polluted environment (Zhu et al. 2022).

Unfortunately, plant-soil feedbacks in soils polluted by heavy metals have not yet been sufficiently investigated. Also, little is known about how heavy metals affect plant-soil feedbacks. Considering the fact that heavy metals can interact with all soil constituents including plant roots, soil particles and soil microbes, it is of great importance to recognize the behavior of heavy metals in polluted soil as well as their influence on plant-soil feedbacks. These feedbacks have long-term, significant and widespread effects on the composition of plants and microbial communities in contaminated soils, and thus their understanding is essential for effective long-term soil remediation.

Plant-soil feedbacks in the heavy metal polluted environment have also important evolutionary implications particularly with regard to plant community composition (Dietterich 2016). For example, negative plant-soil feedbacks that occurs in the heavy metal polluted soils can affect the plant communities by causing rarity of most plant species or suppressing dominant plant species (Maron et al. 2016). On the other hand, positive plant-soil feedbacks alter the plant communities in such a way as to allow dominance of plant species that have a high heavy metal tolerance or to allow new plant species to occupy a heavy metal polluted site (Khan et al. 2023).

Elucidating the connections between plant community composition and plant-soil feedbacks in the heavy metal polluted environments is of great importance to understand how soil physicochemical properties and processes affect plant fitness and genetic-based plant traits related to heavy metal tolerance. In addition, a better understanding of the patterns and driving factors of plant-soil feedbacks would greatly facilitate the restoration of ecosystem multi-functionality in heavy metal polluted soils.

2.4 The Use of Plant-Soil-Microbes Interactions in Phytoremediation

Phytoremediation represents a collection of different techniques that use plants to reduce the concentrations or toxic effects of pollutants in the environment. There are a number of phytoremediation techniques that are appropriate for the remediation of heavy metal polluted soils, including phytoextraction and phytostabilization as the most promising phytoremediation techniques (Sharma et al. 2023).

The success of any phytoremediation technique depends not only on plant species and heavy metal speciation and concentration in soil, but also on the interaction of plants with the surrounding growing medium and soil microbial communities. Numerous studies have shown that plant-soil-microbe interactions play an important role in achieving a successful phytoremediation (Weyens et al. 2015; Rane et al. 2022); however, little is known how these interactions are influenced by heavy metals. In this regard, knowing the behavior of heavy metal in the plant-soil-microbe interactions can help determine the biological effects of heavy metals in soil during phytoremediation processes, which implies the success of any phytoremediation method.

The study of interactions between plants, soil and microbes is also of great importance for the scientific community, including plant physiologists and ecologists. Most scientists agree that interaction of plant with heavy metal tolerant microbes play a crucial role in acclimatizing plants to heavy metal polluted soils (Tiwari and Lata 2018; Narayanan and Ma 2023a, b). These microbes use mechanisms such as biosorption, bioimmobilization, extracellular/intracellular sequestration, production of metal chelators, biofilm production, bioprecipitation, etc., not only to stay alive in heavy metal polluted soils, but also to enhance plant growth and survival under heavy metal stress conditions (Pande et al. 2022).

A list of some plant-beneficial microorganisms that have the ability to reduce heavy metal toxicity in soil is given in Table 2.1.

Although interactions between plants and beneficial rhizosphere microorganisms have been proven to be effective in phytoremediation of soil polluted by heavy metals, the lack of physical contacts between plants, microbes, and heavy metals often reduces the effectiveness of phytoremediation (Jing et al. 2007). In addition,

Table 2.1 List of some plant-beneficial microorganisms that could be used for phytoremediation purposes

Microorganisms	Heavy metal(s) pollution	Mechanisms of heavy metal resistance	References
Arthrobacter sp.	Cr	Reduction	Megharaj et al. (2003)
Bacillus haynesii	Cr	Reduction	Maurya et al. (2022)
Streptomyces tendae F4	Cd	Siderophores	Dimkpa et al. (2009)
Pseudomonas aeruginosa	Cd, Pb	Siderophores	Braud et al. (2010)
Bacilus cereus	Cd, Pb	Bioaugmentation	Nath et al. (2018)
Staphylococus xylosus	As	Biosorption	Aryal et al. (2010)
Cupriavidus sp. strain	Cd	Bioprecipitation	Li et al. (2019)
Bacillus strain MRS-2	Pb	Biosorption	Hoyle-Gardner et al. (2021)
Desulfovibrio desulfuricans	Cu, Cd, Zn	Sequestration	Yue et al. (2015)
Bacillus cereus sys1	Cu, Cd	Biosorption	Sonawdekar and Gupte (2020)
Klebsiella sp.	Cd, Mn	Biosorption	Hou et al. (2015)
Bacillus firmus	Hg	Reduction	Noroozi et al. (2017)
Bacillus subtilis X3	Pb	Bioimmobilization	Qiao et al. (2019)

soil chemical and physical properties strongly influence the association and interactions between plant roots and heavy metal-resistant microbes, indicating that plant-soil-microbe interactions are very complex and dynamic processes (Bara Caracciolo and Terenzi 2021). Plant-soil-microbe interactions are also strongly influenced by environmental factors. Any environmental perturbations such as nutritional shortages, water scarcity, intense light, as well as extreme cold or heat can drastically change the association and relationship between plant roots and microbes and thus their effectiveness in remediation of soils polluted with heavy metals (Cheng et al. 2019). Therefore, a better understanding of this intricate relationship between plants' roots, microbe and soil properties is a key factor for improving phytoremediation.

As shown in Fig. 2.2, plant-soil-microbe interactions are affected by a whole range of factors, including plant characteristics and activity, soil chemical and physical properties, composition and activity of soil microbial communities, etc.

Unfortunately, the increase in anthropogenic activities due to rapid industrialization and modern agricultural practices has drastically increased soil pollution with heavy metals in the past few decades. In order to remediate heavy metal polluted soils and prevent further soil degradation, there is a need to develop more effective techniques to remediate heavy metal-polluted soils, and unraveling plant-soil-microbe

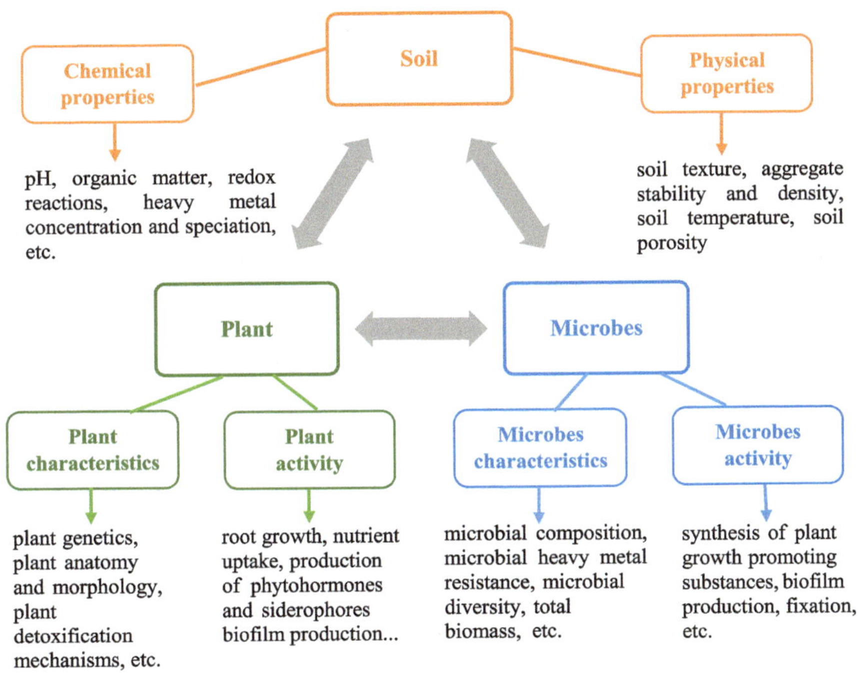

Fig. 2.2 Factors affecting plant-soil-microbe interactions

interactions is vital to achieving these aims. Some of the challenges that phytoremediation researchers have in this regard are: (1) how to achieve a better understanding of the changes in plant-soil-microbe interactions during phytoremediation processes, and (2) how to increase plant-soil-microbe interactions so that beneficial soil microorganisms and heavy metals have closer and more frequent contact with each other. These approaches should also take into account soil variability and complexity, including the effects of soil properties on plant–microbe interactions.

There are numerous studies that identify and describe the beneficial interaction between plants and microorganisms that can be utilized for phytoremediation purposes. For instance, a variety of heavy metal resistant bacteria such as *Pseudomonas aeruginosa* and *Streptomyces tendae* F4 synthesize and release siderophores that promote both plant growth under heavy metal stress conditions and heavy metal uptake by plants, which indicate their capability to enhance phytoextraction (Wang et al. 2011; Singh et al. 2022). Kapahi and Sachdeva (2019) reported that *Pseudomonas aeruginosa* also has the ability to induce precipitation of Cd ions. Furthermore, some microorganisms such as *Bacillus haynesii* and *Cupriavidus pauculus* 1490 have developed a mechanism to synthesize extracellular polymeric substances or biofilm to prevent toxic heavy metals entering the cells (Zeng et al. 2020; Maurya et al. 2022). Extracellular polymeric substances also having a preference for heavy metal binding, reducing their bioavailability, and thus improving phytostabilization. On the other hand, biosurfactants produced by plant-associated microbes can improve phytoextraction by increasing heavy metal mobility and, subsequently, their bioavailability (Lal et al. 2018). Numerous studies have also shown that some plant-associated microbes may prevent heavy metal phytotoxicity and stimulate plant growth through production of phytohormones, siderophores and other plant growth promoting substances (Alves et al. 2022; Gladkov et al. 2023).

Considering all these facts, it becomes obvious that greater clarification of biochemical and molecular mechanisms involved in beneficial plant–microbe-metal interactions are essential for the development of effective and sustainable phytoremediation techniques. Biochemical mechanisms include those by which plant-associated microbes interact with extracellular soluble heavy metal ions, increasing phytoremediation efficiency mainly through heavy metal detoxification, mobilization or immobilization while molecular mechanisms refer to the genetic determinants involved in heavy metal tolerance, which include, for example, the synthesis of stress-related amino acids, proteins or signaling molecules (Ma et al. 2016). Recent advances in various branches of biology such as transcriptomics and proteomics, have helped identify a number of transcription factors, secondary metabolites and stress-inducible proteins involved in heavy metal tolerance, which in turn can be utilized for phytoremediation purposes (Singh et al. 2016).

It is worth mentioning that the biochemical and molecular mechanisms involved in plant–microbe-metal interactions in the heavy metal polluted soils cannot be considered separately because the plant response to heavy metal soil pollution often requires the coordination of complex physiological, biochemical and molecular mechanisms at the whole-plant level, including, among others, changes in global gene expression (Gallego et al. 2012).

A promising development in transgenic research together with advances in understanding the biochemical and molecular mechanisms involved in heavy metal detoxification in plants and plant-assisted microorganisms, as well as their interactions under heavy metals stress conditions, have opened up new possibilities for improving phytoremediation efficiency through genetic engineering. Genetic engineering, also known as genetic modification, refers to all the laboratory-based technologies that allow manipulation, modification or recombination of an organism's genes in order to optimize whole-organism performance or to obtain the desired characteristics in living organisms or their products. In this light, manipulation or modification of any gene responsible for heavy metal detoxification in plants and soil microbes can potentially improve their ability to take up, translocate, immobilize or convert toxic heavy metals into less toxic forms (Priya et al. 2023). This type of approach can potentially be used as a strategy to increase the interaction of plants and microbes with the aim of enhancing the heavy metal detoxification in culture media where it was previously impossible to detoxify heavy metals by natural microrganisms and plants (Fasani et al. 2018). However, plant genetic engineering is a relatively new and developing field, but it undoubtedly provides promising opportunities for progress in the field of phytoremediation.

2.5 Conclusions

Variability in soil properties, especially differences in soil pH, organic matter content, soil texture, and soil structure, density and porosity preclude the ability to definitively predict the behavior and fate of heavy metal ions in soil-plant continuum. This already complex situation is further complicated by differences in remediation efficiencies among plant species. In this light, a better understanding of the patterns and driving factors of plant-soil-microbe interactions as well as the physicochemical processes that influence the behavior of heavy metals in the soils such as complexation, cation exchange, specific adsorption/desorption reactions and precipitation can help to clarify the processes that determine the fate of heavy metals in soil-plant continuum. Consequently, this knowledge would greatly facilitate the restoration of ecosystem multi-functionality in heavy metal polluted soils.

References

Agrahari RK, Singh P, Koyama H, Panda SK (2020) Plant-microbe interactions for sustainable agriculture in the post-genomic era. Curr Genomics 21(3):168–178. https://doi.org/10.2174/1389202921999200505082116

Aigberua AO (2018) Soil organic matter: application for heavy metal source diagnostics in sediments of the Middleton River, Bayelsa State, Nigeria. MOJ Toxicol 1(4):410–415. https://doi.org/10.15406/mojt.2018.04.00138

Alengebawy A, Abdelkhalek ST, Qureshi SR, Wang MQ (2021) Heavy metals and pesticides toxicity in agricultural soil and plants: ecological risks and human health implications. Toxics 9(3):42. https://doi.org/10.3390/toxics9030042

Alsafran M, Saleem MH, Al Jabri H, Rizwan M, Usman H (2023) Principles and applicability of integrated remediation strategies for heavy metal removal/recovery from contaminated environments. J Plant Growth Regul 42:3419–3440. https://doi.org/10.1007/s00344-022-10803-1

Alves ARA, Yin Q, Oliveira RS, Silva EF, Novo LAB (2022) Plant growth-promoting bacteria in phytoremediation of metal-polluted soils: current knowledge and future directions. Sci Total Environ 838:156435. https://doi.org/10.1016/j.scitotenv.2022.156435

Angelaki A, Dionysidis A, Sihag P, Golia EE (2022) Assessment of contamination management caused by copper and zinc cations leaching and their impact on the hydraulic properties of a sandy and a loamy clay soil. Land 11(2):290. https://doi.org/10.3390/land11020290

Angulo-Bejarano PI, Puente-Rivera J, Cruz-Ortega R (2021) Metal and metalloid toxicity in plants: an overview on molecular aspects. Plants (basel) 10(4):635. https://doi.org/10.3390/plants10040635

Aryal M, Ziagova M, Liakopoulou-Kyriakides M (2010) Study on arsenic biosorption using Fe(III)-treated biomass of *Staphylococcus xylosus*. Chem Eng J 2010(162):178–185. https://doi.org/10.1016/j.cej.2010.05.026

Aydinalp C, Marinova S (2003) Distribution and forms of heavy metals in some agricultural soils. Pol J Environ Stud 12(5):629–633

Baluška F, Mancuso S (2020) Plants, climate and humans: plant intelligence changes everything. EMBO Rep 21(3):e50109. https://doi.org/10.15252/embr.202050109

Bararzadeh Ledari M, Saboohi Y, Valero A, Azamian S (2020) Exergy analysis of a bio-system: soil-plant interaction. Entropy (Basel) 23(1):3. https://doi.org/10.3390/e23010003

Barra Caracciolo A, Terenzi V (2021) Rhizosphere microbial communities and heavy metals. Microorganisms 9(7):1462. https://doi.org/10.3390/microorganisms9071462

Beals KK, Moore JAM, Kivlin SN, Bayliss SLJ, Lumibao CY, Moorhead LC, Patel M, Summers JL, Ware IM, Bailey JK, Schweitzer JA (2020) Predicting plant-soil feedback in the field: meta-analysis reveals that competition and environmental stress differentially influence PSF. Front Ecol Evol 8:191. https://doi.org/10.3389/fevo.2020.00191

Biswas B, Qi F, Biswas JK, Wijayawardena A, Khan MAI, Naidu R (2018) The fate of chemical pollutants with soil properties and processes in the climate change paradigm—a review. Soil Syst. 2:51. https://doi.org/10.3390/soilsystems2030051

Borch T, Kretzschmar R, Kappler A, Cappellen PV, Ginder-Vogel M, Voegelin A, Campbell K (2010) Biogeochemical redox processes and their impact on contaminant dynamics. Environ Sci Technol 44(1):15–23. https://doi.org/10.1021/es9026248

Braud A, Geoffroy V, Hoegy F, Mislin GL, Schalk IJ (2010) Presence of the siderophores pyoverdine and pyochelin in the extracellular medium reduces toxic metal accumulation in *Pseudomonas aeruginosa* and increases bacterial metal tolerance. Environ Microbiol Rep 3:419–425. https://doi.org/10.1111/j.1758-2229.2009.00126.x

Caporale AG, Violante A (2016) Chemical processes affecting the mobility of heavy metals and metalloids in soil environments. Curr Pollut Rep 2:15–27. https://doi.org/10.1007/s40726-015-0024-y

Chai YN, Schachtman DP (2022) Root exudates impact plant performance under abiotic stress. Trends Plant Sci 27(1):80–91. https://doi.org/10.1016/j.tplants.2021.08.003

Chen H, Teng Y, Lu S, Wang Y, Wang J (2015) Contamination features and health risk of soil heavy metals in China. Sci Total Environ 512:143–553. https://doi.org/10.1016/j.scitotenv.2015.01.025

Chen Y, Yao Z, Sun Y, Wang E, Tian C, Sun Y, Liu J, Sun C, Tian L (2022) Current studies of the effects of drought stress on root exudates and rhizosphere microbiomes of crop plant species. Int J Mol Sci 23(4):2374. https://doi.org/10.3390/ijms23042374

Cheng YT, Zhang L, He SY (2019) Plant-microbe interactions facing environmental challenge. Cell Host Microbe 26(2):183–192. https://doi.org/10.1016/j.chom.2019.07.009

Chu D (2018) Effects of heavy metals on soil microbial community. IOP Conf Ser: Earth Environ Sci 113:012009. https://doi.org/10.1088/1755-1315/113/1/012009

Chu Z, Lin C, Yang K, Cheng H, Gu X, Wang B, Wu L, Ma J (2022) Lability, bioaccessibility, and ecological and health risks of anthropogenic toxic heavy metals in the arid calcareous soil around a nonferrous metal smelting area. Chemosphere 307(4):136200. https://doi.org/10.1016/j.chemosphere.2022.136200

Dastogeer KMG, Kao-Kniffin J, Okazaki S (2022) Editorial: plant microbiome: diversity, functions, and applications. Front Microbiol 13:1039212. https://doi.org/10.3389/fmicb.2022.1039212

De Long JR, Heinen R, Heinze J, Morriën E, Kenny Png G, Sapsford SJ, Teste FP, Fry EL (2023) Plant-soil feedback: incorporating untested influential drivers and reconciling terminology. Plant Soil 485:7–43. https://doi.org/10.1007/s11104-023-05908-9

Dietterich LH (2016) Plant-soil feedbacks in heavy metal soils. Publicly Accessible Penn Dissertations. https://repository.upenn.edu/edissertations/2257

Dimkpa C, Merten D, Svatoš A, Büchel G, Kothe E (2009) Siderophores mediate reduced and increased uptake of cadmium by *Streptomyces tendae* F4 and sunflower (*Helianthus annuus*), respectively. J Appl Microbiol 107:1687–1716. https://doi.org/10.1111/j.1365-2672.2009.04355.x

Dodgen LK, Ueda A, Wu X, Parker DR, Gan J (2015) Effect of transpiration on plant accumulation and translocation of PPCP/EDCs. Environ Pollut 198:144–153. https://doi.org/10.1016/j.envpol.2015.01.002

Ehrmann J, Ritz K (2014) Plant: soil interactions in temperate multi-cropping production systems. Plant Soil 376:1–29. https://doi.org/10.1007/s11104-013-1921-8

Emamverdian A, Ding Y, Mokhberdoran F, Xie Y (2015) Heavy metal stress and some mechanisms of plant defense response. Sci World J 2015:756120. https://doi.org/10.1155/2015/756120

Fasani E, Manara A, Martini F, Furini A, DalCorso G (2018) The potential of genetic engineering of plants for the remediation of soils contaminated with heavy metals. Plant Cell Environ 41:1201–1232. https://doi.org/10.1111/pce.12963

Forero LE, Kulmatiski A, Grenzer J, Norton JM (2021) Plant-soil feedbacks help explain biodiversity-productivity relationships. Commun Biol 4(1):789. https://doi.org/10.1038/s42003-021-02329-1

Gallego SM, Pena LB, Barcia RA, Azpilicueta CE, Lannone MF, Rosales EP, Zawoznik MS, Groppa MD, Benavides MP (2012) Unravelling cadmium toxicity and tolerance in plants: insight into regulatory mechanisms. Environ Exp Bot 83:33–46. https://doi.org/10.1016/j.envexpbot.2012.04.006

Gladkov EA, Tereshonok DV, Stepanova AY, Gladkova OV (2023) Plant-microbe interactions under the action of heavy metals and under the conditions of flooding. Diversity 15:175. https://doi.org/10.3390/d15020175

Goh CH, Veliz Vallejos DF, Nicotra AB, Nicotra AB, Mathesius U (2013) The impact of beneficial plant-associated microbes on plant phenotypic plasticity. J Chem Ecol 39:826–839. https://doi.org/10.1007/s10886-013-0326-8

Guala SD, Vega FA, Covelo EF (2010) The dynamics of heavy metals in plant-soil interactions. Ecol Model 221:1148–1152. https://doi.org/10.1016/j.ecolmodel.2010.01.003

Hou S, Zheng N, Tang L, Ji X, Li Y (2019) Effect of soil pH and organic matter content on heavy metals availability in maize (*Zea mays* L.) rhizospheric soil of non-ferrous metals smelting area. Environ Monit Assess 191(10):634. https://doi.org/10.1007/s10661-019-7793-5

Hou Y, Cheng K, Li Z, Ma X, Wei Y, Zhang L, Wang Y (2015) Biosorption of cadmium and manganese using free cells of *Klebsiella* sp. isolated from waste water. PLOS ONE 10(10):e0140962. https://doi.org/10.1371/journal.pone.0140962

Hoyle-Gardner J, Jones W, Badisa VLD, Mwashote B, Ibeanusi V, Gaines T, Lowenthal H, Tucker L (2021) Lead metal biosorption and isotherms studies by metal-resistant *Bacillus* strain MRS-2 bacterium. J Basic Microbiol 2021(61):697–708. https://doi.org/10.1002/jobm.202100155

Jacoby R, Peukert M, Succurro A, Koprivova A, Kopriva S (2017) The role of soil microorganisms in plant mineral nutrition-current knowledge and future directions. Front Plant Sci 8:1617. https://doi.org/10.3389/fpls.2017.01617

Järup L (2003) Hazards of heavy metal contamination. Br Med Bull 68:167–182. https://doi.org/10.1093/bmb/ldg032

Jiang O, Li L, Duan G, Gustave W, Zhai W, Zou L, An X, Tang X, Xu J (2023) Root exudates increased arsenic mobility and altered microbial community in paddy soils. J Environ Sci (China) 127:410–420. https://doi.org/10.1016/j.jes.2022.05.036

Jing YD, He ZL, Yang XE (2007) Role of soil rhizobacteria in phytoremediation of heavy metal contaminated soils. J Zhejiang Univ Sci B 8(3):192–207. https://doi.org/10.1631/jzus.2007.B0192

Kandziora-Ciupa M, Nadgórska-Socha A, Barczyk G (2021) The influence of heavy metals on biological soil quality assessments in the Vaccinium myrtillus L. rhizosphere under different field conditions. Ecotoxicology 30:292–310. https://doi.org/10.1007/s10646-021-02345-1

Kapahi M, Sachdeva S (2019) Bioremediation options for heavy metal pollution. J Heal Pollut 9:191203. https://doi.org/10.5696/2156-9614-9.24.191203

Khan IU, Qi S-S, Gul F, Manan S, Rono JK, Naz M, Shi X-N, Zhang H, Dai Z-C, Du D-L (2023) A green approach used for heavy metals 'phytoremediation' via invasive plant species to mitigate environmental pollution: a review. Plants 12:725. https://doi.org/10.3390/plants12040725

Kulmatiski A (2019) Plant-soil feedbacks predict native but not non-native plant community composition: a 7-year common-garden experiment. Front Ecol Evol 7:326. https://doi.org/10.3389/fevo.2019.00326

Lal S, Ratna S, Said OB, Kumar R (2018) Biosurfactant and exopolysaccharide-assisted rhizobacterial technique for the remediation of heavy metal contaminated soil: an advancement in metal phytoremediation technology. Environ Technol Innov 10:243–263. https://doi.org/10.1016/j.eti.2018.02.011

Li F, Zheng Y, Tian J, Ge F, Liu X, Tang Y, Feng C (2019) *Cupriavidus* sp. strain Cd02-mediated pH increase favoring bioprecipitation of Cd^{2+} in medium and reduction of cadmium bioavailability in paddy soil. Ecotoxicol Environ Saf 184:109655. https://doi.org/10.1016/j.ecoenv.2019.109655

Luo L, Guo C, Wang L, Zhang J, Deng L, Luo K, Huang H, Liu Y, Mei X, Zhu S, Yang M (2019) Negative plant-soil feedback driven by re-assemblage of the rhizosphere microbiome with the growth of Panax notoginseng. Front Microbiol 10:1597. https://doi.org/10.3389/fmicb.2019.01597

Ma H-K, Pineda A, van der Wurff AWG, Raaijmakers C, Bezemer TM (2017) Plant-soil feedback effects on growth, defense and susceptibility to a soil-borne disease in a cut flower crop: species and functional group effects. Front Plant Sci 8:2127. https://doi.org/10.3389/fpls.2017.02127

Ma Y, Oliveira RS, Freitas H, Zhang C (2016) Biochemical and molecular mechanisms of plant-microbe-metal interactions: relevance for phytoremediation. Front Plant Sci 7:918. https://doi.org/10.3389/fpls.2016.00918

Madzhieva S, Minkina T, Pinskiy D, Bauer T, Sushkova S (2014) The role of soil's particle-size fractions in the adsorption of heavy metals. Eurasian J Soil Sci 3:197–205. https://doi.org/10.18393/ejss.16003

Manikandan SK, Pallavi P, Shetty K, Bhattacharjee D, Giannakoudakis DA, Katsoyiannis IA, Nair V (2023) Effective usage of biochar and microorganisms for the removal of heavy metal ions and pesticides. Molecules 28(2):719. https://doi.org/10.3390/molecules28020719

Mapaure I, Chimwamurombe PM, Mapani BS, Kamona FA (2011) Impacts of mine dump pollution on plant species diversity, composition and structure of a semiarid savanna in Namibia. Afr J Range Forage Sci 28:149–154. https://doi.org/10.2989/10220119.2011.647753

Maron JL, Laney Smith A, Ortega YK, Pearson DE, Callaway RM (2016) Negative plant-soil feedbacks increase with plant abundance, and are unchanged by competition. Ecology 97(8):2055–2063. https://doi.org/10.1002/ecy.1431

Maurya A, Kumar R, Yadav P, Singh A, Yadav A, Chowdhary P, Raj A (2022) Biofilm formation and extracellular polymeric substance (EPS) production by *Bacillus haynesii* and influence of hexavalent chromium. Bioresour Technol 352:127109. https://doi.org/10.1016/j.biortech.2022.127109

Megharaj M, Avudainayagam S, Naidu R (2003) Toxicity of hexavalent chromium and its reduction by bacteria isolated from soil contaminated with tannery waste. Curr Microbiol 47:51–54. https://doi.org/10.1007/s00284-002-3889-0

Narayanan M, Ma Y (2023a) Metal tolerance mechanisms in plants and microbe-mediated bioremediation. Environ Res 222:115413. https://doi.org/10.1016/j.envres.2023.115413

Narayanan M, Ma Y (2023b) Mitigation of heavy metal stress in the soil through optimized interaction between plants and microbes. J Environ Manag 345:118732. https://doi.org/10.1016/j.jenvman.2023.118732

Nath S, Deb B, Sharma I (2018) Isolation of toxic metal-tolerant bacteria from soil and examination of their bioaugmentation potentiality by pot studies in cadmium- and lead-contaminated soil. Int Microbiol 21:35–45. https://doi.org/10.1007/s10123-018-0003-4

Noroozi M, Amoozegar MA, Pourbabaee AA, Naghavi NS, Nourmohammadi Z (2017) Isolation and characterization of mercuric reductase by newly isolated halophilic bacterium, *Bacillus firmus* MN8. Glob J Environ Sci Manag 3(4):427–436. https://doi.org/10.22034/gjesm.2017.03.04.008

Olaniran AO, Balgobind A, Pillay B (2013) Bioavailability of heavy metals in soil: impact on microbial biodegradation of organic compounds and possible improvement strategies. Int J Mol Sci 14(5):10197–10228. https://doi.org/10.3390/ijms140510197

Olanrewaju OS, Ayangbenro AS, Glick BR, Babalola OO (2019) Plant health: feedback effect of root exudates-rhizobiome interactions. Appl Microbiol Biotechnol 103:1155–1166. https://doi.org/10.1007/s00253-018-9556-6

Pan Y, Kang P, Tan M, Hu J, Zhang Y, Zhang J, Song N, Li X (2022) Root exudates and rhizosphere soil bacterial relationships of *Nitraria tangutorum* are linked to k-strategists bacterial community under salt stress. Front Plant Sci 13:997292. https://doi.org/10.3389/fpls.2022.997292

Pande V, Pandey SC, Sati D, Bhatt P, Samant M (2022) Microbial interventions in bioremediation of heavy metal contaminants in agroecosystem. Front Microbiol 13:824084. https://doi.org/10.3389/fmicb.2022.824084

Priya AK, Muruganandam M, Ali SS, Kornaros M (2023) Clean-up of heavy metals from contaminated soil by phytoremediation: a multidisciplinary and eco-friendly approach. Toxics 11(5):422. https://doi.org/10.3390/toxics11050422

Qayyum S, Khan I, Zhao Y, Maqbool F, Peng C (2016) Sequential extraction procedure for fractionation of Pb and Cr in artificial and contaminated soil. Main Group Met Chem 39(1–2):49–58. https://doi.org/10.1515/mgmc-2016-0001

Qiao W, Zhang Y, Xia H, Luo Y, Liu S, Wang S, Wang W (2019) Bioimmobilization of lead by *Bacillus subtilis* X3 biomass isolated from lead mine soil under promotion of multiple adsorption mechanisms. R Soc Open Sci 6(2):181701. https://doi.org/10.1098/rsos.181701

Ram N, Verloo M (1985) Effect of various organic materials on the mobility of heavy metals in soil. Environ Pollut B 10(4):241–248

Rane NR, Tapase S, Kanojia A, Watharkar A, Salama ES, Jang M, Kumar Yadav K, Amin MA, Cabral-Pinto MMS, Jadhav JP, Jeon BH (2022) Molecular insights into plant-microbe interactions for sustainable remediation of contaminated environment. Bioresour Technol 344(Pt B):126246. https://doi.org/10.1016/j.biortech.2021.126246

Rodrigues CID, Brito LM, Nunes LJR (2023) Soil carbon sequestration in the context of climate change mitigation: a review. Soil Syst 7:64. https://doi.org/10.3390/soilsystems7030064

Sharma JK, Kumar N, Singh NP, Santal AR (2023) Phytoremediation technologies and their mechanism for removal of heavy metal from contaminated soil: an approach for a sustainable environment. Front Plant Sci 14:1076876. https://doi.org/10.3389/fpls.2023.1076876

Shi J, Wu Q, Zheng C, Yang J (2018) The interaction between particulate organic matter and copper, zinc in paddy soil. Environ Pollut B 243:1394–1402. https://doi.org/10.1016/j.envpol.2018.09.085

Singh P, Chauhan PK, Upadhyay SK, Singh RK, Dwivedi P, Wang J, Jain D, Jiang M (2022) Mechanistic insights and potential use of siderophores producing microbes in rhizosphere for mitigation of stress in plants grown in degraded land. Front Microbiol 13:898979. https://doi.org/10.3389/fmicb.2022.898979

Singh S, Parihar P, Singh R, Singh VP, Prasad SM (2016) Heavy metal tolerance in plants: role of transcriptomics, proteomics, metabolomics, and ionomics. Front Plant Sci 6:1143. https://doi.org/10.3389/fpls.2015.01143

Sladkovska T, Wolski K, Bujak H, Radkowski A, Sobol Ł (2022) A review of research on the use of selected grass species in removal of heavy metals. Agronomy 12:2587. https://doi.org/10.3390/agronomy12102587

Sonawdekar S, Gupte A (2020) Biosorption of copper(II) and cadmium(II) by *Bacillus cereus* sys1 isolated from oil-contaminated site. SN Appl Sci 2:1254. https://doi.org/10.1007/s42452-020-3062-z

Sulaiman FR, Hamzah HA (2018) Heavy metals accumulation in suburban roadside plants of a tropical area (Jengka, Malaysia). Ecol Process 7:28. https://doi.org/10.1186/s13717-018-0139-3

Sun R, Gao Y, Yang Y (2022) Leaching of heavy metals from lead-zinc mine tailings and the subsequent migration and transformation characteristics in paddy soil. Chemosphere 291:132792. https://doi.org/10.1016/j.chemosphere.2021.132792

Szabó P, Jordan G, Kocsis T, Posta K, Kardos L, Šajn R, Alijagić J (2022) Biomonitoring and assessment of toxic element contamination in floodplain sediments and soils using fluorescein diacetate (FDA) enzymatic activity measurements: evaluation of possibilities and limitations through the case study of the Drava River floodplain. Environ Monit Assess 194(9):632. https://doi.org/10.1007/s10661-022-10301-7

Tchounwou PB, Yedjou CG, Patlolla AK, Sutton DJ (2012) Heavy metal toxicity and the environment. Exp Suppl 101:133–164. https://doi.org/10.1007/978-3-7643-8340-4_6

Tiwari S, Lata C (2018) Heavy metal stress, signaling, and tolerance due to plant-associated microbes: an overview. Front Plant Sci 9:452. https://doi.org/10.3389/fpls.2018.00452

Viehweger K (2014) How plants cope with heavy metals. Bot Stud 55:35. https://doi.org/10.1186/1999-3110-55-35

Violante A, Cozzolino V, Perelomov L, Caporale AG, Pigna M (2010) Mobility and bioavailability of heavy metals and metalloids in soil environments. J Soil Sci Plant Nutr 10(3):268–292. https://doi.org/10.4067/S0718-95162010000100005

Wang Q, Xiong D, Zhao P, Yu B, Tu B, Wang G (2011) Effect of applying an arsenic-resistant and plant growth-promoting rhizobacterium to enhance soil arsenic phytoremediation by Populus deltoids LH05-17. J Appl Microbiol 111(5):1065–1074. https://doi.org/10.1111/j.1365-2672.2011.05142.x

Weyens N, Thijs S, Popek R, Witters N, Przybysz A, Espenshade J, Gawronska H, Vangronsveld J, Gawronski SW (2015) The role of plant-microbe interactions and their exploitation for phytoremediation of air pollutants. Int J Mol Sci 16(10):25576–25604. https://doi.org/10.3390/ijms161025576

Woch MW, Kapusta P, Stefanowicz AM (2016) Variation in dry grassland communities along a heavy metals gradient. Ecotoxicology 25:80–90. https://doi.org/10.1007/s10646-015-1569-7

Yan A, Wang Y, Tan SN, Mohd Yusof ML, Ghosh S, Chen Z (2020) Phytoremediation: a promising approach for revegetation of heavy metal-polluted land. Front Plant Sci 11:359. https://doi.org/10.3389/fpls.2020.00359

Yu H, Li C, Yan J, Ma Y, Zhou X, Yu W, Kan H, Meng Q, Xie R, Dong P (2023) A review on adsorption characteristics and influencing mechanism of heavy metals in farmland soil. RSC Adv 13(6):3505–3519. https://doi.org/10.1039/d2ra07095b

Yue ZB, Li Q, Li CC, Chen TH, Wang J (2015) Component analysis and heavy metal adsorption ability of extracellular polymeric substances (EPS) from sulfate reducing bacteria. Bioresour Technol 194:399–402. https://doi.org/10.1016/j.biortech.2015.07.042

Zeng W, Zhang S, Xia M, Wu X, Qiu G, Shen L (2020) Insights into the production of extracellular polymeric substances of *Cupriavidus pauculus* 1490 under the stimulation of heavy metal ions. RSC Adv 10(34):20385–20394. https://doi.org/10.1039/c9ra10560c

Zhai Y, Zheng F, Li D, Cao X, Teng Y (2022) Distribution, genesis, and human health risks of groundwater heavy metals impacted by the typical setting of Songnen Plain of NE China. Int J Environ Res Public Health 19(6):3571. https://doi.org/10.3390/ijerph19063571

Zhu S-C, Zheng H-X, Liu W-S, Liu C, Guo M-N, Huot H, Morel JL, Qiu R-L, Chao Y, Tang Y-T (2022) Plant-soil feedbacks for the restoration of degraded mine lands: a review. Front Microbiol 12:751794. https://doi.org/10.3389/fmicb.2021.751794

Chapter 3
Source and Distribution of Heavy Metal and Their Effects on Human Health

Farah Noshin Chowdhury and Md. Mostafizur Rahman

Abstract Being a naturally occurring substance Heavy Metals (HM) are irrevocably a persistent component of the environment. However, anthropogenic activities including excessive use of the HMs in agricultural, industrial, medical, painting, mining and other purposes, as well as subsequent indiscriminate disposal of these products without or with minimal treatment in the environmental matrices has caused a global upsurge of toxic HM accumulation in the nature. Which has undeniably led us on the verge of an impending crisis of possible HM toxicity by the contaminated environment. Sourced both naturally and artificially HMs are widely distributed in the nature and might be changed by environmental process and travel from one environmental compartment to other. They tend to bioaccumulate and biomagnify, in addition to being non-biodegradable thus persistent in the nature. However, they may also exert toxicity on the living organisms including human beings and thus are considered threat to health of the populace. The toxicity of HMs ranges from Neurotoxicity, Nephrotoxicity, Carcinogenicity, Hepatotoxicity, Immunological Toxicity, Cardiovascular toxicity, Dermal Toxicity, Reproductive and Developmental Toxicity and Genotoxicity. This further manifests into a range of diseases arising from these toxic pathways. The nexus of HM source-distribution-health impacts is important to be studied together for devising possible roadmaps of alleviating the present pollution load oh HMs.

Keywords Heavy metals · Source · Distribution · Health · Environment

---

F. N. Chowdhury · Md. M. Rahman (✉)
Laboratory of Environmental Health and Ecotoxicology, Department of Environmental Sciences, Jahangirnagar University, Savar, Dhaka 1342, Bangladesh
e-mail: rahmanmm@juniv.edu

Md. M. Rahman
Department of Environmental Sciences, Jahangirnagar University, Savar, Dhaka 1342, Bangladesh

N. Kumar (ed.), *Heavy Metal Toxicity*, Environmental Science and Engineering,
https://doi.org/10.1007/978-3-031-56642-4_3

3.1 Introduction

The levels of Heavy Metals (HM) in the environment have risen dramatically in recent decades (Zaynab et al. 2022). Metals are naturally occurring elements of the nature. Among the 35 naturally occurring metals, 23 have a high specific density exceeding 5 g/cm^3, atomic number over 20 and an atomic weight larger than 40.04, and are thus classified as HM (Azeh Engwa et al. 2019; Mishra et al. 2019). The 23 HM are, Antimony, Arsenic, Bismuth, Cadmium, Cerium, Chromium, Cobalt, Copper, Gallium, Gold, Iron, Lead, Manganese, Mercury, Nickel, Platinum, Silver, Tellurium, Thallium, Tin, Uranium, Vanadium, And Zinc. Such HMs are widespread in nature. In little concentrations, they are necessary for optimal health, but in high quantities, they can be poisonous or hazardous (Jaishankar et al. 2014). HMs as well as metalloid ions are divided into three categories. The initial group comprise mercury, cadmium, as well as lead, all of which are dangerous at low amounts. The next category of metals tends to be less hazardous (bismuth, indium, arsenic, thallium, and antimony), and the final group encompasses vital metals including zinc, cobalt, copper, iron, and selenium, that are involved in multiple chemical or biochemical reactions within the human body and are hazardous only at high concentrations. As a consequence of sorption, inhaling, and incidents or mistreatment, HM build in soil, human beings, and tissues from animals (Vijay Dahiya 2022; Zaynab et al. 2022). On the average, 10,000–20,000 L of air are breathed every day, and HM are taken in through inhaled particle matter and cause harm to tissues in the lungs (Srivastava et al. 2018).

HMs can be found in the earth's atmosphere, the lithosphere the hydrosphere, as well as biosphere (Azeh Engwa et al. 2019). HM have been consistently present on Earth from its inception, according to recurring biogeochemical processes. The presence of HMs in the ground was caused by the basic weathering process. Metal loading throughout the soil's creation may take place as a consequence of slightly elevated HM concentrations in bedrock (Zaynab et al. 2022). Whilst these HM are found in the environment, humans are exposed to them through a variety of human-caused events. These HM are found in the crust of the earth as ores that are extracted as minerals through mining operations (Azeh Engwa et al. 2019). HM pollution has been caused by anthropogenic activities such as electroplating, metal mining, smelting, pesticide along with fertilizer releases, biological solids, urban sewage/sludge, and the apparel and painting industries (Mishra et al. 2019). Human and anthropogenic influences are the primary causes for the higher toxicity in the environment caused by HMs. Whereas, wind-blown soil particles, wildfires, volcano erupting biogenic procedures, and sea salt are among natural sources of HMs. Mining activities, the use of chemical fertilizers, pesticides, and herbicides as well and crop field watering with factories and waste water are all anthropogenic drivers of HMs pollution. HM ultimately inflict a variety of pressures on living things, adding to higher mortality. Since HM are incapable of biodegradability, they are harmful even in minimal levels (Zaynab et al. 2022).

HM contamination, induced by large scale urbanization and industrialization, is one of the most pressing concerns confronting today's civilization (Jayakumar et al. 2021). HMs enter habitats via a variety of sources, including industrial waste, home effluents, atmospheric outlets, and other metal-based industrial functions, as well as e-waste (Sonone et al. 2021). HMs ubiquitous residential, industrial, agricultural, and technical applications have resulted in their vast distribution in the environment, raising concerns about their potential effects on human health and the natural world (Hassaan et al. 2016). The inappropriate disposal of manufactured waste materials immediately into water bodies and land regions has resulted in a constant increase in HM concentrations and major environmental concerns (Mishra et al. 2019). As a consequence of the overabundance of agricultural chemicals, wastewater from municipalities, industrial wastewater, and sewage in its raw form for agriculture, the food web is continually supplied with HMs (Mitra et al. 2022). HM pollution in agricultural fields/soils contaminates food supplies, eventually entering the human food web. Metals or metalloid concentrations in agricultural soil are quickly increasing, affecting plant development, food safety, and soil microflora (Jayakumar et al. 2021). HM even infiltrate the system of plants and have an impact on the development and growth of plants. Once heavy metals laden vegetables are ingested by humans, they produce a variety of serious health problems (Sandeep et al. 2019). Due to the unseen mechanism of HM toxicity, the issue of HM infiltrating food chains necessitates rigorous evaluations in order to make swift choices to prevent serious health impacts (Balkhair and Ashraf 2016).

HMs are being biomagnified in the food web, that has a deleterious influence on ecology (Zaynab et al. 2022) and the widespread distribution of HM has caused concerns about human health (Srivastava et al. 2018). These HMs are carried and sequestered into bodily cells and tissues, where they bind to proteins as well as nucleic acids, damaging them and altering their biological processes. As a result, HM poisoning can have a variety of effects on the body of an individual (Azeh Engwa et al. 2019). Any necessary or not required trace element that exists in exceeding of acceptable amounts can cause physiological or morphological deviations, as well as genetic alterations, such as reducing or stopping growth or producing mutations. Human intake of HMs is mostly via dietary vegetables, accounting for around 90% of total consumption, with the remaining 10% coming from dermal contact and inhaling contaminated dust. Because of the health concerns associated with the use of chemicals, pesticides, and heavy metals and toxin-contaminated food items, the security of food has grown into a serious public health problem (Mitra et al. 2022). HMs toxicity can drain energy and harm important organs such as the lungs, brain, kidneys, the liver, blood, and others. Long-term exposure causes deteriorating physical, tissue, and neurological processes that resemble illnesses including Alzheimer's, Parkinson's, muscular dystrophy, and multiple sclerosis. Appetite reduction, headaches, hypertension, stomach pain, renal failure, weariness, sleeplessness, arthritis, hallucinations, and vertigo are all symptoms of lead exposure. Acrodynia, often known as pink sickness, is caused by mercury intoxication (Zaynab et al. 2022).

Because of their toxicity, persistence in ambient air, and potential for buildup in the body of humans via bioaccumulation, HMs become prominent environmental contaminants. Toxic HM poisoning of ecosystems on land and in water is a serious environmental hazard with public health implications. HMs may become very hazardous after interacting with various environmental factors such as soil, water, and air, and people and other living species can be subjected to them via the food chain (Mitra et al. 2022). Some HMs are somewhat volatile chemically and have a diameter of less than one micro meter, that attaches to particulate matter hence traverses to a larger extent beyond the site of release. Because of its tiny size, it penetrates the body via the system of respiration and attaches to complex protein molecules, causing denaturing causing death of cells (Srivastava et al. 2018). Since HM are unable to be entirely biodegraded, but must be converted into benign forms (Mishra et al. 2019). Undertaking a combined assessment of the origin, exposure, as well as health threat of pollutants is crucial for devising soil pollution mitigation methods (Yang et al. 2019). A precise grasp of the features that contribute to HM contamination and source attribution, as well as an awareness of the primary elements determining ecological and health concerns, are required (Liang et al. 2022). This chapter thus presents a comprehensive analysis of the sources, distribution and health impacts of HMs on humans.

3.2 Source and Distribution of Heavy Metal

Heavy metal contamination in the environment is constant, hidden, and long-term. Soil qualities are critical in food manufacturing, and HM contamination, alongside their ensuing ingestion and bioaccumulation in crop residues, poses significant health risks, particularly in developing nations (Mitra et al. 2022). The determination of the source is critical in monitoring the environment (Deng et al. 2022). HMs including arsenic, cadmium, lead, and mercury, are unnecessary in animals, plants, or humans because they are extremely toxic. It is necessary to explain the sources and levels of HM in soils in order to define standards of quality and determine threats to human wellness and food safety (Mitra et al. 2022). To determine geographical origins, the standard analytical approach combines statistical analysis methods and elemental recognition of fingerprints. As a way to analyze multivariate models, regression analysis, hierarchical cluster analysis, as well as positive matrix factorization might be used. Remote sensing along with isotope analysis have developed as effective monitoring technologies for concentration dispersion. In regard to remote sensing, multispectral aircraft along with satellite systems were used to swiftly monitor HMs on a broad scale to perform spectral along with morphological analysis. In terms of isotope analysis, the polluted region may have a unique isotope signature, and the significance of releases from people's actions and industrial operations may be reflected in their isotopic fingerprints. Accurate and quick tracing remain significant issues (Deng et al. 2022). The levels of HM are controlled by the kind of soil, plant genetic makeup, and interactions between the two. Mineral fertilizers have higher amounts of HM than organic manure. Toxic metals provide health concerns based on the quantities of these elements in particular media and the period of exposure. Continuous and chronic contact with dangerous metals, even at low levels, can cause health concerns. Heavy metal toxicity constitutes one of the most important abiotic stresses on vegetation, and it depends on the metal's physiochemical characteristics (Mitra et al. 2022). Contamination from heavy metals chains usually consistently adhere to a cyclic pattern throughout the environment, such as industry, the air, water, soil, foods, and humans. HM were shown to be harmful to the health of people as well as the environment (Ahmad Bhat et al. 2019). HMs are distributed in various media throughout the environment these act as the source of the HM as well. This is further elaborated in the following discussion.

3.3 Brief Depiction of HMs Distribution Across Various Media Acting Further as Their Sources

3.3.1 HM in Rocks and Soils

3.3.1.1 Igneous Rock

The primary naturally occurring sources of HM in the natural world are soil and rocks. When magma cools, the fundamental rocks, known as magmatic or igneous stones, consolidate and can be carried to the ground by geological events including volcanism or the movement of plates. HMs are integrated as trace components into the crystal lattices of the principal minerals formed during magma cooling. This is known as isomorphic transposition. The majority of HM are concentrated mostly in leftover magma. Very few heavy metals have the ability to produce their own mineral. Cr crystallizes as chromite, Ni as arsenopyrite, Pb as galena, Zn as sphalerite, Cu as chalcopyrite, Mo as molybdenite, and Fe as pyrite (Bradl 2005).

3.3.1.2 Sedimentary Rock

The process of weathering disintegrates rock into particles known as sediments. The rock is dissolved into ions by the process of chemical weathering. These fragments of sediment are carried to sedimentary basins, where they settle. When fresh sediment accumulates, a process known as diagenesis or lithification occurs. The free particles are subsequently chemically bonded to one another, producing solid rocks. Gypsum ($CaSO_4$-$2H_2O$), anhydrite ($CaSO_4$), calcite ($CaCO_3$), phosphorite ($Ca_3(PO_4)_2$), borate salts, and goethite (FeOOH) constitute these. If they are pierced by ore-beating hydrothermal liquids, they incorporate mineral deposits of various HM (Bradl 2005).

3.3.1.3 Metamorphic Rock

Metamorphosis describes the mechanism through which metamorphic rocks are formed. The rearrangement and the recrystallization of rock constituting materials characterizes metamorphic rocks. Any magmatic, sedimentary in nature, or metamorphic rock is capable of being exposed to elevated temperature and pressure situations such as during plate movement into deeper portions of the earth's crust, resulting in chemical changes. HM may accumulate during the production process or from the parent material (Bradl 2005).

3.3.2 HM in Water and Groundwater

3.3.2.1 Surface Water

Surface water dynamics is highly impacted by the type of soil and rock that the water runs on or across, including rivers, streams, springs, and lakes, and ponds. Temperature, pH, redox value, desorption and adsorption mechanisms from either organic or inorganic matter in suspension or sediments at the bottom, exchange of cations, diluting, evaporated water, and creatures present are the key biological, chemical, and physical variables that determine water content. Water running across limestone ($CaCO_3$), for instance, will produce a pH of around 8, whereas water that passes through granite will create a pH of about 6. If pyrite (FeS_2) exists, oxidation of the material will result in the formation of acid fluids, which then can impact HM dissolution and enhance metal mobility (Bradl 2005).

3.3.2.2 Groundwater

Groundwater is a vital direct supply of water for consumption from wells sunk into the aquifer, as well as water utilized for farming, namely irrigation. HM enter groundwater mostly through agricultural as well as industrial operations, landfilling, mineral extraction, and shipping. HM destiny and transit in groundwater and soil can occur in a variety of ways. Colloids are highly mobile in groundwater as well as soil systems, increasing HM transportation (Bradl 2005).

3.3.3 HMs in Soil

Ions of metal can be bonded to mineral components or organic particles through complexation/sorption processes. Soil and soluble organic material are common contributors to inorganic as well as organic chemicals, and sorption mechanisms can absorb HMs. The resultant complexes are leached and transported into groundwater. HM may be discharged into the air at the surface of soil due to erosion of the surface and colloidal losses (Bradl 2005).

3.3.4 HMs in Air

Heavy metals are registered as gasses, aerosols, and particles in air. Mineral particles, sea salt particulates, extraterrestrial substance, volcanic airborne particles, wildfires, and pollution from industries such as pollutants from transport, the burning of coal, and escape particulate emissions are all sources of HM. For instance, Cadmium air emissions are mostly created by mineral extraction and burning of waste, whereas lead emissions are brought about via the volatilizing of gasoline additives. Nickel is a naturally occurring component of oil that is discharged into the environment upon burning. Typical enriching products in fly ash include molybdenum, copper, zinc, cadmium, lead, arsenic, selenium, antimony, as well as mercury, that were previously bonded in sulfide minerals found in coal and certain ores. Pollutants are emitted at the point of origin and carried by prevalent air flows. On their route to the immission site, they may be diluted, formed precipitates, or altered by chemical processes. The volatile metalloids including Se, Hg, As, and Sb can be transferred in both gas and particulate forms. Cu, Pb, and Zn, on the contrary hand, are only carried as particles. Metals in plants and soils are largely derived from deposited in the atmosphere. In the case of Pb, atmospheric deposition is responsible for more than 90% of total plant absorption, whereas in the case of Cd, atmospheric deposit is responsible for over half of total plant absorption (Bradl 2005).

However, the physical characteristics and chemical makeup of HMs may change as a result of geochemical activity (weathering as well as eruptions of volcanoes) and human actions (mine and farming), which may result in certain HMs release. Furthermore, in nature, these HMs might move and convert with numerous chemical forms. For example, cadmium sulfide in soils might readily be oxidized, and obtain rather great migratory potential into the food chain. Environmental variables including pH, redox capacity, salinity, and temperatures influence HM release. It has been observed that HMs are readily released from the solid phase in naturally contaminated rocks, sediment aquifers, and fragmented bedrocks because of pH variations. The synergistic advantages of pH-dependent activities are related to the significant connection that exists between HMs leakage and pH (Deng et al. 2022). Heavy metals migrate out of the abiotic the setting to living species, where they concentrate in biota at various trophic levels and pollute food webs. Trophic transfer, bio accumulation, and biological magnification of dangerous heavy metals, as well as biotransference, bio concentration, dietary buildup, biodilution, and trophodynamics, have significant consequences for wildlife and human health (Ali and Khan 2019).

After being discharged into the natural world, HMs' temporal and geographical mobility are essentially determined by their chemical/physical characteristics, ambient medium parameters, and contacting material characteristics. Mechanical movement, physical/chemical immigration, and biological movement are the three types of transportation systems. In the context of mechanical migration, mechanical power, which can be separated into fluids and gravitational force, plays an essential role in migratory and the redistribution of HMs. The major mechanical transport is fluid migration, which comprises free diffusion and transport. The migration's distance may be influenced by meteorological/conservancy circumstances, emission levels, and total quantities. Finally, the HM particles appear to have been settled by gravity migration. Chemical and physical modifications in HMs are another way for contaminants to migrate into the surroundings. HMs experience dissolution–precipitation as well as adsorption–desorption in this procedure, and they are additionally adsorbed and moved in tandem with colloids as well as particles (Deng et al. 2022).

Moreover, Geophagy is a frequent practice amongst some ethnic groups, particularly women, in several Nigerian rural areas. The safety of consuming such clays due to their substantial metal content has received little attention. All geophagic clay minerals consist of hydrated silicates of Al, (Na as well as Ca), Al and Mg, or Mg and Fe. Na, Al, Ca, Fe, Mg, Cu, and Zn concentrations are bearable and appear to be a veritable supply of minerals that are missing in the human body. In regard to health risk evaluation, the existence of heavy metals having health risk indexes such as As, Cd, Pb, Se, and Sb makes the geophagic clays dangerous for ingestion. In a comparable manner Al, Fe, and Na are abundant in the clay, causing major health hazards. Consequently, ingesting geophagic clay materials containing heavy metals such as Pb, As, Cd, Se, as well as Sb by pregnant women and children entails a risk of several medical diseases and should be regarded a public health hazard (Lar et al. 2015).

3.4 Sources of HM in the Environment

HMs can be sourced naturally or artificially (Srivastava et al. 2018). They exist inherently in the planet's crust, and their chemical makeup vary between locations, resulting in differences in structure in adjacent concentrations. Some HM are vital in tiny levels to living creatures due to their metabolic activity. When certain HM are released into the environment, their high solubility transforms them into very poisonous and dangerous contaminants. When these metals are discharged into the surroundings, they can leach into subsurface waters and accumulate in aquifers, or they can flow away into water on the surface and the earth's soil causing water and soil contamination. As a result, HM constitute a possible environmental pollutant that might participate in trophic transmission in food chains (Mishra et al. 2019). HM arise naturally as a result of volcanoes, geothermal spring operation, erosion, infiltration, and other processes (Sall et al. 2020). Activities such as volcanic eruptions, jungle wildfires, cyclonic disruptions, and cosmological dust pollute the atmosphere by releasing rare earth metals such as Pb, As, Cd, and Cr in the format of particulates, with volcanic activity accounting for 80% of HM spreading and the other twenty percent coming from additional organic sources along with fires in forests (Srivastava et al. 2018). An investigation in Ghana's biggest manganese mine revealed that underground water transformed from Ca–Mg–SO_4, Ca–Mg–HCO_3–SO_4, Ca–Mg–Na–Cl towards Ca–Na–Cl–HCO_3 water category alongside other combined types, illustrating strong water–rock relationship and the dissolution of rocks in the region pertaining to the Banded Manganese Formation as well as the Birimian metasediments. The water's Hazard Index scores vary between 0.20 and 30.5, indicating that the water is harmful to varying degrees. The mine pits water has incredibly high levels of Fe, Mn, Ni, As, and Sb, resulting in hydromorphic dispersal of the heavy metals to the surrounding water and underground. The risk of cancer in groundwater is modest, but it is substantial in water on the surface and quite high in mine pit water (Ewusi et al. 2022). Rock genesis, erosion, weathering, along with volcanoes are each generally liable for the exposure and release of massive amounts of Al, Cu, Hg, Mn, Ni, as well as Zn into the natural world. HMs such as aluminum, copper, even silver are typically deposited as minerals in various kinds of rocks such as namely igneous hornblende, olivine, augite, along with sedimentary, alongside the rocks containing sufficient quantities of metallic substances for profitable extraction are referred to as ores, leading significant amounts of metals in their most pure obtained forms (Mishra et al. 2019).

Anthropogenic practices that cause contamination include improper industrial waste administration, traffic congestion, the consumption of lead as a fuel anti-knock, cans of aerosol, metallurgy and smelting, wastewater discharge, and building supplies. Trace quantities of HM in fertilizers are significant contributors to heavy metal contamination in diet (Zaynab et al. 2022).

Figure 3.1 depicts a brief pathway from source to HM ingestion in Human Beings. Rapid urbanization and industrialization have culminated in substantial amounts of HM (Liu et al. 2019). The largest single contributor of toxic metals in the natural world comes from a variety of operations and routes, including the ignition and obtaining of toxic waste into the air, as well as excess emissions of wastewaters/effluents comprising detrimental toxic pollutants and chemicals, which pollute surface waters and soils. The continued use of dangerous metals in human-induced processes such as the electroplating process painting, the tanning process, textiles and pigments, paper production, mineral extraction, and others has expanded dramatically, threatening the variety of life upon Earth. The indiscriminate discharge of dangerous metals from various industrial operations, which frequently occurs at excessive quantities (beyond legal limits), poses a toxicity concern (Mishra et al. 2019). HMs such as copper, chromium, and nickel are distributed in mining sectors because they undergo the elevated temperatures refinement process. The coal includes trace amounts of various elements, and transport of coal and mine fires emit particulate matter including heavy metals. The most prevalent emissions from the chromium and lead sectors are secondary pollutants and mixtures of various heavy metals. Increased chromium requirement has been seen in the steel and iron industries. Oil combustion, fossil fuels, and automotive emissions are larger areas for investigation of metal emissions such as lead, nickel, plus arsenic, as well as manufacturing waste dissipates cadmium in the air. Those heavy metals exhibit spatial fluctuation in intensity based on their site

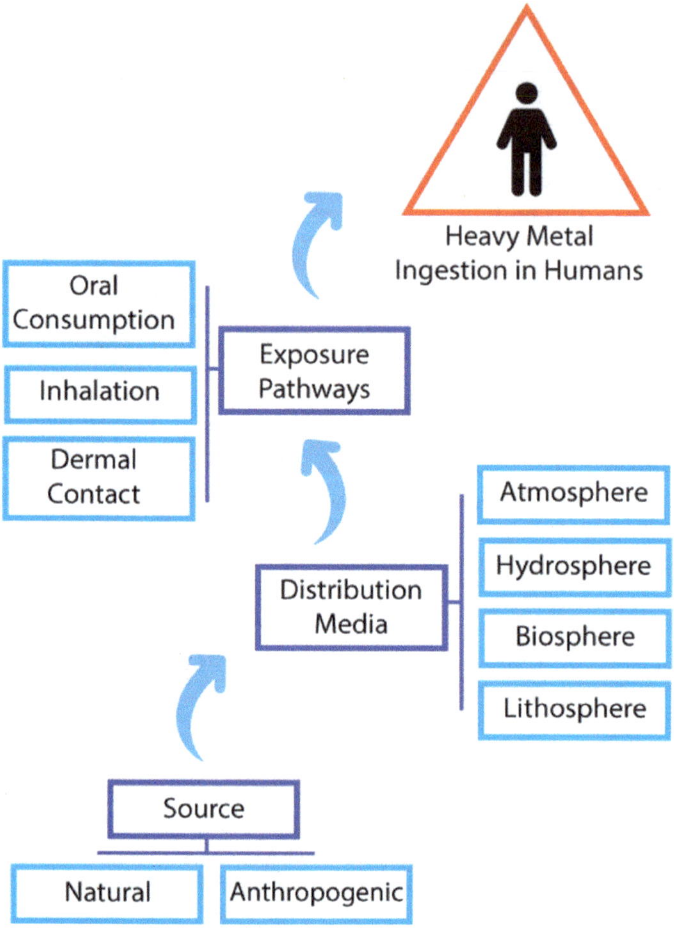

Fig. 3.1 Brief depiction of pathway from source to ingestion of heavy metal in human

of origin and the distance traveled after release (Srivastava et al. 2018). Mercury is released into the environment by sectors such as medication manufacture, paper and pulp preservatives, agriculture, and the chlorine as well as caustic soda enterprises. Cadmium is found in soils and rocks, as well as coal and fertilizers made from minerals. Cadmium is frequently employed in electroplating for a broad range of applications such as batteries, dyes, fabrics, and metal coatings. All of these actions contribute to rising HMs pollution of the environment (Zaynab et al. 2022).

3.5 Source Analysis of HMs Detected in Various Media

HM are typically introduced into populations through water drinking (Chowdhury et al. 2016). Both environmental and human activities were identified as sources of pollution by HMs in samples of water in a research that examined HM concentrations in 8 randomly chosen boreholes in South Africa. The estimated non-carcinogenic consequences using the hazard quotient toxicity possibility, cumulative hazard index, as well as chronic everyday consumption of groundwater via consumption and skin adsorption pathways had been below one, indicating that drinking the water might present little to no essential health risk. Maximum predicted values for a person, however, surpassed the danger threshold for Cr as well as Pb in groundwater (Edokpayi et al. 2018). Water contamination by HM has numerous human basis, like as the combustion of fossil energy sources, exhaust emissions of automobiles, mining, crop production, and combustion of solid and liquid garbage. Heavy metals arise naturally as a result of volcanoes, geothermal spring operation, erosion, infiltration, and other processes. Since majority remain hazardous to people and marine life, water poisoning poses a threat to all living things (Sall et al. 2020). The major cause of HM contamination in humans is polluted drinking water. Heavy metal exposure in the workplace arises as a consequence of employing such metals in a number of manufacturing procedures and/or products, particularly pigments for color and alloys (Fu and Xi 2020).

The source assessment of HM extracted in various seasons collected from the Taizihe River, China of Zn, Cu, Pb, Cr, and Cd demonstrated HM pollution was mostly caused by industry, particularly mining, metal smelting, and electroplating (Chen et al. 2020). According to a research conducted in Nigeria, the primary origin of HM in water and soil may be traced to human endeavors including mining and quarrying. Some researchers have also suggested that geogenic activities are involved. Detail investigation revealed that mining is responsible for 62% of HMs in water and soil, indiscriminate waste disposal is responsible for 23%, and other anthropogenic and geogenic activities are responsible for 15%. The buildup of HMs in soil, vegetation, and water constitutes a major threat to humans, plants, and aquatic species in the research region (Eyankware and Obasi 2021). HM origin detection in a study examining the Beiyun River in China discovered that Factory operations (23.0%) are thought to contribute the most HM to the Beiyun River, next to transportation (17%), agriculture (16%), then fallout from the atmosphere (16%). According to the source-oriented risk evaluation, farms in the Beiyun River provide the most Hazard Index and Cancer Risk, next to industrial operations (Wu et al. 2021). A groundwater investigation found that heavy metal levels were greater throughout the rainy season than throughout the dry season. The primary sources include industrial as well as agricultural operations, as well as natural leakage, and each type of heavy metal could possess distinct origins. The chief contaminants are Fe and Mn, which are mostly caused by the native ecosystem and farming operations. The health concerns posed by heavy metals were distributed as follows: kid > older > adolescent > adulthood.

Carcinogenic heavy metals lead to the primary dangers, with Cr and As being the most significant risk sources (Zhai et al. 2022).

According to the danger of HM buildup in soils and crop yields of vegetables watered with effluent water in Saudi Arabia, sewage irrigation infrastructure might create potential problems with contaminants to groundwater and soil, potentially posing a harm to human beings. In a study that examined Cd, Cr, Cu, Pb, along with Zn. In this investigation, the origin of soil contamination was discovered as irrigation. Sewage irrigation with water is recognized to considerably increase soil HM levels. Particular HM may build up in plants, posing a major threat to the well-being of people when foods made from plants are ingested. Sewerage water and waste from industries are frequently drained to agricultural regions, where they are utilized to grow crops such as vegetables. Such sewage effluents raise HM content of recipient soils, including Fe, Mn, Cu, Zn, Pb, Cr, Ni, Cd, and Co (Balkhair and Ashraf 2016). A study was conducted to address the issue of medicinal plant pollution caused by pollutants in the environment caused by various industrial operations and the atmospheric release of some harmful substances. Plants can take in organic and inorganic substances from all parts of the environment (fluid, dirt, air) and pass them up the food chain to people (Asiminicesei et al. 2020). Vegetables, whether leafy or nonleafy, tend to be excellent HM accumulators. The biological accumulation sequence for nonleafy plants comprised leaf > root > stem > tuber. Because heavy metals possess a significant impact on the nutritional content, plants produced on metal-contaminated earth were deficient in nutrients, moreover consuming of these vegetables could contribute to a lack of nutrients in the population, especially in developing countries where malnutrition is already an issue (Khan et al. 2015).

Moreover, currently, there is a tendency regarding the use of conventional as well as herbal treatments, while there is rising fear that they are contaminated with heavy metals. According to a research of eight distinct kinds of medicinal plants and eight different varieties of herbal distillate offered in Iranian marketplaces, the mean percentage of all metals was within permitted limits. The therapeutic plants had much higher concentrations of Al, As, Cd, Cr, Fe, Mn, Ni, Pb, as well as Zn compared to the herbal distillates (Kohzadi et al. 2019). Lettuce, spinach, radish, along with carrot were produced on manure which was earlier polluted with Cd, Cu, Mn, Pb, and Zn at varying quantities. The results revealed that plant absorption of Cd, Cu, Mn, and Zn increased with increasing soil pollution, however Pb uptake was negligible. Mn_ Zn > Cd > Cu > Pb lowered soil-to-plant transference factor values (Intawongse and Dean 2006). Furthermore, this study discovered that particular plant varieties varied substantially in their metal absorption; for example, spinach acquired an elevated level of Mn and Zn in its tissues, but Cu and Pb levels were very low (Intawongse and Dean 2006).

Source allocation in farmed soil at Lanzhou, China identified four key sources of 8 HMs. Pb, Cd, Zn, and Ni were linked to transportation; Hg was linked to farming and healthcare; Cr, Cu, and Ni obtained from organic sources; and As originated from factories (Chai et al. 2021). The soil HM evaluation in the Yellow River irrigated area assessed the level of Cd, Cu, Pb, Zn, Cr, Ni, as well as As in soil used for agriculture in the Ningxia part. It was revealed that agricultural and commercial sources made up 27.06%, sources from nature for 14.12%, industrial supplies for 23.04%, transportation sources for 21.50%, while Yellow River geological resources for 14.28%. Furthermore, the most major factor producing non-carcinogenic hazard (HI) for young people is a combination of agricultural and industrial sources (55.75%). Two of the primary causes of HI in adulthood were factories and traffic-related factors (factories sources comprised 25.16% and transportation sources contributed for 28.78%). The two primary variables that produce carcinogenic risk (CR) are a combination of sources of farming and manufacturing and natural sources (a combination of sources of farming and factories account for 35.34%, while natural resources accounted for 33.23%) (Shen et al. 2021). Studying HM pollution within the Shi Rivers Basin soils in China reveals that the concentrations of HM could be influenced by naturally occurring sources like parent matter. Human-caused pollution is predominantly impacted by emissions from transportation as well as manufacturing sources, which account for 62.6%, then by agricultural products that involve fertilizers and pesticides, that account for 21.1% (Zhang et al. 2021).

Cropland soils surrounding an e-waste disposal facility in Taizhou, China, were employed in a study to assess the contamination features of HM. High levels of Cd, Cu, Pb, and Zn were found near the e-waste disposal location, according to spatial analysis. In the research region, the principal contributors to HMs are e-waste deconstruction operations, sources from nature, and the atmosphere, which constitute 27, 46, and 27% of HM pollutants, correspondingly. E-waste disposal activities were a prominent element influencing noncarcinogenic danger (Liang et al. 2022). Using an old electronic waste disposal facility in China, researchers analyzed arsenic, cadmium, copper, chromium, nickel, lead, and zinc. The findings revealed that farming soils were primarily polluted with Cd as well as Cu. Parent substance and usage of pesticides, the use of fertilizer, factory release, and automobile emissions contributed around 46.6, 22.2, and 31.2% of metal buildup in soil, sequentially (Yang et al. 2019). Characterization of the concentration and origin of heavy metals in soil is required to define regional quality requirements which enable the identification of polluted sample sites. Heavy metal concentration of soils used for agriculture in a European Mediterranean area was determined by sampling Cd, Co, Cr, Cu, Fe, Mn, Ni, Pb, and Zn origins. Co, Cr, Fe, Mn, Ni, and Zn concentrations were related with the parent rock and belonged to the first primary component known as the lithogenic constituent. Anthropic activity were associated with components such as Cd, Cu, and Pb, which composed the second Cu and Pb and tertiary primary component Cd, dubbed the human-caused components (Micó et al. 2006).

A combined strategy for determining ecological and health dangers to numerous resources of soil heavy metals discovered that in terms of non-carcinogenic exposure, farming procedures had been the biggest origin under wooded areas and agricultural land, whereas factories were the biggest origin under land used for construction. In terms of carcinogenic risk, agricultural operations were the most common source, regardless of land use (Jiang et al. 2020). The occurrence of Cr, Ni, Sb, Ag, Pb, Cu, and Zn in dust and soil were primarily affected by human-caused causes, especially emissions from vehicles, from industry, and native household substances, whereas

Co, Fe, As, Mn, and Cd are from sources that are natural, according to source evaluation and a health risk evaluation of heavy metals pollution in dust from homes and soil surfaces from Nepal. The projected cancerous risk of HMs in dust and dirt surpassed the safe limit for human exposure (Yadav et al. 2019). In a farmland in China Human activities such as mine and smelting, industry, sewage irrigation, urban growth, and the use of fertilizers discharged heavy metals into the soil, resulting in contaminated farming soil (Zhang et al. 2015). Human actions significantly influence heavy metal contamination in soils. In 105 samples of soil obtained from an ordinary rural-industrial community in China, levels of 15 metallic components were found. Seven heavy metals were included in the health risk evaluation: lead, copper, zinc, arsenic, chromium, cadmium, as well as nickel. Overall non-carcinogenic and carcinogenic hazards caused by soil metals were found to be above permissible limits. Cr and As demand special monitoring since their specific health hazards surpass permissible limits. The consumption of homegrown vegetables was mostly to blame for the significant dangers. Incineration of waste and textile/dyeing enterprises were identified as potential sources, as were the natural environment, emissions from vehicles, electroplating businesses, and livestock/poultry rearing (Jiang et al. 2017).

The resource allocation of HM in airborne particles in a municipality revealed five major probable resources, burning of coal, transportation and factories, organic, architecture and home furnishings sources, along with unknown sources, with corresponding percentages of 34.09, 20.72, 18.72, 7.597, and 18.87% (Ma et al. 2021). Analysis of HM contamination features of ambient exposure to PM2.5 in China reveals that Cu, Zn, Cd, Pb, as well as Hg are mostly from human sources. According to the analysis of principal components, the primary sources of heavy metals are car exhaust, waste from chemical plants, and coal-burning operations. In order of decreasing danger level, all 9 heavy metals Cr, Cd, As, Ni, Cu, Pb, Mn, Zn, and Hg were under the permitted threshold (Chen et al. 2015). The 3 origins of heavy metals detected in street dust in a Vietnamese metropolitan region were automotive activities 32.38%, a combined source of automotive and residential operations 26.72%, and a combination of commercial and organic sources 20.23% (Dat et al. 2021). 20 trace elements found in fine particulate materials were identified in a Chinese megacity in order to research the features, origins, and dangers to human health of particle hazardous heavy metals such as As, Cd, Cr, Cu, Mn, Ni, Pb, along with Zn. The findings revealed that Cd may be emitted by metallurgy and mechanical manufacture, but the other metals were mostly impacted by traffic pollution and coal burning (Li et al. 2016). Multiple concentrations of heavy metals were determined in interior dust from recycling electronic waste, countryside, and urban locations in China. Except for Zn, the concentrations of all contaminants in dust collected from the electronic waste recycling region were substantially greater than those in the other sites. Cd and Pb pollution patterns were identical in all three sites, showing that e-waste recycling operations are the primary source of pollution. Cu, Cr, and Zn, on the other hand, were most likely sourced from home items in both rural and urban settings (He et al. 2017). Heavy metals Cu, Ni, Pb, and Zn were detected in deposited dust at the surface in 9 houses in various parts of Pulau Pinang, indicating that there was a predominant cause for the heavy metal pollution, that could be tracked back to car exhaust, street dust, as well as associated activities (Wahab et al. 2012).

A research looked at Cd, Ni, Zn, Pb, along with Fe in Iraqi sandstorms. Human activities were speculated to result in HM in desert dust storms (Kadhum 2020). HM are known to be released during electronic trash or e-waste disposal procedures (Zhang et al. 2019). For Xinqu Park, the main contributor was industrial emissions 42.14%, with coal combustion and transportation emissions contributing 34.03% and 23.83%, correspondingly. During the non-heating season, the principal contributors at Huagong Hospital had been the combustion of coal 45.29%, industrial exhaust 45.29%, and vehicular emissions 10.12% (Sun et al. 2021). HM evaluation of As, Cd, Co, Cr, Cu, Ni, Pb, V, and Zn discovered contributors in the surface ground of an oil field in Iran. Metals such as Co, Cr, and V were found to have natural origins, but As, Cd, Cu, Ni, Pb, and Zn were shown to have manmade origins, such as petroleum leaks and pollution produced by drilling fluid from oil wells. Children have a greater risk of both carcinogenic and non-carcinogenic illnesses than adults. Under experimental circumstances, the carcinogenic risk suggested a danger (Ghorbani et al. 2020).

There were the following sources recognized for HM from a smelter and mine zone in China: air deposits from coal combustion and vehicular emissions (18.0%), naturally occurring sources from the parent substances (34.3%), nonferrous mineral extraction and smelting industry (37.9%), and previous arsenic-related activities (9.8%). The findings emphasize the need of taking legacy As contamination from decommissioned enterprises into account when formulating risk reduction measures in this region (Zhou et al. 2022). At a mining location in Ghana the HM rated by severity of health concerns were as As < Pb < Cr < Cd where study indicates anthropogenic sources for these metals, which may be emitted by gold-mining operations (Hadzi et al. 2019). Urban landfill leachates include hazardous heavy metals that were recently proved to be harmful to people and the natural world. The purpose of this study was to examine the presence of heavy metals in effluents, water from the surface, as well as sediments at non-sanitary dumps in Nigeria, as well as to determine possible health and ecological repercussions of leachate pollution. Pollution indicators indicated that samples of sediment were mild to highly contaminated by chemical pollutants from non-sanitary dumps, and that biological accumulation might represent a problem (Essien et al. 2022). Tailings involving smelting and mining operations can pollute the land with heavy metals and endanger human health. A number of HM (As, Cd, Cr, Cu, Ni, Pb, along with Zn) were discovered in tailings and farming soils near a typical copper tailings lake. The findings revealed that tailings discharge is the primary cause of soil HM pollution, accounting for 35.81% of the total, which was followed by farming operations (19.41%) and traffic emissions (16.31%) (Xiao et al. 2022). Nonferrous metal smelting is an important contributor of manmade HM emissions, resulting in severe environmental contamination that endangers the health of nearby populations (Hu et al. 2021).

The HM Composition in the Hair of Humans study was conducted to evaluate environmental exposure depending on the distance between a subject's home and a pollution source. Hair has been discovered to be an excellent indication of environmental contamination. Sources included proximity to a heat and electricity producing facility, a nonferrous metals plant, or additional manufacturing facilities, in addition to relationships among components in a human organism. Furthermore, statistically significant variations in Ni concentration in male and female hair were discovered. These findings demonstrated that hair mineral composition reflected exposure to environmental factors (Chojnacka et al. 2012). A randomized study of blood from individuals samples in Kerala, South India, was undertaken to determine the hazard of pollution with heavy metals, particularly in connection to the quantity of such metals found in meals and other ambient materials. Except for pure vegetarians, rice remains the staple meal of Kerala, in addition to fish. Heavy metal levels in the blood were studied and food choices such as fish consumption or vegetarian. The concentrations of arsenic, lead, and mercury were found to be substantially below the reference norms, but eating appeared to bring them up, since the quantity of metals in blood differed across vegetarians and non-vegetarians (Jose and Ray 2018).

3.6 Effects of Heavy Metal on Human Health

Certain heavy metals are important as trace elements, however their biotoxicity with regard to human biochemistry is a major issue. Their origins, leaching processes, chemical transformations, and routes of deposition all contribute to pollution of the natural world (Mahurpawar 2015). HM poisoning is recognized as a serious hazard, and it is related with a number of health issues. They can behave as pseudo components of the body at times, but they are also capable of interfering with metabolic processes. Few HM, such as aluminum, may be eliminated by excretion processes, but others accumulate in the human body as well as the food chain, demonstrating a chronic character (Ahmad Bhat et al. 2019). Pb, Cd, and Hg poisoning predominantly affects the kidney, liver, as well as neurological system in people. Methylmercury poisoning throughout childbearing, in especially, has an irreversible effect on the growth of the baby' nervous system. HM can be transferred from mother's body to the embryo via the umbilical cord, and transfer through mother's milk can cause abnormalities in early infancy such as cognitive problems and urinary tract abnormalities. Long-term consequences in numerous tissues, including genetic mutation, cancer risk, teratogenicity, immune toxicity, and hormonal disruption, are more commonly related with heavy metal intake via diet (Russo et al. 2013). Mercury exposure alters the structure of the brain, causing shyness, tremors, cognitive decline, and anger. At greater concentrations, metallic mercury fumes cause lung damage, diarrhea, nausea, nausea, skin rashes, and elevated blood pressure. Depression, cognitive difficulties, tremors, weariness, headache, and hair loss are all symptoms of organic mercury poisoning (Zaynab et al. 2022). Lead accumulates rapidly in the gills as well as livers of fish. Humans often consume fish in the course of their daily diet. Because lead is carried into the bloodstream and absorbed into tissues upon absorption, HMs accumulate in the body (Zaynab et al. 2022). Public health risk assessment approaches, such as the Rabinowitz, Bert, Stern, along with Leggett model, might quantify the risk level of HMs exposure. Health risk assessment HRA, in particular, has been frequently used in the health-related evaluation of HMs contamination (Deng et al. 2022). Moreover, the significance of HMs in generating drug resistance was verified. The specimens collected from an industrialized region's treatment facility had greater HM pollution as well as a greater percentage of substantial correlations, demonstrating that wastewater treated and distributed through the surrounding ecosystem may represent an ongoing risk to human well-being by transmitting antibiotic resistance genes, antibiotics, as well as HMs to the natural world. These findings give light on the influence of industrialization and HM on the spread of resistance to antibiotics (Hubeny et al. 2021).

3.6.1 Exposure and Transportation Pathways of HM for Humans

HMs migrate/assimilate into water used for drinking, agricultural seeds/fruits, and living organisms, eventually entering the human body by consumption. Particularly, HMs accumulate on fine PM and micro plastics in the atmosphere, thereby increasing the risk of inhaling exposure (Deng et al. 2022). Because of the stratum corneum, contact with the skin is regarded as a less relevant mechanism of exposure. Although human exposure doses may be obtained using relevant data, several real exposure parameters on geographical dependency ought to be examined by questionnaires or fieldwork. The harmful effect of HM is strongly linked to their bioavailability (Deng et al. 2022). Certain HM, including lead, cadmium, manganese, and arsenic, may reach the body via the gut path, or via swallowing while eating meals, fruits, and vegetables, or consuming water and other liquids. Others, including lead, are able to reach the body by breathing in, whereas others, such as mercury, can infiltrate via the skin. The majority of HMs are transported to tissues in human beings via the blood. Lead is delivered to the kidneys and liver by the blood cells and then reallocated to one's teeth, bones, as well as hair, primarily as phosphate salt. Cadmium first attaches to blood cells or albumin before binding to kidney and liver tissue. Manganese vapor dissipates over the lung barrier to the Central Nervous System (CNS) after being transported from the bloodstream to the lungs. Manganese salts that are soluble in lipids are dispersed in the gut for fecal waste elimination, whereas fluid-soluble manganese salts are disseminated in the blood plasma as well as kidney for renal clearance. Arsenic circulates in bloodstreams and collects in the heart, lungs, liver, kidney, muscle, as well as neural tissues, in addition to the skin, nails, and hair (Azeh Engwa et al. 2019).

3.6.2 HM Impact on Metabolism

Heavy metals can disrupt the metabolic functioning of the human body in a variety of ways. Furthermore, they may collect in vital human organs that include the brain, heart, liver, as well as kidneys, disrupting normal biological activities. Cadmium, arsenic, aluminum, mercury, and iron are the most common HMs that have an influence on the human body (Fu and Xi 2020). HM deposited in the body disrupt the body's primary metabolic processes and induce an unbalanced level of antioxidants. The activity of numerous hormones as well as the function of essential enzymes are also altered. Changes in glucose, protein, as well as metabolism of lipids enhance the body's vulnerability to diseases. Finally, all of these systems affect the production and application of neurotransmitter in the human organism, which alters a variety of central nervous system activities. As a result, no matter which molecular pathways heavy metals disrupt, they eventually form reactive oxygen species that can induce oxidative damage, which can lead to many types of malignancies, reduced renal function, neurological illnesses, and other endocrine diseases (Fu and Xi 2020).

Because of how they interact with biological components, HMs are hazardous. Furthermore, prolonged or severe exposure will have negative health implications. Reactive Oxygen Species (ROS) production, and activity suppression of antioxidant defense enzymes have all been widely accepted as response mechanisms for HM-induced cell toxicity. The harmful impact is mostly caused by oxidative stress caused by excessive ROS production. On the other hand, HMs or intermediary reactive species produce free radicals. HMs, on the other hand, impair the antioxidant enzyme mechanism, diminish ROS scavengers, and promote oxidation. Enzymatic reacting substances and non-enzymatic antioxidants, in general, affect the dynamic stability of the redox stations. The increased oxidative response is associated with the disruption of biological processes that include cell division, cell cycle, immunological reaction, and death of cells (Deng et al. 2022). HMs cancer-causing potential and genetic mutation are linked to the development of oxidative stress. Some investigations have indicated that cancerous metal ions, including nickel, chromium, cobalt, and arsenic, are involved in redox processes in biological systems. As a result, the free radicals created by these processes oxidize proteins and DNA. Besides the immediate harm to DNA caused by these redox events, the species created have two additional crucial roles that induce cancer in humans. A mechanism that disrupts with DNA repair is considered to be another avenue of heavy metal carcinogenesis (Fu and Xi 2020).

The disruption of redox equilibrium is widely documented as a source of HMs-related toxicity. The cells might repair different damages caused by redox imbalances by activating signaling pathways. The participation of numerous signaling pathways in HMs-induced pathogenic processes has been shown. Initially, the goal is to heal the damage by activating pro-survival pathways. When the level of damage is severe, cells engage death pathways to clean out the injured cells. Several regulatory compounds have been discovered in reaction to HMs-induced oxidative damage (Deng et al. 2022). HM exposure can either stimulate or block the transcription pathway, resulting in a variety of disease situations. Study into cell signaling and

HMs might aid in understanding how HM exposure converts them from molecular to pathological expression. Aside from the oxidative reaction, the toxicity of HMs is intimately connected to their structural features. Cd^{2+}, for example, may disrupt vitamin D metabolism and estrogenic transmission due to its structural similarity to Ca^{2+} (Deng et al. 2022).

3.7 Types of Toxic Impacts of HM on Human Health

3.7.1 Neurotoxicity

Manganese plays a role in a number of bodily physiological activities. Its acute exposure has the potential to be neuroprotective by lowering apoptotic cellular death; however, excessive exposure can cause adverse diseases such as neurological disorders such as Alzheimer's and Parkinson's disease, which result in apoptosis of cells and changes in homeostasis. Continuous Mn buildup increases the generation of ROS, which leads to the breakdown of mitochondria. When arsenic is eaten, it causes cognitive decline in the central nervous system. It's also connected to a variety of neurological ailments, including neurodevelopmental abnormalities, and causes an overabundance of neurodegenerative diseases. Alteration in transmission between neurons and neurotransmitter balance are also caused by arsenic poisoning. Arsenic's neurotoxic effects are also thought to generate a variety of apoptotic pathways. Cadmium neurotoxicity causes neurodegenerative diseases such as amyotrophic lateral sclerosis, Parkinson's disease, Alzheimer's disease, including multiple sclerosis. A number of preclinical investigations have shown that cadmium severely impairs the performance of the PNS (Peripheral Nervous System) and CNS, resulting in a variety of clinical symptoms such as neuropathy in the peripheral regions, olfactory disorders, neurological disruptions, intellectual disabilities, and intellectual disability, as well as motor damage and alterations in behavior. Furthermore, numerous forms of biological activity are altered, including cell differentiation, proliferation, and death. Cadmium neurotoxicity results from brain cell death by apoptosis; there are several apoptosis-inducing variables, including loss of neurogenesis, suppression of neuron expression of genes, epigenetic impact, and endocrine disruption (Mitra et al. 2022).

3.7.2 Nephrotoxicity

Cadmium-induced nephrotoxicity causes clinical signs that include glucosuria, Fanconi-like syndrome, phosphaturia, as well as aminoaciduria. Direct contact with the kidneys impacted the proximal tubular epithelium, leading to a substantial quantity of Cd in urine. Aminoaciduria, 32-microglobulinuria, including glucosuria, and decreased renal tubular phosphate resorption. Overexposure can cause renal tubular acidosis, kidney failure, and hypercalciuria. Lead is harmful to all body parts, but its effects have the biggest impact on the kidneys. Severe lead nephropathy results in proximal tubular dysfunction and Fanconi-like syndrome. Chronic lead nephropathy is distinguished by hyperplasia, interstitial fibrosis, tubule atrophy, kidney damage, and glomerulonephritis. Rapid tubular necrosis, acute dyspnea, changed mental state, stomach discomfort, salivation that is excessive, shakes, nausea, shivers, and hypertension are all symptoms of mercury poisoning (Mitra et al. 2022).

3.7.3 *Carcinogenicity*

Arsenic induces epigenetic alterations, harm to DNA, alterations to the expression of the p53 protein, modifications to histone DNA methylation, and decreased p21 function. Arsenic poisoning increases the risk of developing cancer by binding to proteins that bind to DNA and inhibiting the healing process. By generating ROS, lead induces destruction of the DNA repair system, cellular tumor controlling genes, and the structure of chromosomes and sequencing. Mercury's peroxidative action produces a large amount of oxygen species that are reactive, which can promote protumorigenic transmission and cancer cell proliferation. Nickel acts as a carcinogen via influencing gene regulation, transcription factor administration, and the creation of free radicals (Mitra et al. 2022).

3.7.4 *Hepatotoxicity*

Lead's toxicity to cells in the liver is widely documented. It promotes oxidative stress, which causes liver damage. Organic solvents, when mixed with lead, induce liver damage because they share the same properties as lead. Chronic lead exposure can be harmful to liver cells, causing depletion of glycogen and cellular penetration, which can progress to chronic cirrhosis. Cadmium binds to two human tissues: the kidney's cortex as well as the liver. It concentrates in the liver upon acute exposure and has been associated to a range of hepatic dysfunctions. Cadmium causes oxidative stress as well as hepatocellular damage. Cadmium-induced hepatotoxicity, acute as well as long-term, promotes liver failure and so increases the risk of cancer. Wilson's illness is well recognized for causing copper to build up in the liver. Because higher copper levels can promote oxidative stress, hepatic copper accumulation is not merely pathognomonic, but also harmful. Cholestatic liver disorders are also associated with elevated hepatic copper levels. Numerous investigations have demonstrated that Cr may affect the liver, and histological abnormalities such hepatocyte steatosis, parenchymatous degeneration, and necrosis have previously been documented. Cr hepatotoxicity is related with increased ROS levels, peroxidation of lipids, inhibition of protein, RNA, and DNA synthesis, destruction of DNA, a reduction in antioxidant enzyme activity, mitochondrial malfunction, such as altered the mitochondrial bioenergetics, development arrest, and apoptosis (Mitra et al. 2022).

3.7.5 *Immunological Toxicity*

Rapid and long-term exposure to lead has a number of adverse impacts on the body's immune system and produces a variety of immunological responses, including increased allergies, infectious illnesses, autoimmunity, and cancer. Lead exposure has been related to an increased incidence of lung, stomach, along with bladder cancer in numerous demographic groups. Lead exposure can alter cellular and humoral reactions by altering T-cell function and increasing vulnerability to the onset of autoimmune and hypersensitive. Depending on the exposure settings, occupational and environmental cadmium exposure may have immunosuppressive effects. Humoral immune system reactions are increased at low levels of exposure, but the consequences at greater levels are unknown. In most cases, phagocytosis and host resistance are significantly diminished. Chromium has a number of negative impacts on human immunity. Hexavalent chromium at high dosages inhibits the phagocytic activity of alveolar macrophages as well as the humoral immune system's reaction. Furthermore, chromium causes two forms of hypersensitivity responses: type I (anaphylactic) and type IV (delayed). Many investigations have revealed that chromium exposure can cause contact allergy dermatitis (Mitra et al. 2022).

Here, Fig. 3.2 shows the toxicological health impacts of HM on human beings.

3.7.6 *Cardiovascular Toxicity*

Prolonged lead exposure can result in arteriosclerosis or high blood pressure, as well as thrombosis, atherosclerosis, and heart illness. Long-term exposure also raises the pressure in the arteries. Contact with low to average levels of cadmium causes diabetes, high blood pressure, carotid atherosclerosis, peripheral artery disease, persistent renal disease, coronary artery disease, stroke, and cardiac arrest. Cadmium was associated to an elevated likelihood of cardiovascular mortality in prospective studies. Mercury levels in hair have been associated to elevated oxidized levels of LDL in atherosclerotic lesions, a condition known as acute coronary syndrome, and atherosclerosis. Cobalt poisoning induces transient systolic cardiac depression, which may be separated from other types of cardiomyopathies. Cobalt cardiomyopathy may be gradual and lethal (Mitra et al. 2022). Several additional organs are also impacted as a result of tiny particles entering the alveoli of the pulmonary system and traveling through the circulation. A number of studies have been published that focus upon the younger generation, which is more vulnerable to toxicants as a result of the general impact caused by HMs (Srivastava et al. 2018).

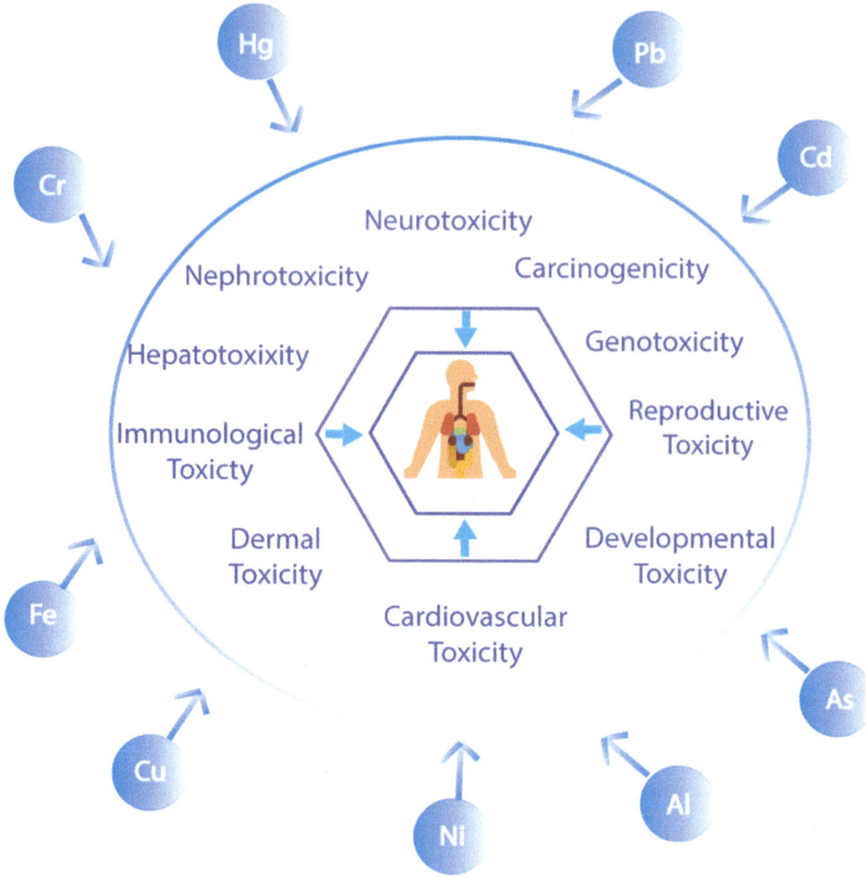

Fig. 3.2 Brief depiction of toxicological health impact on humans due to heavy metals

3.7.7 Skin Toxicity

Chronic contact with arsenic can cause a variety of skin problems, notably hyperkeratosis, hyperpigmentation, and multiple kinds of skin cancer. The most common skin alteration produced by arsenic exposure is hyperpigmentation. Bowen's disease may be caused by arsenic exposure. Arsenic hyperkeratosis mostly affects the soles and palms, although it may additionally impact the legs, toes, hands, arms, and palm dorsum. Some hyperkeratotic and Bowen's disease sores have a chance to progress to invasive cancer. Contact dermatitis, systemic contact dermatitis, including cancer of the skin are all caused by chromium exposure. Many skin illnesses, notably acrodynia or pink disease, are brought on by mercury and mercury-containing substances. Inflammation may occur in people who have been tattooed using the colors red cadmium sulfide and mercury sulfide. The most common cause of problems with the skin is mercury toxicity (Mitra et al. 2022).

3.7.8 Reproductive and Developmental Toxicity

Arsenic is a recognized human reproductive poison. Inorganic arsenic inhibits male fertility by lowering testicular weight and the amount of sperm found in the epididymis. In addition to reducing the generation of sperm, inorganic arsenic exposure alters testosterone and gonadotropin quantities and disrupts the steroidogenesis process. Arsenic intake is linked to an increased risk of endometrial cancer in women. Arsenic exposure while pregnant then impairs endometrial angiogenesis, a process essential for embryo growth. Due to their contact with heavy metals that include lead, cadmium, mercury, and other pollutants, over ten percent of women have a risk of sterility. The Female infertility is primarily caused by an imbalance in hormones, which is compounded by heavy metal poisoning's disruption of the endocrine system (Mitra et al. 2022). Heavy metals constitute endocrine disruptors, interfering with activities regulated through the organism's natural hormones. Exposure to heavy metals has been linked to male sterility in several studies (López-Botella et al. 2021).

3.7.9 Genotoxicity

Arsenic gene toxicity causes DNA alterations such as chromosomal anomalies, alteration, micronuclei formation, elimination, and sister chromatid interchange. Regardless of the fact that arsenic exerts no immediate impact on DNA, it influences the capacity for mutation of various other carcinogens. When subjected to UV light, arsenic, for instance, was found to be more mutagenic within human cells. The genetically toxic impacts of chemical chemicals on humans produce alterations in genetic material predominantly through a pair of mechanisms: teratogenesis and carcinogenesis. In human and cultivated cells, Cr(VI) produces an extensive variety of genetic material structural alterations, involving DNA chromosomal protein crosslinks, inter-DNA strand crosslinks, as well as nucleotide strand breaking (Mitra et al. 2022).

3.8 Health Risk Analysis of HM from Various Sources

Immune deficiency osteoporosis, dementia, and organ dysfunction are all possible health problems induced by long-term or short-term HM exposure. HM concentrations in various human organs including blood have been linked to estrogen-dependent disorders that include breast cancer, cancer of the endometrium, endometriosis, including premature abortions, alongside pre-term deliveries, stillbirths, and hypotrophy (Rzymski et al. 2015). A research was conducted to examine the amounts of Cd, Cr, Pb, Cu, Ni, Co, Mn, Fe, and Zn in biological specimens from humans such as urine, blood, human hair, and nails, as well as the antioxidant reaction in blood samples from persons exposed to heavy metals. The findings revealed metal-specific disparities in concentrations across exposure groups throughout the locations examined. HM concentrations in samples of blood appeared to be in that sequence Pb > Cd > Ni > Co > Cr. Cu > Pb > Cr > Ni > Co was sequenced in samples of urine. In nail samples, the sequence was Pb > Ni > Cr > Co > Cd > Mn; whereas for hair specimens, the ordering was Pb > Ni > Cr > Co > Cd > Mn. Cr > Mn > Cu > Co. While, there was a substantial reduction in antioxidant enzyme activity HM levels were rising. This was the first research to reveal biological evidence of harmful changes in people, as a result of environmental prolonged exposure (Bibi et al. 2016).

Elemental absorption in teeth dentine serves as a bioindicator because it occurs throughout the process of creation and mineralization stages and is preserved to a considerable extent over long periods of time. The intake comprises both necessary elements derived from geogenic food sources and non-essential elements derived from environmental shocks. Most samples showed the presence of the following HMs in descending order: As, Mn, Ba, Cu, Cr, Pb, Zn, Hg, Sb, Al, Sr, Sn. The levels have been found to rise with aging. Among the ethnic groupings, the dental surfaces of ethnic Chinese had somewhat higher levels of metal compared to those of Indians and Malays, with female dentine having higher concentrations over male dentine. Molars had higher quantities of Hg, Cu, and Sn, whereas incisors had greater levels of Pb, Sr, Sb, and Zn. With higher heavy metal levels of concentration in dental dentine indicating contamination from manufacturing emissions and urban development, it is clear that human teeth dentine may offer historical information on exposure, serving as a credible bio-indicator of environmental contamination (Asaduzzaman et al. 2017).

3.9 Health Risk from Consumption of Food Containing HM

HMs are widespread in our food. In little concentrations, they are necessary for optimal health, but in high quantities, they can be poisonous or hazardous (Jaishankar et al. 2014). HM pollute the food chain due to potential bioaccumulation, no degradability, as well as excessive proportions in which they exist, and so constitute a source of toxicity to humans and the overall ecological function (Nkwunonwo et al. 2020). A research measured Cu, Zn, Cd, and Pb levels, as well as Pb isotopic arrangements, in vegetables, rice, along with human hair in regions near a smelter in China. Cd and Pb levels were found to be substantially higher in rice along with vegetable samples. Both the age of the rice along with the source proved major variables in the enriching of the levels of HM in human hair. Pb in human hair was mostly derived from dietary consumption. According to a non-carcinogenic risk assessment, Cd exposure was the primary health concern for area populations. Crops grown near the smelter were heavily polluted with Cd, and human contact was linked to food consumption (Hu et al. 2021). Food crops polluted with hazardous HM such as Cd,

Pb, and Cr were investigated. Greater levels of Cd, Cr, Ni, Mn, including Pb were found in wastewater-irrigated topsoil and crops for food. Cd levels in soil, as well as Pb, Cd, and Cr levels in irrigated water as well as agricultural crops, above the allowable limit. Human health may be jeopardized by sewage irrigation for food plants. Pb, Cd, as well as Mn in agricultural crops with health risk indexes greater than one provide a possible health danger (Khan et al. 2013).

3.9.1 Water

The actions of humans have had a significant impact on the world's water resources in recent years, resulting in major supplies and potable quality problems. The levels of HM in water for drinking are greater in many places of the world than various international guideline limits. It is vital to discuss HM contamination in drinking water, their integration within the food chain, and their consequences as a worldwide concern to human health. It is known that millions of people suffer from chronic HM poisoning, which has grown into a worldwide health emergency, and that roughly 1.6 million kids die each year from illnesses caused by polluted drinking water (Fernández-Luqueño et al. 2013). There additionally exists proof of HM in water that people consume, which is causing harm to human health via food chain pollution (Fernández-Luqueño et al. 2013). HM-contaminated water causes significant disease and mortality globally (Fu and Xi 2020). Immediate contact with HM in water used for drinking beyond the acceptable limit can be harmful to human health. A study assessed the risks of HM in water for drinking taken from Anambra State, Nigeria's drinking water. A total of 81 water samples were obtained from boreholes, wells, sachets, harvested rain, and streams. The chronic daily consumption indices in the study locations were greatest for Cd, with Cd > Hg > Pb in borehole, well, sachet water, and stream water samples. It was more prevalent among children than adults. In the analyzed locations, children had a greater chance of developing cancer than adults. Cadmium was a significant risk factor (Emmanuel et al. 2022).

3.9.2 Rice

Excessive amounts of HM deposition in Iranian grown rice indicate that there is a substantial danger towards the health of frequent rice users. These findings demonstrated that rice within this area might be an important food source of Pb and Cd exposure for the general population (Vatanpour et al. 2020). Cd level in rice in China had the greatest biological buildup coefficient and proved slightly more than the highest permitted level in 7.5% of specimens, although the remaining metals remained less than their respective allowable values. Heavy metal exposure (particularly As exposure) from rice eating poses considerable carcinogenic and noncarcinogenic hazards to adults, as well as a noncarcinogenic hazard to children, with a moderate carcinogenic risk to children. Natural causes of Cr and Ni, including lithogenic materials and soil parent components, farming operations with Cd, Cu, and Zn, particularly high levels of pesticide and application of fertilizers, and industrial operations with Hg, Pb, and As), such as car exhaust, burning of coal, as well as those associated with the clothing and chemical industries, have been recognized as major sources (Ren et al. 2021). A research was planned to look at the presence of HM in quantities in rice produced near mining regions in India. Apart from Pb, Ni, and Zn in select areas, the metal contents were less than the Indian maximum permissible amounts for food. HM in rice may be traced back to soil, irrigation water, and air dust deposit. Based on projected daily intake and the intended hazard quotient, the high prospective health effects of exposure to metals from rice eating were demonstrated. HM consumption for local individuals were greater than the tolerated daily allowances for Cr, Fe, Ni, and V. The major components contributing to the probable non-carcinogenic hazard were Pb, Cu, and Cr (Giri and Singh 2017). Some studies regarding HM in Food material is tabulated in Table 3.1.

Table 3.1 Heavy metals detected in food and relevant findings

Food material	Heavy metal detected	Country	Findings	References
Packaged tuna	Hg, Pb, Cd	Italy and others	Cd and Hg within limits, Pb exceeded limits in almost 10% of the samples	Russo et al. (2013)
Okra vegetable crop	Cd, Cr, Cu, Pb, Zn, Ni	Saudi Arabia	Pb, Ni, Cd, Cr were above acceptable levels	Balkhair and Ashraf (2016)
Vegetables grown with wastewater	Cr^{2+}, Pb^{2+}, and Cd^{2+}, Cd and Mn	Pakistan	Cr^{2+}, Pb^{2+}, and Cd^{2+} above legal limits, Cd and Mn constituted a serious health danger	Mahmood and Malik (2014)
Drinking water (from boreholes, wells, sachets, harvested rain, and streams)	Pb, Cd, Hg	Anambra State, Nigeria	The chronic daily consumption indices in the study locations were greatest for Cd, with Cd > Hg > Pb in borehole, well, sachet water, and stream water samples. It was more prevalent among children than adults. Children had a greater chance of developing cancer than adults. Cadmium was a significant risk factor	Emmanuel et al. (2022)

(continued)

Table 3.1 (continued)

Food material	Heavy metal detected	Country	Findings	References
Onion	Cd, Pb	Iran	Found in all samples	Shokri et al. (2022)
Rice	Cd, As etc.	China	Cd level had the greatest biological buildup coefficient and proved slightly more than the highest permitted level in 7.5% of specimens, exposure (particularly As poses considerable carcinogenic and noncarcinogenic hazards to adults, as well as a noncarcinogenic hazard to children)	Ren et al. (2021)
Dairy products	Cadmium, copper, lead, and zinc	Romania	After consuming dairy goods, there was a moderate to high risk of nephrotoxicity, liver toxicity, hematotoxicity, cardiac toxicity, and reproductive toxicity	Năstăsescu et al. (2020)
Wine	Cr, Fe, Cu, Mn, Ni, and Zn	–	Throughout the winemaking procedure, Pb, Cu, and Cr levels declined in white wines, but only the copper and lead levels fell in red wines. During winemaking, Ni and Fe levels in white wines increased gradually, whereas Cr, Ni, and Fe levels in red wines increased gradually	Dumitriu (Gabur) et al. (2021)
Cabbages and carrots	Cd, Pb, Zn, Fe, Ni, Cu	Ghana	HMs were detected in both urban and rural samples	Mensah et al. (2009)
Multifloral honey	As, Cd, Co, Cr, Ni, Pb, Zn	Saudi Arabia	Highest concentration of zinc, lowest concentrations of copper	Bazeyad et al. (2019)
Brown rice from HM mining area	Pb, Cd, As, Mn, Sb	China	All were detected, Cd contributed most to cancer risk	Fan et al. (2017)
Tropical fish and shellfish	V, Cr, Mn, Ni, Cu, Zn, As, Se, Mo, Ag, Cd, Sb, Ba, and Pb	Bangladesh	Cu proved the most abundant metal. Ni, As, Ag, and Sb were found in lesser amounts, whereas Cr, Mn, Zn, and Se had been found at higher concentrations. Mn, Zn, and Pb were found in higher amounts than recommended	Ahmed et al. (2015)

(continued)

Table 3.1 (continued)

Food material	Heavy metal detected	Country	Findings	References
Spring water	Pb, Fe, Cd, Cr	Nigeria	HM levels surpassed the standard for drinking water quality	Anyanwu and Onyele (2018)
Raw milk	Cu, Pb, Cd, Cr	Pakistan	The average amounts of Cu, Pb, and Cd were found to be significantly greater than Guidelines. Seasonal variations were observed	Younus et al. (2016)
Gallus or broiler chicken	Cr, Cu, Pb, Cd, As, Fe, Zn	–	HMs were below the permissible level of quantification	Thirulogachandar et al. (2014)
Drinking water from a dam and a river	Fe, Mn, Cu, Pb, Cd, Ni, Co, Zn	Nigeria	HMs were detected in risky levels from both sites especially the dam water	Ekanem et al. (2016)
Seven commercial fish species	Mn, Cu, Zn, Cd, Pb	Ghana	Concentration below threshold	Kwaansa-Ansah et al. (2019)
Cultured fish	Cd, Cu, Zn	Egypt	HM were under permitted levels in fish musculature	El Bahgy et al. (2021)
Rice	Various HMs	India	Except for Pb, Ni, and Zn in select areas, the metal contents were less than the Indian maximum permissible amounts for food	Giri and Singh (2017)
Rice	Pb, Cd, Zn, Cr, Ni, Mn	Nigeria	Daily consumption of lead and mercury from field rice were greater than the acceptable range	Ihedioha et al. (2016)
Lettuce, spinach, radish, carrot	Cd, Cu, Mn, Zn, Pb	–	Plant absorption of Cd, Cu, Mn, and Zn increased with increasing soil pollution, however Pb uptake was negligible Particular plant varieties varied substantially in their metal absorption	Intawongse and Dean (2006)

(continued)

Table 3.1 (continued)

Food material	Heavy metal detected	Country	Findings	References
Vegetables including; celery, cabbages, carrots, asparagus lettuces, cowpeas, tomatoes, and cayenne pepper etc.	Pb, Cd, Cr, Hg, As	China	HM were found in all of them	Liu et al. (2013)
Marine fish	Cd, Cr, Cu, Zn, Pb, As	China	Cd, Cr, Cu, Zn, Pb, and As in the muscle tissues of the tested marine fishes remained undetectable. But were found in gills and liver	Liu et al. (2015)
Twenty-four types of fruits and vegetables	Pb, Cd, As, Hg	India	The average levels for the selected HM in vegetables and fruits appeared as follows: Pb > Hg > Cd > As. Garlic has the largest HM buildup, followed by potato. Vegetables accumulated more metals than fruits	Mawari et al. (2022)
Vegetables irrigated with wastewater and sewage sludge	Cd, Cu, Pb, Zn	Zimbabwe	HMs were detected in high concentrations	Muchuweti et al. (2006)
Rice	Mn, Pb, Ni, Cr	Thailand	Except for Mn, the quantities of most metals within roots appeared high. Pb levels in all rice samples were above the threshold	Neeratanaphan et al. (2017)
Five fish species	Cd, As, Pb, Cr, Zn, Ni, Se, Cu, Mo, MN, Sb, Ba, Ag, V	Bangladesh	Mn, Zn, Se, Pb in all fish species were above threshold	Kawser Ahmed et al. (2016)
Milk	Cr, Fe, Cu, Mn, Cd, Pb	Bangladesh	HM were within safe limit	Muhib et al. (2016)

3.9.3 Milk and Dairy Products

The existence of HMs in agricultural inputs such as feed, fertilizer, water, and the surroundings causes residues to be excreted in animals' milk. Since consuming HM-contaminated milk causes substantial health risks to consumers. A study was carried out to assess the content of HM in untreated raw milk obtained from Pakistan at various seasons. Cu, Pb, Cd, and Cr were all examined. The average amounts of Cu, Pb, and Cd were found to be significantly greater than the International Dairy Federation Guidelines. Heavy metal levels varied dramatically between samples throughout a season. There was a substantial change in mean HM concentrations among summer and winter for each kind of milk supply (Younus et al. 2016). HM poisoning has been related to a variety of ailments, but the severity of the condition doubles several folds when these contaminants are detected in milk, which happens to be a staple diet for the vulnerable population (Chandrakar et al. 2018). The presence of pollutants can affect the health advantages of cattle milk. Cadmium, copper, lead, and zinc concentrations in both cheese and milk were measured. After consuming dairy goods, there was a moderate to high risk of nephrotoxicity, liver toxicity, hepatotoxicity, cardiac toxicity, and reproductive toxicity (Năstăsescu et al. 2020). In Pakistan, irrigation with wastewater is commonly used to raise feed for animals for cattle husbandry. HM risk evaluation in unprocessed milk from buffalo fed on wastewater-irrigated farmland in Pakistan showed the overall target health quotient levels of heavy metals for raw milk from buffalo varied from 6.92 to 42.44, with the greatest in wastewater-irrigated croplands and the least at tubewell water sites, indicating a substantial carcinogenic danger to the exposed populace. Polluted fodder, such as maize as well as brassica plants cultivated on polluted soil, is a typical cause of HM being found in raw milk (Iqbal et al. 2020).

3.9.4 Fish and Sea Food

Toxicity to HM is to blame for the degeneration of aquaculture populations, creating physical abnormalities in creatures and contaminating the aquatic environment. These poisonous HM cause a variety of fish ailments. Because fish are consumed by people, it has an indirect impact on them. The infiltration of HM into water bodies and aquatic ecosystems has a significant influence on the food chain (Sonone et al. 2021). Fish is a good source of nutrients, but its nutritional value can be influenced by the habitat in which it lives. Fish cannot avoid the negative impacts of these toxins in their watery environment. HM poisoning from fish eating can cause mental and CNS destruction, decreased energy levels, as well as harm to the chemical makeup of the blood, lungs, kidneys, bones, liver, and other essential organs. Exposure for a long time may cause physical, muscular, and neurological conditions such as Parkinson's disease, dystrophy of the muscles, and multiple sclerosis to worsen. Allergies are frequent and can potentially lead to cancer (David and Isangedighi 2019). Fish is a wonderful source of nutrition for mankind, but it can be hazardous to their health due to the existence of xenobiotics that comprise HMs. The levels of mercury, lead, as well as Cadmium in tuna packed in various manners from different nations were studied. Concentration of Hg and Cd were below the limit set by the European Commission but for almost 10% of the materials tested, Pb values surpassed European guidelines. Hg, Pb, and Cd thresholds in the freshly caught tuna tissue are 1.0, 0.3, and 0.1 mg/kg at wet weight, correspondingly. Among the most important biochemical concerns to the security of food in the seafood industry is the possible existence of HM in seafood. The steady rise in pollution from the oceans, the internationalization of marketplaces with regard to of both ingredient manufacturing and processing plant setting, and expanding customer awareness to food security problems render pollution from HMs a critical factor for fish industry development and competitiveness. HM pollution can be caused by human actions, but it can also be caused by natural geological causes (Russo et al. 2013). Evaluating the possibility of health threats associated with 5 heavy metals identified in 7 most consumed fish species, including *Cirrhinus mrigala, Cirrhinus reba, Catla catla, Lebio rohita, Crossocheilus latius, Clupisoma garua*, and *Mystus tengara* gathered from market places in, India. Cr, Cd, and lead levels were above the permissible limits, whereas Zn was found to be within permitted levels at all sample locations. HM buildup was found to be greatest in the livers of all examined fish species. In chosen fish tissues, the metals were found in the following sequence: Zn > Pb > Cu > Cd > Cr (Maurya et al. 2019).

Pollution of fish from freshwater with poisonous metalloids as well as HM is a serious public health concern. Metal-contaminated consumption of fish consequently poses a threat to human health (Ali and Khan 2018). The possible dangers to people's health from heavy metal transference rely on the quantity of fish ingested. Interdisciplinary research on the biological accumulation of xenobiotics including potentially hazardous metals as well as metalloids in fish from freshwater is underway (Ali and Khan 2018). In Ghana, the muscles of frequently eaten *Trachurus trachurus, Lutjanus fulgens, Lutjanus goreensis, Acanthocybium solandri, Pagellus bellottii, Scomber colias*, as well as *Dentex congoensis* varieties of fish were assessed for health risk along with HM buildup. The results demonstrated that the quantities of Mn, Cu, Zn, Cd, and Pb were within the allowable limits stipulated by numerous health institutes. The health risk evaluation utilizing non-carcinogenic as well as cancer factor effects suggests that fish consumption has no negative health effects (Kwaansa-Ansah et al. 2019). V, Cr, Mn, Ni, Cu, Zn, As, Se, Mo, Ag, Cd, Sb, Ba, and Pb buildup together with adverse health danger were studied in fish in Bangladesh: *Ailia coila, Gagata youssoufi*, along with *Mastacembelus pancalus*; one of them crustacean, *Macrobrachium rosenbergii*; and a single *Gastropoda, Indoplanorbis exustus*. Cu proved the most abundant metal. Ni, As, Ag, and Sb were found in lesser amounts, whereas Cr, Mn, Zn, and Se had been found at higher concentrations. Mn, Zn, and Pb were found in higher amounts than recommended. The target hazard quotient indicated that there was no possible health danger. However, the cumulative impact, assessed danger for *M. pancalus* ingestion. intake at the current quantity is not dangerous, but continuous intake for over seventy years increases the risk of cancer (Ahmed et al. 2015).

3.9.5 Vegetable and Corn

A research to look at the potential negative health effects linked with eating okra vegetable crop that has been polluted with hazardous HM produced the crop on land watered with purified wastewater. Ni, Pb, Cd, and Cr contents in edible parts were over the allowed levels in 90, 28, 83, and 63% among the specimens, correspondingly. The following HM were found in the edible parts: Chromium > Zinc > Nickel > Cadmium > Manganese > Lead > Copper > Iron. The result of the Health Risk Index measured more than one, indicating a possible health risk. a higher level of bio-contamination than earlier findings in locations across the world. The findings suggested a possible route for human susceptibility to HM toxicity caused by the indirect consumption of vegetables produced on HM polluted soil that was watered by water that was contaminated resources. The okra analyzed failed to be acceptable for consumption by humans, particularly direct eating. The levels of HM were substantially greater in plants cultivated in wastewater-irrigated soils compared to crops cultivated in baseline soil. The use of both processed and unprocessed water for watering enhanced the contamination of edible crops with Cd, Pb, and Ni, posing a possible health concern in the long run. Different vegetables were shown to collect and translocate varying quantities of metals through the soil onto their tissues. Most agricultural and vegetable species grown on HM-polluted soil are unable to be ingested to prevent uptake of these metals. Cucumber collects the fewest metals and metalloids, hence it may be less hazardous to one's health to consume it rather than okra (Balkhair and Ashraf 2016). The study to assess the effect of HM from onion intake on human health found that all samples analyzed contained HM (Shokri et al. 2022). Public health concerns from five distinct HMs have been investigated in 9 distinct Asian, European, and African regions. The levels of heavy metals in wheats appeared greater than in corns, suggesting that HMs remained simpler to build up in wheats compared in corns. In diverse exposure methods, the human hazards caused by the cancerous metals As and Cd were much greater compared to those caused by noncarcinogenic metals. The human health hazards produced by wheat intake comprised 67.75–97.52% and 69.52–98.04%, accordingly, of the overall human health hazards for adults and children. Adults faced greater overall human health hazards than children (Wang et al. 2019).

3.9.6 Other Food Products

Evaluation of HM inside Gallus or Broiler Chicken as well as pollution due to chicken consumption of water, foods, premixes, and additional sources such as air, water from the ground, dirt, for 7 HMs that is, Chromium, Copper, Lead, Cadmium, Arsenic, Iron, and Zinc revealed that for every sample, heavy metals were below the permissible level of quantification (Thirulogachandar et al. 2014). Ingestion of polluted medicinal plants as well as herbal medical goods is thought to be a possible cause of HM poisoning (Ezeabara et al. 2014). A high level of chemical contaminants, particularly HM, has an impact on honey quality. According to a research to evaluate the quantities of HM in honey specimens from Saudi Arabia, it should not be considered a risk for human health for these samples (Bazeyad et al. 2019). HM pollution in wine may offer over time health and dangers to the environment. HM that are regularly discovered in the winemaking phases of white and red wines varietals include Cr-Chromium, Fe-Iron, Cu-Copper, Mn-Manganese, Ni-Nickel, and even Zn-Zinc. Wine and samples of waste water were obtained at eight distinct phases of the winemaking process. The results revealed that throughout the winemaking procedure, Pb, Cu, and Cr levels declined in white wines, but only the copper and lead levels fell in red wines. During winemaking, Ni and Fe levels in white wines increased gradually, whereas Cr, Ni, and Fe levels in red wines increased gradually. The target hazard quotients indicated that moderate wine drinking has no harmful effects on human health (Dumitriu (Gabur) et al. 2021).

3.10 HM Health Impact from Other Sources

3.10.1 Soil

Heavy metal pollution of soil can provide dangers and hazards to persons and the environment, whereas harmful metals in soils can substantially impede organic pollutant biodegradation. In a research about an agricultural region near a former metallurgical treatment industrial plant, the pollutants studied included beryllium, cadmium, chromium, nickel, and lead. The findings indicate that the potential harm to human health is greater than what is tolerable (Cocârţă et al. 2016). The results of a study in agricultural soil in China revealed that average Hg concentrations surpassed the national contamination risk assessment value. Based on the findings of the health risk evaluation, the cancerous and non-carcinogenic dangers of heavy metals for human beings are negligible (Chai et al. 2021). HM poisoning of soil from unprocessed urban and commercial wastewater discharge is a severe concern to human health. Irrigated soil, water, and crops were tested, and Cr^{2+}, Pb^{2+}, and Cd^{2+} levels in vegetables grown with wastewater above legal limits, whereas the Health Risk Index was highest in *Spinacia oleracea* along with *Brassica campestris* grown with wastewater. With regard to Cd and Mn, *S. oleracea, B. campestris*, and *Coriandrum sativum* constituted a serious health danger (Mahmood and Malik 2014). According to the risk evaluations in the Shi Rivers Basin soils in China, the HM offer greater carcinogenic and additional health hazards for children than to grownups, and consumption is the primary way that these chemicals infiltrate the body. Elder and child non-carcinogenic risks are at tolerable limits. The primary origin of the non-carcinogenic danger is transporting and processing in factories (Zhang et al. 2021).

A quantitative evaluation of human health hazards under diverse usages of land determined by soil HM contamination sources revealed that human activities considerably enhanced nickel, copper, and lead. Farming operations provided the greatest to soil HM pollution (33.72%) of the four pollutant sources. Although the non-carcinogenic as well as carcinogenic hazards to children remained greater compared to those to adults, the health hazards for both cohorts followed the same pattern across land-use categories. Agricultural operations were the leading cause of cancer risk, contributing to 37, 41, and 38% of the risk of cancer in building, related to agriculture, and forest areas, correspondingly. Industrial emissions caused non-carcinogenic dangers, while manufacturing activity was next only to farming operations in terms of carcinogenic risk. Dangerous HM sources endanger one's health (Lu et al. 2021). HMs such as As, Cd, Cr, Ni, along with Pb were tested in soil in an industrial area in Iran and shown to provide insignificant carcinogenic along with non-carcinogenic hazards to both adults and kids (Mohammadi et al. 2020). A research from a South Indian metropolitan region indicates that city soils has grown into one of the primary ecological and human health issues. Six of the HM that demonstrated no carcinogenic danger to humans in the researched region are entirely comprised Cr and Pb, whereas Cr controls the carcinogenic risk, and the remainder of the metals provide no conceivable risk for local people (Adimalla et al. 2020). Humans, particularly children, face possible health concerns in the Farming Soils of China's industrial districts along the Jinxi River and around Lake Qingshan. Furthermore, As was discovered to lead to a greater severity of cancer risks (Abdelhafez and Li 2015). HMs in topsoil harmed human health via breathing, swallowing, and exposure to the skin. The health risk evaluation from soils HM resources under diverse land use can offer an essential reference point for avoiding and managing soil contaminants at the source. In one research, soil supplies from residential properties, forests, and agricultural land were used as examples, and four main sources were statistically examined. The most significant cause of cancer risk was emissions from factories. In terms of non-cancer danger industrial discharge remained the most important source in household and forests, whereas agricultural operations accounted for farm land. Children's risk trends were comparable to those of adults (Huang et al. 2018).

3.10.2 Water Bodies

Health risk evaluation in the Beiyun River in China discovered that the long-term persistence and buildup of HM in rivers constitute a significant hazard to the resilience of environments and the health of people. The findings revealed that both the Hazard Index as well as Cancer Risk for As and Ni for indigenous peoples were severe (Wu et al. 2021). The evaluation of eight HMs Cd, Cr, As, Fe, Mn, Cu, Zn, and Pb within groundwater shallowly situated in an extracting area, China found Fe, As, and Cr levels greatly above World Health Organization acceptable water for consumption limits. In grownups, the overall cancer risks to health of Cr and As indicated 2.49×10^{-3} as well as 3.43×10^{-4}, correspondingly, above the upper limit of tolerable risk value (1×10^{-4}). Evaluation of exposure associated health risks HM ions in groundwater are mostly consumed via drinking water intake and external exposure (such as washing), which are the two principal routes of entrance into the bodies of humans (Jiang et al. 2021). The evaluation of HMs contact, carcinogenic as well as non-carcinogenic health concerns in groundwater near mine in Iran showed cancer risk levels (sum of As, Cd, as well as Cr) for kids and adults across all communities were over the limit. The risk of cancer values for children as well as adults via the oral route were discovered in the following order: As > Cr > Cd (Shams et al. 2022).

The health hazards associated with heavy metals extracted in various seasons collected from the Taizihe River, China demonstrated that the mean concentrations of Pb, Cr, and Cd was not substantially different across seasons. The yearly average hazards to human health from Cd and Cr were greater than the maximum permissible risk threshold. Zn, Pb, and Cu risk levels for human health were within the suggested threshold (Chen et al. 2020). Fe, Mn, Zn, Cr, and Pb heavy metal contamination in the the upper region Ganga River, India, was assessed. According to the health risk evaluation, exposure via ingesting and cutaneous channels presently offers no substantial hazard to human health (Prasad et al. 2020). The human health danger of several HMs in a Nigerian rural spring was evaluated to determine its appropriateness for human use. Lead (Pb), iron (Fe), cadmium (Cd), as well as chromium (Cr) levels surpassed the Nigerian Standard for Drinking Water Quality along with World Health Organization requirements. The health risk evaluation for all locations found no especially harmful individual heavy metal, but overall combined effect, reflected by the hazard index, is cause for worry (Anyanwu and Onyele 2018). A research investigated the human health hazards posed by heavy metal pollution (As, Cd, Cr, Cu, Hg, Pb, Ni, as well as Zn) in the upper levels of the water in the Houjing River, the most heavily polluted river within southern Taiwan. The Adapted Risk Assessments Modeling Framework was used to model the health hazards associated with heavy metal exposure in three situations: fish consumption, cutaneous water contact, and inadvertent water intake while swimming. On the "adverse effect" threshold, copper showed the greatest non-carcinogenic danger, whereas Ni along with Cr exhibited the greatest cancer-causing potential at the "unacceptable" level. Furthermore, the cumulative dangers of ingesting fish were much larger than those of swimming situations (Hoang et al. 2021).

3.10.3 Airborne Sources

HM found in city airborne particles may offer significant health hazards. Oral bio accessibility metrics for V, Co, Ni, As, Mo, Cr, Mn, Cu, Zn, as well as Pb in city dust varied from 1.563% through 76.51%. Following adjusting for bio accessibility, HM possessed lesser non-carcinogenic alongside carcinogenic threats, alongside average hazard index measures below the recommended value and average cumulative cancer risk metrics surpassing the preventative criteria for adults as well as children. Vehicle and emissions from industry, as well as coal burning, led the noncarcinogenic and cancerous hazards generated by HM in city dust (Ma et al. 2021). A research investigation of HM in PM_{10} within Agra, India found that carcinogenic hazards of As, Cd, Co, Cr, Pb, and Ni via contact with the skin and ingesting exposure remained within safe limits for adults as well as kids. Co posed a possible carcinogenic risk exclusively to adults, but Cr posed a danger towards kids and adults by inhalation (Sah et al. 2019). The erosive influence of wind might convey tainted soil from mining activity. As one of the primary dust storm causes, contamination with Cu, Zn, Cd, and Pb found in the top soil of a zinc-lead mining in Iran. This study's significant contamination level indicated the possibility of serious health hazards in the mine region. Furthermore, analytical methodologies demonstrated a substantial interrelationship amongst Pb, Zn, and Cd in the residential region and their similar anthropogenic origin (Abouian Jahromi et al. 2020). HM found in city dusts (road-dust as well as house-dust) in Egypt demonstrated consequences for human health Geographically, for majority heavy metals, the significant concentrations were reported in areas associated with increasing anthropogenic activities, strong traffic, and high human density. analysis divided the examined metals in 2 categories as: naturally source Co, Cr, Mn, Ni, and Fe and human inputs Cd, Cu, Pb, and Zn. The noncancerous hazards provided by the metals under consideration varied as 0.0001 (Cd) through 0.15 (Pb) and proved minimal. The carcinogenic risk of Pb for children exposed to Alexandria home dust surpassed the guideline values and was deemed unacceptable, although the cancerous hazards associated with the other metals investigated were deemed tolerable for both groups. Children are at greater danger than grownups (Jadoon et al. 2021).

HMs found in city road dusts in Kolkata and Bengaluru, India, have consequences for human health since they are a possible source of exposure and a channel for exposure to humans. In other situations, the bioaccessibility of road particles at the level of ingesting was greater than 50%, with zinc along with lead being accessible. Yet, being exposed to lead to children in Kolkata has been demonstrated to be a major risk factor. The danger pathways dominated by ingestion were much bigger than cutaneous and inhalation methods (Chenery et al. 2020). $PM_{2.5}$ was examined for heavy metals from two locations in China in order to evaluate their origins and human health concerns throughout warming and non-heating seasons in a research. During the heating season, coal combustion accounted for 59.64% of heavy metal pollution at Huagong Hospitals, with industry and transportation emissions contributing 21.06, 19.30%, correspondingly (Sun et al. 2021). HM depositing occurs in indoor dust. Previous research has demonstrated that heavy metal concentrations in indoor dust are altered by local activity by humans, and that air transmission can be hazardous to human health. Furthermore, because of their hand-to-mouth behavior and fast bodily development, children are especially vulnerable to HMs (Tan et al. 2016). Rapid urbanization and industrialization have culminated in substantial amounts of atmospheric particulates comprising different HM, which has a negative impact on the health of people. PM health risk assessment necessitates an accurate estimate of bioaccessibility rather than total concentration of HM in PM. A study found that the leachable portion of particle-bound As, Pb, Cr, Mn, Cd, Cu, Ni, and Zn in lung fluid differed dramatically between particles originating from various airborne pollution pathways, such as the combustion of coal, biomass burning, escapee dust, roadway dust, constructing dust, cement, as well as soil. Furthermore, the bioaccessibility of HM in biomass burning, cement, and soil particulates was substantially reliant on pollution inputs; particle Cu, Ni, Pb, and Cd were indications of bioaccessibility source dependency (Liu et al. 2019).

3.10.4 Mines and Similar Factories

Metalliferous extraction of minerals, including mines for coal, are primarily accountable for the unearthing of metals including as As, Cd, Fe, and thus on, and these activities are creating continuous soil pollution due to routine practices in the coalfield. HM pollution into the environment is also a major source of metal transit from mining sites to other businesses (Mishra et al. 2019). A conventional smelter and mine zone in South China evaluated for 8 heavy metals Cd, Hg, As, Pb, Cr, Ni, Cu, along with Zn in the province of Hunan, China. Overall non-carcinogenic risk for health led the overall hazard index 84.8% for older people, for children 84.7% while the cancerous health risk led the entire carcinogenic risk index 69% for older people, 68.8% for minors. Ingestion by mouth (89.4–95.2%) was the most common exposure pathway for children as well as adults. Children exhibited higher non-carcinogenic plus carcinogenic risks than adults (Zhou et al. 2022). Metal extraction and waste disposal produce regional heavy metal pollution, which is of great concern due to the possible harm to surrounding inhabitants (Fan et al. 2017). A study on soil as well as brown rice specimens from 3 HM mining regions in China to examine the possible health hazards to local residents for Pb, Cd, As, Mn, and Sb discovered that Cd contributed the most substantially to the risk of cancer, encompassing around 99.77% of this risk. There are possible noncarcinogenic along with carcinogenic health concerns for nearby residents, and frequent surveillance of pollution is urgently needed to preserve human health (Fan et al. 2017). Noncarcinogenic risk was primarily brought about by Cr in soil within a steel manufacturing region in China evaluated for As, Cd, Cr, Cu, Pb, Ni, Zn, and Hg, and children suffer higher risks from the presence of HMs in soils. The mean cancer risk rates for Ni, Cr, Cd, as well as As all fell lower than the threshold levels (Gao and Wang 2018).

3.10.5 Additional Sources

Pb and As levels in lipstick along with eye pencil from Bangladesh were determined to be within the permitted level, however Cr concentrations in several samples above what is permitted in cosmetics. Cd was not found in any of the samples tested; nevertheless, additional elements including Co, Ni, Cu, Mn, Zn, and Fe were identified in significant amounts. The cancer risk numbers for Cr and Ni suggest that lipsticks may cause cancer if they infiltrate the consumer's body (Baroi et al. 2023). The human dangers and source recognition of 6 prevalent HMs; Pb, Cr, As, Cu,

Zn, and Ni, were examined in fireworks dust obtained in Beijing to comprehend the effect of fireworks upon HM in dust. The dangers of HM in dust raised because of firework burning (Fu et al. 2021). HM found in medical prosthesis, water sources, and cigarettes can accumulate in tissues as well as blood, causing hazardous consequences. A study found that cadmium plus nickel can have delayed consequences in cells from humans in vitro, that are indicative of instability in the genome. The consequences can be seen even at levels when no acute hazardous effects can be seen. Instabilities in the genome can be seen in the offspring of cells several generations after exposure by the ongoing production of cytogenetic aberrations and delayed cell death (Coen et al. 2001). HMs as well as metalloids move into surroundings, posing environmental and human health hazards. A research on coal-mine brownfield sites found that As, Cd, Cr, Cu, Hg, Mn, Ni, Pb, Sc, Ti, as well as Zn were abundant in bulk soils. Consumption was the primary mode of getting exposed to soil-associated HMs and metalloids. The overall non-carcinogenic hazards to kids and adults were less than the safe threshold of one, while the total cancerous hazards were greater than the maximum permissible level. Cd and Ni were the primary contributors to the overall carcinogenic risk from ingesting and dermal access (Li et al. 2017). HMs in garbage dumps constitute a significant environmental and health concern. The sequence of movement as well as bioavailability of the following metals in soil near 2 municipal landfills in Nigeria is: Cd > Pb > Cu > Zn > Fe. The degree of cancer risk exceeds the levels declared unacceptable by the US Environmental Protection Agency (Ogunbanjo et al. 2016).

3.11 A Comprehensive Analysis of HMs Sources and Their Health Impacts

Heavy metal poisoning has shown to be a significant hazard, with multiple medical conditions linked with it. Even if these metals serve no biological purpose, their toxic effects are still present in a manner that is damaging to the body of a person and its appropriate operating. They may be disruptive with metabolic processes (Jaishankar et al. 2014). The toxicology of HMs is primarily determined by their respective oxidation state, and this is accountable for physiological and bio-toxic consequences. When these elements enter live beings, they mix coupled with enzymes, proteins, as well as DNA molecules to generate extremely stable bio-toxic compounds, which impair their correct functioning and prevent them from participating in bioreactions. Arsenic, chromium, cadmium, and lead are extremely toxic, causing cancerous, mutagenic, as well as genotoxic consequences (Mishra et al. 2019). A comprehensive analysis of some common HMs is deliberated in the following section.

3.11.1 Mercury

Mercury is a prevalent heavy metal contaminant discovered in the natural world, mostly in the elemental, inorganic, as well as organic forms. The toxicity of these three forms varies, with the element being the least poisonous. Despite the fact that mercury is present at low amounts in the natural world, its toxicity, perseverance, and biological accumulation may constitute a health risk to humans. The element mercury oxidizes in the atmosphere to generate inorganic mercury. Metallic mercury is often ingested through the airways in its vapor state. Due to its soluble nature, elemental mercury moves rapidly through cell membranes and the blood brain, eventually getting to the target region. When elemental mercury enters the circulation, it is readily converted into inorganic Hg. In contrast to the organic form, which predominantly impacts the brain and nervous system, inorganic Mercury is non-regularly distributed, collecting mostly in the kidneys and inducing acute renal failure. Mercury is a leading cause of autoimmune disorders and antinuclear antibodies in those who have been subjected to inorganic mercury. Hg is also a known cause of Alzheimer's and Parkinson's disease. As a result, every kind of mercury are highly poisonous to the brain, nervous system, and the gastrointestinal tract. Mercury exposure enhances the generation of free radicals, ROS, and superoxide anions. Mercury-induced toxicity is characterized by the suppression of antioxidant defense mechanisms, alterations in the oxidant-antioxidant balance, and a rise in ROS. Methylmercury is a neurotoxin that causes peroxidation of lipids, damage to mitochondria, microtubule impairment, and neurotoxic molecule buildup (Fu and Xi 2020).

Farming, city sewer releases, mineral extraction, combustion, and factory discharges are all major contributors of mercury contamination. Mercury exists primarily in the following forms: metallic components, inorganic salts, along with organic molecules, each with varying toxicity and bioavailability. The aforementioned kinds of mercury are commonly found in water resources such as rivers, lakes, and seas, in which they are picked up by microbes and converted into methyl mercury inside the microorganism, creating severe disruption. Human contact with methyl mercury is mostly from ingestion of this infected aquatic species. Mercury is extensively utilized in barometers, pyrometers, hydrometers, mercurial arc lamps, as a catalyst, in the pulp and paper factories, battery packs, and tooth amalgams (Jaishankar et al. 2014). Hg are commonly employed in research instruments and supplies such as thermometer and barometric instruments, arterial blood pressure measuring devices, and fluorescent light bulbs. It is mostly Hg is a well-known dangerous metal that often appears to exist in mercuric, mercurous, elemental, and alkylated form (methyl/ethyl Hg) (Mishra et al. 2019). Mercury toxicity can harm the neurological system, particularly in children. It is one among the most dangerous metals, causing sickness and even death at low concentrations and over extended periods of exposure. Mercury toxicity is subsequently solubilized in plasma, blood, as well as hemoglobin. It can penetrate the placenta and affect the fetus in pregnant women. The risk remains after birth since breast milk can become infected. Mercury is highly neurotoxic and bioaccumulative, and it exerts cytotoxic effects on central nervous system cells called stem cells. Depending on the quantity of pollution and the length of exposure, mercury produces an endocrine effect (Sall et al. 2020).

3.11.2 Chromium

Chromium is found in petroleum as well as coal, in addition to chromium steel, pigment oxidizing agents, chemical fertilizers, catalysts, oil well drilling process, alongside metal plating tanning facilities. Chromium is widely utilized in a variety of sectors, including wood conservation, the electroplating process metallurgy, paints and color manufacture, chemical manufacturing, tanners and the manufacture of paper and pulp. These businesses contribute significantly to chromium contamination, which has a negative impact on biological as well as ecological species. Human actions such as disposing of waste and application of fertilizer may result in the accidental discharge of chromium into the natural world (Azeh Engwa et al. 2019). Because of the substantial usage of Cr salts throughout the chrome tanning process, the leather goods sector is principally responsible for Cr pollution in the environment. HM contamination into soils is additionally exacerbated by the increasing use of chemical fertilizers, pesticides, alongside fungicides in farming operations. Fertilizers contain As, Cr, Cd, Pb, Zn, Ni, Fe, Mo, along with Mn, which are essential for plant development and advancement, but excessive amounts are harmful (Mishra et al. 2019).

Chromium toxicity differs greatly depending on its chemical makeup type (particles, nanoparticle, ion, oxide, hydroxide, valence 2 to valence 6). Chromium species can cause breaks in double-strands and trigger DNA damage, implying that they are the ultimate genetically toxic agent. Plants have a mechanism that regulates their chromium intake. Yet, when the level of chromium in the soil becomes too high, genetic alterations occur in plants. Chromium may lead to respiratory issues, a reduced capacity to fight infections, genetic abnormalities, infertility, and tumor growth in animals. Chronic chromium contamination in humans can cause skin lesions as well as mucous membrane damage, in addition to respiratory difficulties and potentially bronchopulmonary cancer (Sall et al. 2020). Chromium is the world's sixth most prevalent element. The most frequent varieties of Cr being trivalent in nature $Cr + 3$ and hexavalent $Cr + 6$, which are both harmful to humans. Organic sources of chromium include the combustion of oil and coal, petroleum coming from ferro chromate refractory substance, pigmented oxidants, catalysts, chromium steel, chemical fertilizers, oil well drilling process, and metal plating tanning facilities. Chromium is discharged into the environment by humans via sewerage and fertilizers. Cr(III) is stationary and not soluble in water in its reduced form, but Cr(VI) is migratory in its oxidized state. Although the oxidative form of Cr(III) is not considered a necessary pollutant of ground water, Cr(VI) has been demonstrated to be hazardous to people. Chromium is widely utilized in metallurgy, electroplating, painting and pigments manufacture, the tanning process, wood conservation, chemical manufacturing, and paper and pulp manufacturing. These fields contribute significantly to chromium contamination. A variety of industrial as well as agricultural methods raise the hazardous level in the environment, raising concerns about chromium contamination. Numerous toxic heavy metals and chemicals are discharged into waterways by tanneries. Cr(III) oxidizes to Cr(VI) in the presence of sufficient oxygen in the natural world, which is exceedingly poisonous and readily soluble in water. Cr(VI) is classified as a category 1 human carcinogens due to its mutagenic effects (Jaishankar et al., 2014).

3.11.3 Lead

Among the most harmful HMs is lead. Chronic toxicity is generally referred as a poisoning from lead. It can have catastrophic consequences for the neurological, hematopoietic, as well as cardiovascular systems of humans. Lead can induce neurological, hematological, and renal problems at large levels. It may lead to neurological abnormalities in youngsters, as well as psychological issues and learning issues. Lead poisoning reduces fertility, causes fetal mortality along with spontaneous abortion, and causes neurological, heart, and digestive problems. It may also be mutagenic and cancerous (Sall et al. 2020). Lead may be discovered in both organic and inorganic substances, as well as trace levels in food, water, and soil. The greatest amount of lead contamination in the environment is created by the application of paint containing lead on the outside of houses. Lead poisoning occurs in over 900 jobs, involving refinery, smelting, mining, the welding process, and painting, as well as the production of glass for stained glass and ceramics glass. Inorganic compounds made from lead and elemental lead, like lead dust, can enter the body via the digestive and respiratory systems. The existence of lead in the surroundings influences its toxicity. Organic lead substances can reach the brain via the skin and cause a toxin to form in the brain's central nervous system. Lead gets absorbed through the plasma once the fluid outside of the cell is in a condition of and collects in both hard and soft tissues. Ultimately, lead is mostly eliminated via the urine and digestive systems. The harm induced by lead to biological systems involves two distinct pathways due to the generation of free radicals. The first process is the direct creation of reactive oxygen species, which leads to a drop in mitochondrial viability and death via cytochrome c release. The second process is accomplished by the ingestion of cell antioxidants. Previous research has shown that lead inhibits mitochondrial oxidative phosphorylation. The primary mechanism of lead poisoning is oxidative stress (Fu and Xi 2020).

Lead initially was identified in gasoline and home paint, but has now spread to lead bullets, plumbing fixtures, pewter pitcher, batteries for storage, toys, even taps. Lead performs no biological activities. Lead toxicity in live cells is caused by an ionic process as well as oxidative stress (Jaishankar et al. 2014). Pb is widely employed in automobile exhaust, hair colors, pigments, the vermillion, collyrium, art pottery glazing, and enameled pottery. Transportation on roads and cars are the primary sources of Pb leaks into the natural world. Human activities such as burning coal, leaded fuels, and the manufacturing of pyrometallurgical non-ferrous metals all impact significantly to Pb-related contamination (Mishra et al. 2019).

3.11.4 Cadmium

Cadmium is a HM found organically in ores which is commonly utilized as a stabilizer in a variety of goods including color pigments, alloys, and various other related items. Smoke respiration, the nickel–cadmium batteries sector, the electroplating process and pigments for paint are all sources of workplace cadmium exposure. Consumption of polluted food and water causes oral cadmium exposure. This is the most critical channel for people who do not smoke and individuals who have never been exposed to cadmium at work. Exposure to cadmium has been shown in studies to impair the key antioxidant enzymes. Cadmium exposure has been linked with elevated lipid peroxidation in several brain areas, including the cerebellum along with cortex. Cadmium interferes with the number of enzymes that keep the redox reaction going. Additionally, as oxidative stress is thought to be among of the linked processes of genotoxicity, it is additionally believed to be a possible source of neuronal toxicity. Cadmium-induced oxidative stress can damage and impair the proper function of organelles, as well as generate mutations in DNA that result in alterations in gene expression and, eventually, apoptosis. Furthermore, cadmium can disrupt the repair process that repairs DNA damage, resulting in the buildup of DNA damage, generating genetic mutations and consequently cancer, and impeding DNA damage repair (Fu and Xi 2020).

Cadmium ranks ninth among heavy metals in terms of toxicity. It is an output of zinc manufacturing that individuals may come into contact with at work or in their surroundings. Once ingested by people, this metal will build up within the body over time. It is found in tobacco smoke as well as batteries that recharge and specific metals. Cadmium is utilized in alkaline batteries mainly as an electrode section, with the remainder employed in coatings, pigments, platings, as well as a plastic a stabilizing agent Humans can be exposed to the metal largely by breathing and consumption, resulting in acute and chronic toxicity. Cadmium in the natural world will be present in soils and sediments over decades. Plants progressively absorb the metals that collect in their cells and concentrate them down the food chain, eventually reaching the human body. Cadmium is mostly detected in vegetables and fruits owing to its rapid transfer from soil to plant. Cadmium is an extremely harmful and unnecessary heavy metal. The underlying cause of cadmium poisoning is not well known (Jaishankar et al. 2014). The itai-itai illness in Japan was brought on by over time cadmium exposure (Deng et al. 2022). Cadmium is nearly as hazardous as mercury and lead. identified as a human cancer agent (or combination). Indeed, several investigations done in workplaces and related to respiratory exposures have demonstrated a considerable increase in cancer-related mortality when cadmium is present. Cadmium can cause renal issues and raise blood pressure in people. Furthermore, inhaling it is hazardous (Sall et al. 2020).

3.11.5 Nickel

Despite nickel is one of the lightweight heavy metals in its metal state, extended interaction with the pores and mucous membranes of the skin may result in irritation and, in rare cases, allergy. The most prevalent side effect of being exposed to nickel is contact dermatitis. Furthermore, consumption of nickel salts can cause diarrhea, vomiting, and nausea, whilst persistent inhalation of nickel in the monoxide or metal form can cause asthmatic and dysfunction of the tubes. Furthermore, some nickel chemical compounds, like tetra-carbonyl-nickel, are exceedingly poisonous and carcinogenic (lung cancer). As a result, nickel has the potential to cause cancer (Sall et al. 2020). Nickel ranks as the twenty-fourth most plentiful element on the planet, accounting for 3% of all the planet's crust. Nickel is extensively present in water for drinking all throughout the world. Nickel is thought to be crucial for a variety of human bodily processes. Excessive exposure, on the other hand, may result in toxicity. Water and food polluted with nickel-containing substances constitute the two primary sources of oral nickel intake. Ni has various applications in business and industry, making it available to these personnel. Nickel emissions occur naturally as a result of wildfires, lava emissions, and wind debris. Additional sources of nickel discharged into the atmosphere comprise the combustion of coal and the burning of waste, while smoking is a further cause of exposure to nickel for people. Nickel bonds to mercaptan in the body of a person to produce the nickel-mercaptan complexes. The combination then interacts with oxygen in molecules to form free radicals, resulting in nickel toxicity. Nickel attaches to thiol proteins, causing glutathione levels to drop, resulting in nickel toxin in the body. Because of decreased elimination of calcium in urine, physicochemical qualities of the body of an individual could shift as a consequence of nickel exposure. Additionally, nickel exposure reduces nitrogen retention. Prolonged nickel exposure reduces nicotinamide levels, which impairs the oxidative phosphorylation process. Due to nickel exposure, cultivated red blood cells from humans were discovered to be under stress from oxidative damage. Nickel is involved in the development of blood poisoning (Fu and Xi 2020).

3.11.6 Arsenic

Human Beings exposure to arsenic can occur from consuming water polluted with industrial or farming chemical residues. The smoke or arsenic mist might also pollute water bodies. Poisoning can occur by consuming pesticide-contaminated meals or arsenic-rich foods. Surprisingly, consuming dairy products is also an origin of poisoning from arsenic, as contaminated milk includes arsenic-containing water. Besides these routes, AS can enter the human body by breathing and contact with the skin. The ingestion of acids and unrefined metals is also a major cause of arsenic poisoning. Inorganic arsenic is harmful both acutely and chronically. More than 140 million people globally drink water with arsenic that surpasses the World Health

Organization's ten part per billion standards. Many enzymes are inhibited by arsenic. The primary metabolic routes are assumed to be methyl metallic oxide along with glutathione conjugation. Arsenic not only harms the skin and causes bladder cancer, but it also harms the liver, kidneys, along with lungs. Arsenic has been linked to cancer, the syndrome of metabolism, and other metabolic disorders and symptoms. Nevertheless, the precise process through which arsenic causes sickness is unknown. It is thought that reactive oxygen species, changed DNA methylation, damage to DNA, and inhibition of enzymes all play a role (Fu and Xi 2020).

Arsenic is widely accessible as oxides or sulfides, or as a salt containing iron, calcium, sodium, copper, and other elements. Arsenic is the tenth most prevalent chemical on the planet, and its inorganic varieties, including arsenite and arsenate compounds, are toxic to both the environment and living things. Arsenic can enter the body through natural processes, industrial processes, or accidental sources. Water for drinking may become polluted as a result of the use of arsenical insecticides, natural deposits of minerals, or improper disposal of arsenical substances. Arsenic acts as a protoplastic toxin because it mainly impacts the sulphydryl category of cells, enabling respiration in cells, enzymes, as well as mitosis to fail (Jaishankar et al. 2014). Humans come into exposure to arsenic through industrial processes that include smelting and microelectronics, in timber preservation agents, insecticides, weed killers, fungicides, and paints, can pollute drinking water (Azeh Engwa et al. 2019).

3.11.7 Aluminum

Aluminum is the 3rd most plentiful element in the planet's crust. Aluminum is found organically in the atmosphere, water, and soil. Aluminum mining and processing raises its concentration in the environment. Many parameters, particularly water pH as well as organic matter concentration, have a significant impact on aluminum toxicity. Its toxicity rises as the pH decreases. The dispersion of hazardous aluminum ions induced by changes in soil and water pH caused by rain that is acidic and rising acidity of the atmosphere around them has a negative impact on the ecosystem. This manifests as forest dryness, plant poisoning, agricultural decline or failure, aquatic animal mortality, and numerous abnormalities in the operation of both human and animal ecosystems. Crop cultivation was limited to 67 percent of the globe's total acid soil area because of aluminum toxicity (Jaishankar et al. 2014). Most physical and biological processes are disrupted by aluminum. The precise process of aluminum absorption via the gastrointestinal system is unknown. Aluminum toxicity is most likely caused by the interaction of aluminum with plasma membranes, apoplastic or symplastic sites. In individuals, Mg^{2+} as well as Fe^{3+} are substituted by Al^{3+}, causing several disruptions in interaction between cells, cellular development, and secretory processes. Aluminum-induced alterations in neurons are comparable to the degeneration lesions seen in Alzheimer patients. Neurotoxicity consequences such as neuronal shrinkage in the locus ceruleus, substantia nigra, and striatum are the most serious sequelae of aluminum toxicity (Jaishankar et al. 2014).

3.11.8 Iron

The 2nd most prevalent metal in the crust of the planet is iron. Iron is a critical element for the development and longevity of nearly all living creatures. Numerous individuals were subjected to high quantities of iron in their drinking water since the collected groundwater surpassed the allowable limit. When ingested iron fails to attach to protein, a wide spectrum of damaging free radicals are generated. This moving free iron has a corrosive effect on the gut and bodily fluids. These liberated irons enter heart, liver, along with brain cells, raising metabolic acidification and causing damage to mitochondria, microsomes, and various other organelles in cells. Iron toxicity in cells has resulted in damage to tissues, damage to cells, a mutation, and malignant changes (Jaishankar et al. 2014). Iron toxicity among individuals is commonly linked to vomiting, circulatory collapse, and diarrhea (Abdulrahman et al. 2012).

3.11.9 Copper

Copper is often remobilized by acid–base exchange of ions or oxidation. Copper is required for life for humans, but excessive amounts can result in anemia, adrenal hyperactivity and deficiency, allergic reactions, thinning hair, arthritis, autism, carcinoma, depressive disorders, raised cholesterol, diabetes, dyslexia, or failure to prosper, tiredness, anxieties, fractures of the bones, migraines, cardiac arrest, irritability, high blood pressure, diseases, anxiety disorders, strokes, decay of the teeth, and vitamin C along with other vitamin deficiencies (Abdulrahman et al. 2012).

3.12 Future Research and Concluding Remarks

When HM poisoning occurs, it is a chemically critical condition. Toxicology, if undiagnosed or improperly handled, can cause substantial disease and diminished quality of life, ultimately leading to death. Environmental regulation, comprehensive planning, technical techniques for enhancing the standard of waste discharges, and ecological monitoring systems are recommended solutions for combating this threat (David and Isangedighi 2019). It is also advised that, for best management and avoidance of additional pollution, the origin and levels of HM regarding the possible contamination of food items from the region's water, soil, and sand be investigated (Shokri et al. 2022). Recognizing the links between recurrent interaction and HMs is likely the most successful approach to tackle and mitigate the impact to human health (Deng et al. 2022). It is opined that, co-exposure to various HM in drinking water requires more research. Low-cost arsenic removal solutions require immediate attention (Chowdhury et al. 2016). Human Beings bio monitoring is a well-known approach for evaluating human susceptibility to environmental contaminants (Bibi et al. 2016). Toenail and Hair is an especially helpful tool for assessing heavy metal load and contamination in the body (Ab Razak et al. 2015; Chojnacka et al. 2012) this aspect need to be studied more.

Assessing the relationship between contamination of soil with over time use of land is a very effective method for regulating resources by designating danger zones as heavy metal levels that exceed the associated control limit (Lin et al. 2011). In order to secure the long-term growth of local farming, equivalent measures must be implemented in future manufacturing and agricultural operations. The following are specific measures or policies: (1) minimize the application of synthetic fertilizers and herbicides; (2) eliminate or enhance releases of pollutants industrial processes; (3) use more sustainable automobile devices and fuels; (4) immobilize and stabilize HMs in river deposits; and (5) preserve an appropriate secure distance among farmland and factories (Shen et al. 2021). To reduce non-cancer risk, pesticide usage must be reduced over time and strictly monitored (Yang et al. 2019). Urban or commercial waste ought not to be emptied into rivers or agricultural land unless first being treated to avoid metals entering the food chain. In besides addressing the waste that reaches the farms, other methods of cleansing up the currently polluted substrates are required. Continuous monitoring of soil, plant, and water quality, as well as prohibiting metals from infiltrating vegetables, are required for the avoidance of possible human health problems (Balkhair and Ashraf 2016). Utilizing total amounts of particle-bound HM as opposed to bioaccessible quantities results in exaggeration of the health threat posed by source particles and incorrect identification of high-risk contaminants and priority metal contaminants in the originating particles (Liu et al. 2019; Ma et al. 2021). Thus, this aspect should be taken care of when quantifying HM associated risks and this would lead to correct estimation and proper alleviation technique selection for HM contamination.

References

Ab Razak NH, Praveena SM, Aris AZ, Hashim Z (2015) Drinking water studies: a review on heavy metal, application of biomarker and health risk assessment (a special focus in Malaysia). J Epidemiol Glob Health 5(4):297–310. https://doi.org/10.1016/j.jegh.2015.04.003

Abdelhafez AA, Li J (2015) Environmental monitoring of heavy metal status and human health risk assessment in the agricultural soils of the Jinxi River Area, China. Human Ecol Risk Assess 21(4):952–971. https://doi.org/10.1080/10807039.2014.947851

Abdulrahman I, Akan JC, Chellube M, Waziri M (2012) Levels of heavy metals in human hair and nail samples from Maiduguri Metropolis, Borno State, Nigeria. World Environ 2(4):81–89. https://doi.org/10.5923/j.env.20120204.05

Abouian Jahromi M, Jamshidi-Zanjani A, Khodadadi Darban A (2020) Heavy metal pollution and human health risk assessment for exposure to surface soil of mining area: a comprehensive study. Environ Earth Sci 79(14):1–18. https://doi.org/10.1007/s12665-020-09110-3

Adimalla N, Chen J, Qian H (2020) Spatial characteristics of heavy metal contamination and potential human health risk assessment of urban soils: a case study from an urban region of South India. Ecotoxicol Environ Saf 194(126):110406. https://doi.org/10.1016/j.ecoenv.2020.110406

Ahmad Bhat S, Hassan T, Majid S (2019) Heavy metal toxicity and their harmful effects on living organisms—a review. Int J Med Sci Diagnosis Research (IJMSDR) 3(1), 106–122. www.ijmsdr.com

Ahmed MK, Baki MA, Islam MS, Kundu GK, Habibullah-Al-Mamun M, Sarkar SK, Hossain MM (2015) Human health risk assessment of heavy metals in tropical fish and shellfish collected from the river Buriganga, Bangladesh. Environ Sci Pollut Res 22(20):15880–15890. https://doi.org/10.1007/s11356-015-4813-z

Ali H, Khan E (2018) Bioaccumulation of non-essential hazardous heavy metals and metalloids in freshwater fish. Risk to human health. Environ Chem Lett 16(3):903–917. https://doi.org/10.1007/s10311-018-0734-7

Ali H, Khan E (2019) Trophic transfer, bioaccumulation, and biomagnification of non-essential hazardous heavy metals and metalloids in food chains/webs—concepts and implications for wildlife and human health. Hum Ecol Risk Assess 25(6):1353–1376. https://doi.org/10.1080/10807039.2018.1469398

Anyanwu ED, Onyele OG (2018) Human health risk assessment of some heavy metals in a rural. Afr J Environ Nat Sci Res 1(1):15–23. www.abjournals.org

Asaduzzaman K, Khandaker MU, Binti Baharudin NA, Amin YBM, Farook MS, Bradley DA, Mahmoud O (2017) Heavy metals in human teeth dentine: a bio-indicator of metals exposure and environmental pollution. Chemosphere 176:221–230. https://doi.org/10.1016/j.chemosphere.2017.02.114

Asiminicesei DM, Vasilachi IC, Gavrilescu M (2020) Heavy metal contamination of medicinal plants and potential implications on human health. Revista de Chimie 71(7):16–36. https://doi.org/10.37358/RC.20.7.8222

Azeh Engwa G, Udoka Ferdinand P, Nweke Nwalo F, Unachukwu M (2019) Mechanism and health effects of heavy metal toxicity in humans. In: Poisoning in the modern world—new tricks for an old dog?, June. https://doi.org/10.5772/intechopen.82511

Balkhair KS, Ashraf MA (2016) Field accumulation risks of heavy metals in soil and vegetable crop irrigated with sewage water in western region of Saudi Arabia. Saudi J Biol Sci 23(1):S32–S44. https://doi.org/10.1016/j.sjbs.2015.09.023

Baroi A, Siddique MAB, Akbor MA, Chowdhury FN, Jamil MAR, Uddin MK, Rahman MM (2023) Exposure and health risks of metals in imported and local brands' lipsticks and eye pencils from Bangladesh. Environ Sci Pollut Res 30(16):46222–46233. https://doi.org/10.1007/s11356-023-25416-8

Bazeyad AY, Al-Sarar AS, Rushdi AI, Hassanin AS, Abobakr Y (2019) Levels of heavy metals in a multifloral Saudi honey. Environ Sci Pollut Res 26(4):3946–3953. https://doi.org/10.1007/s11356-018-3909-7

Bibi M, Hashmi MZ, Malik RN (2016) The level and distribution of heavy metals and changes in oxidative stress indices in humans from Lahore district, Pakistan. Human Exp Toxicol 35(1):78–90. https://doi.org/10.1177/0960327115578063

Bradl HB (2005) Chapter 1 Sources and origins of heavy metals. Interface Sci Technol 6(C):1–27. https://doi.org/10.1016/S1573-4285(05)80020-1

Chai L, Wang Y, Wang X, Ma L, Cheng Z, Su L (2021) Pollution characteristics, spatial distributions, and source apportionment of heavy metals in cultivated soil in Lanzhou, China. Ecol Indic 125:107507. https://doi.org/10.1016/j.ecolind.2021.107507

Chandrakar C, Kumar Jaiswal S, Chaturvedani AK, Sahu SS, Ika M, Wasist U (2018) A review on heavy metal residues in indian milk and their impact on human health. Int J Curr Microbiol Appl Sci 7(05):1260–1268. https://doi.org/10.20546/ijcmas.2018.705.152

Chen P, Bi X, Zhang J, Wu J, Feng Y (2015) Assessment of heavy metal pollution characteristics and human health risk of exposure to ambient PM2.5 in Tianjin, China. Particuology 20:104–109. https://doi.org/10.1016/j.partic.2014.04.020

Chen Q, Lu Z, Yan D, Wang Q, Xin S (2020) Source analysis and health risk of heavy metals in the different seasons from Taizihe River, China. Acta Ecologica Sinica 40(1):64–71. https://doi.org/10.1016/j.chnaes.2018.12.004

Chenery SRN, Sarkar SK, Chatterjee M, Marriott AL, Watts MJ (2020) Heavy metals in urban road dusts from Kolkata and Bengaluru, India: implications for human health. Environ Geochem Health 42(9):2627–2643. https://doi.org/10.1007/s10653-019-00467-4

Chojnacka K, Saeid A, Michalak I, Mikulewicz M (2012) Effects of local industry on heavy metals content in human hair. Pol J Environ Stud 21(6):1563–1570

Chowdhury S, Mazumder MAJ, Al-Attas O, Husain T (2016) Heavy metals in drinking water: occurrences, implications, and future needs in developing countries. Sci Total Environ 569–570:476–488. https://doi.org/10.1016/j.scitotenv.2016.06.166

Cocârţă DM, Neamţu S, Reşetar Deac AM (2016) Carcinogenic risk evaluation for human health risk assessment from soils contaminated with heavy metals. Int J Environ Sci Technol 13(8):2025–2036. https://doi.org/10.1007/s13762-016-1031-2

Coen N, Mothersill C, Kadhim M, Wright EG (2001) Heavy metals of relevance to human health induced genomic instability. J Pathol 195(3):293–299. https://doi.org/10.1002/path.950

Dat ND, Nguyen VT, Vo TDH, Bui XT, Bui MH, Nguyen LSP, Nguyen XC, Tran ATK, Nguyen TTA, Ju YR, Huynh TMT, Nguyen DH, Bui HN, Lin C (2021) Contamination, source attribution, and potential health risks of heavy metals in street dust of a metropolitan area in Southern Vietnam. Environ Sci Pollut Res 28(36):50405–50419. https://doi.org/10.1007/s11356-021-14246-1

David GS, Isangedighi IA (2019) Heavy metals contamination in fish: effects on human health. J Aquat Sci Mar Biol 2(4):7–12

Deng H, Tu Y, Wang H, Wang Z, Li Y, Chai L, Zhang W, Lin Z (2022) Environmental behavior, human health effect, and pollution control of heavy metal(loid)s toward full life cycle processes. Eco-Environ Health 1(4):229–243. https://doi.org/10.1016/j.eehl.2022.11.003

Dumitriu (Gabur) GD, Teodosiu C, Morosanu I, Plavan O, Gabur I, Cotea VV (2021) Heavy metals assessment in the major stages of winemaking: chemometric analysis and impacts on human health and environment. J Food Compos Anal 100(April):103935. https://doi.org/10.1016/j.jfca.2021.103935

Edokpayi JN, Enitan AM, Mutileni N, Odiyo JO (2018) Evaluation of water quality and human risk assessment due to heavy metals in groundwater around Muledane area of Vhembe District, Limpopo Province, South Africa. Chem Cent J 12(1):1–16. https://doi.org/10.1186/s13065-017-0369-y

Ekanem EUO, Garba IH, Harami A, Akan JC (2016) Health risk assessment for exposure to some selected heavy metals via drinking water from Dadinkowa Dam and River Gombe Abba in Gombe State. World J Anal Chem 4(1):1–5. https://doi.org/10.12691/wjac-4-1-1

El Bahgy HEK, Elabd H, Elkorashey RM (2021) Heavy metals bioaccumulation in marine cultured fish and its probabilistic health hazard. Environ Sci Pollut Res 28(30):41431–41438. https://doi.org/10.1007/s11356-021-13645-8

Emmanuel UC, Chukwudi MI, Monday SS, Anthony AI (2022) Human health risk assessment of heavy metals in drinking water sources in three senatorial districts of Anambra State, Nigeria. Toxicol Rep 9(April):869–875. https://doi.org/10.1016/j.toxrep.2022.04.011

Essien JP, Ikpe DI, Inam ED, Okon AO, Ebong GA, Benson NU (2022) Occurrence and spatial distribution of heavy metals in landfill leachates and impacted freshwater ecosystem: an environmental and human health threat. PLoS ONE 17(2 February):1–18. https://doi.org/10.1371/journal.pone.0263279

Ewusi A, Sunkari ED, Seidu J, Coffie-Anum E (2022) Hydrogeochemical characteristics, sources and human health risk assessment of heavy metal dispersion in the mine pit water–surface water–groundwater system in the largest manganese mine in Ghana. Environ Technol Innov 26:102312. https://doi.org/10.1016/j.eti.2022.102312

Eyankware MO, Obasi PN (2021) A holistic review of heavy metals in water and soil in Ebonyi SE, Nigeria; with emphasis on its effects on human, plants and aquatic organisms. World News Nat Sci 38(June):1–19

Ezeabara C, Okanume O, Emeka A, Okeke C, Mbaekwe E (2014) Heavy metal contamination of herbal drugs: implication for human health—a review. Int J Trop Dis Health 4(10):1044–1058. https://doi.org/10.9734/ijtdh/2014/11481

Fan Y, Zhu T, Li M, He J, Huang R (2017) Heavy metal contamination in soil and brown rice and human health risk assessment near three mining areas in Central China. J Healthc Eng 2017. https://doi.org/10.1155/2017/4124302

Fernández-Luqueño F, López-Valdez F, Gamero-Melo P, Luna-Suárez S, Aguilera-González E, Martínez A, García-Guillermo M, Hernández-Martínez G, Herrera-Mendoza R, Álvarez-Garza M, Pérez-Velázquez I (2013) Heavy metal pollution in drinking water—a global risk for human health: a review. Afr J Environ Sci Technol 7(7):567–584. https://doi.org/10.5897/AJEST12.197

Fu H, Yang Z, Liu Y, Shao P (2021) Ecological and human health risk assessment of heavy metals in dust affected by fireworks during the Spring Festival in Beijing. Air Qual Atmos Health 14(1):139–148. https://doi.org/10.1007/s11869-020-00920-9

Fu Z, Xi S (2020) The effects of heavy metals on human metabolism. Toxicol Mech Methods 30(3):167–176. https://doi.org/10.1080/15376516.2019.1701594

Gao J, Wang L (2018) Ecological and human health risk assessments in the context of soil heavy metal pollution in a typical industrial area of Shanghai, China. Environ Sci Pollut Res 25(27):27090–27105. https://doi.org/10.1007/s11356-018-2705-8

Ghorbani MR, Ghanavati N, Babaenejad T, Nazarpour A, Payandeh K (2020) Assessment of the potential ecological and human health risks of heavy metals in Ahvaz oil field, Iran. PLoS ONE 15(11 November). https://doi.org/10.1371/journal.pone.0242703

Giri S, Singh AK (2017) Human health risk assessment due to dietary intake of heavy metals through rice in the mining areas of Singhbhum Copper Belt, India. Environ Sci Pollut Res 24(17):14945–14956. https://doi.org/10.1007/s11356-017-9039-9

Hadzi GY, Ayoko GA, Essumang DK, Osae SKD (2019) Contamination impact and human health risk assessment of heavy metals in surface soils from selected major mining areas in Ghana. Environ Geochem Health 41(6):2821–2843. https://doi.org/10.1007/s10653-019-00332-4

Hassaan MA, El Nemr A, Madkour FF (2016) Environmental assessment of heavy metal pollution and human health risk. Am J Water Sci Eng 2(3):14–19. https://doi.org/10.11648/j.ajwse.20160203.11

He CT, Zheng XB, Yan X, Zheng J, Wang MH, Tan X, Qiao L, Chen SJ, Yang ZY, Mai BX (2017) Organic contaminants and heavy metals in indoor dust from e-waste recycling, rural, and urban areas in South China: spatial characteristics and implications for human exposure. Ecotoxicol Environ Saf 140(January):109–115. https://doi.org/10.1016/j.ecoenv.2017.02.041

Hoang HG, Chiang CF, Lin C, Wu CY, Lee CW, Cheruiyot NK, Tran HT, Bui XT (2021) Human health risk simulation and assessment of heavy metal contamination in a river affected by industrial activities. Environ Pollut 285(March):117414. https://doi.org/10.1016/j.envpol.2021.117414

Hu Y, Wang C, Song Z, Chen M, Ding L, Liang X, Bi X, Li Z, Li P, Zheng W (2021) Heavy metal in rice and vegetable and human exposure near a large pb/zn smelter in central China. Int J Environ Res Public Health 18(23). https://doi.org/10.3390/ijerph182312631

Huang J, Guo S, Zeng G, Li F, Gu Y, Shi Y, Shi L, Liu W, Peng S (2018) A new exploration of health risk assessment quantification from sources of soil heavy metals under different land use. Environ Pollut 243:49–58. https://doi.org/10.1016/j.envpol.2018.08.038

Hubeny J, Harnisz M, Korzeniewska E, Buta M, Zieliński W, Rolbiecki D, Giebułtowicz J, Nałęcz-Jawecki G, Płaza G (2021) Industrialization as a source of heavy metals and antibiotics which can enhance the antibiotic resistance in wastewater, sewage sludge and river water. PLoS ONE 16(6 June):1–24. https://doi.org/10.1371/journal.pone.0252691

Ihedioha JN, Ujam OT, Nwuche CO, Ekere NR, Chime CC (2016) Assessment of heavy metal contamination of rice grains (Oryza sativa) and soil from Ada field, Enugu, Nigeria: estimating the human healtrisk. Hum Ecol Risk Assess 22(8):1665–1677. https://doi.org/10.1080/10807039.2016.1217390

Intawongse M, Dean JR (2006) Uptake of heavy metals by vegetable plants grown on contaminated soil and their bioavailability in the human gastrointestinal tract. Food Addit Contam 23(1):36–48. https://doi.org/10.1080/02652030500387554

Iqbal Z, Abbas F, Ibrahim M, Qureshi TI, Gul M, Mahmood A (2020) Human health risk assessment of heavy metals in raw milk of buffalo feeding at wastewater-irrigated agricultural farms in Pakistan. Environ Sci Pollut Res 27(23):29567–29579. https://doi.org/10.1007/s11356-020-09256-4

Jadoon WA, Abdel-Dayem SMMA, Saqib Z, Takeda K, Sakugawa H, Hussain M, Shah GM, Rehman W, Syed JH (2021) Heavy metals in urban dusts from Alexandria and Kafr El-Sheikh, Egypt: implications for human health. Environ Sci Pollut Res 28(2):2007–2018. https://doi.org/10.1007/s11356-020-08786-1

Jaishankar M, Tseten T, Anbalagan N, Mathew BB, Beeregowda KN (2014) Toxicity, mechanism and health effects of some heavy metals. Interdiscip Toxicol 7(2):60–72. https://doi.org/10.2478/intox-2014-0009

Jayakumar M, Surendran U, Raja P, Kumar A, Senapathi V (2021) A review of heavy metals accumulation pathways, sources and management in soils. Arab J Geosci 14(20). https://doi.org/10.1007/s12517-021-08543-9

Jiang C, Zhao Q, Zheng L, Chen X, Li C, Ren M (2021) Distribution, source and health risk assessment based on the Monte Carlo method of heavy metals in shallow groundwater in an area affected by mining activities, China. Ecotoxicol Environ Saf 224:112679. https://doi.org/10.1016/j.ecoenv.2021.112679

Jiang HH, Cai LM, Wen HH, Hu GC, Chen LG, Luo J (2020) An integrated approach to quantifying ecological and human health risks from different sources of soil heavy metals. Sci Total Environ 701:134466. https://doi.org/10.1016/j.scitotenv.2019.134466

Jiang Y, Chao S, Liu J, Yang Y, Chen Y, Zhang A, Cao H (2017) Source apportionment and health risk assessment of heavy metals in soil for a township in Jiangsu Province, China. Chemosphere 168:1658–1668. https://doi.org/10.1016/j.chemosphere.2016.11.088

Jose A, Ray JG (2018) Toxic heavy metals in human blood in relation to certain food and environmental samples in Kerala, South India. Environ Sci Pollut Res 25(8):7946–7953. https://doi.org/10.1007/s11356-017-1112-x

Kadhum SA (2020) A preliminary study of heavy metals pollution in the sandy dust storms and its human risk assessment from middle and south of Iraq. Environ Sci Pollut Res 27(8):8570–8579. https://doi.org/10.1007/s11356-019-07380-4

Kawser Ahmed M, Baki MA, Kundu GK, Saiful Islam M, Monirul Islam M, Muzammel Hossain M (2016) Human health risks from heavy metals in fish of Buriganga river, Bangladesh. SpringerPlus 5(1). https://doi.org/10.1186/s40064-016-3357-0

Khan A, Khan S, Khan MA, Qamar Z, Waqas M (2015) The uptake and bioaccumulation of heavy metals by food plants, their effects on plants nutrients, and associated health risk: a review. Environ Sci Pollut Res 22(18):13772–13799. https://doi.org/10.1007/s11356-015-4881-0

Khan MU, Malik RN, Muhammad S (2013) Human health risk from Heavy metal via food crops consumption with wastewater irrigation practices in Pakistan. Chemosphere 93(10):2230–2238. https://doi.org/10.1016/j.chemosphere.2013.07.067

Kohzadi S, Shahmoradi B, Ghaderi E, Loqmani H, Maleki A (2019) Concentration, source, and potential human health risk of heavy metals in the commonly consumed medicinal plants. Biol Trace Elem Res 187(1):41–50. https://doi.org/10.1007/s12011-018-1357-3

Kwaansa-Ansah EE, Nti SO, Opoku F (2019) Heavy metals concentration and human health risk assessment in seven commercial fish species from Asafo Market, Ghana. Food Sci Biotechnol 28(2):569–579. https://doi.org/10.1007/s10068-018-0485-z

Lar UA, Agene JI, Umar AI (2015) Geophagic clay materials from Nigeria: a potential source of heavy metals and human health implications in mostly women and children who practice it. Environ Geochem Health 37(2):363–375. https://doi.org/10.1007/s10653-014-9653-0

Li H, Ji H, Shi C, Gao Y, Zhang Y, Xu X, Ding H, Tang L, Xing Y (2017) Distribution of heavy metals and metalloids in bulk and particle size fractions of soils from coal-mine brownfield and implications on human health. Chemosphere 172(30):505–515. https://doi.org/10.1016/j.chemosphere.2017.01.021

Li, Y., Zhang, Z., Liu, H., Zhou, H., Fan, Z., Lin, M., Wu, D., & Xia, B. (2016). Characteristics, sources and health risk assessment of toxic heavy metals in PM2.5 at a megacity of southwest China. Environ Geochem Health 38(2):353–362. https://doi.org/10.1007/s10653-015-9722-z

Liang Q, Tian K, Li L, He Y, Zhao T, Liu B, Wu Q, Huang B, Zhao L, Teng Y (2022) Ecological and human health risk assessment of heavy metals based on their source apportionment in cropland soils around an e-waste dismantling site, Southeast China. Ecotoxicol Environ Saf 242(June):113929. https://doi.org/10.1016/j.ecoenv.2022.113929

Lin YP, Cheng BY, Chu HJ, Chang TK, Yu HL (2011) Assessing how heavy metal pollution and human activity are related by using logistic regression and kriging methods. Geoderma 163(3–4):275–282. https://doi.org/10.1016/j.geoderma.2011.05.004

Liu JL, Xu XR, Ding ZH, Peng JX, Jin MH, Wang YS, Hong YG, Yue WZ (2015) Heavy metals in wild marine fish from South China Sea: levels, tissue- and species-specific accumulation and potential risk to humans. Ecotoxicology 24(7–8):1583–1592. https://doi.org/10.1007/s10646-015-1451-7

Liu X, Song Q, Tang Y, Li W, Xu J, Wu J, Wang F, Brookes PC (2013) Human health risk assessment of heavy metals in soil-vegetable system: a multi-medium analysis. Sci Total Environ 463–464:530–540. https://doi.org/10.1016/j.scitotenv.2013.06.064

Liu X, Ouyang W, Shu Y, Tian Y, Feng Y, Zhang T, Chen W (2019) Incorporating bioaccessibility into health risk assessment of heavy metals in particulate matter originated from different sources of atmospheric pollution. Environ Pollut 254:113113. https://doi.org/10.1016/j.envpol.2019.113113

López-Botella A, Velasco I, Acién M, Sáez-Espinosa P, Todolí-Torró JL, Sánchez-Romero R, Gómez-Torres MJ (2021) Impact of heavy metals on human male fertility—an overview. Antioxidants 10(9):1–24. https://doi.org/10.3390/antiox10091473

Lu Q, Wang S, Bai X, Liu F, Li C, Deng Y, Tian S (2021) Quantitative assessment of human health risks under different land uses based on soil heavy metal pollution sources. Hum Ecol Risk Assess 27(2):327–343. https://doi.org/10.1080/10807039.2019.1710811

Ma J, Yan Y, Chen X, Niu Z, Yu R, Hu G (2021) Incorporating bioaccessibility and source apportionment into human health risk assessment of heavy metals in urban dust of Xiamen, China. Ecotoxicol Environ Saf 228:112985. https://doi.org/10.1016/j.ecoenv.2021.112985

Mahmood A, Malik RN (2014) Human health risk assessment of heavy metals via consumption of contaminated vegetables collected from different irrigation sources in Lahore, Pakistan. Arab J Chem 7(1):91–99. https://doi.org/10.1016/j.arabjc.2013.07.002

Mahurpawar M (2015) Effects of heavy metals on human healtheffects of heavy metals on human health. Int J Res GRANTHAALAYAH 3(9SE):1–7. https://doi.org/10.29121/granthaalayah.v3.i9se.2015.3282

Maurya PK, Malik DS, Yadav KK, Kumar A, Kumar S, Kamyab H (2019) Bioaccumulation and potential sources of heavy metal contamination in fish species in River Ganga basin: possible human health risks evaluation. Toxicol Rep 6(May):472–481. https://doi.org/10.1016/j.toxrep.2019.05.012

Mawari G, Kumar N, Sarkar S, Daga MK, Singh MM, Joshi TK, Khan NA (2022) Heavy metal accumulation in fruits and vegetables and human health risk assessment: findings from Maharashtra, India. Environ Health Insights 16. https://doi.org/10.1177/11786302221119151

Mensah E, Kyei-Baffour N, Ofori E, Obeng G (2009) Influence of human activities and land use on heavy metal concentrations in irrigated vegetables in Ghana and their health implications. In: Appropriate technologies for environmental protection in the developing world—selected papers from ERTEP 2007, pp 9–14. https://doi.org/10.1007/978-1-4020-9139-1_2

Micó C, Recatalá L, Peris M, Sánchez J (2006) Assessing heavy metal sources in agricultural soils of an European Mediterranean area by multivariate analysis. Chemosphere 65(5):863–872. https://doi.org/10.1016/j.chemosphere.2006.03.016

Mishra S, Bharagava RN, More N, Yadav A, Zainith S, Mani S, Chowdhary P (2019) Heavy metal contamination: an alarming threat to environment and human health. Environ Biotechnol Sustain Future 103–125. https://doi.org/10.1007/978-981-10-7284-0_5

Mitra S, Chakraborty AJ, Tareq AM, Emran TB, Nainu F, Khusro A, Idris AM, Khandaker MU, Osman H, Alhumaydhi FA, Simal-Gandara J (2022) Impact of heavy metals on the environment and human health: novel therapeutic insights to counter the toxicity. J King Saud Univ Sci 34(3):101865. https://doi.org/10.1016/j.jksus.2022.101865

Mohammadi AA, Zarei A, Esmaeilzadeh M, Taghavi M, Yousefi M, Yousefi Z, Sedighi F, Javan S (2020) Assessment of heavy metal pollution and human health risks assessment in soils around an industrial zone in Neyshabur, Iran. Biol Trace Elem Res 195(1):343–352. https://doi.org/10.1007/s12011-019-01816-1

Muchuweti M, Birkett JW, Chinyanga E, Zvauya R, Scrimshaw MD, Lester JN (2006) Heavy metal content of vegetables irrigated with mixtures of wastewater and sewage sludge in Zimbabwe: implications for human health. Agr Ecosyst Environ 112(1):41–48. https://doi.org/10.1016/j.agee.2005.04.028

Muhib MI, Chowdhury MAZ, Easha NJ, Rahman MM, Shammi M, Fardous Z, Bari ML, Uddin MK, Kurasaki M, Alam MK (2016) Investigation of heavy metal contents in cow milk samples from area of Dhaka, Bangladesh. Int J Food Contamin 3(1):1–10. https://doi.org/10.1186/s40550-016-0039-1

Năstăsescu V, Mititelu M, Goumenou M, Docea AO, Renieri E, Udeanu DI, Oprea E, Arsene AL, Dinu-Pîrvu CE, Ghica M (2020) Heavy metal and pesticide levels in dairy products: evaluation of human health risk. Food Chem Toxicol 146:111844. https://doi.org/10.1016/j.fct.2020.111844

Neeratanaphan L, Khamma S, Benchawattananon R, Ruchuwararak P, Appamaraka S, Intamat S (2017) Heavy metal accumulation in rice (Oryza sativa) near electronic waste dumps and related human health risk assessment. Hum Ecol Risk Assess 23(5):1086–1098. https://doi.org/10.1080/10807039.2017.1300856

Nkwunonwo UC, Odika PO, Onyia NI (2020) A review of the health implications of heavy metals in food chain in Nigeria. Sci World J 2020. https://doi.org/10.1155/2020/6594109

Ogunbanjo O, Onawumi O, Gbadamosi M, Ogunlana A, Anselm O (2016) Chemical speciation of some heavy metals and human health risk assessment in soil around two municipal dumpsites in Sagamu, Ogun state, Nigeria. Chem Speciat Bioavailab 28(1–4):142–151. https://doi.org/10.1080/09542299.2016.1203267

Prasad S, Saluja R, Joshi V, Garg JK (2020) Heavy metal pollution in surface water of the Upper Ganga River, India: human health risk assessment. Environ Monit Assess 192(11). https://doi.org/10.1007/s10661-020-08701-8

Ren Y, Lin M, Liu Q, Zhang Z, Fei X, Xiao R, Lv X (2021) Contamination assessment, health risk evaluation, and source identification of heavy metals in the soil-rice system of typical agricultural regions on the southeast coast of China. Environ Sci Pollut Res 28(10):12870–12880. https://doi.org/10.1007/s11356-020-11229-6

Russo R, Voi AL, De Simone A, Serpe FP, Anastasio A, Pepe T, Cacace D, Severino L (2013) Heavy metals in canned tuna from Italian markets. J Food Prot 76(2):355–359. https://doi.org/10.4315/0362-028X.JFP-12-346

Rzymski P, Tomczyk K, Rzymski P, Poniedziałek B, Opala T, Wilczak M (2015) Impact of heavy metals on the female reproductive system. Ann Agric Environ Med 22(2):259–264. https://doi.org/10.5604/12321966.1152077

Sah D, Verma PK, Kandikonda MK, Lakhani A (2019) Pollution characteristics, human health risk through multiple exposure pathways, and source apportionment of heavy metals in PM10 at Indo-Gangetic site. Urban Clim 27(November 2018):149–162. https://doi.org/10.1016/j.uclim.2018.11.010

Sall ML, Diaw AKD, Gningue-Sall D, Efremova Aaron S, Aaron JJ (2020) Toxic heavy metals: impact on the environment and human health, and treatment with conducting organic polymers, a review. Environ Sci Pollut Res 27(24):29927–29942. https://doi.org/10.1007/s11356-020-09354-3

Sandeep G, Vijayalatha KR, Anitha T (2019) Heavy metals and its impact in vegetable crops. Int J Chem Stud 7(1):1612–1621

Shams M, Tavakkoli Nezhad N, Dehghan A, Alidadi H, Paydar M, Mohammadi AA, Zarei A (2022) Heavy metals exposure, carcinogenic and non-carcinogenic human health risks assessment of groundwater around mines in Joghatai, Iran. Int J Environ Anal Chem 102(8):1884–1899. https://doi.org/10.1080/03067319.2020.1743835

Shen W, Hu Y, Zhang J, Zhao F, Bian P, Liu Y (2021) Spatial distribution and human health risk assessment of soil heavy metals based on sequential Gaussian simulation and positive matrix factorization model: a case study in irrigation area of the Yellow River. Ecotoxicol Environ Saf 225:112752. https://doi.org/10.1016/j.ecoenv.2021.112752

Shokri S, Abdoli N, Sadighara P, Mahvi AH, Esrafili A, Gholami M, Jannat B, Yousefi M (2022) Risk assessment of heavy metals consumption through onion on human health in Iran. Food Chem X 14(July 2021):100283. https://doi.org/10.1016/j.fochx.2022.100283

Sonone SS, Jadhav S, Sankhla MS, Kumar R (2021) Water contamination by heavy metals and their toxic effect on aquaculture and human health through food chain. Lett Appl NanoBioSci 10(2):2148–2166. https://doi.org/10.33263/LIANBS102.21482166

Srivastava A, Siddiqui NA, Koshe RK, Singh VK (2018) Human health effects emanating from airborne heavy metals due to natural and anthropogenic activities: a review. 279–296. https://doi.org/10.1007/978-981-10-7122-5_29

Sun S, Zheng N, Wang S, Li Y, Hou S, Song X, Du S, An Q, Li P, Li X, Hua X, Dong D (2021) Source analysis and human health risk assessment based on entropy weight method modification of pm2.5 heavy metal in an industrial area in the northeast of China. Atmosphere 12(7). https://doi.org/10.3390/atmos12070852

Tan SY, Praveena SM, Abidin EZ, Cheema MS (2016) A review of heavy metals in indoor dust and its human health-risk implications. Rev Environ Health 31(4):447–456. https://doi.org/10.1515/reveh-2016-0026

Thirulogachandar A, Rajeswari M, Ramya S (2014) Assessment of heavy metals in gallus and their impacts on human. Int J Sci Res Publ 4(6):1–8. www.ijsrp.org

Vatanpour N, Feizy J, Hedayati Talouki H, Es'haghi Z, Scesi L, Malvandi AM (2020) The high levels of heavy metal accumulation in cultivated rice from the Tajan river basin: health and ecological risk assessment. Chemosphere 245. https://doi.org/10.1016/j.chemosphere.2019.125639

Vijay Dahiya (2022) Heavy metal toxicity of drinking water: a silent killer. GSC Biol Pharm Sci 19(1):020–025. https://doi.org/10.30574/gscbps.2022.19.1.0107

Wahab NAA, Darus FM, Isa N, Sumari SM, Hanafi NFM (2012) Heavy metal concentration of settled surface dust in residential building. Malays J Anal Sci 16(1):18–23

Wang S, Wu W, Liu F (2019) Assessment of the human health risks of heavy metals in nine typical areas. Environ Sci Pollut Res 26(12):12311–12323. https://doi.org/10.1007/s11356-018-04076-z

Wu H, Xu C, Wang J, Xiang Y, Ren M, Qie H, Zhang Y, Yao R, Li L, Lin A (2021) Health risk assessment based on source identification of heavy metals: a case study of Beiyun River, China. Ecotoxicol Environ Saf 213. https://doi.org/10.1016/j.ecoenv.2021.112046

Xiao M, Xu S, Yang B, Zeng G, Qian L, Huang H, Ren S (2022) Contamination, source apportionment, and health risk assessment of heavy metals in farmland soils surrounding a typical copper tailings pond. Int J Environ Res Public Health 19(21). https://doi.org/10.3390/ijerph192114264

Yadav IC, Devi NL, Singh VK, Li J, Zhang G (2019) Spatial distribution, source analysis, and health risk assessment of heavy metals contamination in house dust and surface soil from four major cities of Nepal. Chemosphere 218:1100–1113. https://doi.org/10.1016/j.chemosphere.2018.11.202

Yang S, He M, Zhi Y, Chang SX, Gu B, Liu X, Xu J (2019) An integrated analysis on source-exposure risk of heavy metals in agricultural soils near intense electronic waste recycling activities. Environ Int 133(June):105239. https://doi.org/10.1016/j.envint.2019.105239

Younus M, Abbas T, Zafar M, Raza S, Khan A, Saleem AH, Idrees MA, Nisa QU, Akhtar R, Saleem G (2016) Assessment of heavy metal contamination in raw milk for human consumption. South Afr J Anim Sci 46(2):166–169. https://doi.org/10.4314/sajas.v46i2.7

Zaynab M, Al-Yahyai R, Ameen A, Sharif Y, Ali L, Fatima M, Khan KA, Li S (2022) Health and environmental effects of heavy metals. J King Saud Univ Sci 34(1):101653. https://doi.org/10.1016/j.jksus.2021.101653

Zhai Y, Zheng F, Li D, Cao X, Teng Y (2022) Distribution, genesis, and human health risks of groundwater heavy metals impacted by the typical setting of songnen plain of NE China. Int J Environ Res Public Health 19(6). https://doi.org/10.3390/ijerph19063571

Zhang T, Ruan J, Zhang B, Lu S, Gao C, Huang L, Bai X, Xie L, Gui M, Qiu R (2019) Heavy metals in human urine, foods and drinking water from an e-waste dismantling area: Identification of exposure sources and metal-induced health risk. Ecotoxicol Environ Saf 169(October 2018):707–713. https://doi.org/10.1016/j.ecoenv.2018.10.039

Zhang X, Zhong T, Liu L, Ouyang X (2015) Impact of soil heavy metal pollution on food safety in China. PLoS ONE 10(8):1–14. https://doi.org/10.1371/journal.pone.0135182

Zhang Y, Wang S, Gao Z, Zhang H, Zhu Z, Jiang B, Liu J, Dong H (2021) Contamination characteristics, source analysis and health risk assessment of heavy metals in the soil in Shi River Basin in China based on high density sampling. Ecotoxicol Environ Saf 227:112926. https://doi.org/10.1016/j.ecoenv.2021.112926

Zhou L, Zhao X, Meng Y, Fei Y, Teng M, Song F, Wu F (2022) Identification priority source of soil heavy metals pollution based on source-specific ecological and human health risk analysis in a typical smelting and mining region of South China. Ecotoxicol Environ Saf 242(8):113864. https://doi.org/10.1016/j.ecoenv.2022.113864

Chapter 4
Impact of Heavy Metals Contamination on Human Health

Meena Yadav and Ahmad Pervez

Abstract We reviewed the recent literature on toxic impacts of heavy metals or 'toxic' metals, such as, Arsenic, Cadmium, Chromium, Lead, Mercury, Copper, Manganese and Zinc on human health. The toxic levels of heavy metals impact major physiological processes, such as immune system, nervous system, cardio-vascular system, hepato-urinary system, respiratory system etc. resulting in compromised physical and mental health.

Keywords Anthropogenic activities · Genotoxicity · Health hazards · Immunity · Mutagenicity

4.1 Introduction

Heavy metal (HM) pollution has become a serious concern for the environment as well as biodiversity, primarily due to rapid industrialization and other anthropogenic activities. HMs are naturally present in the biosphere, i.e., in water, soil and rocks. The major heavy metals present in the environment are lead (Pb), chromium (Cr), antimony (Sb), zinc (Zn), arsenic (As), copper (Cu), cadmium (Cd), manganese (Mn) and mercury (Hg). More than the permissible levels of these heavy metals become toxic for the environment and biodiversity (Mitra et al. 2022). HMs enter the environment (air, water and soil) through several sources, which can be natural or anthropogenic. Natural sources of HMs include weathering, rock formation, erosion, and volcanic eruptions while anthropogenic sources include mining, smelting, excessive use of fertilizers and pesticides, industrial effluents, municipal waste (sewage) and traffic emissions (Yadav and Sharma 2023). Due to anthropogenic sources, the levels of

M. Yadav (✉)
Department of Zoology, Maitreyi College, University of Delhi, New Delhi, India
e-mail: myadav@maitreyi.du.ac.in

A. Pervez
Biocontrol Laboratory, Department of Zoology, Sri Dev Suman Uttarakhand University, Pt. L.M.S. Campus Rishikesh, Rishikesh, Uttarakhand, India
e-mail: ahmadpervez@yahoo.com

N. Kumar (ed.), *Heavy Metal Toxicity*, Environmental Science and Engineering,
https://doi.org/10.1007/978-3-031-56642-4_4

HMs increase to toxic levels in the soil and water bodies. High levels of HMs have been reported from many countries such as India, China, Italy, Korea, Germany, Turkey, Bangladesh, Iran, etc. (Yadav and Sharma 2023). The toxicity due to HMs has become a serious public health problem for humans. HMs are known to show plethora of toxic impacts on human health. Thallium (Tl) is more toxic than common HMs and causes alopecia in humans (Mitra et al. 2022), exposure to high levels of Sb and Cr promotes carcinogenicity (Sun et al. 2015), and Pb causes intellectual abnormalities in children (Hou et al. 2013). It is well-known that Hg toxicity causes *Minamata* disease which is characterized by loss of sensation in feet and hands, difficulty in coordination of movements in hands and feet, impairment in hearing, difficulty in maintaining equilibrium, slurred and unclear speech, narrowing of visual field, tremors in hands and feet and erratic ocular movement (Minamata disease 2007). Similarly, Cd toxicity causes *itai-itai* (*itai* means '*ouch*' in Japanese) disease which can be identified by severe pain due to osteomalacia (Horiguchi 2014). HMs cause many other toxic impacts in the human body such as neurotoxicity, hepatotoxicity, nephrotoxicity, and skin toxicity (Mitra et al. 2022). HMs from the soil and water bodies find their way to humans through the food chain, where they undergo bioaccumulation, biomagnification and biotransformation causing severe health problems. Owing to the numerous toxic effects of these HMs on human health, an overview of the impact of these HMs is presented in the light of recent literature.

4.2 Impact of Heavy Metal Contamination on Human Health

The two most common routes of HM exposure in humans are food and water. Excessive exposure/intake of HMs is associated with an increase in the risk of several diseases. Chronic intake of high concentrations of HMs has been associated with diseases such as cancer, gangrene, cardiac arrest, liver disease, hyperkeratosis, vomiting, nausea, hypertension, kidney malfunction, abdominal pain, gliomas, headache, damage to nerves, diarrhea, lung diseases and fatal organ failure (Egbueri and Agbasi 2022; Ukah et al. 2020). The elevated levels of HMs impact almost all the physiological systems such as the immune system, cardiovascular system, nervous system, kidneys, development, reproductive system, etc. (Kabir et al. 2021; Mitra et al. 2022; Naija and Yalcin 2023; Wang et al. 2006; Zheng et al. 2023) (Fig. 4.1). For example, Cd is a carcinogen and also causes bone disease, kidney disease, and cardiovascular disease (Mitra et al. 2022). Similarly, Hg is neurotoxic, nephrotoxic as well as hepatotoxic in humans (Emeny et al. 2019; Kim et al. 2015b; Oulhote et al. 2017) (Table 4.1).

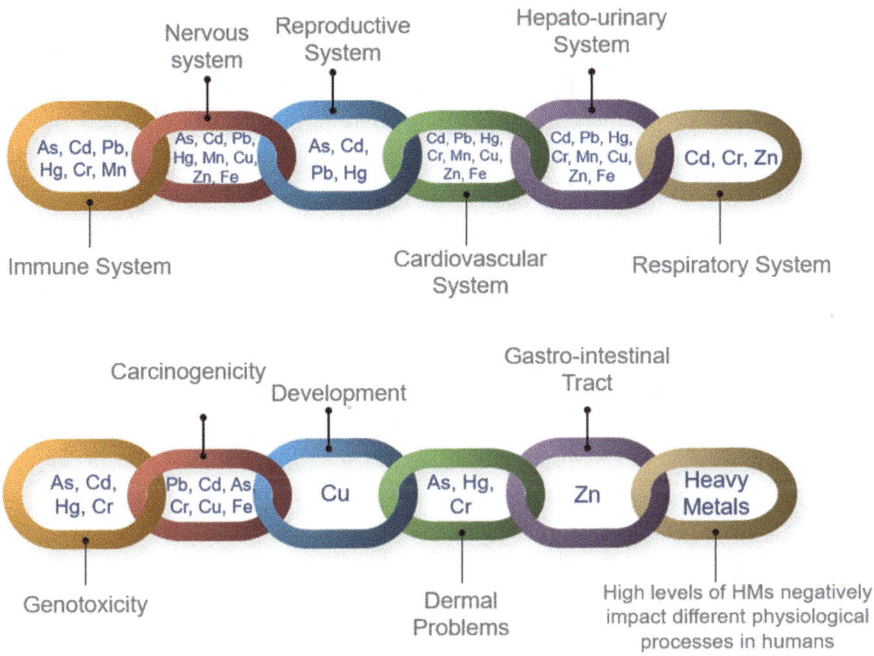

Fig. 4.1 Summary of various physiological processes that are negatively impacted by high levels of HMs

4.2.1 Health Impacts of Arsenic

Contaminated food and water are the major sources of As exposure for humans. The toxicity of As depends on its state i.e., inorganic As is more toxic than organic As, and As(III) is more toxic than As(V) (Tsai et al. 2021). More than permissible levels of As in drinking water cause developmental abnormalities, neurobehavioral disorders, cardiovascular diseases, hearing problems, anemia, eosinophilia, leukopenia, and carcinoma (Mishra et al. 2019).

4.2.1.1 Immune System

Higher As levels impact the immune system in children as well as adults. Arsenic exposure during early infancy increases the risk and severity of infections (Farzan et al. 2016). Toxic As levels reduce the $CD4^+/CD8^+$ T-cells ratio in children leading to immunosuppression. An increased As level in drinking water increases the $CD8^+$ T-cells in mothers, while decreasing $CD4^+$ T-cells and B-cells in the cord blood (Kile et al. 2014). High As levels decrease the memory T cells in the umbilical cord, thereby compromising the immunity of the fetus (Nygaard et al. 2017). Children born to mothers exposed to As contaminated drinking water during pregnancy are

Table 4.1 Major toxic effects of HMs on human health

Type of toxicity	HM	Diseases/Symptoms	References
Neurotoxicity	Cd	Vertigo, headache, vision disorders, peripheral neuropathy, slowing of vasomotor functions, reduction in ability to concentrate, tremors, sweating, dermographia, decreased intelligence	Naija and Yalcin (2023)
		Neurodegenerative defects, amyotrophic lateral sclerosis, Alzheimer disease, Parkinson's disease	Okuda et al. (1997) Branca et al. (2018)
		Neural cell death, apoptosis, impairment of neurogenesis, endocrine disruption, epigenetic effect	Wang and Du (2013)
	Hg	Cerebral palsy, deafness, muscle weakness, ataxia, tremors, spasticity, numb limbs, restlessness, anxiety, depression, sleep disturbance, sensory disturbances, dysarthria, constriction of visual field	
	Mn	Alzheimer's and Parkinson's disease leading to apoptosis and disturbance in homeostasis	Goldhaber (2003)
		Mitochondrial dysfunctions	Harischandra et al. (2019)
	Ar	Cognitive impairment, autophagy, alterations in synaptic transmission and balance in neurotransmitter	Garza-Lombó et al. (2019)
	Cu	Wilson's disease	Mitra et al. (2022)
	Zn	Lethargy, lightheadedness, staggering, anxiety, depression, comatose and somnolence Mitochondrial disruption, neuronal damage	Nriagu (2007) Inoue et al. (2015)
Nephrotoxicity	Cd	Glucosuria, phosphaturia, aminoaciduria Fanconi-like syndrome	Reyes et al. (2013)
		Hypercalciuria	Friberg et al. (2019)

(continued)

Table 4.1 (continued)

Type of toxicity	HM	Diseases/Symptoms	References
	Pb	Fanconi-like syndrome, chronic nephropathy hyperplasia, interstitial fibrosis, renal failure, glomerulonephritis	Mitra et al. (2022)
	Hg	Acute dyspnea, abdominal pain, altered mental status, profuse salivation, vomiting, chills, tremors, hypotension	Lentini et al. (2017)
Hepatotoxicity	Pb	Liver damage	Malaguarnera et al. (2012)
		Chronic cirrhosis, glycogen depletion, cellular infiltration	Hegazy and Fouad (2014)
	Cd	Liver failure, cancer	Hyder et al. (2013)
	Cu	Cholestatic liver diseases	Yu et al. (2019)
	Cr	Steatosis of hepatocytes, necrosis, parenchymatous degeneration	Hasanein and Emamjomeh (2019)
Immunological toxicity	Pb	Allergies, autoimmunity, infectious diseases, cancer	Hsiao et al. (2011)
		Indication of MHC activity. Producing B and T-cells, hypersensitivity	Kasten-Jolly et al. (2010)
	Cd	Immunosuppression	Mitra et al. (2022)
	Cr	Allergic contact dermatitis	Leroyer et al. (1998)
Cardiovascular toxicity	Pb	Arteriosclerosis, thrombosis, hypertension, atherosclerosis	Vaziri (2002, 2008)
	Cd	Hypertension, diabetes mellitus, carotid atherosclerosis, peripheral arterial disease, myocardial infarction and stroke	Mitra et al. (2022)
		Cardiovascular death	Tellez-Plaza et al. (2013)
	Hg	Atherosclerotic lesions, atherosclerosis, acute coronary failure	Yoshizawa et al. (2002)
		Coronary heart disease, carotid artery stenosis, cardiovascular disease, myocardial infarction	Kulka (2016)
Reproductive toxicity	Ar	Neural tube anomalies	Wang et al. (2006)
		Varied testosterone and gonadotropin levels, impairment of steroidogenesis process	Kim and Kim (2015)

(continued)

Table 4.1 (continued)

Type of toxicity	HM	Diseases/Symptoms	References
		Endometrial cancer	Salnikow and Zhitkovich (2008)
		Endometrial angiogenesis, endometriosis	Milton et al. (2017)
Genotoxicity	Ar	Chromosomal abnormalities, mutation	Roy et al. (2018)
		Oxidative stress, disruption of DNA repair	Pierce et al. (2012)
	Hg	Teratogenic effects	Young et al. (2008)
		Carcinogenesis	Crespo-López et al. (2009)
	Cr	Carcinogenesis	Thompson et al. (2012)
		Crosslinks between DNA and chromosomal protein along with inter-DNA strand cross-links, nucleotide strand breakage	Fang et al. (2014)

at higher risk of developing acute respiratory infections, during the first 4–5 years of their life (Ahmed et al. 2020), and allergic rhinitis (Ruan et al. 2022). Children drinking As contaminated water show lower Hb levels and decreased neutrophils and lymphocytes but increased levels of eosinophils (Chatterjee et al. 2018).

4.2.1.2 Nervous System

Arsenic causes cognitive impairment leading to neurodegenerative diseases, when present in toxic levels in the human body. It disturbs neurotransmitter balance and synaptic transmission. As also shows its effects via apoptosis using caspase-mediated MAPK signaling pathways and intracellular calcium upturn (Garza-Lombó et al. 2019).

4.2.1.3 Reproductive System

Arsenic is a known toxin for the reproductive system in humans. It causes neural tube anomalies in animals (Wang et al. 2006), and reduces weights of testes, accessory sex organs, and sperm numbers in epididymis in humans. Inorganic As also interferes with steroidogenesis and causes variations in gonadotropin and testosterone levels (Kim and Kim 2015). In females, it increases the risk of endometrial cancer, and during pregnancy alters endometrial angiogenesis which leads to conditions such as endometriosis, prematurity, subfertility, sterility and spontaneous abortions (Salnikow and Zhitkovich 2008; Milton et al. 2017).

4.2.1.4 Genotoxicity and Mutagenicity

Arsenic is genotoxic as it alters genomes of organisms by inducing chromosomal abnormalities, production of micronuclei, mutations, and sister chromatid exchange (Roy et al. 2018). It disrupts DNA repair mechanisms (Pierce et al. 2012) by attaching to the DNA binding proteins. Due to this, the DNA repair process slows down, thereby increasing the risk of cancer (García-Esquinas et al. 2013). It damages the DNA and causes epigenetic alterations by DNA methylation, histone modifications, altering expression of p53, and reducing p21 expression (Park et al. 2015). Numerous genes responsible for regulating cancer, apoptosis, growth and signaling events are greatly affected by As-induced chromosomal abnormalities (Adetutu et al. 2013).

4.2.1.5 Other Health Issues

The systemic effects of As are caused due to the compounds containing As that cause cancer, skin and circulatory system problems (Scragg 2006). Chronic exposure leads to several skin problems such as hyperpigmentation, various types of skin cancers including Bowen's disease, and hyperkeratosis (Huang et al. 2019).

4.2.2 Health Impacts of Cadmium

Cadmium occurs naturally in the Earth's crust with its concentration ranging between 0.1 and 0.2 ppm (Sharma et al. 2015). In crops, it accumulates in the edible parts of fruits, seeds and vegetables and enters human systems through food-chain (Chavez et al. 2015). Humans are exposed to Cd through food ingestion (90%), inhalation, smoking, and drinking water (Järup and Åkesson 2009; Satarug et al. 2010). Higher Cd levels increase the risk of eczema, asthma, and food allergies in children (Pesce et al. 2021). They also cause many other health conditions such as kidney damage, high blood pressure, and elevated risks of certain cancers such as endometrial, bladder, prostate, lung and breast (Adams et al. 2014; Satarug et al. 2010). Exposure to Cd can impact bones causing osteoporosis and osteomalacia due to disturbance in Vitamin D uptakes in kidneys. Cd toxicity leads to the well-known *Itai-Itai* disease which is most commonly found in postmenopausal and multiparous women, especially those whose diets lack calcium and vitamin D (Woolf 2022). Various physiological systems are under direct threat of Cd toxicity.

4.2.2.1 Immune System

The number of blood cells is affected by toxic levels of Cd exposure. For example, toxic Cd levels decrease the number of T-cells, which causes immunosuppression (Nygaard et al. 2017), but increase the number of neutrophils in children. Cd exposure increases the risk of chronic inflammation in children due to increased levels of IL-1β (Zhang et al. 2020).

4.2.2.2 Nervous System

Toxic levels of Cd have been observed in multiple neurodegenerative diseases such as Parkinson's disease, Alzheimer's disease, amyotrophic lateral sclerosis and multiple sclerosis (Branca et al. 2018; Charkiewicz et al. 2023; Min and Min 2016; Okuda et al. 1997). These diseased conditions might be due to impact of Cd on the peripheral nervous system (PNS) and the central nervous system (CNS) which is manifested in the form of peripheral neuropathy, learning disabilities, neurological disturbances, olfactory dysfunctions, impairment of motor functions, mental retardation and behavioral changes (Marchetti 2014; Miura et al. 2013). However, Cd levels in brain are influenced by multiple factors such as permeability of the blood–brain-barrier (BBB), increase in age and increase in neurodegenerative diseases. So, it is difficult to say with certainty whether Cd concentration increases due to the progression of these neurological diseases, or it contributes to the development of these diseases in humans (Ruczaj and Brzóska 2023). Cellular activities such as cell division, cellular differentiation, and apoptosis are also affected due to Cd toxicity owing to various mechanisms such as impairment of neurogenesis, endocrine disruption, inhibition of neuronal gene expression, and epigenetic effects (Wang and Du 2013).

4.2.2.3 Reproductive System

Cd interferes with the reproductive system in numerous ways. It alters cell adhesion in testes and interferes with the normal migration of germ cells, which is necessary for spermatogenesis. It reduces testicular growth rate, retards sperm count and reduces the motility of sperm (Charkiewicz et al. 2023; Pizent et al. 2012), all of which contribute to lowering of semen quality. Cd easily crosses the barrier to the fetus and brings about teratogenic effects. It causes impaired spermatogenesis, steroidogenesis, and damage to Leydig cells, Sertoli cells, and seminiferous tubules, all of which compromise the reproductive functions of the males. It also alters functions of the prostate gland. All these disturbances lead to male infertility (Charkiewicz et al. 2023).

4.2.2.4 Cardio-vascular and Respiratory System

The cardiovascular system is also severely affected by Cd toxicity, as toxic levels inhibit the endothelial nitric oxide synthase and suppresses acetylcholine-induced vascular relaxation leading to hypertension (Gallagher and Meliker 2010). Cd can affect glucose metabolism by interfering with the pathways in various organs, like the liver, pancreas and adrenal gland. It increases the blood glucose levels, and induces diabetes in mice, so it provides evidence that Cd exposure increases the risk for people with Diabetes Mellitus (Li et al. 2022). During the occurrence of *Itai-Itai* disease, Cd can induce anemia by suppressing the erythropoietin production, which disturbs iron transport to the duodenum causing iron-deficient anemia (Fujiwara et al. 2020). Lungs are also susceptible to Cd toxicity and chronic exposure may cause progressive pulmonary fibrosis thereby damaging the lungs (Li et al. 2017). Low to moderate exposure to Cd can also lead to diabetes, peripheral arterial disease, carotid atherosclerosis, chronic kidney disease, ischemic stroke, myocardial infarction and heart failure (Charkiewicz et al. 2023; Mitra et al. 2022).

4.2.2.5 Hepato-urinary System

Kidneys are highly susceptible to Cd toxicity, as about 30% of Cd tends to be deposited in the kidney tubules causing tubular damage (Akesson et al. 2005). Diabetics are more prone to this damage. In the kidney, it causes hypercalciuria, glycosuria, aminoaciduria, renal tubular acidosis and renal failure (Friberg et al. 2019; Jacquillet et al. 2007). It also causes proteinuria, increased levels of creatinine and elevated excretion of calcium and phosphorus (Charkiewicz et al. 2023). The liver is also affected as excess Cd accumulates in the liver after acute exposure and causes hepatic dysfunctions resulting in liver failure and also increases the risk of liver cancer (Hyder et al. 2013).

4.2.2.6 Genotoxic and Carcinogenic Effects

Cd can exhibit genotoxic and mutagenic effects due to its low affinity with DNA. It induces mutations by two mechanisms i.e., by inducing production of reactive oxygen species (ROS), and by inhibiting DNA repair of exogenous and endogenous lesions, both of which may become causes of cancer (Filipic et al. 2006). It can even cause DNA breaks and cell death in the proximal tubule cells (Wang et al. 2017). It can mutate immediate early response and stress response genes along with transcriptional and translational factors (Rani et al. 2014). Cd interferes with the DNA repair processes, like nucleotide excision repair and disturbs the removal of DNA lesions (Schwerdtle et al. 2010). In low concentrations, it interferes with oxidative DNA base repair and DNA alkylation damage (Fatur et al. 2003). Thus, Cd may show carcinogenicity through various mechanisms.

4.2.3 Health Impacts of Lead

Lead toxicity is a major concern in children due to environmental pollution in developing nations. Numerous sources generate this toxicity, like painting of home, battery recycling, arms industry, smoking, printing books, leaded automobile fuel, and smelting along with food, water and milk contamination.

4.2.3.1 Immune System

Humans are exposed to Pb from various sources such as gasoline and paints, and excessive Pb exposure triggers the production of T-cells, B-cells, and higher major histocompatibility complex (MHC) activity (Kasten-Jolly et al. 2010), thereby activating the immune system. So, it increases cellular and humoral responses and makes the individual prone to developing autoimmunity and hypersensitivity (Mishra 2009). Due to higher Pb exposure children develop asthma, infections and influenza (Oktapodas Feiler et al. 2020). Pb toxicity increases the number of neutrophils, monocytes, eosinophils, basophils (Rawat et al. 2021), and decreases the population of natural killer cells (Zhang et al. 2016). Altered levels of blood cells disturbs the functioning of the immune system.

4.2.3.2 Nervous System

Neurophysiological functions are affected due to Pb exposure. Higher levels of Pb is neurotoxic and causes neurocognitive symptoms and psychiatric disturbances (Debnath et al. 2019). Lead, in the blood, above 10 μg/dL disturbs cognitive functions in humans. Pb toxicity largely affects brain development, and therefore is more threatening to the developing fetus and growing children who are more prone to its neurotoxic effects than adults (Rehman et al. 2018). Pb shows negative impacts by blocking N-methyl-D-aspartate receptor, which plays a major role in the maturation of brain plasticity causing a break in the understanding of newly learned knowledge thereby hindering the cognitive functions of a developing child (Sanders et al. 2009). Lead can cross the endothelial cells at blood–brain barrier disturbing the brain fitness.

4.2.3.3 Reproductive System

Lead induces toxicity in the reproductive systems of both males and females. It crosses the placenta causing intrauterine deaths, prematurity, and low birth weight. Constant Pb exposure can inhibit ovulation, menstruation, and follicular growth, and slowdown ovum implantation (Vigeh et al. 2011). Pb damages testes in humans by inducing alterations of seminiferous tubules, blood–testis barrier, Sertoli cells, and sperm loss. Furthermore, it alters the development and function of Leydig cells and induces Leydig cell tumors leading to male infertility (Rajendar et al. 2012).

4.2.3.4 Cardio-vascular System

Lead induces fatalistic responses to the cardio-vascular system by blocking blood vessels leading to instant heart attack and death. So, rising blood Pb levels increase cardiovascular morbidity and mortality (Akil and Ahmad 2011). Both chronic and acute Pb toxicity may lead to vascular damage and cardiac dysfunction (Flora et al. 2012). Lead toxicity is so severe to the heart that even low levels of Pb can lead to cardiac problems and high blood pressure (Kim et al. 2015a). Chronic Pb exposure can cause atherosclerosis, arteriosclerosis, hypertension, thrombosis and cardiac disease. Pb exerts its effects by several mechanisms such as reduction in NO availability, increased oxidative stress, elevated prostaglandin levels (vasoconstrictor), altered renin-angiotensin system, lowered prostaglandin levels (vasodilator), increased inflammation, and disrupted Ca^{2+} signaling in vascular smooth muscles (Vaziri 2002, 2008).

4.2.3.5 Hepato-urinary System

Chronic Pb exposure generates reactive oxygen species (ROS) which damage liver cells and cause cellular infiltration and glycogen depletion resulting in chronic liver cirrhosis (Hegazy and Fouad 2014). Exposure to Pb can affect the kidneys leading to nephrotoxicity which is manifested as hyperplasia, tubule atrophy, interstitial fibrosis, glomerulonephritis and renal failure. Proximal tubular dysfunction induced by excessive Pb exposure causes Fanconi-like syndrome i.e., malabsorption of electrolytes and other substances (Harischandra et al. 2019). Kidney is meant to eliminate Pb from the blood but it becomes a victim of Pb toxicity, as Pb is absorbed in the renal tubules by the proximal tubular cells and binds with proteins, and hence accumulates in the kidney (Kwon et al. 2015).

4.2.3.6 Carcinogenicity

Higher exposure to Pb increases the risk of cancers such as stomach, lung and bladder (Rousseau et al. 2007). It induces cancer by damaging the DNA repair mechanisms, chromosomal structure and sequence, and tumor regulatory genes (Harischandra et al. 2019). Impairment of transcription by shifting zinc from some regulatory proteins is another mechanism by which Pb causes cancer (Silbergeld et al. 2000).

4.2.4 Health Impacts of Mercury

Mercury (Hg) is a highly toxic and persistent element that exists in liquid form at room temperature. It is found in three forms in the environment, i.e., elemental Hg, organic Hg and inorganic Hg. Aquatic anaerobic microorganisms convert Hg to methyl Hg

(MeHg), which easily enters the human food chain via fish and seafood, and gets bioaccumulated. Humans are highly susceptible to Hg exposure and associated health risks due to feeding on aquatic foods. Mercury poisoning, due to the exposure of Hg and its compounds, is also known as *mercurialism* or *hydrargyria*. Infantile acrodynia in children, exhibiting pink discoloration of the hands and feet, is a classic example of mercury poisoning (Omkar and Pervez 2017), not forgetting the famous Minamata disease caused due to Hg poisoning. Some of the examples of impacts of mercury toxicity on various physiological systems are discussed.

4.2.4.1 Immune System

Hg toxicity reduces the levels of total white blood cells, lymphocytes and monocytes (Oulhote et al. 2017), thereby compromising the immunity of the individual. High Hg levels increase the risk of asthma, wheezing and airway hyperresponsiveness (Kim et al. 2015b). Hg impairs the immune system of children and causes many disorders. Maternal exposure to Hg during pregnancy increases the risk of respiratory infections for children of 9–12 years of age (Emeny et al. 2019).

4.2.4.2 Nervous System

The CNS is the primary target of Hg toxicity. Under the neurotoxic effects, exposure of Hg vapour impairs a variety of cognitive, sensory and motor functions of CNS. Hg is a potent neurotoxin that easily penetrates the placental and blood–brain barrier and negatively affects growth in humans (Ismanto et al. 2023). Hg harms blood-brain barrier and facilitates other toxic substances to cause injury to brain. Prenatal or postnatal exposure of Hg can risk the induction of autism spectrum disorders in children (Ismanto et al. 2023).

Since Hg poisoning disturbs the CNS, it causes depression, fear, memory failure, hallucination, tremors, weight loss, drowsiness, weakness, and insomnia (Verma et al. 2018). It can also jeopardize the senses, like sight, smell and hearing. Some of its compounds, such as methyl mercury, have the potential to damage both the cerebellum and cerebrum due to which numerous problems arise such as ataxia, neurasthenia, loss of hearing and vision, difficulty in locomotion, unconsciousness and finally death (Hong et al. 2012). CNS toxicity due to Hg also includes erethism with signs of shyness, nervousness, lack of concentration, insomnia, emotional lability, memory impairment, encephalopathy, peripheral neuropathy, depression, symptoms as in Parkinson's disease and impotency. Hg-induced toxicity results in lesions in the cerebellum, which leads to speech disorders (dysarthria) and balance impairment (ataxia). Further toxicity can impair vision, weaken muscles, cause cramps, disability in hearing, and disturbance of other sensory senses (Hong et al. 2012).

4.2.4.3 Cardio-vascular System

Hg exposure causes atherosclerosis, and increases the risk of cardiovascular disease, coronary heart disease, acute myocardial infarction, coronary dysfunction and carotid artery stenosis. This is because Hg inactivates paraoxonase, an extracellular antioxidant enzyme associated with HDL dysfunction (Kulka 2016). Exposure of methylmercury is linked with reduced sympathetic low-frequency along with parasympathetic high-frequency modulation of heart rate variability (Grandjean et al. 2004).

4.2.4.4 Hepato-urinary System

While acute exposure of Hg to the kidney causes acute tubular necrosis, chronic exposure injures the epithelium and induces necrosis in the *pars recta* of the proximal tubule. Acute exposure may be manifested as mental impairment, acute dyspnea, abdominal pain, excess saliva production, vomiting, chills, tremors, and hypotension while chronic exposure of Hg leads to tubular failure, higher excretion of albumin and retinol-binding protein in urine, and nephritic state with membranous nephropathy (Lentini et al. 2017). Hg toxicity also causes proteinuria (excretion of high levels of proteins in urine), and kidney failure.

4.2.4.5 Reproductive and Genotoxicity

Excess Hg causes subfertility or infertility in humans by affecting various parameters of reproductive fitness. In males, it impacts semen quality and induces DNA damage in sperm. In females, Hg is associated with longer TTP (TTP = thrombotic thrombocytopenic purpura; a disorder characterized by microvascular thrombosis involving arterioles and capillaries) and causes spontaneous abortions, fetal malformations and menstrual disorders. One possibility that can explain these effects could be that Hg might act as endocrine disruptor thereby causing hormonal disorders that impact the functions of testes and ovaries (Henriques et al. 2019). Excess Hg generates ROS which damages proteins, DNA and lipids and thus, increases cancer risk (Reczek and Chandel 2017; Zefferino et al. 2017).

4.2.4.6 Dermal Problems

Exposure to Hg and Hg-containing products also adversely affects the skin and causes *acrodynia* (the pink disease) in which skin becomes pink when exposed to Hg-containing compounds (Horowitz et al. 2002).

4.2.5 Health Impacts of Chromium

Chromium is an abundant heavy metal that is accumulating in the environment due to anthropogenic activities such as plating, tanning, nuclear plants, coal-based power plants, chromite mining, refractory industries, etc. Six forms of Cr are highly prevalent, among which Cr(III) and Cr(VI) are the major ones that highly contribute to environmental toxicity (Mohanty et al. 2023). The hexavalent form of chromium is more toxic to human beings than Cr(III), as the former is water soluble and can enter the biological membrane causing numerous carcinogenic and mutagenic effects (Cervantes et al. 2001).

4.2.5.1 Immune System

High levels of Cr(VI) reduce the phagocytic action of alveolar macrophages, and humoral immune response. It induces type I and type IV hypersensitivity reactions, and induces allergic contact dermatitis (Leroyer et al. 1998; Mitra et al. 2022).

4.2.5.2 Cardio-vascular System

Chromium toxicity manifesting in the chromium contamination-susceptible areas can be examined through haematological parameters. Cr(VI) is permissible in the blood serum between 50–500 ng/ml. Anything in excess creates problem, as Cr(VI) disseminates into erythrocytes thereby reducing the RBC count, packed cell volume and percent haemoglobin in the blood (Ray 2016). A lethal exposure of Cr(VI) can even cause hemorrhages along with hypochromic and microcytic anemia (Nigam et al. 2014). Cr(VI) even produces ROS in RBCs which can elevate oxidative stress conditions by damaging lipids and proteins of the cell leading to cancers (Ray 2016).

4.2.5.3 Hepato-urinary System

Cr poisoning is also evident as histopathological changes in the liver and kidneys. Acute exposure to Cr(VI) can cause hepatic necrosis and renal tubular necrosis leading to kidney failure (Zheng et al. 2020). Cr(VI) causes many histopathological symptoms in the liver such as parenchymatous degeneration, steatosis of hepatocytes, necrosis, oxidative stress, lipid peroxidation, hampering nucleic acid and protein synthesis, DNA damage, mitochondrial dysfunction, arrest of cell growth and apoptosis (Hasanein and Emamjomeh 2019).

4.2.5.4 Respiratory System

Cr(VI) can induce toxicity after being absorbed into the body via oral, dermal and pulmonary routes. It can cause allergies, and asthma-related symptoms like breathlessness or shortness of breath. It also exhibits symptoms, like pneumonia, bronchitis, rhinitis, itching and irritation in nasal tract. The acute forms of these symptoms can lead to the destruction of mucous membrane and explosion of septum, which will culminate in ulcers and cancers in the respiratory tract (Mohanty et al. 2023). Chronic respiratory effects like macrophage infiltration, hyperplasia, fibrosis and pulmonary inflammation are also evident after chromium poisoning (Mohanty et al. 2023).

4.2.5.5 Mutagenic and Carcinogenic Effects

Hexavalent chromium is highly carcinogenic, especially to industrial workers and occupational toxicity leads to lung, nasal, sinus, bladder, glandular and stomach cancer (Straif et al. 2018). Such tumours and cancers are largely due to the cross-links, standard breaks, chromosomal aberration, and cell cycle disruption leading to DNA damage and fatal health issues (Xie et al. 2004). Cr(VI) may even accelerate cell apoptosis, disturb the cell cycle regulation and damage immune cells and the cytoskeleton (Nigam et al. 2014). Cr(VI) after crossing the cell membrane can even get converted into Cr(III) causing mild genomic instability and can even lead to major damages like cell cycle arrest and apoptosis (Ray 2016).

4.2.5.6 Dermal and Optic Effects

Exposure to Cr(VI) results in an increase in skin sensitization causing irritation and skin burning. Minor symptoms, such as burning and itchy skin, rashes, and redness in the skin are also evident. Further increase in the duration of exposure can cause skin ulcers and an immune-hampering disease known as allergic contact dermatitis (Hessel et al. 2021). It also causes skin cancer, contact dermatitis, and systemic contact dermatitis (Mitra et al. 2022). Toxic Cr derivatives, like chromic acid or chromates can permanently damage the naked eye if you come in direct contact. Cr(VI) trioxide has corrosive effects like rashes and burning sensations. Thus, people should be made aware of possible deleterious effects of chromium-loaded dust, air and smoke oozing out from chromium contaminant industrial sources (Mitra et al. 2022).

4.2.6 Health Impacts of Manganese

Manganese (Mn) is an essential element for human health, which despite being a heavy metal acts as a co-factor in numerous enzymatic pathways for homeostasis, regulation of blood sugar, maintenance of nerve impulses and functions, etc. (Aschner et al. 2007). However, its higher levels can lead to Mn toxicity, which disturbs various systems in humans and other life forms.

4.2.6.1 Immune System

Although Mn is a trace element and necessary for important physiological functions in humans, higher levels show toxicity. If levels of Mn are high in maternal serum, there is an increased chance of developing childhood eczema (Pesce et al. 2021). Antibody production requires nominal amounts of Mn which is also critical for the host innate immune response (Chen et al. 2020). However, excessive amounts of Mn significantly decrease C3 (a core protein in the complement system), total blood lymphocytes, and $CD4^+$ and $CD8^+$ lymphocytes (Chen et al. 2020; Nakata et al. 2006; Antonini et al. 2012).

4.2.6.2 Nervous System

Mn in moderate amounts has a neuroprotective role as it reduces apoptosis, but in higher concentrations, it may cause complications in neurons which may lead to conditions such as Alzheimer's and Parkinson's disease (Goldhaber 2003). Studies have also shown that the heavy exposure of Mn can induce signs and symptoms akin to those of Parkinson's disease (Rutchik et al. 2012a, b). Further, it can also alter the levels of neurotransmitters and metabolites (Gwiazda et al. 2007). Mn homeostasis is disrupted in case of excessive Mn exposure and it happens due to two events—down-regulation of receptors for metal uptake and upregulation of receptors for discharge of Mn from the cell. However, Mn accumulation leads to increased formation of reactive oxygen species (ROS) which in turn, through the *cytochrome c-Caspase 3-PKCδ* pathway leads to DNA fragmentation and apoptosis (Harischandra et al. 2019).

4.2.6.3 Cardio-vascular System

High levels of exposure to Mn perturbs cardiovascular function (O'Neal and Zheng 2015). An overdose of Mn (5–10 mg Mn/kg) can decrease the heart rate and blood pressure. Precisely, it lowers the mean diastolic and systolic blood pressure and elevates the levels of diastolic hypotension (Jiang and Zheng 2005). Mn exposure may inhibit myocardial contraction, induces hypotension and dilates blood vessels, thereby imposing heavy impact on the cardiac function.

4.2.6.4 Hepato-urinary System

Liver is the main storage organ for manganese, and the clearance of excess Mn is directly linked to the liver through hepatobiliary excretion (Chua and Morgan 1997). Thus, a heavy impact of Mn can severely damage the liver. Numerous chronic liver diseases may lead to the accumulation of excess Mn in the brain leading to hepatic encephalopathy, which could be highly lethal (Long et al. 2009; Squitti et al. 2009).

4.2.7 Health Impacts of Copper

Copper (Cu) is an important and essential trace mineral, which is responsible for our mental and physical health. The daily copper consumption by humans, if exceeds more than 2–3 mg then it may lead to copper poisoning or Copperiedus (Omkar and Pervez 2017). This copper toxicity can cause oxidative stress and tissue damage (Cecconi et al. 2002). Approximately 95% of Cu in plasma is bound to ceruloplasmin. Cases of copper poisoning through ingestion of liquids in the form of beverages and water are frequent. Victims of Cu toxicity show symptoms like abdominal pain, vomiting, nausea, headache, diarrhea, breathing problems, hemolytic anemia, tachycardia, kidney failure and even death (Srivastava et al. 2005). Seemingly, Cu poisoning perturbs numerous hormonal components, which leads to severe disturbance of endocrinological system (Yang et al. 2004). Any perturbation in Cu homeostasis is connected with numerous diseases and tissue damage (Bleackley and Macgillivray 2011).

4.2.7.1 Nervous System

In excessive concentrations, Cu interferes with neurodevelopment (Prohaska 2000). Excessive accumulation of Cu causes Wilson's disease characterized by neurobehavioral abnormalities as observed in people with schizophrenia. Dysregulation of Cu can cause amyloid plaque formation and oxidative stress, as manifested in Alzheimer's disease (Cheignon et al. 2018). Cu accumulation is also linked to Huntington's disease, Parkinson's disease, and autism spectrum disorder (Bulcke et al. 2017).

4.2.7.2 Cardio-vascular System

High levels of Cu in serum increases the risk for cardiovascular disease, such as cardiac arrhythmia (Salustri et al. 2010; Bost et al. 2016). Cu poisoning decreases the number of RBCs, leading to anemia (Singh et al. 2011).

4.2.7.3 Hepato-urinary System

Cu induced toxicity can disrupt absorption and excretion of Cu, which is connected with severe disorders, like Menkes disease and Wilson's disease (White et al. 2009). Chronic Cu toxicity causes liver cirrhosis and damages other organs, which is a symptom of the Wilson's disease along with increased levels cause oxidative stress (Yu et al. 2019). Elevated levels of Cu in the body damage liver as well as the kidney (Singh et al. 2011).

4.2.7.4 Carcinogenicity and Mutagenicity

Although little is known about Cu accumulation to be carcinogenic (ECI 2008), there are evidences that Cu could be a tumor promoter through alteration of oxidative phosphorylation (Ishida et al. 2013).

4.2.7.5 Developmental Health Impacts

Cu toxicity has been associated with development defects such as slow development, delayed bone growth, and lighter body weight (Singh et al. 2011). Insufficient uptake of Cu during development affects the activities of the cuproenzymes that leads to increased production of ROS, disturbed iron metabolism, reduced availability of NO, and disturbed cell signaling. Cu deficiency causes Menkes disease characterized by neurological problems, growth retardation, and hypopigmentation, while Cu accumulation, in liver and brain, causes Wilson disease that leads to liver cirrhosis and hepatitis, and neurological symptoms similar to those as observed in Parkinson's disease (Bleackley and Macgillivray 2011). Cu shows minor toxic effects on GI tract and leads to vomiting, diarrhea and stomach aches (Singh et al. 2011).

4.2.8 Health Impacts of Zinc

Zn is a multipurpose element, largely needed for the human health. It is an essential micronutrient, and an anti-inflammatory agent that plays an important role in functioning of various human systems. It provides therapeutic benefits in healing severe diseases, like atherosclerosis, autoimmune diseases, Alzheimer's disease, several malignancies, cancer, depression, aging, diabetes, and Wilson's disease. However, high levels of Zn can cause imbalance in the regulation of numerous pathways and body functions. This leads to zinc-toxicity, which can harm various human physiological systems.

4.2.8.1 Gastro-intestinal System

Zinc salts are corrosive and can induce severe injury to the alimentary canal. This exhibits symptoms, like mouth burning, vomiting, esophagitis, gastrointestinal hemorrhage, gastritis and pancreatitis (Nriagu 2007). Food and beverages preserved in galvanized zinc containers can cause diarrhea and gastrointestinal distress. Ingestion of high levels of zinc salts, like zinc sulphate, can cause gastrointestinal ulcerations and burns. This is followed by significant fall in percent hemoglobin and hematocrit levels leading to intravascular hemolysis (Hussain et al. 2022).

4.2.8.2 Cardio-vascular System

Humans exposed to high zinc levels show symptoms of hypertension and premature atrial beats, and hypovolemic shock (Nriagu 2007). High doses of zinc may also upset in the serum ferritin, and serum lipid profile. Salts of zinc, for e.g., zinc chloride, can decline blood platelet count and can cause microcytic anemia (Jeng and Chen 2022).

4.2.8.3 Respiratory System

Certain daily use substances contain zinc salts, which are deleterious, if taken orally. A mild intake of correction fluid by a patient suffering from asthma could result in an acute episode of bronchospasms, and oropharyngeal inflammation leading to dysphonia (Nriagu 2007).

4.2.8.4 Hepato-urinary System

High levels of zinc in children can cause chronic cholestatic liver disease leading to biliary cirrhosis. High doses of zinc may also cause renal failure and mild albuminuria (Nriagu 2007).

4.2.8.5 Nervous System

Zn-induced neurotoxicity causes numerous brain disorders, such as stroke, seizures and brain injury (Morris and Levenson 2017). Intake of high levels of zinc can lead to several neurological disorders, such as, lethargy, lightheadedness, staggering, anxiety, depression, comatose and somnolence (Nriagu 2007). Increased levels of synaptic Zn can dysregulate Ca2+, produce ROS and lead to mitochondrial disruption, neuronal damage and death (Inoue et al. 2015). Excess Zn can inhibit gamma-aminobutyric acid receptors (GABAAR) and glycine receptors (GlyR), which are known to mitigate neurotoxicity (Kuenzel et al. 2016). Excess Zn also inhibits cellular energy production causing neuronal death (He and Aizenman 2010).

4.2.9 *Health Impacts of Iron*

Iron is an important essential element in human body, which is largely needed for the growth, development, maintenance and cell division. A normal man of 70 kg contains approximately 4.0 g of Fe, of which, about 70, 25 and 5% is found in hemoglobin, iron stores (in the form of ferritin and hemosiderin) and muscle myoglobin (along with tissue enzymes and plasma transferrin), respectively. However, industrialization, high consumption of red meat, hereditary hemochromatosis, and intake of Fe due to cultural practices have resulted in overloading of Fe in the iron stores, which may be 20–30 folds greater than the normal capacity. Overload of iron leads to iron deposits in various tissues causing tissue damage, organ failure and its functional impairment. Acute toxicity of Fe also results in skin pigmentation, hepatic dysfunction, hepatomegaly, liver fibrosis, hepatoma, heart failure, diabetes mellitus, testicular atrophy, arthropathy and hereditary hemochromatosis (Brissot et al. 2012; Hsu et al. 2022). Secondary Fe overload can lead to Thalassemia and sideroblastic anemias (Taher and Cappellini 2018).

4.2.9.1 Nervous System

Several neural problems are associated with misregulation of Fe metabolism in the Central Nervous System (Ponka 2004). Fe overloads are usually reported in the brains of persons suffering from Parkinson's disease, Alzheimer's disease, Huntington's disease or Hallervorden–Spatz syndrome. An autosomal recessive disorder known as Aceruloplasminemia is due to the Fe overload in brain and visceral organs (Nittis and Gitlin 2002). Such patients also show the symptoms of diabetes mellitus, anemia, and retinal degeneration. An extremely rare disease, neuroferritinopathy is also related to Fe overload, as free Fe triggers the formation of ROS causing neurodegeneration and neuronal cell death (Curtis et al. 2001). Neuroferritinopathy also exhibits symptoms of chorea, dystonia, and bradykinesia.

4.2.9.2 Cardio-vascular System

Due to iron overload, Fe gets deposited in coronary plaques linked with elevated cholesterol levels in persons with atherosclerotic lesions (Stadler et al. 2004). These deposits of plaques of symptomatic patients exhibit higher Fe concentrations with high risk of cap rupture than the asymptomatic patients (Gustafsson et al. 2014). Fe overload is also known to cause myocardial ischemia, calcific valvular and vascular diseases, heart failure, pulmonary arterial hypertension and arrhythmogenesis (Li and Zhang 2021). An Fe-dependent apoptosis known as ferroptosis is distinct from apoptosis, necroptosis, and other types of cell deaths. Ferroptosis has been found in several cardiovascular diseases, such as atherosclerotic cardiovascular disease, cardiomyopathy, and myocardial reperfusion injury (Li and Zhang 2021).

4.2.9.3 Hepato-urinary System

Fe overload disturbs liver functioning and causes chronic liver diseases, such as dysmetabolic iron overload syndrome (Milic et al. 2016). High iron stores are also linked with a decline in insulin secretion with increased hepatic insulin resistance (Datz et al. 2013).

4.2.9.4 Carcinogenicity

Fe overload carries the risk for carcinogenesis (Huang 2003). Hereditary hemochromatosis due to Fe accumulation in parenchymal tissue can lead to hepatocellular carcinoma (Deugnier and Turlin 2001). Other cancerous forms, including melanoma, skin, esophageal and acute myeloid leukemia are also linked with hereditary hemochromatosis.

4.3 Conclusion

Pollution of water, soil and air due to heavy metals has been on the rise in recent years, majorly due to anthropogenic activities. With their increasing use in industries, agriculture, cosmetics, and other fields, there is no way to escape from the ill-effects of these HMs. The higher levels of HMs impact all living organisms, including humans. Although, trace amounts of some of the HMs are necessary for normal physiological functions but their higher levels disturb these processes which are manifested as different types of diseases/disorders. The severity of the symptoms depends upon the type of HM taken up by humans due to various routes of exposure, and the amount of exposure. They act by upregulating or downregulating the signaling pathways. Looking at the diverse toxic impacts of the HMs, there is need to take two major steps i.e., to reduce the levels of these HMs at contaminated sites (remediation), and to understand their mechanism of action so that appropriate measures can be taken to counter their adverse effects on living organisms.

References

Adams SV, Quraishi SM, Shafer MM, Passarelli MN, Freney EP, Chlebowski RT et al (2014) Dietary cadmium exposure and risk of breast, endometrial, and ovarian cancer in the Women's Health Initiative. Environ Health Persp 122(6): 594–600. https://doi.org/10.1289%2Fehp.1307054

Adetutu A, Oyewo EB, Adesokan AA (2013) Protective effects of Vernonia amygdalina against sodium arsenite-induced genotoxicity in rat. Pharm Res 5(3):207–211. https://doi.org/10.4103%2F0974-8490.112431

Ahmed SM, Branscum A, Welch BM, Megowan M, Bethel JW, Odden MC, Kile ML (2020) A prospective cohort study of in utero and early childhood arsenic exposure and infectious disease in 4-to 5-year-old Bangladeshi children. Environ Epidemiol 4(2):e086. https://doi.org/10.1097%2FEE9.0000000000000086

Akesson A, Lundh T, Vahter M, Bjellerup P, Lidfeldt J, Nerbrand C et al (2005) Tubular and glomerular kidney effects in Swedish women with low environmental cadmium exposure. Environ Health Perspect 113:1627–1631. https://doi.org/10.1289/ehp.8033

Akil L, Ahmad HA (2011) Relationships between obesity and cardiovascular diseases in four Southern states and Colorado. J Health Care Poor Underserved 22:61–72. https://doi.org/10.1353/hpu.2011.0166

Antonini JM, Zeidler-Erdely PC, Young SH, Roberts JR, Erdely A (2012) Systemic immune cell response in rats after pulmonary exposure to manganese-containing particles collected from welding aerosols. J Immunotoxicol 9(2):184–192. https://doi.org/10.3109/1547691x.2011.650733

Aschner M, Guilarte TR, Schneider JS, Zheng W (2007) Manganese: recent advances in understanding its transport and neurotoxicity. Toxicol Appl Pharmacol 221:131–147. https://doi.org/10.1016/j.taap.2007.03.001

Bleackley MR, MacGillivray RT (2011) Transition metal homeostasis: from yeast to human disease. Biometals 24:785–809. https://doi.org/10.1007/s10534-011-9451-4

Bost M, Houdart S, Oberli M, Kalonji E, Huneau JF, Margaritis I (2016) Dietary copper and human health: current evidence and unresolved issues. J Trace Elem Med Biol 35:107–115. https://doi.org/10.1016/j.jtemb.2016.02.006

Branca JJV, Morucci G, Pacini A (2018) Cadmium-induced neurotoxicity: still much ado. Neural Regen Res 13:1879–1882. https://doi.org/10.4103/1673-5374.239434

Brissot P, Ropert M, Le Lan C, Loréal O (2012) Non-transferrin bound iron: a key role in iron overload and iron toxicity. Biochim Biophys Acta 1820(3):403–410

Bulcke F, Dringen R, Scheiber IF (2017) Neurotoxicity of copper. Neurotox Metals 313–343. https://doi.org/10.1007/978-3-319-60189-2_16

Cecconi I, Scaloni A, Rastelli G, Moroni M, Vilardo PG, Costantino L et al (2002) Oxidative modification of aldose reductase induced by copper ion. Definition of the metal–protein interaction mechanism. J Biol Chem 277(44):42017–42027. https://doi.org/10.1074/jbc.M206945200

Cervantes C, Campos-García J, Devars S, Gutiérrez-Corona F, Loza-Tavera H, Torres-Guzmán JC, Moreno-Sánchez R (2001) Interactions of chromium with microorganisms and plants. FEMS Microbiol Rev 25(3):335–347. https://doi.org/10.1111/j.1574-6976.2001.tb00581.x

Charkiewicz AE, Omeljaniuk WJ, Nowak K, Garley M, Nikliński J (2023) Cadmium toxicity and health effects—a brief summary. Molecules 28(18):6620. https://doi.org/10.3390/molecules28186620

Chatterjee D, Adak S, Banerjee N, Bhattacharjee P, Bandyopadhyay AK, Giri AK (2018) Evaluation of health effects, genetic damage and telomere length in children exposed to arsenic in West Bengal, India. Mutat Res Genet Toxicol Environ Mutagen 836:82–88. https://doi.org/10.1016/j.mrgentox.2018.06.012

Chavez E, He ZL, Stoffella PJ, Mylavarapu RS, Li YC, Moyano B, Baligar VC (2015) Concentration of cadmium in cacao beans and its relationship with soil cadmium in southern Ecuador. Sci Total Environ 533:205–214. https://doi.org/10.1016/j.scitotenv.2015.06.106

Cheignon C, Tomas M, Bonnefont-Rousselot D, Faller P, Hureau C, Collin F (2018) Oxidative stress and the amyloid beta peptide in Alzheimer's disease. Redox Biol 14:450–464. https://doi.org/10.1016/j.redox.2017.10.014

Chen X, Liu Z, Ge X, Luo X, Huang S, Zhou Y et al (2020) Associations between manganese exposure and multiple immunological parameters in manganese-exposed workers healthy cohort. J Trace Elem Med Biol 59:126454. https://doi.org/10.1016/j.jtemb.2020.126454

Chua AC, Morgan EH (1997) Manganese metabolism is impaired in the Belgrade laboratory rat. J Comp Physiol B 167(5):361–369. https://doi.org/10.1007/s003600050085

Crespo-López ME, Macêdo GL, Pereira SI, Arrifano GP, Picanco-Diniz DL, do Nascimento JLM, Herculano AM (2009) Mercury and human genotoxicity: critical considerations and possible molecular mechanisms. Pharmacol Res 60:212–220. https://doi.org/10.1016/j.phrs.2009.02.011

Curtis AR, Fey C, Morris CM, Bindoff LA, Ince PG, Chinnery PF et al (2001) Mutation in the gene encoding ferritin light polypeptide causes dominant adult-onset basal ganglia disease. Nat Genet 28:350–354

Datz C, Felder TK, Niederseer D, Aigner E (2013) Iron homeostasis in the metabolic syndrome. Eur J Clin Invest 43(2):215–224

Debnath B, Singh WS, Manna K (2019) Sources and toxicological effects of lead on human health, India. J Med Spec 10(2):66–71. https://doi.org/10.4103/INJMS.INJMS_30_18

Deugnier Y, Turlin B (2001) Iron and hepatocellular carcinoma. J Gastroenterol Hepatol 16:491–494

Egbueri J, Agbasi J (2022) Data-driven soft computing modeling of groundwater quality parameters in Southeast Nigeria: comparing the performances of different algorithms. Environ Sci Pollut Res 29:38346–38373. https://doi.org/10.1007/s11356-022-18520-8

Emeny RT, Korrick SA, Li Z, Nadeau K, Madan J, Jackson B et al (2019) Prenatal exposure to mercury in relation to infant infections and respiratory symptoms in the New Hampshire Birth Cohort Study. Environ Res 171:523–529. https://doi.org/10.1016/j.envres.2019.01.026

Fang Z, Zhao M, Zhen H, Chen L, Shi P, Huang Z, Shi X (2014) Genotoxicity of tri- and hexavalent chromium compounds in vivo and their modes of action on DNA damage in vitro. PLoS ONE 9(8):e103194. https://doi.org/10.1371/journal.pone.0103194

Farzan SF, Li Z, Korrick SA, Spiegelman D, Enelow R, Nadeau K et al (2016) Infant infections and respiratory symptoms in relation to in utero arsenic exposure in a US cohort. Environ Health Perspect 124(6):840–847. https://doi.org/10.1289/ehp.1409282

Fatur T, Lah TT, Filipic M (2003) Cadmium inhibits repair of UV-, methyl methanesulfonate- and N-methyl-N-nitrosourea-induced DNA damage in Chinese hamster ovary cells. Mutat Res 529:109–116. https://doi.org/10.1016/s0027-5107(03)00112-x

Filipic M, Fatur T, Vudrag M (2006) Molecular mechanisms of cadmium induced mutagenicity. Hum Exp Toxicol 25:67–77. https://doi.org/10.1191/0960327106ht590oa

Flora G, Gupta D, Tiwari A (2012) Toxicity of lead: a review with recent updates. Interdiscip Toxicol 5:47–58. https://doi.org/10.2478/v10102-012-0009-2

Friberg LT, Elinder GG, Kjellstrom T, Nordberg GF (eds) (2019) Cadmium and health: a toxicological and epidemiological appraisal: Volume 2: Effects and response, vol 1. CRC Press. https://doi.org/10.1201/9780429260599

Fujiwara Y, Lee JY, Banno H, Imai S, Tokumoto M, Hasegawa T et al (2020) Cadmium induces iron deficiency anemia through the suppression of iron transport in the duodenum. Toxicol Lett 332:130–139. https://doi.org/10.1016/j.toxlet.2020.07.005

Gallagher CM, Meliker JR (2010) Blood and urine cadmium, blood pressure, and hypertension: a systematic review and metaanalysis. Environ Health Perspect 118(12):1676–1684. https://doi.org/10.1289/ehp.1002077

García-Esquinas E, Pollán M, Umans JG, Francesconi KA, Goessler W, Guallar E et al (2013) Arsenic exposure and cancer mortality in a US-based prospective cohort: the strong heart study. Cancer Epidemiol Biomarkers Prev 22(11):1944–1953. https://doi.org/10.1158/1055-9965.epi-13-0234-t

Garza-Lombó C, Pappa A, Panayiotidis MI, Gonsebatt ME, Franco R (2019) Arsenic-induced neurotoxicity: a mechanistic appraisal. J Biol Inorg Chem 24(8):1305–1316. https://doi.org/10.1007/s00775-019-01740-8

Goldhaber SB (2003) Trace element risk assessment: essentiality vs. toxicity. Regul Toxicol Pharmacol 38(2):232–242. https://doi.org/10.1016/s0273-2300(02)00020-x

Grandjean P, Murata K, Budtz-Jorgensen E, Weihe P (2004) Cardiac autonomic activity in methylmercury neurotoxicity: 14-year follow-up of a Faroese birth cohort. J Pediatr 144:169–176. https://doi.org/10.1016/j.jpeds.2003.10.058

Gustafsson H, Hallbeck M, Norell M, Lindgren M, Engström M, Rosén A et al (2014) Fe(III) distribution varies substantially within and between atherosclerotic plaques. Magn Reson Med 71:885–892. https://doi.org/10.1002/mrm.24687
Gwiazda R, Lucchini R, Smith D (2007) Adequacy and consistency of animal studies to evaluate the neurotoxicity of chronic low-level manganese exposure in humans. J Toxicol Environ Health A 70(7):594–605. https://doi.org/10.1080/10937400600882897
Harischandra DS, Ghaisas S, Zenitsky G, Jin H, Kanthasamy A, Anantharam V, Kanthasamy AG (2019) Manganese-induced neurotoxicity: new insights into the triad of protein misfolding, mitochondrial impairment, and neuroinflammation. Front Neurosci 13:654. https://doi.org/10.3389/fnins.2019.00654
Hasanein P, Emamjomeh A (2019) Beneficial effects of natural compounds on heavy metal-induced hepatotoxicity. Diet Interv Liver Dis Foods Nutr Diet Suppl 2019:345–355. https://doi.org/10.1016/B978-0-12-814466-4.00028-8
He K, Aizenman E (2010) ERK signaling leads to mitochondrial dysfunction in extracellular zinc-induced neurotoxicity: ERK1/2 activation in zinc neurotoxicity. J Neurochem 114:452–461. https://doi.org/10.1111/j.1471-4159.2010.06762.x
Hegazy AMS, Fouad UA (2014) Evaluation of lead hepatotoxicity; histological, histochemical and ultrastructural study. Forensic Med Anat Res 02(03):70–79. https://doi.org/10.4236/fmar.2014.23013
Henriques MC, Loureiro S, Fardilha M, Herdeiro MT (2019) Exposure to mercury and human reproductive health: a systematic review. Reprod Toxicol 85:93–103. https://doi.org/10.1016/j.reprotox.2019.02.012
Hessel EV, Staal YC, Piersma AH, den Braver-Sewradj SP, Ezendam J (2021) Occupational exposure to hexavalent chromium. Part I. Hazard assessment of non-cancer health effects. Regul Toxicol Pharmacol 126:105048. https://doi.org/10.1016/j.yrtph.2021.105048
Hong YS, Kim YM, Lee KE (2012) Methylmercury exposure and health effects. J Prev Med Public Health 45(6):353–363. https://doi.org/10.3961/jpmph.2012.45.6.353
Horiguchi H (2014) *Itai Itai* disease. Encyclopedia of toxicology, 3rd edn. In: Reference module in biomedical sciences, pp 1–2. https://doi.org/10.1016/B978-0-12-386454-3.00033-6
Horowitz Y, Greenberg D, Ling G, Lifshitz M (2002) Acrodynia: a case report of two siblings. Arch Dis Child 86(6):453–453. https://doi.org/10.1136/adc.86.6.453
Hou S, Yuan L, Jin P, Ding B, Qin N, Li L et al (2013) A clinical study of the effects of lead poisoning on the intelligence and neurobehavioral abilities of children. Theor Biol Med Model 10(1):1–9. https://doi.org/10.1186/1742-4682-10-13
Hsiao C-L, Wu K-H, Wan K-S (2011) Effects of environmental lead exposure on T-helper cell-specific cytokines in children. J Immunotoxicol 8(4):284–287. https://doi.org/10.3109/1547691X.2011.592162
Hsu CC, Senussi NH, Fertrin KY, Kowdley KV (2022) Iron overload disorders. Hepatol Comm 6(8):1842–1854
Huang HW, Lee CH, Yu HS (2019) Arsenic-induced carcinogenesis and immune dysregulation. Int J Environ Res Public Health 16(15):2746. https://doi.org/10.3390%2Fijerph16152746
Huang X (2003) Iron overload and its association with cancer risk in humans: evidence for iron as a carcinogenic metal. Mutat Res 533:153–171
Hussain S, Khan M, Sheikh TMM, Mumtaz MZ, Chohan TA, Shamim S, Liu Y (2022) Zinc essentiality, toxicity, and its bacterial bioremediation: a comprehensive insight. Front Microbiol 13:900740. https://doi.org/10.3389/fmicb.2022.900740
Hyder O, Chung M, Cosgrove D, Herman JM, Li Z, Firoozmand A et al (2013) Cadmium exposure and liver disease among US adults. J Gastrointest Surg 17(7):1265–1273. https://doi.org/10.1007/s11605-013-2210-9
Inoue K, O'Bryant Z, Xiong ZG (2015) Zinc-permeable ion channels: effects on intracellular zinc dynamics and potential physiological/pathophysiological significance. Curr Med Chem 22(10):1248–1257. https://doi.org/10.2174%2F0929867322666150209153750

Ishida S, Andreux P, Poitry-Yamate C, Auwerx J, Hanahan D (2013) Bioavailable copper modulates oxidative phosphorylation and growth of tumors. Proc Natl Acad Sci USA 110(48):19507–19512. https://doi.org/10.1073%2Fpnas.1318431110

Ismanto A, Hadibarata T, Widada S, Indrayanti E, Ismunarti DH, Safinatunnajah N et al (2023) Groundwater contamination status in Malaysia: level of heavy metal, source, health impact, and remediation technologies. Bioprocess Biosyst Eng 46(3):467–482. https://doi.org/10.1007/s00449-022-02826-5

Jacquillet G, Barbier O, Rubera I, Tauc M, Borderie A, Namorado MC et al (2007) Cadmium causes delayed effects on renal function in the offspring of cadmium-contaminated pregnant female rats. Am J Physiol Renal Physiol 293(5):F1450–F1460. https://doi.org/10.1152/ajprenal.00223.2007

Järup L, Åkesson A (2009) Current status of cadmium as an environmental health problem. Toxicol Appl Pharmacol 238(3):201–208. https://doi.org/10.1016/j.taap.2009.04.020

Jeng SS, Chen YH (2022) Association of zinc with anemia. Nutrients 14(22):4918. https://doi.org/10.3390%2Fnu14224918

Jiang Y, Zheng W (2005) Cardiovascular toxicities upon manganese exposure. Cardiovasc Toxicol 5:345–354. https://doi.org/10.1385/ct:5:4:345

Kabir MT, Uddin MS, Zaman S, Begum Y, Ashraf GM, Bin-Jumah MN et al (2021) Molecular mechanisms of metal toxicity in the pathogenesis of Alzheimer's disease. Mol Neurobiol 58:1–20. https://doi.org/10.1007/s12035-020-02096-w

Kasten-Jolly J, Heo Y, Lawrence DA (2010) Impact of developmental lead exposure on splenic factors. Toxicol Appl Pharmacol 247(2):105–115. https://doi.org/10.1016/j.taap.2010.06.003

Kile ML, Houseman EA, Baccarelli AA, Quamruzzaman Q, Rahman M, Mostofa G et al (2014) Effect of prenatal arsenic exposure on DNA methylation and leukocyte subpopulations in cord blood. Epigenetics 9(5):774–782. https://doi.org/10.4161%2Fepi.28153

Kim HC, Jang TW, Chae HJ, Choi WJ, Ha MN, Ye BJ et al (2015a) Evaluation and management of lead exposure. Ann Occup Environ Med 27:30. https://doi.org/10.1186/s40557-015-0085-9

Kim KN, Bae S, Park HY, Kwon HJ, Hong YC (2015b) Low-level mercury exposure and risk of asthma in school-age children. Epidemiology 26(5):733–739. https://doi.org/10.1097/ede.0000000000000351

Kim YJ, Kim JM (2015) Arsenic toxicity in male reproduction and development. Dev Reprod 19(4):167–180. https://doi.org/10.12717%2FDR.2015.19.4.167

Kuenzel K, Friedrich O, Gilbert DF (2016) A recombinant human pluripotent stem cell line stably expressing halide-sensitive YFP-I152L for GABAAR and GlyR-targeted high-throughput drug screening and toxicity testing. Front Mol Neurosci. https://doi.org/10.3389/fnmol.2016.00051

Kulka M (2016) A review of paraoxonase 1 properties and diagnostic applications. Pol J Vet Sci 19(1):225–232. https://doi.org/10.1515/pjvs-2016-0028

Kwon SY, Bae ON, Noh JY, Kim K, Kang S, Shin YJ et al (2015) Erythrophagocytosis of lead-exposed erythrocytes by renal tubular cells: possible role in lead-induced). nephrotoxicity. Environ Health Perspect 123:120–127. https://doi.org/10.1289%2Fehp.1408094

Lentini P, Zanoli L, Granata A, Signorelli SS, Castellino P, Dell'Aquila R (2017) Kidney and heavy metals—the role of environmental exposure. Mol Med Rep 15(5):3413–3419. https://doi.org/10.3892/mmr.2017.6389

Leroyer C, Dewitte JD, Bassanets A, Boutoux M, Daniel C, Clavier J (1998) Occupational asthma due to chromium. Respiration 65(5):403–405. https://doi.org/10.1159/000029303

Li FJ, Surolia R, Li H, Wang Z, Liu G, Liu RM et al (2017) Low-dose cadmium exposure induces peribronchiolar fibrosis through site-specific phosphorylation of vimentin. Am J Physiol Lung Cell Mol Physiol 313:L80–L91. https://doi.org/10.1152/ajplung.00087.2017

Li M, Wang S, Liu X, Sheng Z, Li B, Li J et al (2022) Cadmium exposure decreases fasting blood glucose levels and exacerbates type-2 diabetes in a mouse model. Endocrine 76(1):53–61

Li S, Zhang X (2021) Iron in cardiovascular disease: challenges and potentials. Front Cardiovasc Med 8:707138

Long LL, Li XR, Huang ZK, Jian YM, Fu SX, Zheng W (2009) Relationship between changes in brain MRI and 1H-MRS, severity of chronic liver damage, and recovery after liver transplantation. Exp Biol Med 234:1075–1085. https://doi.org/10.3181/0903-rm-118

Malaguarnera G, Cataudella E, Giordano M, Nunnari G, Chisari G, Malaguarnera M (2012) Toxic hepatitis in occupational exposure to solvents. World J Gastroenterol 18:2756–2766. https://doi.org/10.3748/wjg.v18.i22.2756

Marchetti C (2014) Interaction of metal ions with neurotransmitter receptors and potential role in neurodiseases. Biometals 27:1097–1113. https://doi.org/10.1007/s10534-014-9791-y

Milic S, Mikolasevic I, Orlic L, Devcic E, Starcevic-Cizmarevic N, Stimac D et al (2016) The role of iron and iron overload in chronic liver disease. Med Sci Monit 22:2144–2151

Milton AH, Hussain S, Akter S, Rahman M, Mouly TA, Mitchell K (2017) A review of the effects of chronic arsenic exposure on adverse pregnancy outcomes. Int J Environ Res Public Health 14(6):556. https://doi.org/10.3390/ijerph14060556

Min JY, Min KB (2016) Blood cadmium levels and Alzheimer's disease mortality risk in older US adults. Environ Health 15:69. https://doi.org/10.1186/s12940-016-0155-7

Minamata Disease (2007) Minamata disease: its history and lessons. https://minamata195651.jp/pdf/kyoukun_en/kyoukun_eng_all.pdf

Mishra KP (2009) Lead exposure and its impact on immune system: a review. Toxicol in Vitro 23(6):969–972. https://doi.org/10.1016/j.tiv.2009.06.014

Mishra S, Bharagava RN, More N, Yadav A, Zainith S, Mani S, Chowdhary P (2019) Heavy metal contamination: an alarming threat to environment and human health. In: Sobti R, Arora N, Kothari R (eds) Environmental biotechnology: for sustainable future. Springer, Singapore, pp 103–125. https://doi.org/10.1007/978-981-10-7284-0_5

Mitra S, Chakraborty AJ, Tareq AM, Emran TB, Nainu F, Khusro A et al (2022) Impact of heavy metals on the environment and human health: novel therapeutic insights to counter the toxicity. J King Saud Univ Sci 34(3):101865. https://doi.org/10.1016/j.jksus.2022.101865

Miura S, Takahashi K, Imagawa T, Uchida K, Saito S, Tominaga M, Ohta T (2013) Involvement of TRPA1 activation in acute pain induced by cadmium in mice. Mol Pain 9:1744–8069. https://doi.org/10.1186/1744-8069-9-7

Mohanty S, Benya A, Hota S, Kumar MS, Singh S (2023) Eco-toxicity of hexavalent chromium and its adverse impact on environment and human health in Sukinda Valley of India: a review on pollution and prevention strategies. Environ Chem Ecotoxicol 5:46–54. https://doi.org/10.1016/j.enceco.2023.01.002

Morris DR, Levenson CW (2017) Neurotoxicity of zinc. In: Aschner M, Costa LG (eds) Neurotoxicity of metals. Springer, Cham, pp 303–312. https://doi.org/10.1007/978-3-319-60189-2_15

Naija A, Yalcin HC (2023) Evaluation of cadmium and mercury on cardiovascular and neurological systems: effects on humans and fish. Toxicol Rep 10:498–508. https://doi.org/10.1016/j.toxrep.2023.04.009

Nakata A, Araki S, Park SH, Park JT, Kim DS, Park HC, Yokoyama K (2006) Decreases in $CD8^+$ T, naive ($CD4^+$ $CD45RA^+$) T, and B ($CD19^+$) lymphocytes by exposure to manganese fume. Ind Health 44(4):592–597. https://doi.org/10.2486/indhealth.44.592

Nigam A, Priya S, Bajpai P, Kumar S (2014) Cytogenomics of hexavalent chromium (Cr^{6+}) exposed cells: a comprehensive review. Indian J Med Res 139(3):349–370

Nittis T, Gitlin JD (2002) The copper–iron connection: hereditary aceruloplasminemia. Semin Hematol 39:282–289

Nriagu J (2007) Zinc toxicity in humans. School of Public Health, University of Michigan, pp 1–7. https://web.archive.org/web/20170808155353id_/http://www.extranet.elsevier.com/homepage_about/mrwd/nvrn/Zinc%20Toxicity%20in%20Humans.pdf

Nygaard UC, Li Z, Palys T, Jackson B, Subbiah M, Malipatlolla M et al (2017) Cord blood T cell subpopulations and associations with maternal cadmium and arsenic exposures. PLoS ONE 12(6):e0179606. https://doi.org/10.1371/journal.pone.0179606

O'Neal SL, Zheng W (2015) Manganese toxicity upon overexposure: a decade in review. Curr Environ Health Rep 2:315–328. https://doi.org/10.1007/s40572-015-0056-x

Oktapodas Feiler M, Caserta MT, van Wijngaarden E, Thevenet-Morrison K, Hardy DJ, Zhang YV et al (2020) Environmental lead exposure and influenza and respiratory syncytial virus diagnoses in young children: a test-negative case-control study. Int J Environ Res Public Health 17(20):7625. https://doi.org/10.3390/ijerph17207625

Okuda B, Iwamoto Y, Tachibana H, Sugita M (1997) Parkinsonism after acute cadmium poisoning. Clin Neurol Neurosurg 99:263–265. https://doi.org/10.1016/s0303-8467(97)00090-5

Omkar, Pervez A (2017) Concepts of toxicology, 3rd edn. Vishal Publishing Co., Jalandhar, p 295

Oulhote Y, Shamim Z, Kielsen K, Weihe P, Grandjean P, Ryder LP, Heilmann C (2017) Children's white blood cell counts in relation to developmental exposures to methylmercury and persistent organic pollutants. Reprod Toxicol 68:207–214. https://doi.org/10.1016/j.reprotox.2016.08.001

Park YH, Kim D, Dai J, Zhang Z (2015) Human bronchial epithelial BEAS-2B cells, an appropriate in vitro model to study heavy metals induced carcinogenesis. Toxicol Appl Pharmacol 287(3):240–245. https://doi.org/10.1016/j.taap.2015.06.008

Pesce G, Sesé L, Calciano L, Travert B, Dessimond B, Maesano CN et al (2021) Foetal exposure to heavy metals and risk of atopic diseases in early childhood. Pediatr Allergy Immunol 32(2):242–250. https://doi.org/10.1111/pai.13397

Pierce BL, Kibriya MG, Tong L, Jasmine F, Argos M, Roy S et al (2012) Genome-wide association study identifies chromosome 10q24. 32 variants associated with arsenic metabolism and toxicity phenotypes in Bangladesh. PLoS Genet 8(2):e1002522. https://doi.org/10.1371/journal.pgen.1002522

Pizent A, Tariba B, Zivkovic T (2012) Reproductive toxicity of metals in men. Arh Hig Rada Toksikol 63(Suppl 1):35–46. https://doi.org/10.2478/10004-1254-63-2012-2151

Ponka P, Beaumont C, Richardson DR (1998) Function and regulation of transferrin and ferritin. Semin Hematol 35:35–54

Ponka P (2004) Hereditary causes of disturbed iron homeostasis in the central nervous system. Ann NY Acad Sci 1012(1):267–281.

Prohaska JR (2000) Long-term functional consequences of malnutrition during brain development: copper. Nutrition 7(16):502–504. https://doi.org/10.1016/s0899-9007(00)00308-7

Rajendar B, Bharavi K, Rao GS, Kishore PVS, Kumar PR, Kumar CSVS, Kumar DS (2012) Protective effect of alpha-tocopheral on biochemical and histological alterations induced by cadmium in rat testes. Indian J Physiol Pharmacol 55:213–220

Rani A, Kumar A, Lal A, Pant M (2014) Cellular mechanisms of cadmium-induced toxicity: a review. Int J Environ Health Res 24:378–399. https://doi.org/10.1080/09603123.2013.835032

Rawat PS, Singh S, Zahid M, Mehrotra S (2021) An integrated assessment of lead exposure in children: correlation with biochemical and haematological indices. J Trace Elem Med Biol 68:126835. https://doi.org/10.1016/j.jtemb.2021.126835

Ray RR (2016) Adverse hematological effects of hexavalent chromium: an overview. Interdiscip. Toxicol 9(2):55–65. https://doi.org/10.1515%2Fintox-2016-0007

Reczek CR, Chandel NS (2017) The two faces of reactive oxygen species in cancer. Annu Rev Cancer Biol 1:79–98. https://doi.org/10.1146/annurev-cancerbio-041916-065808

Rehman K, Fatima F, Waheed I, Akash MS (2018) Prevalence of exposure of heavy metals and their impact on health consequences. J Cell Biochem 119:157–184. https://doi.org/10.1002/jcb.26234

Reyes JL, Molina-Jijón E, Rodríguez-Muñoz R, Bautista-García P, Debray-García Y, Namorado MDC (2013) Tight junction proteins and oxidative stress in heavy metals-induced nephrotoxicity. Biomed Res Int 1–14. https://doi.org/10.1155/2013/730789

Rousseau MC, Parent ME, Nadon L, Latreille B, Siemiatycki J (2007) Occupational exposure to lead compounds and risk of cancer among men: a population-based case-control study. Am J Epidemiol 166(9):1005–1014. https://doi.org/10.1093/aje/kwm183

Roy JS, Chatterjee D, Das N, Giri AK (2018) Substantial evidences indicate that inorganic arsenic is a genotoxic carcinogen: a review. Toxicol Res 34:311–324. https://doi.org/10.5487%2FTR.2018.34.4.311

Ruan F, Zhang J, Liu J, Sun X, Li Y, Xu S, Xia W (2022) Association between prenatal exposure to metal mixtures and early childhood allergic diseases. Environ Res 206:112615. https://doi.org/10.1016/j.envres.2021.112615

Ruczaj A, Brzóska MM (2023) Environmental exposure of the general population to cadmium as a risk factor of the damage to the nervous system: a critical review of current data. J Appl Toxicol 43(1):66–88. https://doi.org/10.1002/jat.4322

Rutchik JS, Zheng W, Jiang YM, Mo XE (2012a) How does an occupational neurologist assess welders and steelworkers for a manganese-induced movement disorder? An international team's experiences in Guangxi, China, part I. J Occup Environ Med 54(11):1432–1434. https://doi.org/10.1097%2FJOM.0b013e318216d0df

Rutchik JS, Zheng W, Jiang YM, Mo XE (2012b) How does an occupational neurologist assess welders and steelworkers for a manganese-induced movement disorder? An international team's experiences in Guangxi, China, part II. J Occup Environ Med 54(12):1562–1564. https://doi.org/10.1097/jom.0b013e318216d01b

Salnikow K, Zhitkovich A (2008) Genetic and epigenetic mechanisms in metal carcinogenesis and cocarcinogenesis: nickel, arsenic, and chromium. Chem Res Toxicol 21(1):28–44. https://doi.org/10.1021/tx700198a

Salustri C, Barbati G, Ghidoni R, Quintiliani L, Ciappina S, Binetti G et al (2010) Is cognitive function linked to serum free copper levels? A cohort study in a normal population. Clin Neurophysiol 121(4):502–507. https://doi.org/10.1016/j.clinph.2009.11.090

Sanders T, Liu Y, Buchner V, Tchounwou PB (2009) Neurotoxic effects and biomarkers of lead exposure: a review. Rev Environ Health 24:15–45. https://doi.org/10.1515%2Freveh.2009.24.1.15

Satarug S, Garrett SH, Sens MA, Sens DA (2010) Cadmium, environmental exposure, and health outcomes. Environ Health Perspec 118(2):182–190. https://doi.org/10.1289/ehp.0901234

Schwerdtle T, Ebert F, Thuy C, Richter C, Mullenders LHF, Hartwig A (2010) Genotoxicity of soluble and particulate cadmium compounds: impact on oxidative DNA damage and nucleotide excision repair. Chem Res Toxicol 23:432–442. https://doi.org/10.1021/tx900444w

Scragg A (2006) Environmental biotechnology, 2nd edn. Oxford University Press, Oxford

Sharma H, Rawal N, Mathew BB (2015) The characteristics, toxicity and effects of cadmium. Int J Nanotechnol Nanosci 3:1–9

Silbergeld EK, Waalkes M, Rice JM (2000) Lead as a carcinogen: experimental evidence and mechanisms of action. Am J Ind Med 38(3):316–323. https://doi.org/10.1002/1097-0274(200009)38:3%3C316::aid-ajim11%3E3.0.co;2-p

Singh R, Gautam N, Mishra A, Gupta R (2011) Heavy metals and living systems: an overview. Indian J Pharmacol 43(3):246–253. https://doi.org/10.4103/0253-7613.81505

Squitti R, Gorgone G, Panetta V, Lucchini R, Bucossi S, Albini E et al (2009) Implications of metal exposure and liver function in Parkinsonian patients resident in the vicinities of ferroalloy plants. J Neural Transm 116:1281–1287. https://doi.org/10.1007/s00702-009-0283-0

Srivastava A, Peshin SS, Kaleekal T, Gupta SK (2005) An epidemiological study of poisoning cases reported to the National Poisons Information Centre, All India Institute of Medical Sciences, New Delhi. Hum Exp Toxicol 24:279–285. https://doi.org/10.1191/0960327105ht527oa

Stadler N, Lindner RA, Davies MJ (2004) Direct detection and quantification of transition metal ions in human atherosclerotic plaques: evidence for the presence of elevated levels of iron and copper. Arterioscler Thromb Vasc Biol 24:949–954. https://doi.org/10.1161/01.ATV.0000124892.90999.cb

Straif K, Loomis D, Guha N, Hall AL (2018) Identifying occupational carcinogens: an update from the IARC monographs. Occup Environ Med 75:543–544. https://doi.org/10.1136/oemed-2018-105189

Sun H, Brocato J, Costa M (2015) Oral chromium exposure and toxicity. Curr Environ Heal Rep 2(3):295–303. https://doi.org/10.1007/s40572-015-0054-z

Taher AT, Cappellini MD (2018) How I manage medical complications of β-thalassemia in adults. Blood 132(17):1781–1791

Tellez-Plaza M, Guallar E, Howard BV, Umans JG, Francesconi KA, Goessler W et al (2013) Cadmium exposure and incident cardiovascular disease. Epidemiology 24(3):421–429. https://doi.org/10.1097/EDE.0b013e31828b0631

Thompson CM, Fedorov Y, Brown DD, Suh M, Proctor DM, Kuriakose L et al (2012) Assessment of Cr(VI)-induced cytotoxicity and genotoxicity using high content analysis. PLoS ONE 7(8):e42720. https://doi.org/10.1371/journal.pone.0042720

Tsai TL, Lei WT, Kuo CC, Sun HL, Su PH, Wang SL (2021) Maternal and childhood exposure to inorganic arsenic and airway allergy—a 15-year birth cohort follow-up study. Environ Int 146:106243. https://doi.org/10.1016/j.envint.2020.106243

Ukah B, Ameh P, Egbueri J, Unigwe CO, Ubido OE (2020) Impact of effluent-derived heavy metals on the groundwater quality in ajao industrial area, Nigeria: an assessment using entropy water quality index (EWQI). Int J Energy Water Res 4:231–244. https://doi.org/10.1007/s42108-020-00058-5

Vaziri ND (2002) Pathogenesis of lead-induced hypertension: role of oxidative stress. J Hypertens Suppl 20(3):S15-20

Vaziri ND (2008) Mechanisms of lead-induced hypertension and cardiovascular disease. Am J Physiol Heart Circ Physiol 295(2):H454–H465. https://doi.org/10.1152/ajpheart.00158.2008

Verma RK, Sankhla MS, Kumar R (2018) Mercury contamination in water & its impact on public health. Int J Forensic Sci 1(2):72–78.

Vigeh M, Smith DR, Hsu PC (2011) How does lead induce male infertility? Iran J Reprod Med 9:1–8

Wang A, Holladay SD, Wolf DC, Ahmed SA, Robertson JL (2006) Reproductive and developmental toxicity of arsenic in rodents: a review. Int J Toxicol 25(5):319–331. https://doi.org/10.1080/10915810600840776

Wang B, Du Y (2013) Cadmium and its neurotoxic effects. Oxid Med Cell Longev 2013:898034. https://doi.org/10.1155/2013/898034

Wang XY, Yang H, Wang MG, Yang DB, Wang ZY, Wang L (2017) Trehalose protects against cadmium-induced cytotoxicity in primary rat proximal tubular cells via inhibiting apoptosis and restoring autophagic flux. Cell Death Dis 8(10):e3099–e3099

White MK, Johnson EM, Khalili K (2009) Multiple roles for Pur-α in cellular and viral regulation. Cell Cycle 8(3):414–420. https://doi.org/10.4161/cc.8.3.7585

Woolf AD (2022) Itai-Itai disease—Japan, 1955. In: History of modern clinical toxicology. Academic Press, pp 109–120

Xie H, Holmes AL, Wise SS, Gordon N, Wise JP (2004) Lead chromateinduced chromosome damage requires extracellular dissolution to liberate chromium ions but does not require particle internalization or intracellular dissolution. Chem Res Toxicol 17:1362–1367

Yadav M, Sharma P (2023) Current eco-friendly and sustainable methods for heavy metals remediation of contaminated soil and water: special emphasis on use of genetic engineering and nanotechnology. Pollution 9(3):1028–1048. https://doi.org/10.22059/poll.2023.345177.1522

Yang CC, Wu ML, Deng JF (2004) Prolonged hemolysis and methemoglobinemia following organic copper fungicide ingestion. Vet Hum Toxicol 46:321–323

Yoshizawa K, Rimm EB, Morris JS, Spate VL, Hsieh CC, Spiegelman D, Stampfer MJ (2002) Mercury and the risk of coronary heart disease in men. N Engl J Med 347:1755–1760. https://doi.org/10.1056/NEJMoa021437

Young HA, Geier DA, Geier MR (2008) Thimerosal exposure in infants and neurodevelopmental disorders: an assessment of computerized medical records in the Vaccine Safety Datalink. J Neurol Sci 271(1–2):110–118. https://doi.org/10.1016/j.jns.2008.04.002

Yu L, Liou IW, Biggins SW, Yeh M, Jalikis F, Chan LN, Burkhead J (2019) Copper deficiency in liver diseases: a case series and pathophysiological considerations. Hepatol Commun 3(8):1159–1165. https://doi.org/10.1002%2Fhep4.1393

Zefferino R, Piccoli C, Ricciardi N, Scrima R, Capitanio N (2017) Possible mechanisms of mercury toxicity and cancer promotion: Involvement of gap junction intercellular communications and inflammatory cytokines. Oxid Med Cell Longev 2017:7028583. https://doi.org/10.1155/2017/7028583

Zhang Y, Huo X, Cao J, Yang T, Xu L, Xu X (2016) Elevated lead levels and adverse effects on natural killer cells in children from an electronic waste recycling area. Environ Pollut 213:143–150. https://doi.org/10.1016/j.envpol.2016.02.004

Zhang Y, Huo X, Lu X, Zeng Z, Faas MM, Xu X (2020) Exposure to multiple heavy metals associate with aberrant immune homeostasis and inflammatory activation in preschool children. Chemosphere 257:127257. https://doi.org/10.1016/j.chemosphere.2020.127257

Zheng K, Zeng Z, Tian Q, Huang J, Zhong Q, Huo X (2023) Epidemiological evidence for the effect of environmental heavy metal exposure on the immune system in children. Sci Total Environ 868:161691. https://doi.org/10.1016/j.scitotenv.2023.161691

Zheng X, Li S, Li J, Lv Y, Wang X, Wu P (2020) Hexavalent chromium induces renal apoptosis and autophagy via disordering the balance of mitochondrial dynamics in rats. Ecotoxicol Environ Safety 204:111061. https://doi.org/10.1016/j.ecoenv.2020.111061

Chapter 5
Microbial Remediation of Heavy Metal Contamination in Soils

Lamya Muazzeda Medha and Tofazzal Islam

Abstract Natural ecosystems face pollution from organic contaminants, toxic wastes, and heavy metals, posing threats to human, animal, plant, and microbial well-being. Despite good intentions, many countries' chemical and physical treatment approaches are time-consuming, costly, and limited in efficacy, especially in regions affected by metal contamination. In response to these challenges, bioremediation using microorganisms has emerged as a viable and environmentally sustainable approach for remediating heavy metal-contaminated soils and streams. Microbial remediation, gaining popularity for its cost-effectiveness and efficacy, relies on beneficial microorganisms like fungi and bacteria. Whether naturally occurring or genetically engineered, these microorganisms can sequester, convert, or detoxify heavy metals, facilitating ecosystem restoration and safeguarding living organisms. This chapter emphasizes the concepts, applications, and future potential of microbial remediation as a means to mitigate environmental pollution and promote sustainable, nature-based remediation of heavy metal pollutants.

Keywords Bioremediation · Biosorption · Biotransformation · In-situ bioremediation · Ex-situ bioremediation

5.1 Introduction

The term "heavy metals" (HM) refers to a class of elements that can be found in the crust of our planet (Yabe et al. 2010). However, it is important to note that various human activities, including but not limited to industrial discharges, mining operations, and agricultural practices, have the potential to lead to the dispersion of heavy metals into the surrounding atmosphere (Sarwar et al. 2017; Sarker et al. 2022, 2023). In numerous countries, soil heavy metal levels exceed permissible limits (Su 2014). This situation is primarily attributed to unregulated urban growth,

L. M. Medha · T. Islam (✉)
Institute of Biotechnology and Genetic Engineering (IBGE), Bangabandhu Sheikh Mujibur Rahman Agricultural University, Gazipur 1706, Bangladesh
e-mail: tofazzalislam@bsmrau.edu.bd

N. Kumar (ed.), *Heavy Metal Toxicity*, Environmental Science and Engineering,
https://doi.org/10.1007/978-3-031-56642-4_5

industrialization, population increases, and inadequate governance. As a result, it presents a concealed risk to the chain of human sustenance (Khalid et al. 2017; Rakib et al. 2022).

Metallic elements include not only the heavy metals and transition metals but also certain lanthanides, metalloids, and actinides. According to Afal and Wiener (2014), these heavy metals have a density of 3.5–7 g/cm^3, atomic weights of 22.98 to less than 40, and atomic numbers below 2. In pure form, they have high electrical and thermal conductivity. Solubility, binding with inorganic soil components, attachment to organic matter, and precipitation from solids are all processes that affect the metal content of soil. Approximately 30 of nature's 92 elements, including beryllium (Be), boron (B), lithium (Li), aluminum (Al), titanium (Ti), vanadium (V), chromium (Cr), manganese (Mn), cobalt (Co), nickel (Ni), copper (Cu), arsenic (As), selenium (Se), strontium (Sr), molybdenum (Mo), palladium (Pd), silver (Ag), cadmium (Cd), tin (Sn), antimony (Sb), tellurium (Te), cesium (Cs), barium (Ba), tungsten (W), platinum (Pt), gold (Au), mercury (Hg), lead (Pb), and bismuth (Bi) can have adverse effects on human health (Mingho 2005).

Soil contamination by heavy metals pertains to the accumulation of hazardous elements such as Hg, Cd, Pb, Cr, and other highly perilous substances in the soil, surpassing their inherent baseline levels (Abd-Alla et al. 2012). These metallic elements do not undergo natural degradation and can accumulate significantly through biological mechanisms, hence presenting considerable hazards to human well-being (Li et al. 2015). In contemporary times, the discharge of considerable amounts of heavy metals as a consequence of industrial operations and mining, subsequently leading to their accumulation in the soil, has led to heightened levels of heavy metals in it. Furthermore, it is worth noting that the extensive utilization of pesticides and fertilizers could potentially be a contributing factor to the increased levels of heavy metals found in soil (Malidareh et al. 2014).

Biogeochemical processes and marine productivity can be indirectly affected by heavy metal contamination (Bong et al. 2010). The presence of this substance has been seen to have negative consequences on soil health, namely by negatively impacting the respiration, metabolic processes, and activities of soil microorganisms. Additionally, it has been found to impede the conversion of organic carbon into bio-carbon. In addition, the presence of heavy metals in the surrounding environment might result in physiological impairment and nutritional deficiencies in plants. Moreover, there is a possibility of these heavy metals being transmitted to species at higher trophic levels, thereby presenting substantial and irreversible risks to human well-being (Jin et al. 2018). The ability of these toxic heavy metals to accumulate within living beings can lead to the development of a range of metabolic and physiological diseases (Ahemad and Malik 2011). According to previous research (Kumar and Bharadvaja 2020), heavy metals exhibit significant toxicity, carcinogenicity, and mutagenicity, even when present at extremely low quantities. According to previous research (Rehan and Alsohim 2019), the presence of these entities has the potential to interfere with the proper functioning and operational processes of crucial cellular components. Due to their propensity for bioaccumulation and resistance to biodegradation, the issue of heavy metal pollution has become a matter of significant environmental importance (Joutey et al. 2013; Igiri et al. 2018). This phenomenon poses dangers to both human welfare and the overall health of ecosystems (Rehan and Alsohim 2019). Heavy metals are known for their inherent persistence and resistance (Njoku et al. 2020), which poses a significant threat to environmental sustainability and the survival of species (Gupta et al. 2016). In light of the detrimental effects that heavy metals have on ecosystems and their dependent creatures; it is very important to promptly address and mitigate the excessive pollution caused by these toxic elements (Akan et al. 2010).

Abiotic (chemical and physical) and biotic (biological) approaches are used to remediate heavy metal-contaminated media (Dhankhar and Hooda 2011; Siddiquee et al. 2015). Most standard heavy metal removal technologies are too expensive for large-scale adoption and require regular monitoring and control (Siddiquee et al. 2015). The limitations of conventional physical–chemical procedures are being addressed by ecologically friendly technologies using diverse species (Alves et al. 2014; Gupta et al. 2016; Siddiquee et al. 2015). These eco-friendly cleanup methods are called bioremediation. Microorganisms are ubiquitous in myriad of environment that play vital role in geochemical nutrient cycling, promote plant growth (Islam et al. 2023) and detoxification of pollutants including heavy metals (Sarker et al. 2023). Elite strains of microorganisms such as fungi and bacteria can be used for bioremediation of soil contaminants and boost crop yields with little input (Stefanescu 2015; Tang et al. 2021). Environmental heavy metal removal relies on microorganisms (Stefanescu 2015). Heavy metal bioremediation by microbes is becoming successful (Kumar Mishra 2017; Sarker et al. 2023). Natural attenuation through bioaccumulation, biotransformation, or biosorption has made certain micro-organisms effective heavy metal bioremediators (Pandey et al. 2005; Kim et al. 2015; Singh et al. 2013; Sinha et al. 2009; Stefanescu 2015).

Mostly degradative plasmids and spores determine microorganism decomposition (Jin et al. 2018). Microorganisms eliminate heavy metals by removing metal ions from their cells, forming metal-ion complexes, and transforming dangerous metal ions into non-toxic, harmless forms (Akhtar et al. 2013). In the bioremediation of metal-contaminated regions, temperature, pH, low-weight organic acids, and humic acids all affect how heavy metals change, are transported, reach oxidation states, and are accessible to microorganisms (Igiri et al. 2018). Biological and green remediation of heavy metal contaminated water and soils has recently been reviewed (Sarker et al. 2023). This chapter comprehensively reviews current knowledge and critically discusses on remediation heavy metal contamination in soils by utilizing naturally occurring and genetically engineered microorganisms.

5.2 Sources of Pollution with Heavy Metals

There are two main places where heavy metals come from: naturally and because of people. Man-made pollution, which is caused by things like the metals' solubility, instability and bioavailability (Abdu et al. 2011a), is more common and damaging.

Anthropogenic heavy metal pollution is caused by things that people do, like mining, burning, using mineral fertilizers, applying sewage sludge, and industrial processes. The environment also gets heavy metals from things like painting, smoking, metallurgy, aerosol cans, and sewage release (Abdu et al. 2011b). When making plastics, coloring is done by adding dyes that contain heavy metals. Heavy metals like cadmium are often used to coat different things and in glowing dials and rubber curing processes, which are all examples of human-made pollution (Kirkham 2006). Also, normal wear and tear on car parts can release cadmium, which can get into the sewers through waste or into the air as particulate matter (Kirkham 2006).

Heavy metals come from nature mostly through weathering and pedogenesis. For example, chemical weathering can break down mineral ores and release heavy metals (Abdu 2011; Abdu et al. 2011b). Inclusion and adsorption are two ways that heavy metals can be found in rocks (Sposito 2008). Acid rain, dew, dust storms in the air, wildfires, and volcanic events are all natural sources of heavy metal pollution (Naidu et al. 1997). When the parent material has a lot of heavy metals, pedogenic pollution may be worse than pollution from people (Brown et al. 1999).

5.3 Impacts of Heavy Metal

5.3.1 Impacts on the Soil

The presence of elevated levels of heavy metals has the potential to cause disturbances in soil microbial communities and enzyme activity, resulting in a significant reduction ranging from 10 to 50 times (Chander et al. 1995). Heavy metals, including cadmium, lead and arsenic have the ability to last in soil for prolonged durations, mostly as a result of industrial emissions and mining residue (Ramos et al. 1994). The presence of heavy metals has been found to have an impact on several soil properties, such as pH and organic matter concentration. Additionally, it has been observed that heavy metals can impede the decomposition of organic matter and the recycling of nutrients by soil microbes (Chao et al. 2014).

5.3.2 Impacts on the Plants

Some plants can tolerate heavy metal poisoning with limited growth and metabolic adaptations (Cheng 2003). Reduced agricultural yields and food chain disruptions result from their biochemical and physiological abnormalities (Chibuike and Obiora 2014). Oxidative stress from these metals inhibits enzyme activity and damages cell structures (Assche and Clijsters 1990; Jadia and Fluker 2009). They can also deplete plant nutrients (Taiz and Zeiger 2002). *Brassica juncea* plants tolerate heavy metals, however lead exposure reduced their water content (Zaier et al. 2010). Heavy metal exposure reduces plant growth, development, metabolism, and physicochemical activity (Servilia et al. 2005). According to Tchounwou et al. (2012) mercury, cadmium, arsenic, chromium and lead are particularly detrimental to plant growth and photosynthetic pigments.

5.3.3 Impacts on Human

Environmental heavy metals can harm humans through inhalation, skin absorption, and ingestion. According to McLaughin et al. (2000) and Ling et al. (2007), high metal concentrations in soil can affect people and ecosystems. Consuming contaminated groundwater and plants harms human health and reduces agricultural output (Wuana and Okieimen 2011). Heavy metal exposure can originate from various sources, including water, soil, crops, fish, toys, and animals. Oral exposure to Cd and Cr beyond acceptable levels can pose considerable noncarcinogenic risks to adolescents and young people.

5.4 Microbiological Interaction with HM in Bioremediation

Bioremediation employs microorganisms, green plants, and vegetation to mitigate and cleanse pollution caused by heavy metals. It utilizes microorganisms that can thrive in heavy metal-contaminated environments and exhibit significant metal absorption (Shakoori and Rehman 2004). Microbial biomass, including bacteria, fungi, and yeast, has been frequently utilized in bioremediation efforts (Feng and Aldrich 2004; Rehman et al. 2007; Morales-Barrera and Cristiani-Urbina 2008). Due to its economic feasibility, high efficiency in removing HMs, and widespread availability in contaminated soil and water, bioremediation is unique among physicochemical methods for HM pollution. Microorganisms are key to HM bioremediation. They help dissolve HMs and oxidize and reduce transition HMs. This eco-friendly technique uses microbial metabolism to remove HM from the environment. Converting poisonous heavy metals into harmless ions is its goal. Bioremediation uses *Arthrobacter, Aspergillus, Alcaligenes, Azotobacter, Burkholderia, Fusarium, Bacillus, Mucor, Pseudomonas, Penicillium, Streptomyces, Rhizopus, Stenotrophomonas, Talaromyces, Thiobacillus,* and *Natrialba, Trichoderma* and *Haloferax,* extreme-environment archaea, are effective biological remediation resources. These prospective bacteria can be exploited in soil (Manoj et al. 2020; Sreedevi et al. 2022). To tackle the problem of hazardous substances, two discrete methodologies are utilized: in-situ remediation and ex situ remediation. In-situ remediation entails the introduction of a biological agent to the contaminated site or the establishment of favorable conditions for native organisms to handle the contaminants. An alternative approach is ex-situ remediation, where the polluted substance is moved to a different site for the purpose of treatment (Kumar et al.

2011a, b, 2016; Raghunandan et al. 2014, 2018). The detoxification process can be facilitated by microorganisms via a variety of mechanisms, the most prominent of which is the utilization of toxic metals as a nutritional source (Sun et al. 2020). So microbial remediation requires additional investigation and research.

5.5 Bioremediation Types

There are various forms of bioremediation techniques that are employed for the purpose of environmental restoration. Bioremediation can be categorized into two distinct techniques based on the specific location where the bioremediation process takes place.

5.5.1 In-Situ Bioremediation

This approach aligns with the treatment of contaminated surfaces in their current locations. It relies on the direct detoxification of sorbed and dissolved pollutants by microorganisms. It can be implemented in both saturated and unsaturated soils. In comparison to the ex-situ strategy, in-situ bioremediation is an effective method for remediating chemicals that are organic (Brar et al. 2006). Furthermore, in-situ bioremediation has also been extended to address the remediation of toxic as well inorganic metals. Furthermore, the use of chemotactic properties possessing microorganisms to assist migration into polluted areas and then breakdown dangerous compounds is expected to improve safety (Kulshreshtha et al. 2014).

In addition, stimulation of heavy metal reduction at the site serves to minimize the potential for downstream transportation of contaminants. One of the primary challenges encountered in the field of in-situ bioremediation pertains to the identification and selection of either a single organism or a combination of organisms which has the inherent capability to effectively mitigate the toxic effects of specific metals. In laboratory experiments, it was observed that the combination of Fe^{3+} and sulfate-depleting bacteria have enzymatic capability to digest few heavy metals, including Co(III), Tc(VIII), U(VI), and Cr(VI) (Tebo and Obraztsova 1998; Gorby et al. 1998; Lloyd et al. 2000). Furthermore, it was observed that species belonging to the *Geobacteraceae* family had a significant presence during the stimulation phase aimed at reducing Fe^{3+} levels. Additionally, these members were identified throughout the stimulation process targeting the reduction of U(VI) in a contaminated Aquifer. The *Geobacteraceae* group has been recognized as having a significant impact on the stabilization of pollutants and the reduction of metals in subterranean environments (He et al. 2019). The subsequent methods are employed for the purpose of "in-situ" bioremediation.

5.5.1.1 Biosparging

Biosparging is a method by which air is injected through a conduit beneath the water table, resulting in an increase in oxygen concentrations that encourage the growth of native microorganisms (Jain et al. 2012). Significantly, it is different from bioventing through the introduction of air into the saturated region, thus promoting the transfer of volatile organic compounds to the unsaturated region. The efficacy of this procedure is contingent upon the biodegradability of the impurities and the properties of the soil. An aspect that distinguishes this system is its cost-effectiveness and its capacity for adaptable design (Atlas and Philp 2005).

5.5.1.2 Bioventing

Bioventing, as described by Jain et al. (2012), is a remediation methodology wherein a regulated quantity of oxygen is introduced into the unsaturated soil to induce the degradation of contaminants by indigenous soil microorganisms. In addition to the aforementioned elements, moisture and nutrients are introduced in accordance with the guidelines delineated by Philp and Atlas (2005). The efficiency of bioventing in promoting anaerobic biodegradation has been significantly improved through the addition of nitrogen to oxygen, as evidenced by the research conducted by Shah et al. (2001) and Mihopoulos et al. (2000, 2002).

5.5.1.3 Bioaugmentation

Bioaugmentation commences with the utilization of exogenous microbial strains, which may be genetically modified or naturally present, in order to purify polluted water or soil. Frequently, a collection of microorganisms endowed with the requisite enzymes and degradation pathways is utilized in this manner. Bioaugmentation is a process utilized to purify chlorinated ethenes from groundwater, municipal wastewater, and soil. The chlorinated ethenes are converted into innocuous chloride and ethylene (Jain et al. 2012).

5.5.1.4 Natural or Intrinsic Bioremediation

Bioremediation by intrinsic method involves enhancement of natural microorganisms through the provision of necessary nutrition and oxygen to mitigate heavy metal contamination, without any intervention of genetically engineered microorganisms (Riseh et al. 2023).

5.5.1.5 Engineered Bioremediation

Engineered bioremediation encompasses the deliberate alteration of environmental factors to boost the proliferation of the introduced microorganisms, hence accelerating the bioremediation process (Sayler and Ripp 2000; Ripp et al. 2000; Abo-Alkasem et al. 2023).

5.5.1.6 Advantages of In-Situ Bioremediation

In-situ bioremediation is a highly cost-efficient and versatile technique for remediating contaminated sites. It leverages naturally occurring biological processes and microorganisms to break down and neutralize contaminants, significantly reducing the need for expensive equipment and extensive site excavation. This cost-effectiveness makes it a economically viable option, especially for remediating large, contaminated areas, even in remote or challenging locations where transporting contaminated materials is logistically complex and costly. A notable strength of in-situ bioremediation is its ability to address a wide range of pollutants, both organic and inorganic. It achieves this by harnessing the diverse metabolic activities of various microorganisms to degrade or transform contaminants into less harmful forms. This adaptability allows it to effectively treat complex mixtures of contaminants commonly found in polluted sites, ensuring a comprehensive and thorough remediation process. In-situ bioremediation's reliance on natural biological processes, coupled with its cost efficiency and versatility, makes it a valuable tool in the effort to rehabilitate polluted sites and contribute to environmental protection and sustainability (Abo-Alkasem et al. 2022).

5.5.1.7 Challenges Facing In-Situ Bioremediations

In-situ bioremediation encounters significant challenges that can impede its effectiveness in addressing environmental contamination. One primary obstacle arises from elevated levels of heavy metals, which can disrupt the metabolic processes of native bacteria, limiting their ability to facilitate remediation. These heavy metals, when present in substantial quantities, act as impediments, inhibiting the metabolic activity of indigenous microorganisms crucial for bioremediation. This interference poses a significant limitation to the applicability of in-situ bioremediation in areas contaminated with such metals. Furthermore, the microbial metabolic activities involved in bioremediation can unintentionally transform contaminants into more hazardous and mobile intermediates, worsening the contamination issue. To prevent these adverse transformations, a comprehensive understanding of microbial ecology and specific bioremediation processes is essential. In-situ bioremediation may not be suitable for addressing resistant pollutants, as some contaminants resist microbial degradation. In such cases, alternative remediation methods must be considered. However, in-situ bioremediation remains an ideal choice for lightly contaminated environments, where the challenges posed by heavy metals, contaminant transformation, and resistant pollutants are less pronounced, making it a valuable tool for environmental remediation in specific contexts (Kulshreshtha et al. 2014).

5.5.2 *Ex-Situ Bioremediation*

The fundamental principle underlying ex-situ bioremediation technique involves the remediation of a polluted site by the process of excavating soil to facilitate decomposition by microorganisms. This strategy employed a total of five techniques.

5.5.2.1 Slurry-Phase

The slurry-phase procedure includes digging up polluted soil and then combining it with water before moving it to a bioreactor to remove stones. The kind and concentration of contaminants, soil conditions, and biodegradation rate determine the amount of water needed. After filtration, the soil is dried then returned. EPA (2003) states that liquids receive further treatment.

5.5.2.2 Solid-Phase

Solid-phase method involves excavation, piles of soil containing municipal, agricultural, and organic wastes, and oxygen supply through pipes to boost microbial respiration and activity. Solid-phase bioremediation is time-consuming and space-intensive (Hyman and Dupont 2001).

5.5.2.3 Landfarming

This method involves stimulating indigenous organisms by supplying excavated soil with minerals and nutrients, then tilling it periodically to promote biodegradation (Abo-Alkasem et al. 2023).

5.5.2.4 Soil Biopiles

Landfarming bioremediation involves injection of air through soil using perforated pipes and above ground piles (Verma 2021).

This technique is preferrable for treating extreme environments and low molecular weight compounds by limiting volatilization (Gomez and Sartaj 2014) due to its low cost, full control over nutritional feed, temperature and aeration (Whelan et al. 2015).

5.5.2.5 Composting Bioremediation

Composting bioremediation and landfarming bioremediation share similarities in their approach to excavating contaminated soil and stimulation of indigenous microbes by air injection and nutrient feeding. However, they differ in the supplementation of the soil with supplements such as corncobs, hay, and straw. These additives aid in distribution of oxygen, moisture content maintenance, and turning frequency. It should be noted that the employing composting for the biological degradation of volatile contaminants is unfavorable due to the periodic turning required during composting (Hobson et al. 2005).

5.5.2.6 Benefits of Ex-Situ Bioremediation

Ex situ bioremediation offers several advantages in the field of environmental remediation. Firstly, it allows for precise control and regulation of the biological degradation process, ensuring that the treatment is conducted efficiently and according to specific requirements. Additionally, this method exhibits the remarkable capacity to successfully eliminate a wide spectrum of pollutants, making it a versatile and effective solution for tackling various forms of contamination. Perhaps one of its most significant benefits is the notable reduction in the time required to complete the treatment process. By taking the contaminated material out of its natural environment and into a controlled setting, ex situ bioremediation accelerates the degradation of pollutants, leading to faster and more efficient cleanup efforts. These advantages collectively make ex situ bioremediation a valuable and practical approach in addressing environmental contamination, offering precision, versatility, and expedited remediation (Abo-Alkasem et al. 2022).

5.5.2.7 Challenges Facing Ex-Situ Bioremediation

Ex-situ bioremediation encounters specific challenges that limit its applicability. Firstly, it is primarily effective for the degradation of chlorinated hydrocarbons, restricting its usefulness for a broader range of contaminants. Additionally, certain soil types may require supplementary treatments, such as the incorporation of impermeable substrates, to optimize the remediation process. These challenges underscore the importance of carefully considering the suitability of ex-situ bioremediation for addressing specific pollution scenarios, as it may not be universally applicable and may necessitate additional interventions to achieve successful remediation outcomes (Abo-Alkasem et al. 2022).

5.6 Mechanism of Action of Microbial Remediation

In general, the majority of Heavy Metals (HMs) are typically regarded as hazardous, but various living organisms have developed strategies to adapt to and detoxify them, promoting cellular rejuvenation (de Alencar et al. 2017). Bioremediation encompasses methods like detoxification, biosorption, degradation, and the conversion of highly toxic HMs into less harmful forms. Additionally, it involves retaining HMs within cells through bioaccumulation, retaining them outside cells through precipitation as insoluble compounds, and generating metabolites that dissolve and bind metal compounds, reducing their leaching potential (Jasrotia et al. 2017). Furthermore, the bioremediation of HMs can be facilitated by plant growth-promoting microbes (PGPMs), including bacteria, mycorrhizae, and fungi (Rai et al. 2020; Raklami et al. 2022). Among these PGPMs, bacteria and mycorrhizae are capable of releasing bioactive substances, including metal chelators, which assist in the processes of biosorption and bioaccumulation (Raklami et al. 2022). Moreover, certain beneficial bacteria and fungi can expedite the microbial precipitation of HMs and the release of metal ions through biosurfactants, contributing to the reclamation of areas contaminated with trace metals (Manoj et al. 2020). However, it's important to note that the effectiveness of these potential mechanisms is greatly influenced by the characteristics of the organisms involved and the specific type and concentration of HMs (Akhtar et al. 2020; Sharma 2020). Figure 5.1 illustrates the various methods through which microorganisms remove HMs.

5.6.1 Biosorption

An organism forms a complex with a metal during this process, producing a non-toxic compound. In order to ascertain an effective biosorption mechanism, it is imperative to examine various critical factors: the nature of the biosorbent, its capacity to adsorb metals, the rate at which adsorption occurs, the biosorbent's regenerative capability, the percentage of metal recovered, the economic viability of the biosorption process, and the flexibility of the method utilized to separate biosorbent-metal complex (Bae et al. 2001, 2002, 2003). Bioremediation methods that utilize the biosorption mechanism can be classified into two main categories (Fig. 5.2).

5.6.1.1 Metabolism-Independent Biosorption

Metabolism-independent biosorption relies on cell physical and chemical properties, whether alive or non-living. This category includes numerous mechanisms:

Periplasmic components attract metal ions, causing **adsorption or extracellular sequestration**. Metal adsorption is influenced by EPS on bacterial cell walls. EPS contains polysaccharides, mucopolysaccharides, and proteins. Carboxyl, hydroxyl,

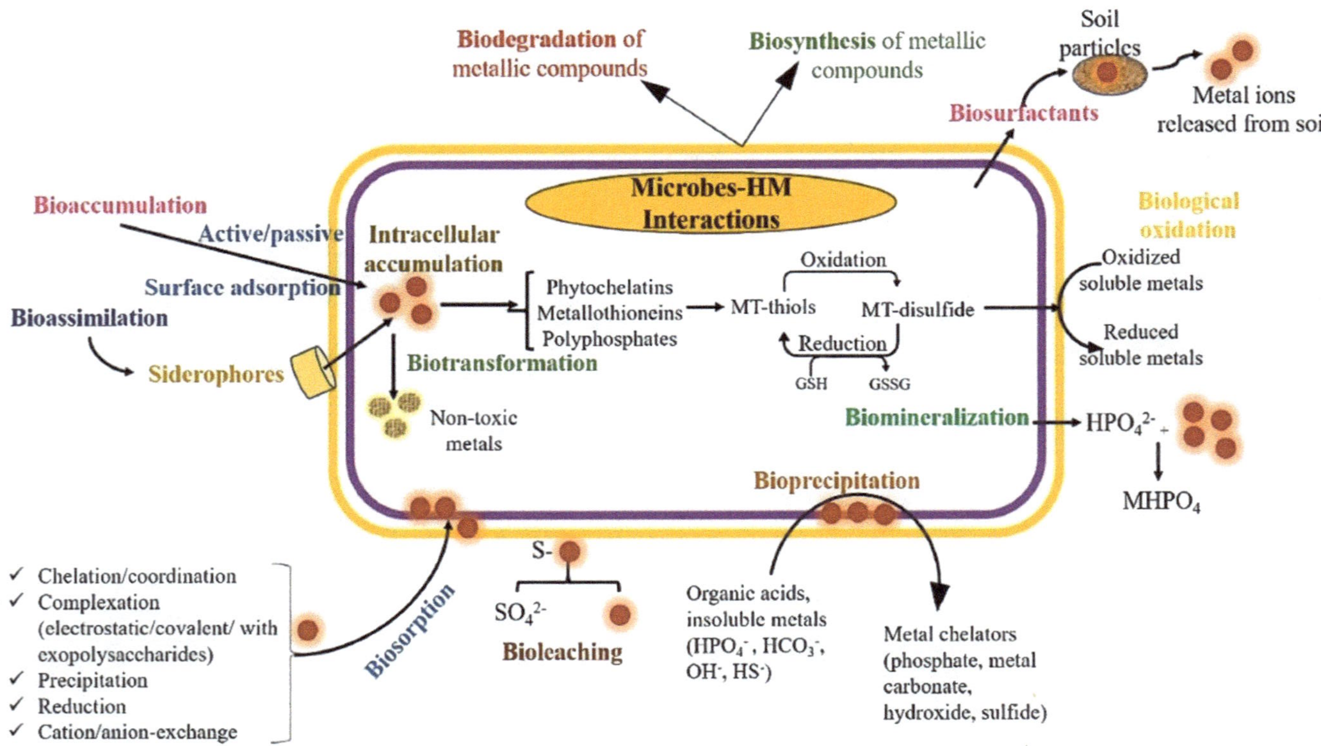

Fig. 5.1 Various bioremediation mechanisms of microorganisms including bioaccumulation, biosorption, biomineralization, bioleaching, bioprecipitation, redox, and biodegradation during potential microbes-HM interaction. Red round shapes represent the heavy metals, MT-Metallothionein, GSH-Glutathione, and GSSG-Glutathione disulfide (Sarker et al. 2023)

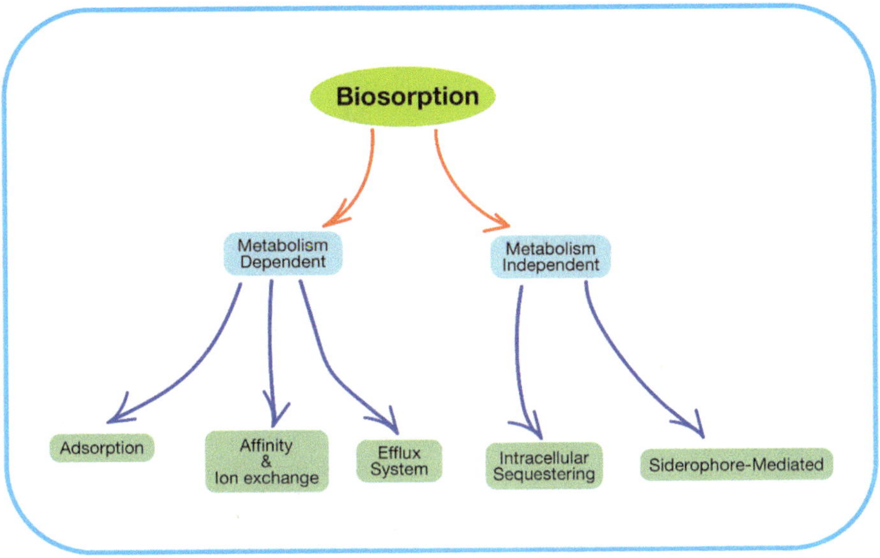

Fig. 5.2 Mechanism of bioremediation action adapted from Abo-Alkasem et al. (2023)

amine, and phosphoric functional groups help sequester heavy metals (Guiné et al. 2006).

5.6.1.2 Ion Exchange and Affinity

It includes biosorbent (cellular component) interacting with metal ions. Cunninghamella can bind textile wastewater heavy metals (Tigini et al. 2010). By exchanging ions, Saccharomyces cerevisiae may eliminate Zn(II) and Cd(II).

5.6.1.3 Extracellular Sequestration, or Efflux

It is one of the main ways organisms fight against heavy metal toxicity. This involves an exterior protective barrier and metal ion ejection from cytoplasm to periplasm (Dixit et al. 2015). Transformation and efflux are bacteria's main heavy metal resistance mechanisms, according to Ma et al. (2016).

5.6.2 Metabolism-Dependent Biosorption

Sequestration of heavy metals relies on metabolism-dependent biosorption, which is linked to viable microorganisms' metabolic activity (Ramasamy and Banu 2007).

TEM investigation of *Salipaludibacillus agaradhaerens* strain NRC-R cells showed intracellular sequestering or bioaccumulation, which involves intricate cell-metal interactions in the cytoplasm. Chromium buildup in the cell was confirmed by TEM and EDX analysis (Abo-Alkasem et al. 2022). Mishra and Malik (2013) characterize metal accumulation as adhesion to the cell surface, delayed penetration into the periplasm, and eventual cytoplasm, similar to food uptake. Cysteine-rich proteins helped cadmium-tolerant Pseudomonas putida adsorb Zn, Cd, and Cu, while *Rhizobium leguminosarum* used glutathione to sequester Cd. Using their strong cell walls as ligands, fungi help decontaminate inorganic metals.

5.6.3 Biotransformation

In order to alter the oxidation state of metals, biotransformation relies on the redox mechanism, an integral component of microbial metabolism. The natural process by which metals decompose has been observed in the transformations of chromium (as documented in the investigations of Abo-Alkasem et al. 2022, 2023), selenium (as examined in the research of Lloyd et al. 2001), uranium (as investigated by Chang et al. 2001), and mercury (as documented in the findings of Brim et al. 2000). Siderophore-mediated biosorption (chelation) is another key process besides metabolism-dependent biosorption. *Clostridium* spp., which uses sulfate-reducing bacteria, produces these chemicals in aerobic soils (Vara Prasad and Oliveira Freitas 2003).

5.6.4 Oxidation and Reduction Mechanisms

Microorganisms act as oxidizing agents, releasing electrons that interact with anions in polluted soil, similar to the anaerobic detoxification of organic substances by Lovley and Phillips (1988). In 2000, Spormann and Widdel found that iron (III) accelerates deterioration. The soil can be dug and inoculated with the right microbial consortia or reduced directly in a bioreactor using a pump and treat technique. In 2022, Abo Elsoud and Abo-Alkasem noted that sulfate-reducing bacteria, which are essential to ecological equilibrium, can reduce sulfate directly or indirectly through biofilms. Asha and Sandeep (2013) noted that the indirect technique is cheaper and greener. Uranium cleanup is successful with *Desulfosphorosinus* spp. Clostridium spp. serves as a relevant example for the utilisation of sulfate-reducing bacteria, as demonstrated by Vara Prasad and Oliveira Freitas (2003).

5.6.5 Volatilization or Methylation of Metals

The process involves methylation by microbes, which results in the volatilization of metals, depicts a vital role in remediating metal. Ramasamy and Banu (2007) revealed the ability of several bacterial species, including Pseudomonas, Escherichia, Clostridium, and Bacillus, to convert mercury (II), selenium, arsenic, and lead into methylated compounds in gaseous form.

5.6.6 Bioleaching

Bioleaching and bio-oxidation are components of biomining (Brunetti et al. 2012; Volesky and Holan 1995; Rahman and Sathasivam 2015). Bioleaching involves the mobilization of positive heavy metal ions from insoluble ores via biological dissolution or complexation. Microbial metabolism generates organic acids with low molecular weights, which have the ability to dissolve heavy metals and soil particles containing heavy metal minerals (Fig. 5.3). In nutrient-rich environments, microorganisms have the ability to emit organic acids and increase Cd leaching, as stated by Chanmugathas and Bollag (1998). The rate of leaching increased from 9% in the absence of nutrients to 36% when glucose and other nutrients were added (Chanmugathas and Bollag 1998). Research findings suggest that specific bacteria, including Citrobacter, are capable of generating unbound inorganic phosphate, which can aggregate to form an impermeable metal phosphate layer—as illustrated (Marchenko et al. 2015). Certain microorganisms, such as *Citrobacter*, are capable of generating unbound inorganic phosphate, which can accumulate to form an impermeable metal phosphate layer capable of entrapping hazardous metals (Fig. 5.3) (Marchenko et al. 2015).

Heavy metal valence is altered via redox reactions facilitated by prokaryotic microorganisms, which impacts both mobility and toxicity (Fig. 5.3) (Gavrilescu 2004). Hg^{2+}, for example, can disperse via aerobic bacterial conversion to Hg0; tobacco smoke is also capable of doing so (Danis et al. 2008). Water-soluble and toxic Cr^{6+} can be reduced by Coryne bacterium and other microorganisms. Valence of heavy metals (Fig. 5.3) can influence their mobility and toxicity (Gavrilescu 2004), thereby affecting their activity. Dead Bacillus licheniformis R08 is capable of converting Pb^{2+} to Pb0, whereas microbes such as Coryne bacteria can convert hazardous Cr^{6+} to Cr^{3+} (Goyal et al. 2003) (Effect and mechanism).

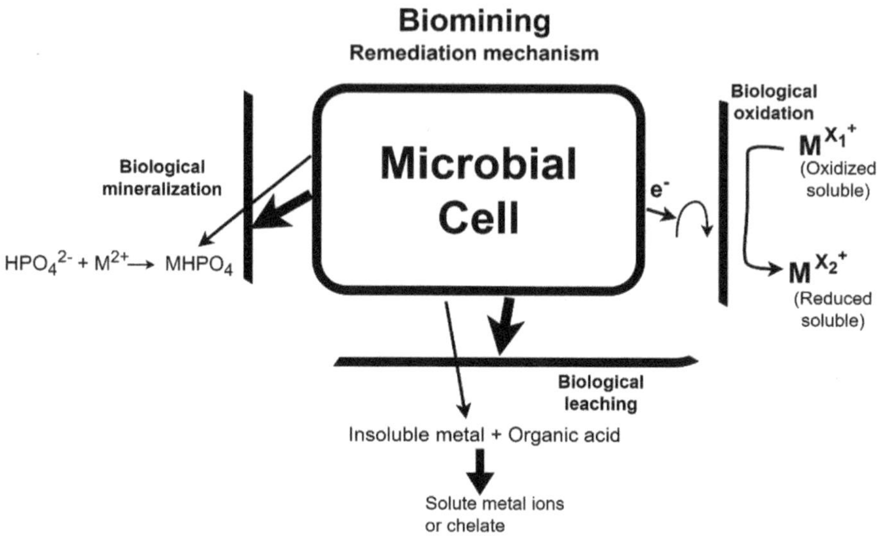

Fig. 5.3 Biomining mechanism of microbes. Adapted from Jin et al. (2018)

5.6.7 Plant-Microbial Remediation

Mycorrhizal fungi and other rhizosphere microorganisms can improve plant heavy metal absorption (Bojórquez et al. 2016). Joner and Leyval (1997) found that mycorrhizal plants receive up to 90, 127, and 131% more Cd than non-mycorrhizal plants at 1, 10, and 100 mg/kg soil Cd^{2+} concentrations.

Mycorrhizal fungi boost plant root surface area by extending their mycelia into the soil (Trellu et al. 2016). Bissonnette et al. (2010) found that mycorrhiza inoculation enhances the absorption of Cu^{2+}, Cd^{2+}, and Zn^{2+}. Endophytic mycorrhiza can aid host plants in resisting heavy metal ions. Plant-endophytic mycorrhiza synergize by acidifying, producing chelating agents, iron carriers, organic acids, and activating metal phosphates. The fungal cell wall's mucus can join with polyphosphate and organic acid ions in the fungal tissue to combine heavy metal ions and restrict movement when soil heavy metal levels are hazardous. Chen et al. (1997) found that arbuscular mycorrhizal fungi absorbed 1.6, 2.8, and 13.3% of their dry weights in Mn^{2+}, Zn^{2+}, and Cd^{2+}. The amount and nature of root exudates alter after fungi infect plant roots, impacting heavy metal oxidation near the rhizosphere (Mathew 2005; Niu et al. 2011). Plant mycorrhizae also bind heavy metals to cell walls to inhibit transmission to plants (Cervantes et al. 2001).

5.7 Microbes Involved in Bioremediation

A key aspect in bioremediation is the type of micro-flora, which utilizes gene and metabolic diversity to reduce toxicity of contaminants (Abdul and Fatema 2000; Juwarkar et al. 2010; Alvarez et al. 2017). Santos et al. (2011) suggest that bioremediation relies on indigenous or derived strains from different sites. Asha and Sandeep (2013) suggest that using indigenous microbial strains avoids the requirement for monitoring. Energy and carbon sources are essential for bioremediation (Vidali 2001). A list of microorganisms used in bioremediation of heavy metals is given in Table 5.1.

5.8 Factors Affecting the Bioremediation Process

Focusing on the approach and ignoring bioremediation elements can reduce its efficacy. As noted by Bandowe et al. (2014), metal ion concentration, valence state, bioavailability, redox potential, low molecular weight organic acids, and environmental variables like temperature and pH can affect bioremediation success.

5.8.1 Substrate Concentration

Quantifying bio-sorbent buildup is crucial to bioremediation success. Usually, two models are used. In 2004, Acar and Malkoc proposed the Langmuir model, which assumes single-layer adsorption. The Freundlich model, introduced by Febrianto et al. (2009), focuses on adsorption equilibrium. A key premise is that adsorption efficiency increases with heavy metal content up to a threshold.

5.8.2 Type of the Substrate

The type of substrate plays a pivotal role in determining the adsorption efficiency, as it is influenced by various factors such as the specific characteristics of the soil, the type of heavy metal present, and the incorporation of soil additives. It is essential to recognize that the interaction between soil and heavy metals significantly constrains the mobility of these elements, as well as their susceptibility to microbial adsorption, as highlighted in the study conducted by Hu et al. (2010). Furthermore, it is worth noting that higher concentrations of $FeSO_4.7H_2O$, exceeding 20 g per liter, have been found to exert a detrimental effect on the leaching of copper (Cu) and zinc (Zn), as documented by Tyagi et al. (2014).

Table 5.1 Different highly efficient microorganisms used in alleviating heavy metals from contaminated environments

Microbial species	Studied HM	Removal efficiency	Removal mechanisms	References
Bacillus cereus	Cr	81% biosorption	Biosorption	Nayak et al. (2018)
Ochrobactrum MT180101	Cu	90% removal	Biosorption	Torres (2020)
B. thuringiensis	Hg	62% metal sorption	Biosorption	Saranya et al. (2019)
Burkholderia sp. Z-90	Pb, Mn, As, Zn, Cu, Cd	24–52% bioleaching	Combined bioleaching and flocculation	Yang et al. (2018)
Cellulosimicrobium sp. (KX710177)	Pb	Nearly complete remediation (99%)	Biosorption	Bharagava and Mishra (2018)
Pseudomonas aeruginosa ATCC 9027	Cu, and Zn	A wide range of removal (18–65%)	Soil washing	Mulligan et al. (2001)
P. aeruginosa FZ-2	Hg	Above 99%	Biosorption	Hwang and Jho (2018)
P. aeruginosa	Zn, Ni, Cd, Pb, Sr and Ba	Adsorption varied 7–62%	Adsorption	Elouzi et al. (2012)
Bacillus sp. HIP3	Cd, Cu, Cr, Zn	Metal chelation was evident up to 13%	Metal chelation	Hisham et al. (2019)
Pseudomonas sp. LKS06	Cd, Pb	N/A	Biosorption	Huang and Lin (2013)
Bacillus sp. MSI 54	Pb, Hg, Cd and Mn	Precipitation achieved 75%–99%	Co-precipitation	Ravindran et al. (2020)
Citrobacter freundii MG812314.1	Fe, Cu, Cd, Zn, Al, Mn	Precipitation documented 40–80%	Precipitation	Gomaa and El-Meihy (2019)
Desulfovibrio desulfuricans	Zn	100% removal	Biosorption	Mwandira et al. (2020)
Alcaligenes sp. MMA	Zn, and Cd	Accumulation of metal nearly 70%	Bioaccumulation	Sodhi et al. (2020)
Oceanobacillus profundus	Zn, and Pb	Metal sorption 54–97%	Biosorption	Abd El-Motaleb et al. (2020)
Klebsiella sp. USL2D	Pb	97% efficiency	Biosorption	Orji et al. (2021)
Streptomyces sp.	Pb	83% removal	Biosorption	Hamdan et al. (2021)
Klebsiella sp. USL2D	Pb	97% efficiency	Biosorption	Orji et al. (2021)

(continued)

Table 5.1 (continued)

Microbial species	Studied HM	Removal efficiency	Removal mechanisms	References
Klebsiella sp. USL2D	Pb	97% efficiency	Biosorption	Orji et al. (2021)
P. azotoformans JAW1	Cd, and Pb	Metal sorption 78–98%	Biosorption	Orji et al. (2021), Choinska-Pulit et al. (2018)
Paecilomyces sp. *Aspergilus niger* *Penicillium* sp.	Co	70–93% removal	Biosorption	Gonzalez et al. (2019)
Acinetobacter sp. B9	Ni	69% removal	Biosorption	Bhattacharya and Gupta (2013)
Chlorella sp. (immobilized)	Cd	More than 92% removal	Biosorption	Shen et al. (2018)
Sterigmatomyces halophilus	Pb, and Zn	57–90% effectiveness	Biosorption	Bano et al. (2018)
Maugeotia genuflexa *Ulothrix cylindricum*	As	Metal sorption up to 98%	Biosorption	Leong and Chang (2020)
Scenedesmus quadricauda	Cr	98–100% removal	Biosorption	Daneshvar et al. (2019), Khoubestani et al. (2015)
Aphanothess sp.	Pb	Higher efficiency (99.9%)	Biosorption	Keryanti and Mulyono (2021)
C. sorokiniana	Cd	Metal accumulation 65%	Bioaccumulation	Leon-Vaz et al. (2021)
Phormidium sp.	Pb	Above 92%	Biosorption	Das et al. (2016)
C. kessleri	Pb	Efficiency 97%	Biosorption	Sultana et al. (2020)

5.8.3 pH

pH plays a critical role in influencing the activities of both metals and microbes, exerting a significant impact on various biological and chemical processes. When the pH is not appropriately adjusted, it can lead to a reduction in enzyme activity, alter the surface charges of microbial organisms, and disrupt the mobility of metal ions, as noted by Bandowe et al. in their 2014 study and Galiulin and Galiulina (2008). According to research findings, it has been observed that raising the pH to a level of 5.5 has a beneficial effect on the adsorption of Pb^{2+} and Zn^{2+} ions. This adjustment in pH enhances the ability to remove these metal ions from a given system, as reported by Rodríguez-Tirado et al. in their 2012 study and further supported by the findings of Wierzba (2015). However, it is important to note that exceeding this pH level can have a counterproductive effect, diminishing the efficiency of metal removal, and therefore, careful pH control is crucial in optimizing the removal of these heavy metal ions from the environment.

5.8.4 Temperature

Temperature plays a pivotal role in the intricate processes of microbial growth, metabolic activities, and enzyme function, thereby exerting a significant influence on the intricate dynamics of metal diffusion between various layers and their bioavailability within ecosystems, as highlighted by Fang et al. in their 2011 study. Remarkably, the optimal temperature for the biodegradation of $Cd^{2+}$was determined to be 35 °C in the context of *Bacillus jeotgali*, as reported by Chanmugathas and Bollag (1988). In contrast, for the efficient biodegradation of Zn^{2+}, a slightly lower temperature of 30 °C emerged as the most favorable condition according to the same study. These findings underscore the nuanced relationship between temperature and the microbial-mediated transformation of heavy metals, shedding light on the intricate interplay of environmental factors in metal bioavailability.

5.9 Biotechnological Approaches for Enhanced Bioremediation

Biotechnology possesses the remarkable capability to engineer a single microorganism, equipping it with a comprehensive suite of enzymes or a specific breakdown pathway designed for the purpose of environmental remediation, as highlighted in the work of Dangi et al. (2019). This entails the genetic modification of microbial strains, endowing them with the capacity to efficiently degrade hydrocarbons. For instance, a multi-plasmid-carrying Pseudomonas strain has the potential to oxidize a wide range of hydrocarbon types, including polyaromatic, terpenic, and aliphatic compounds. This genetic modification is achieved through the utilization of recombinant DNA techniques, which enable the exchange of genetic material between different microorganisms, leading to the creation of genetically modified microorganisms (GMMs) or genetically engineered microorganisms (GEMs).

In 1981, the USA patented two genetically modified bacterial strains, *Pseudomonas aeruginosa* (NRRLB-5472) and *Pseudomonas putida* (NRRLB-5473), which were initially developed in the early 1970s by Chakrabarty. These strains were engineered to possess genes responsible for the degradation of highly toxic compounds such as camphor, naphthalene, and salicylate, making them efficient at toxin degradation. Additionally, the genes responsible for polychlorinated biphenyl (PCB) degradation, initially located in *R. eutropha* A5, *A. denitrificans* JB1, and

Achromobacter sp. LBS1C1, were successfully transferred to the heavy metal-resistant strain *R. eutropha* CH34 through natural conjugation, enabling the removal of PCBs. Moreover, to address heavy metal contamination, an *Arabidopsis thaliana* gene was introduced into *Mesorhizobium huakuii* subsp. *Rengei* strain B3 to produce phytochelatin synthase (PCS), fostering a symbiotic relationship between *M. huakuii* subsp. *Rengei* strain B3 and *Astragalus sinicus*. This allowed for the accumulation of Cd^{2+} and the release of phytochelatins, thanks to a bacteroid-specific promoter (the nif H gene). Furthermore, genetic engineering has been employed for heavy metal degradation in various environmental contexts, including chromium degradation from industrial wastewater using *Alcaligenes eutrophus* AE104 (pEBZ141) and the construction of a recombinant photosynthetic bacterium, *Rhodopseudomonas palustris*, for the simultaneous expression of the Hg transport system and metallothionein, enabling Hg^{2+} degradation in heavy metal-contaminated wastewater.

In practical applications, GEMs have proven highly effective in bioremediation efforts targeting soil, activated sludge, and groundwater. By harnessing genetic engineering and plasmid breeding, scientists can efficiently develop microorganisms endowed with catalytic capabilities for the removal of pollutants (Sayler and Ripp 2000; Ripp et al. 2000). The field of genetic engineering holds great promise for the advancement of bioremediation, a sentiment strongly supported by microbiologists and molecular biologists, as exemplified in the information presented in Table 5.2 (Verma and Kuila 2019). Genome editing of plants and microorganisms using CRISPR-Cas technology has emerged as a powerful tool for highly efficient bioremediation of heavy metals (Islam 2019).

5.10 Problems That Are Faced by Genetically Engineered Microbes in Bioremediation

Genetic engineering has advanced bioremediation efforts, but practical challenges persist. Engineered strains may not be the primary contributors to bioremediation, as natural microorganisms like *Rhodococcus* and *Pseudomonas* often outperform them in specific conditions. Translating lab successes to real-world applications remains difficult. To address this, "Suicidal genetically engineered microorganisms" (S-GEMS) have been used to enhance safety and efficacy in polluted sites. *Pseudomonas fluorescens* HK44 has been field-tested, but ecological risks still accompany genetically engineered microorganisms' release. Harsh environmental factors pose a major challenge for engineered microorganisms. Bacteria like *E. coli, Bacillus subtilis,* and *P. putida* are utilized in field conditions, focusing on molecular properties. Adaptations are necessary to address real-world challenges. Ensuring environmental safety for field release is paramount. Engineered bacteria must exhibit resilience and avoid horizontal gene transfer when introduced into harsh environments to prevent impacting native microflora. Engineered bacteria are often tailored for in vitro bioremediation to sidestep field complexities. The deliberate release

Table 5.2 GEMs that are used in heavy metal bioremediation

Microbes	Modified gene expression	Heavy metals	References
Sphingomonas desiccabilis and *Bacillus Idriensis strains*	Ars M gene over expression	Arsenic	Liu et al. (2011)
B. subtilis BR151 (pTOO24)	Luminescent Cadmium sensors	Cadmium	Ivask et al. (2011)
Methylococcus capsulatus (Bath)	CrR genes for Cr (VI) reductase activity	Chromium (VI)	Hasin et al. (2010)
Caulobacter crescentus JS4022/p723-6H	RsaA-6His fusion protein	Cadmium (II)	Patel et al. (2010)
Pseudomonas strain K-62	MerE protein encoded by transposon Tn21	Mercury	Kiyono et al. (2009)
Achromobacter sp. AO22	Mercury reductase expressing mer gene	Mercury	Ng et al. (2009)
E. coli strain	Metallothionein	Arsenic	Singh et al. (2008)
E. coli strain	AsIII S-adenosylmethionine methyltransferase gene	Arsenic	Yuan et al. (2008)
Pseudomonas fluorescens OS8; *Escherichia coli* MC1061; *Bacillus subtilis* BR151; *Staphylococcus aureus* RN4220	MerR/CadC/ZntR/Pmer/PcadA/PzntA	Cadmium, lead, mercury, zinc	Bondarenko et al. (2008)
E. coli strain	PCS gene expression (SpPCS)	Cd^{2+}	Kang et al. (2007)
E. coli JM109	Cadmium transport system and metallothionein	Cadmium	Deng et al. (2007)
P. putida 06909	Expression of metal binding peptide EC20	Cadmium	Wu et al. (2006)
Pseudomonas K-62	Expression of mercury transport system and Organomurcuriallyase	Mercury	Kiyono and Pan-Hou (2006)
E. coli	SE5000 Nickel transport system and metallothionein	Nickel	Deng et al. (2005)
E. coli JM109	Hg^{2+} transporter and metallothionein	Mercury	Zhao et al. (2005)
E. coli strain	Over expression of Serin acetyl transferase	Nickel and cobalt	Freeman et al. (2005)
Acidithiobacillus ferrooxidans	Mercury ion transporter gene expression	Mercury	Sasaki et al. (2005)

(continued)

Table 5.2 (continued)

Microbes	Modified gene expression	Heavy metals	References
E. coli	Metalloregulatory protein ArsR (over expressing ELP153AR)	Arsenic	Kostal et al. (2004)
Escherichia coli and Moraxella sp.	Expressing EC20	Mercury and cadmium	Bae et al. (2003)
Mesorhizobium huakuii B3	Phytochelatin synthase (PCS) gene expression	Cd^{2+}	Sriprang et al. (2003)
E. coli strain	Organomurcuriallyase gene expression	Mercury	Murtaza et al. (2002)
P. fluorescens 4F39	Nickel transport system	Nickel	Lopez et al. (2002)
Deinococcus radiodurans	Hg (II) resistance gene (merA)	Mercury (Radioactive waste from nuclear weapons)	Brim et al. (2000)

of engineered bacteria raises concerns about adverse effects on native microorganisms, making the survival and behavior of genetically engineered bacteria in harsh environments a significant field concern.

5.11 Conclusion and Future Perspectives

Microbial remediation stands out as a promising solution for combating soil pollution, driven by advancements in genetic engineering and the formulation of microbial consortia. Unlike complex and environmentally burdensome traditional methods, bioremediation, leveraging microorganisms, provides a quicker and more eco-friendly approach. Genetic engineering plays a pivotal role in modifying microorganisms, enhancing their capacity to detoxify heavy metals and thereby increasing the precision and efficiency of bioremediation. This makes it a potent tool for remediating contaminated sites. Moreover, biological processes, especially in in situ applications, offer greater accuracy and specificity, reducing the necessity for soil excavation and transportation, ultimately minimizing waste and industrial pollution. To fully exploit the potential of bioremediation, the focus should center on developing genetically modified microbes (GEMs) using CRISPR-Cas genome editing capable of adapting to the harshest conditions, particularly in heavy metal-contaminated areas. This adaptability broadens the scope of contaminants that can be effectively treated, rendering bioremediation a versatile solution. However, the environmental safety of GEMs remains a concern, necessitating comprehensive studies and rigorous safety protocols to mitigate potential risks to ecosystems. An equally exciting avenue is the creation of microbial consortia—communities comprising diverse microbial species that collaborate to remediate soil. By assembling consortia with complementary metal tolerance and remediation abilities, a broader range of heavy metal pollutants can be effectively addressed. Understanding the intricate interactions of these microbial species in soil is crucial for optimizing bioremediation and tailoring solutions to specific contamination scenarios. In conclusion, microbial remediation, propelled by genetic engineering and microbial consortia, offers an eco-friendly, efficient, and versatile approach to tackle soil pollution. Ongoing research and stringent safety measures are imperative to fully realize the potential of this innovative technology while ensuring its environmental safety.

References

Abd El-Motaleb M, El-Sabbagh S, Mohamed W, Wafy K (2020) Biosorption of Cu^{2+}, Pb^{2+} and Cd^{2+} from wastewater by dead biomass of Streptomyces cyaneus Kw42. Int J Curr Microbiol Appl Sci 9:422–435

Abd-Alla MH, Morsy FM, El-Enany AWE, Ohyama T (2012) Isolation and characterization of a heavy-metal-resistant isolate of Rhizobium leguminosarum bv. viciae potentially applicable for biosorption of Cd^{2+} and Co^{2+}. Int Biodeterior Biodegrad 67:48–55

Abdu N (2011) Availability, transfer and balances of heavy metals in urban agriculture of West Africa. Kassel University Press GmbH

Abdu N, Abdulkadir A, Agbenin JO, Buerkert A (2011a) Vertical distribution of heavy metals in wastewater-irrigated vegetable garden soils of three West African cities. Nutr Cycl Agroecosyst 89:387–397

Abdu N, Agbenin JO, Buerkert A (2011b) Geochemical assessment, distribution, and dynamics of trace elements in urban agricultural soils under long-term wastewater irrigation in Kano, northern Nigeria. J Plant Nutr Soil Sci 174(3):447–458

Abdul RS, Fatima Q (2000) Cadmium resistant bacteria from industrial effluents and their role in environmental cleanup. Pak J Zool 32(2):165–178

Abo Alkasem MI, El-Abd MAEH, Maany DA, Ibrahim ASS (2023) Hexavalent chromium reduction by a potent novel Haloalkaliphilic Nesterenkonia sp strain NRC-Y isolated from hypersaline soda lakes. Egypt J Chem 66(6):463–473

Abo-Alkasem MI, Hassan NMH, Abo Elsoud MM (2023) Microbial bioremediation as a tool for the removal of heavy metals. Bull Natl Res Centre 47(1):1–15

Abo-Alkasem MI, Maany DA, El-Abd MA, Ibrahim AS (2022) Bioreduction of hexavalent chromium by a novel haloalkaliphilic Salipaludibacillus agaradhaerens strain NRC-R isolated from hypersaline soda lakes. 3 Biotech 12:1–15

Abo Elsoud MM, Abo-Alkasem MI (2022) Environmental sulfate-reducing microorganisms. Appl Microbes Environ Microb Biotechnol 625–654

Acar FN, Malkoc E (2004) The removal of chromium (VI) from aqueous solutions by Fagus orientalis L. Biores Technol 94(1):13–15

Afal A, Wiener SW (2014) Metal toxicity. Med Medscape 81(4):338–347

Ahemad M, Malik A (2011) Bioaccumulation of heavy metals by zinc resistant bacteria isolated from agricultural soils irrigated with wastewater. Bacteriol J 2(1):12–21

Akan JC, Abdulrahman FI, Sodipo OA, Ochanya AE, Askira YK (2010) Heavy metals in sediments from river Ngada, Maiduguri Metropolis, Borno State, Nigeria. J Environ Chem Ecotoxicol 2(9):131–140

Akhtar FZ, Archana KM, Krishnaswamy VG, Rajagopal R (2020) Remediation of heavy metals (Cr, Zn) using physical, chemical and biological methods: a novel approach. SN Appl Sci 2:1–14

Akhtar MS, Chali B, Azam T (2013) Bioremediation of arsenic and lead by plants and microbes from contaminated soil. Res Plant Sci 1(3):68–73

Alvarez A, Saez JM, Costa JSD, Colin VL, Fuentes MS, Cuozzo SA, Benimeli CS, Polti MA, Amoroso MJ (2017) Actinobacteria: current research and perspectives for bioremediation of pesticides and heavy metals. Chemosphere 166:41–62

Alves MB, Medeiros FCM, Sousa MH, Rubim JC, Suarez AZ (2014) Cadmium and tin magnetic nanocatalysts useful for biodiesel production. J Braz Chem Soc 25(12):2304–2313

Asha LP, Sandeep RS (2013) Review on bioremediation-potential tool for removing environmental pollution. Int J Basic Appl Chem Sci 3(3):21–33

Atlas RM, Philp J (2005) Bioremediation. Applied microbial solutions for real-world environmental cleanup. ASM Press

Bae W, Mehra RK, Mulchandani A, Chen W (2001) Genetic engineering of Escherichia coli for enhanced uptake and bioaccumulation of mercury. Appl Environ Microbiol 67(11):5335–5338

Bae W, Mulchandani A, Chen W (2002) Cell surface display of synthetic phytochelatins using ice nucleation protein for enhanced heavy metal bioaccumulation. J Inorg Biochem 88(2):223–227

Bae W, Wu CH, Kostal J, Mulchandani A, Chen W (2003) Enhanced mercury biosorption by bacterial cells with surface-displayed MerR. Appl Environ Microbiol 69(6):3176–3180

Bandowe BAM, Bigalke M, Boamah L, Nyarko E, Saalia FK, Wilcke W (2014) Polycyclic aromatic compounds (PAHs and oxygenated PAHs) and trace metals in fish species from Ghana (West Africa): bioaccumulation and health risk assessment. Environ Int 65:135–146

Bano A, Hussain J, Akbar A, Mehmood K, Anwar M, Hasni MS, Ullah S, Sajid S, Ali I (2018) Biosorption of heavy metals by obligate halophilic fungi. Chemosphere 199:218–222

Bharagava RN, Mishra S (2018) Hexavalent chromium reduction potential of Cellulosimicrobium sp. isolated from common effluent treatment plant of tannery industries. Ecotoxicol Environ Saf 147:102–109

Bhattacharya A, Gupta A (2013) Evaluation of Acinetobacter sp. B9 for Cr (VI) resistance and detoxification with potential application in bioremediation of heavymetals-rich industrial wastewater. Environ Sci Pollut Res 20:6628–6637

Bissonnette L, St-Arnaud M, Labrecque M (2010) Phytoextraction of heavy metals by two Salicaceae clones in symbiosis with arbuscular mycorrhizal fungi during the second year of a field trial. Plant Soil 332:55–67

Bojórquez C, Frías Espericueta MG, Voltolina D (2016) Removal of cadmium and lead by adapted strains of Pseudomonas aeruginosa and Enterobacter cloacae. Revista Internacional de Contaminación Ambiental 32(4):407–412

Bondarenko O, Rolova T, Kahru A, Ivask A (2008) Bioavailability of Cd, Zn and Hg in soil to nine recombinant luminescent metal sensor bacteria. Sensors 8:6899–6923

Bong CW, Malfatti F, Azam F, Obayashi Y, Suzuki S (2010) The effect of zinc exposure on the bacteria abundance and proteolytic activity in seawater. Interdisciplinary studies on environmental chemistry—biological responses to contaminants. Terrapub, Tokyo, pp 57–63

Brar SK, Verma M, Surampalli RY, Misra K, Tyagi RD, Meunier N, Blais JF (2006) Bioremediation of hazardous wastes—a review. Pract Period Hazard Toxic Radioact Waste Manag 10(2):59–72

Brim H, McFarlan SC, Fredrickson JK, Minton KW, Zhai M, Wackett LP, Daly MJ (2000) Engineering Deinococcus radiodurans for metal remediation in radioactive mixed waste environments. Nat Biotechnol 18(1):85–90

Brown GK, MacCarthy, Leenheer JA (1999) Simultaneous determination of Ca, Cu, Ni, Zn and Cd binding strengths with fulvic acid fractions by Schubert's method. Anal Chim Acta 402(1–2):169–181

Brunetti G, Farrag K, Soler-Rovira F, Nigro F, Senesi N (2012) The effect of compost and Bacillus licheniformis on the phytoextraction of Cr, Cu, Pb and Zn by three brassicaceae species from contaminated soils in the Apulia region, Southern Italy. Geoderma 170:322–330

Cervantes C, Campos-García J, Devars S, Gutiérrez-Corona F, Loza-Tavera H, Torres-Guzmán JC, Moreno-Sánchez R (2001) Interactions of chromium with microorganisms and plants. FEMS Microbiol Rev 25(3):335–347

Chander KPCB, Brookes C, Harding SA (1995) Microbial biomass dynamics following addition of metal-enriched sewage sludges to a sandy loam. Soil Biol Biochem 27(11):1409–1421
Chang YJ, Peacock AD, Long E, Stephen JR, McKinley JP, Macnaughton SJ, Hussain AA, Saxton AM, White DC (2001) Diversity and characterization of sulfate-reducing bacteria in groundwater at a uranium mill tailings site. Appl Environ Microbiol 67(7):3149–3160
Chanmugathas, Bollag JM (1988) A column study of the biological mobilization and speciation of cadmium in soil. Arch Environ Contam Toxicol 17:229–237
Chao S, LiQin J, WenJun Z (2014) A review on heavy metal contamination in the soil worldwide: situation, impact and remediation techniques. Environ Skept Crit 3(2):24–38
Chen TB, Wong JWC, Zhou HY, Wong MH (1997) Assessment of trace metal distribution and contamination in surface soils of Hong Kong. Environ Pollut 96(1):61–68
Cheng S (2003) Heavy metal pollution in China: origin, pattern and control. Environ Sci Pollut Res 10:192–198
Chibuike GU, Obiora SC (2014) Heavy metal polluted soils: effect on plants and bioremediation methods. Appl Environ Soil Sci
Choinska-Pulit A, Sobolczyk-Bednarek J, Łaba W (2018) Optimization of copper, lead and cadmium biosorption onto newly isolated bacterium using a Box-Behnken design. Ecotoxicol Environ Saf 149:275–283
Daneshvar E, Zarrinmehr MJ, Kousha M, Hashtjin AM, Saratale GD, Maiti A, Vithanage M, Bhatnagar A (2019) Hexavalent chromium removal from water by microalgal-based materials: adsorption, desorption and recovery studies. Bio Technol 293:122064
Dangi AK, Sharma B, Hill RT, Shukla (2019) Bioremediation through microbes: systems biology and metabolic engineering approach. Crit Rev Biotechnol 39(1):79–98
Danis U, Nuhoglu A, Demirbas A (2008) Ferrous ion-oxidizing in Thiobacillus ferrooxidans batch cultures: influence of pH, temperature and initial concentration of Fe^{2+}. Fresenius Environ Bull 17(3):371–377
Das D, Chakraborty S, Bhattacharjee C, Chowdhury R (2016) Biosorption of lead ions (Pb^{2+}) from simulated wastewater using residual biomass of microalgae. Desalin Water Treat 57(10):4576–4586
de Alencar FLS, Navoni JA, do Amaral VS (2017) The use of bacterial bioremediation of metals in aquatic environments in the twenty-first century: a systematic review. Environ Sci Pollut Res 24:16545–16559
Deng X, Li QB, Lu YH, Sun DH, He N (2005) Genetic engineering of Escherichia coli SE5000 and its potential for Ni^{2+} bioremediation. Proc Biochem 40:425–430
Deng X, Yi XE, Liu G (2007) Cadmium removal from aqueous solution by gene modified Escherichia coli JM109. J Hazard Mater 139:340–344
Dhankhar R, Hooda A (2011) Fungal biosorption—an alternative to meet the challenges of heavy metal pollution in aqueous solutions. Environ Technol 32(5):467–491
Dixit R, Wasiullah X, Malaviya D, Pandiyan K, Singh UB, Sahu A, Shukla R, Singh BP, Rai JP, Sharma K, Lade H (2015) Bioremediation of heavy metals from soil and aquatic environment: an overview of principles and criteria of fundamental processes. Sustainability 7(2):2189–2212
Elouzi AA, Akansha AA, Elgerbi AM, El-BaseirEl Gammudi BA M (2012) Removal of heavy metal contamination by bio-surfactants (rhamnolipids). J Chem Pharmaceut Res 4:4337–4341
EPA (2003) EPA protocol for the review of existing national primary drinking water regulations, united states environmental protection agency office of water office of ground water and drinking water standards and risk management division targeting and analysis branch1200 Pennsylvania avenue, NW (4607M) Washington, DC 20460
Fang L, Zhou C, Cai C, Rong X, Dai K, Liang W, Gu JD, Huang Q (2011) Binding characteristics of copper and cadmium by cyanobacterium Spirulina platensis. J Hazard Mater 190(1–3):810–815
Febrianto J, Kosasih AN, Sunarso J, Ju YH, Indraswati N, Ismadji S (2009) Equilibrium and kinetic studies in adsorption of heavy metals using biosorbent: a summary of recent studies. J Hazard Mater 162(2–3):616–645

Feng D, Aldrich C (2004) Adsorption of heavy metals by biomaterials derived from the marine alga Ecklonia maxima. Hydrometallurgy 73(1–2):1–10

Freeman JL, Persans MW, Nieman K, Salt DE (2005) Nickel and cobalt resistance engineered in Escherichia coli by overexpression of serine acetyltransferase from the nickel hyperaccumulator plant Thlaspi goesingense. Appl Environ Microbiol 71:8627–8633

Galiulin RV, Galiulina RA (2008) Removing heavy metals from soil with plants. Her Russ Acad Sci 78(2):141–143

Gavrilescu M (2004) Removal of heavy metals from the environment by biosorption. Eng Life Sci 4(3):219–232

Gomaa EZ, El-Meihy RM (2019) Bacterial biosurfactant from Citrobacter freundii MG812314.1 as a bioremoval tool of heavy metals from wastewater. Bull Natl Res Cent 43:69

Gomez F, Sartaj M (2014) Optimization of field scale biopiles for bioremediation of petroleum hydrocarbon contaminated soil at low temperature conditions by response surface methodology (RSM). Int Biodeterior Biodegrad 89:103–109

Gonzalez CJF, Rodríguez Perez AS, Vargas Morales JM, Martínez Juarez VM, Rodríguez IA, Cuello CM, Fonseca GG, Escalera Chavez ME, Munoz Morales A (2019) Bioremoval of cobalt (II) from aqueous solution by three different and resistant fungal biomasses. Bioinorgan Chem Appl 1–8

Gorby YA, Caccavo F, Bolton H (1998) Microbial reduction of cobaltIIIEDTA-in the presence and absence of manganese (IV) oxide. Environ Sci Technol 32(2):244–250

Goyal N, Jain SC, Banerjee UC (2003) Comparative studies on the microbial adsorption of heavy metals. Adv Environ Res 7(2):311–319

Guiné V, Spadini L, Sarret G, Muris M, Delolme C, Gaudet JP, Martins JM (2006) Zinc sorption to three gram-negative bacteria: combined titration, modeling, and EXAFS study. Environ Sci Technol 40(6):1806–1813

Gupta A, Joia J, Sood A, Sood R, Sidhu C, Kaur G (2016) Microbes as potential tool for remediation of heavy metals: a review. J Microb Biochem Technol 8(4):364–372

Hamdan AM, Abd-El-Mageed H, Ghanem N (2021) Biological treatment of hazardous heavy metals by Streptomyces rochei ANH for sustainable water management in agriculture. Sci Rep 11:9314

Hasin AA, Gurman SJ, Murphy LM, Perry A, Smith TJ, Gardiner E (2010) Remediation of chromium (VI) by a methane-oxidizing bacterium. Environ Sci Tech 44:400–405

He Y, Gong Y, Su Y, Zhang Y, Zhou X (2019) Bioremediation of Cr (VI) contaminated groundwater by Geobacter sulfurreducens: environmental factors and electron transfer flow studies. Chemosphere 221:793–801

Hisham NHB, Ibrahim MF, Ramil N, Abd-Aziz S (2019) Production of biosurfactant produced from used cooking oil by Bacillus sp. HIP3 for heavy metals removal. Molecules 24:2617

Hobson AM, Frederickson J, Dise NB (2005) CH_4 and N_2O from mechanically turned windrow and vermicomposting systems following in-vessel pre-treatment. Waste Manag 25(4):345–352

Hu N, Luo Y, Song J, Wu L, Zhang H (2010) Influences of soil organic matter, pH and temperature on Pb sorption by four soils in Yangtze River Delta. Acta Pedol Sin 47(2):246–252

Huang W, Lin ZN (2013) Biosorption of Cd(II)/Pb(II) from aqueous solution by biosurfactant-producing bacteria: isotherm kinetic characteristic and mechanism. Colloids Surf B Biointerfaces 105:113

Hwang SK, Jho EH (2018) Heavy metal and sulfate removal from sulfate-rich synthetic mine drainages using sulfate reducing bacteria. Sci Total Environ 635:1308–1316

Hyman M, Dupont RR (2001) Groundwater and soil remediation: process design and cost estimating of proven technologies. American Society of Civil Engineers

Igiri BE, Okoduwa SI, Idoko GO, Akabuogu EP, Adeyi AO, Ejiogu IK (2018) Toxicity and bioremediation of heavy metals contaminated ecosystem from tannery wastewater: a review. J Toxicol 27:2018:2568038

Islam T, Fatema HMN, Gupta DR, Mahmud NU, Sakif TI, Sharpe AG (2023) Improvement of growth, yield and associated bacteriome of rice by the application of probiotic *Paraburkholderia* and *Delftia*. Front Microbiol 14:1212505. https://doi.org/10.3389/fmicb.2023.1212505

Ivask A, Dubourguier HC, Pollumaa L, Kahru A (2011) Bioavailability of cd in 110 polluted top soils to recombinant bioluminescent sensor bacteria: effect of soil particulate matter. J Soils Sediments 11:231–237

Jadia CD, Fulekar MH (2009) Phytoremediation of heavy metals: recent techniques. African J Biotechnol 8(6)

Jain AN, Udayashankara TH, Lokesh KS (2012) Review on bioremediation of heavy metals with microbial isolates and amendments on soil residue. Int J Sci Res 6:2319–7064

Jasrotia S, Kansal A, Mehra A (2017) Performance of aquatic plant species for phytoremediation of arsenic-contaminated water. Appl Water Sci 7:889–896

Jin Y, Luan Y, Ning Y, Wang L (2018) Effects and mechanisms of microbial remediation of heavy metals in soil: a critical review. Appl Sci 8(8):1336

Joner EJ, Leyval C (1997) Uptake of 109Cd by roots and hyphae of a Glomus mosseae/Trifolium subterraneum mycorrhiza from soil amended with high and low concentrations of cadmium. New Phytol 135(2):353–360

Joutey TN, Bahafid W, Sayel H, Ghachtouli EIN (2013) Biodegradation: involved microorganisms and genetically engineered microorganisms. INTECH 28:8–320

Juwarkar AA, Singh SK, Mudhoo A (2010) A comprehensive overview of elements in bioremediation. Rev Environ Sci Bio/technol 9:215–288

Kang SH, Singh S, Kim JY, Lee W, Mulchandani A, Chen W (2007) Bacteria metabolically engineered for enhanced phytochelatin production and cadmium accumulation. Appl Environ Microbiol 73:6317–6320

Keryanti K, Mulyono EWS (2021) Determination of optimum condition of lead (Pb) biosorption using dried biomass microalgae Aphanothece sp. Periodica Polytechnica Chem Eng 65(1):116–123

Khalid S, Shahid M, Niazi NK, Murtaza B, Bibi I, Dumat C (2017) A comparison of technologies for remediation of heavy metal contaminated soils. J Geochem Explor 182:247–268

Khoubestani SR, Mirghaffari N, Farhadian O (2015) Removal of three and hexavalentchromium from aqueous solutions using a microalgae biomass-derived biosorbent. Environ Prog Sustain Energy 34:949–956

Kim IH, Choi JH, Joo JO, Kim YK, Choi JW, Oh BK (2015) Development of a microbe-zeolite carrier for the effective elimination of heavy metals from seawater. J Microbiol Biotechnol 25(9):1542–1546

Kirkham MB (2006) Cadmium in plants on polluted soils: Effects of soil factors, hyperaccumulation, and amendments. Geoderma 137(1–2):19–32

Kiyono M, Pan-Hou H (2006) Genetic engineering of bacteria for environmental remediation of mercury. J Health Sci 52:199–204

Kiyono M, Sone Y, Nakamura R, Pan-Hou H, Sakabe K (2009) The Mer E protein encoded by transposon Tn21 is a broad mercury transporter in Escherichia coli. FEBS Lett 583:1127–1131

Kostal JRY, Wu CH, Mulchandani A, Chen W (2004) Enhanced arsenic accumulation in engineered bacterial cells expressing ArsR. Appl Environ Microbiol 70:4582–4587

Kulshreshtha A, Agrawal R, Barar M, Saxena S (2014) A review on bioremediation of heavy metals in contaminated water. IOSR J Environ Sci Toxicol Food Technol 8(7):44–50

Kumar Mishra G (2017) Microbes in heavy metal remediation: a review on current trends and patents. Recent Pat Biotechnol 11(3):188–196

Kumar A, Bisht BS, Joshi VD, Dhewa T (2011a) Review on bioremediation of polluted environment: a management tool. Int J Environ Sci 1(6):1079–1093

Kumar A, Chanderman A, Makolomakwa M, Perumal K, Singh S (2016) Microbial production of phytases for combating environmental phosphate pollution and other diverse applications. Crit Rev Environ Sci Technol 46(6):556–591

Kumar L, Bharadvaja N (2020) Microbial remediation of heavy metals. Microbial Bioremed Biodegrad 49–72

Kumar R, Bhatia D, Singh R, Rani S, Bishnoi NR (2011b) Sorption of heavy metals from electroplating effluent using immobilized biomass Trichoderma viride in a continuous packed-bed column. Int Biodeterior Biodegrad 65(8):1133–1139
Leong YK, Chang JS (2020) Bioremediation of heavy metals using microalgae: recent advances and mechanisms. Bioresour Technol 303:122886
Leon-Vaz A, Leon R, Giraldez I, Vega JM, Vigara J (2021) Impact of heavy metals in the microalga Chlorella sorokiniana and assessment of its potential use in cadmium bioremediation. Aquat Toxicol (NY) 239:105941
Li J, Yu H, Luan Y (2015) Meta-analysis of the copper, zinc, and cadmium absorption capacities of aquatic plants in heavy metal-polluted water. Int J Environ Res Public Health 12(12):14958–14973
Ling TY, Liew CF, Modingin A (2007) Optimization of oxidation pond efficiency in animal farm wastewater treatment. Universiti Malaysia Sarawak
Liu S, Zhang F, Chen J, Sun GX (2011) Arsenic removal from contaminated soil via biovolatilization by genetically engineered bacteria under laboratory conditions. J Environ Sci 23:1544–1550
Lloyd JR, Sole VA, Van Praagh CVG, Lovley DR (2000) Direct and Fe (II)-mediated reduction of technetium by Fe (III)-reducing bacteria. Appl Environ Microbiol 66(9):3743–3749
Lloyd JR, Mabbett AN, Williams DR, Macaskie LE (2001) Metal reduction by sulphate-reducing bacteria: physiological diversity and metal specificity. Hydrometallurgy 59(2–3):327–337
Lopez A, Lazaro N, Morales S, Margues AM (2002) Nickel biosorption by free and immobilized cells of pseudomonas fluorescens 4F39: a comparative study. Water Air Soil Pollut 135:157–172
Lovley DR, Phillips EJ (1988) Novel mode of microbial energy metabolism: organic carbon oxidation coupled to dissimilatory reduction of iron or manganese. Appl Environ Microbiol 54(6):1472–1480
Ma Y, Oliveira RS, Freitas H, Zhang C (2016) Biochemical and molecular mechanisms of plant-microbe-metal interactions: relevance for phytoremediation. Front Plant Sci 7:918
Malidareh HB, Mahvi AH, Yunesian M, Alimohammadi M, Nazmara S (2014) Effect of fertilizer application on paddy soil heavy metals concentration and groundwater in North of Iran. Middle-East J Sci Res 20(12):1721–1727
Manoj SR, Karthik C, Kadirvelu K, Arulselvi I, Shanmugasundaram T, Bruno B, Rajkumar M (2020) Understanding the molecular mechanisms for the enhanced phytoremediation of heavy metals through plant growth promoting rhizobacteria: a review. J Environ Manag 254:109779
Marchenko AM, Pshinko GN, Demchenko VY, Goncharuk VV (2015) Leaching heavy metal from deposits of heavy metals with bacteria oxidizing elemental sulphur. J Water Chem Technol 37:311–316
Mathew AM (2005) Phytoremediation of heavy metal contaminated soil. Oklahoma State University
McLaughlin MJ, Zarcinas BA, Stevens DP, Cook N (2000) Soil testing for heavy metals. Commun Soil Sci Plant Anal 31(11–14):1661–1700
Mihopoulos G, Sayles GD, Suidan MT, Shah J, Bishop DF (2000) Vapor phase treatment of PCE in a soil column by lab-scale anaerobic bioventing. Water Res 34(12):3231–3237
Mihopoulos G, Suidan MT, Sayles GD, Kaskassian S (2002) Numerical modeling of oxygen exclusion experiments of anaerobic bioventing. J Contam Hydrol 58(3–4):209–220
Mingho Y (2005) Environmental Toxicology: biological and health effects of pollutants, Chap 12, 2nd edn. CRC Press LLC, BocaRaton. ISBN 1-56670-670-2
Mishra A, Malik A (2013) Recent advances in microbial metal bioaccumulation. Crit Rev Environ Sci Technol 43(11):1162–1222
Morales-Barrera L, Cristiani-Urbina E (2008) Hexavalent chromium removal by a Trichoderma hamatum fungal strain isolated from tannery effluent. Water Air Soil Pollut 187:327–336
Mulligan CN, Yong RN, Gibbs BF (2001) Heavy metal removal from sediments by biosurfactants. J Hazard Mater 85:111–125
Murtaza I, Dutt A, Ali A (2002) Biomolecular engineering of Esherichia coli organomercuriallyase gene and its expression. Ind J Biotech 1:117–120

Mwandira W, Nakashima K, Kawasaki S, Arabelo A, Banda K, Nyambe I, Chirwa M, Ito M, Sato T, Igarashi T, Nakata H (2020) Biosorption of Pb (II) and Zn (II) from aqueous solution by Oceanobacillus profundus isolated from an abandoned mine. Sci Rep 10(1):21189

Naidu R, Kookana RS, Sumner ME, Harter RD, Tiller KG (1997) Cadmium sorption and transport in variable charge soils: a review. J Environ Qual 26(3):602–617

Nayak AK, Panda SS, Basu A, Dhal NK (2018) Enhancement of toxic Cr (VI), Fe, and other heavy metals phytoremediation by the synergistic combination of native Bacillus cereus strain and Vetiveria zizanioides L. Int Phytorem 20(7):682–691

Ng SP, Davis B, Polombo EA, Bhave M (2009) Tn 5051 like mer containing transposon identified in a heavy metal tolerant strain Achromobacter sp. AO22. BMC Res Notes 7:2–38

Niu ZX, Sun LN, Sun TH (2011) The bioadsorption of cadmium and lead by bacteria in root exudates culture. Soil Sedim Contamin: Int J 20(8):877–891

Njoku KL, Asunmo MO, Ude EO, Adesuyi AA, Oyelami AO (2020) The molecular study of microbial and functional diversity of resistant microbes in heavy metal contaminated soil. Environ Technol Innov 17:100606

Orji OU, Awoke JN, Aja M, Aloke C, Obasi OD, Alum EU, Udu-Ibiam OE, Oka GO (2021) Halotolerant and metalotolerant bacteria strains with heavy metals biorestoration possibilities isolated from Uburu Salt Lake, Southeastern, Nigeria. Heliyon 7:e07512

Pandey G, Paul D, Jain RK (2005) Conceptualizing suicidal genetically engineered microorganisms for bioremediation applications. Biochem Biophy Res Commun 327:637–639

Patel J, Zhang Q, Michael R, McKay L, Vincent R, Xu Z (2010) Genetic engineering of Caulobacter crescentus for removal of cadmium from water. Appl Biochem Biotechnol 160:232–243

Philp JC, Atlas RM (2005) Bioremediation of contaminated soils and aquifers. In: Atlas RM, Philp JC, (eds) Bioremediation: applied microbial solutions for real-world environmental cleanup. American Society for Microbiology (ASM) Press, Washington, pp 139–236

Raghunandan K, Mchunu S, Kumar A, Kumar KS, Govender A, Permaul K, Singh S (2014) Biodegradation of glycerol using bacterial isolates from soil under aerobic conditions. J Environ Sci Health Part A 49(1):85–92

Raghunandan K, Kumar A, Kumar S, Permaul K, Singh S (2018) Production of Gellan gum, an exopolysaccharide, from biodiesel-derived waste glycerol by Sphingomonas spp. 3 Biotech 8:1–13

Rahman MS, Sathasivam KV (2015) Heavy metal adsorption onto Kappaphycus sp. from aqueous solutions: the use of error functions for validation of isotherm and kinetics models. BioMed Res Int

Rai PK, Kim KH, Lee SS, Lee JH (2020) Molecular mechanisms in phytoremediation of environmental contaminants and prospects of engineered transgenic plants/microbes. Sci Total Environ 705:135858

Rakib MRJ, Rahman MA, Onyena AP, Kumar R, Sarker A, Hossain MB, Islam ARMT, Islam MS, Rahman MM, Jolly YN, Idris AM (2022) A comprehensive review of heavy metal pollution in the coastal areas of Bangladesh: abundance, bioaccumulation, health implications, and challenges. Environ Sci Pollut Res 29(45):67532–67558

Raklami A, Meddich A, Oufdou K, Baslam M (2022) Plants—microorganisms-based bioremediation for heavy metal cleanup: recent developments, phytoremediation techniques, regulation mechanisms, and molecular responses. Int J Mol Sci 23(9):5031

Ramasamy K, Kamaludeen, Banu SP (2007) Bioremediation of metals: microbial processes and techniques. Environ Bioremed Technol 173–187

Ramos L, Hernandez LM, Gonzalez MJ (1994) Sequential fractionation of copper, lead, cadmium and zinc in soils from or near Donana National Park, vol 23, no 1. American Society of Agronomy, Crop Science Society of America, and Soil Science Society of America, pp 50–57

Ravindran A, Sajayan A, Priyadharshini GB, Selvin J, Kiran GS (2020) Revealingthe efficacy of thermostable biosurfactant in heavy metal bioremediation and surface treatment in vegetables. Front Microbiol 11:222

Rehan M, Alsohim AS (2019) Bioremediation of heavy metals. In: Environmental chemistry and recent pollution control approaches, pp 145–158
Rehman A, Farooq H, Shakoori AR (2007) Copper tolerant yeast, Candida tropicalis, isolated from industrial effluents: its potential use in wastewater treatment. Pak J Zool 39(6):405
Ripp S, Nivens DE, Ahn Y, Werner C, Jarrell J, Easter J, Burlage R, Sayler GS (2000) Controlled field release of a bioluminescent genetically engineered microorganism for bioremediation process monitoring and control. Environ Sci Tech 34:846–853
Riseh RS, Vazvani MG, Hajabdollahi N, Thakur VK (2023) Bioremediation of heavy metals by Rhizobacteria. Appl Biochem Biotechnol 195(8):4689–4711
Rodríguez-Tirado V, Green-Ruiz C, Gómez-Gil B (2012) Cu and Pb biosorption on Bacillus thioparans strain U3 in aqueous solution: Kinetic and equilibrium studies. Chem Eng J 181:352–359
Santos HF, Carmo FL, Paes JE, Rosado AS, Peixoto RS (2011) Bioremediation of mangroves impacted by petroleum. Water Air Soil Pollut 216:329–350
Saranya K, Sundaramanickam A, Shekhar S, Swaminathan S (2019) Biosorption of mercury by Bacillus thuringiensis (CASKS3) isolated from mangrove sediments of southeast coast India. Indian J Geo-Mar Sci 48:143–150
Sarker A, Kim JE, Islam ARMT, Bilal M, Rakib MRJ, Nandi R, Rahman MM, Islam T (2022) Heavy metals contamination and associated health risks in food webs—a review focuses on food safety and environmental sustainability in Bangladesh. Environ Sci Pollut Res 29(3):3230–3245
Sarker A, Masud MAA, Deepo DM, Das K, Nandi R, Ansary MWR, Islam ARMT, Islam T (2023) Biological and green remediation of heavy metal contaminated water and soils: a state-of-the-art review. Chemosphere 332:138861
Sarwar N, Imran M, Shaheen MR, Ishaque W, Kamran MA, Matloob A, Rehim A, Hussain S (2017) Phytoremediation strategies for soils contaminated with heavy metals: modifications and future perspectives. Chemosphere 171:710–721
Sasaki Y, Minakawa T, Miyazaki A, Silver S, Kusano T (2005) Functional dissection of a mercuric ion transporter Mer C from Acidithiobacillus ferrooxidans. Biosci Biotech Biochem 69:1394–1402
Sayler GS, Ripp S (2000) Field applications of genetically engineered microorganisms for bioremediation processes. Curr Opin Biotechnol 11:286–289
Servilia O, Foca N, Airinei A (2005) Effects of heavy metals on plant growth and photosynthetic activity. Not Bio Sci 107–110
Shah JK, Sayles GD, Suidan MT, Mihopoulos Kaskassian S (2001) Anaerobic bioventing of unsaturated zone contaminated with DDT and DNT. Water Sci Technol 43(2):35–42
Shakoori AR, Rehman A (2004) Multiple metal resistance in the ciliate protozoan, Vorticella microstoma, isolated from industrial effluents and its potential in bioremediation of toxic wastes. Bull Environ Contam Toxicol 72:1046–1051
Sharma I (2020) Bioremediation techniques for polluted environment: concept, advantages, limitations, and prospects. In: Trace metals in the environment-new approaches and recent advances. IntechOpen
Shen Y, Zhu W, Li H, Ho SH, Chen J, Xie Y, Shi X (2018) Enhancing cadmium bioremediation by a complex of water-hyacinth derived pellets immobilized with Chlorella sp. Bio Technol 257:157–163
Siddiquee S, Rovina K, Azad SA, Naher L, Suryani S, Chaikaew JJMBT (2015) Heavy metal contaminants removal from wastewater using the potential filamentous fungi biomass: a review. J Microb Biochem Technol 7(6):384–395
Singh N, Verma T, Gaur R (2013) Detoxification of hexavalent chromium by an indigenous facultative anaerobic Bacillus cereus strain isolated from tannery effluent. African J Biotechnol 12(10)
Singh S, Kang SH, Mulchandani A, Chen W (2008) Bioremediation: environmental cleanup through pathway engineering. Curr Opin Biotechnol 19:437–444

Sinha RK, Valani D, Sinha S, Singh S, Herat S (2009) Bioremediation of contaminated sites: a low-cost nature's biotechnology for environmental clean up by versatile microbes, plants & earthworms. Solid Waste Manag Environ Remed 978-1
Sodhi KK, Kumar M, Singh DK (2020) Multi-metal resistance and potential of Alcaligenes sp. MMA for the removal of heavy metals. SN Appl Sci 2:1–13
Spormann AM, Widdel F (2000) Metabolism of alkylbenzenes, alkanes, and other hydrocarbons in anaerobic bacteria. Biodegradation 11:85–105
Sposito G (2008) The chemistry of soils. Oxford University Press
Sreedevi R, Suresh K, Jiang G (2022) Bacterial bioremediation of heavy metals in wastewater: a review of processes and applications. J Water Process Eng 48:102884
Sriprang R, Hayashi M, Ono H, Takagi M, Hirata K, Murooka Y (2003) Enhanced accumulation of Cd^{2+} by a Mesorhizobium sp. transformed with a gene from Arabidopsis thaliana coding for phytochelatin synthase. Appl Environ Microbiol 69:1791–1796
Stefanescu IA (2015) Bioaccumulation of heavy metals by Bacillus megaterium from phosphogypsum waste. Sci Stud Res Chem Chem Eng Biotechnol Food Ind 16(1):93
Su C (2014) A review on heavy metal contamination in the soil worldwide: situation, impact and remediation techniques. Environ Skept Crit 3(2):24
Sultana N, Hossain SM, Mohammed ME, Irfan MF, Haq B, Faruque MO, Razzak SA, Hossain MM (2020) Experimental study and parameters optimization of microalgae based heavy metals removal process using a hybrid response surfacemethodology-crow search algorithm. Sci Rep 10:15068
Sun Y, Zhou S, Pan SY, Zhu S, Yu Y, Zheng H (2020) Performance evaluation and optimization of flocculation process for removing heavy metal. Chem Eng J 385:123911
Taiz L, Zeiger E (2002) Plant physiology, 3rd edn. Sinauer Associates, Massachusetts
Tang X, Huang Y, Li Y, Wang L, Pei X, Zhou D, He, Hughes SS (2021) Study on detoxification and removal mechanisms of hexavalent chromium by microorganisms. Ecotoxicol Environ Saf 208:111699
Tchounwou B, Yedjou CG, Patlolla AK, Sutton DJ (2012) Heavy metal toxicity and the environment. Molecular, clinical and environmental toxicology: volume 3: environmental toxicology, pp 133–164
Tebo BM, Obraztsova AY (1998) Sulfate-reducing bacterium grows with Cr (VI), U (VI), Mn (IV), and Fe (III) as electron acceptors. FEMS Microbiol Lett 162(1):193–199
Tigini V, Prigione V, Giansanti M, Pannocchia A, Varese GC (2010) Fungal biosorption, an innovative treatment for the decolourisation and detoxification of textile effluents. Water 2(3):550–565
Tofazzal Islam TI (2019) CRISPR-Cas technology in modifying food crops. Abstr CABI Rev 1–16. https://doi.org/10.1079/PAVSNNR201914050
Torres E (2020) Biosorption: a review of the latest advances. Processes 8(12):1–23
Trellu C, Mousset E, Pechaud Y, Huguenot D, van Hullebusch ED, Esposito G, Oturan MA (2016) Removal of hydrophobic organic pollutants from soil washing/flushing solutions: a critical review. J Hazard Mater 306:149–174
Tyagi S, Kumar V, Singh J, Teotia B, Sharma S (2014) Bioremediation of pulp and paper mill effluent by dominant aboriginal microbes and their consortium. Int J Environ Res 8(3):561–568
Van Assche F, Clijsters H (1990) Effects of metals on enzyme activity in plants. Plant Cell Environ 13(3):195–206
Vara Prasad MN, de Oliveira Freitas HM (2003) Metal hyperaccumulation in plants: biodiversity prospecting for phytoremediation technology. Electron J Biotechnol 6(3):285–321
Verma A (2021) Bioremediation techniques for soil pollution: an introduction. Biodegradation technology of organic and inorganic pollutants
Verma S, Kuila A (2019) Bioremediation of heavy metals by microbial process. Environ Technol Innov 14:100369
Vidali M (2001) Bioremediation. An overview. Pure Appl Chem 73(7):1163–1172
Volesky B, Holan ZR (1995) Biosorption of heavy metals. Biotechnol Prog 11(3):235–250

Whelan MJ, Coulon F, Hince G, Rayner J, McWatters R, Spedding T, Snape I (2015) Fate and transport of petroleum hydrocarbons in engineered biopiles in polar regions. Chemosphere 131:232–240

Wierzba S (2015) Biosorption of lead (II), zinc (II) and nickel (II) from industrial wastewater by Stenotrophomonas maltophilia and Bacillus subtilis. Pol J Chem Technol 17(1):79–87

Wu CH, Wood TK, Mulchandani A, Chen W (2006) Engineering plant-microbe symbiosis for rhizoremediation of heavy metals. Appl Environ Microbiol 72:1129–1134

Wuana RA, Okieimen FE (2011) Heavy metals in contaminated soils: a review of sources, chemistry, risks and best available strategies for remediation. Int Sch Res Not 2011

Yabe J, Ishizuka M, Umemura T (2010) Current levels of heavy metal pollution in Africa. J Vet Med Sci 72(10):1257–1263

Yang Z, Shi W, Yang W, Liang L, Yao W, Chai L, Gao S, Liao Q (2018) Combination of bioleaching by gross bacterial biosurfactants and flocculation: a potential remediation for the heavy metal contaminated soil. Chemosphere 206:83–91

Yuan CG, Lu XF, Qin J, Rosen BP, Le XC (2008) Volatile arsenic species released from Escherichia coli expressing the AsIII S-adenosyl methionine methyltransferase gene. Environ Sci Technol 42:3201–3206

Zaier H, Mudarra A, Kutscher D, De La Campa MF, Abdelly C, Sanz-Medel A (2010) Induced lead binding phytochelatins in Brassica juncea and Sesuvium portulacastrum investigated by orthogonal chromatography inductively coupled plasma-mass spectrometry and matrix assisted laser desorption ionization-time of flight-mass spectrometry. Anal Chim Acta 671(1–2):48–54

Zhao XW, Zhou MH, Li QB, Lu YH, He N, Sun DH, Deng X (2005) Simultaneous mercury bioaccumulation and cell propagation by genetically engineered Escherichia coli. Proc Biochem 40:1611–1616

Chapter 6
Role of Plants in Heavy Metals Remediation

Senad Murtić

Abstract Soil pollution by heavy metals has become one of the significant environmental concerns facing the world today. This not only reduces soil fertility but also affects human health through food chain. Thus, to safeguard soil resources, it is necessary to seek effective approaches for preventing soil degradation and restoring soils polluted by heavy metals. Phytoremediation is a very promising remediation technique that reduces or stabilizes heavy metals and some other soil pollutants using plants. There are five basic types of phytoremediation, each with a different mechanism for remediating heavy metal-polluted soils: phytoextraction, phytostabilization, phytovolatization, phytodegradation and phytofiltration. However, their successful application is usually limited due to slow and incomplete remediation by plants. Therefore, improving plant performance is a crucial factor for developing highly effective phytoremediation. The efficiency of phytoremediation can be improved through using organic or inorganic amendments, chelating and acidifying agents, using microbial assisted or transgenic plant-mediated phytoremediation and applying plant growth regulators. Besides, the combination of different phytoremediation approaches could offer new opportunities to improve the effectiveness of phytoremediation.

Keywords Health · Heavy metals · Phytoremediation · Plant performance · Soil pollution

6.1 Introduction

Heavy metals are generally known as environmental contaminants due to their toxicity and persistence in the environment. The toxicity of each heavy metal greatly depends on its chemical form and concentration. Other factors that influence heavy

S. Murtić (✉)
Department of Plant Physiology, Faculty of Agriculture and Food Science, University of Sarajevo, Sarajevo, Bosnia and Herzegovina
e-mail: murticsenad@hotmail.com

N. Kumar (ed.), *Heavy Metal Toxicity*, Environmental Science and Engineering,
https://doi.org/10.1007/978-3-031-56642-4_6

metal toxicity include the exposure route, period of exposure, genetic factors and health status of exposed individuals (Mitra et al. 2022).

Arsenic (As), cadmium (Cd), lead (Pb) and chromium (Cr) are among the heavy metals that are of high public health concern because they are highly toxic even at low concentrations. On the other hand, heavy metals such as iron (Fe), copper (Cu), manganese (Mn) and zinc (Zn) are recognized as essential elements for the maintenance of various physiological functions in plants, humans and animals; however, they are also toxic in excessive concentrations (Hejna et al. 2021).

Unfortunately in past decades, the rapid industrialization and intensive agriculture have led to heavy metals wide distribution in the environment especially in the soils which are mainly responsible for safe food production. Soil pollution by heavy metals not only reduces soil productivity but also influences human health through the food chain. In this light, soil pollution by heavy metals has become one of the significant environmental concerns facing the world today (Alsafran et al. 2023).

In order to maintain soil productivity, new remediation techniques have been continuously developed for preventing soil degradation and restoring soils polluted by heavy metals. Today, there are many different techniques for the removal of heavy metal from polluted soils. The main three techniques are physical, chemical and biological soil remediation techniques. The main factors that can influence the choice of remediation technique are: cost, implementation time, commercial availability, and public acceptability. Choice of remediation technique also strongly depends on type of environment and concentration of pollutant (Azubuike et al. 2016).

Physical remediation techniques include soil replacement, soil isolation, soil washing, aeration, electro-kinetic remediation and permeable barrier systems, while chemical remediation techniques include immobilization, chemical precipitation, solvent extraction, ion exchange, coagulation and flocculation. Despite their relatively high efficiency, most of these techniques are expensive and destructive to environment, and therefore not well accepted by the public (Liu et al. 2018).

Biological soil remediation techniques i.e. bioremediations refers to the removal of heavy metals and other pollutants from polluted soils using biological processes. These techniques are not harmful to the environment and retain the soil fertility during the remediation process. Also, they are safer and more cost-effective as compared to physical and chemical remediation techniques. On the other hand, these techniques have some disadvantages. Among those are the slow processing times, relatively low efficiency and non-biodegradability of heavy metals (Nnaji et al. 2023).

Based on the place where the biological process is implemented, biological soil remediation techniques can by divided into ex-situ and in-situ bioremediation techniques. Ex-situ bioremediation involves excavating affected soil from its original location to treat it, while in-situ bioremediation refers to treatment at the site of pollution. Choice of bioremediation treatments mainly depends on the cost of treatment, soil chemical and physical properties, and chemical form and concentration of pollutants. In general, in-situ bioremediation treatments are more cost-effective than the ex-situ bioremediation treatments (Alori et al. 2022).

Common ex situ soil bioremediation techniques include biopile treatment, land farming, composting i.e. solid-phase soil treatment and treatment with slurry containing bioremediators i.e. slurry phase soil treatment.

A biopile is an ex situ bioremediation technique in which excavated soils are mixed with different organic amendments (saw dust, straw, bark or wood chips) and placed onto a treatment area equipped with leachate collection systems and some form of aeration to optimize pollutant removal (Sharma 2020).

Land farming refers to the bioremediation technique in which excavated soils are transported to the land farming site and periodically turned over (tilled) to aerate the mixture (Chemlal et al. 2013). The aeration stimulates microbial activity to aerobically degrade or immobilize the pollutants in the soil. This technique can also be performed directly on the polluted site (in situ).

Composting refers to the use of bacteria and fungi in mature, cured compost to convert pollutants in the soil to humus and inert byproducts. Compost used in bioremediation is specially made to treat specific soil pollutants. Composting can bc implemented for both ex-situ and in-situ treatment; however, ex situ composting is usually more practicable and predictable as compared to in situ composting (Babalola et al. 2012).

Slurry phase soil bioremediation treatment refers to the controlled treatment of excavated soil in a bioreactor. In this type of treatment, the excavated polluted soil without stones and plant debris is mixed with water to a predetermined concentration dependent upon the soil chemical properties and pollutant concentration. This provides suitable conditions for the soil microorganisms to transform, immobilize or break down the pollutants contained in the soil (Shrivastava 2022).

In general, ex situ soil bioremediation techniques have a higher remediation potential as compared to in situ soil bioremediation techniques, but they also have lots of disadvantages including high costs, high energy consumption and potential negative impacts of soil transportation and disposal on the environment (Crawford 2006).

With in-situ soil bioremediation techniques, heavy metals and other pollutants are removed right at the site of pollution. Common in situ soil bioremediation techniques include bioventing, bioslurping, biosparging, bioaugmentation and phytoremediation (Bala et al. 2022).

Bioventing is an in situ remediation technique that stimulates existing soil microorganisms to remove or reduce pollutants in soil by supplying air or oxygen under pressure into the unsaturated soil zone (Anekwe and Isa 2021). Bioventing primarily stimulates the microbial degradation of polycyclic aromatic hydrocarbons such as crude oil products, but also assists in the microbial degradation of volatile organic pollutants because the movement of vapor is slow through biologically active soil (Chaudhary and Kim 2019).

Bioslurping involves the application of two remedial techniques bioventing and vacuum-enhanced extraction in order to remediate hydrocarbon-polluted soils. The role of bioventing is to stimulate the aerobic biodegradation of hydrocarbons in the unsaturated zone, while vacuum-enhanced extraction lifts groundwater pollutants from the area just above the saturated zone (Beretta et al. 2018).

Biosparging is an in-situ bioremediation technique that uses soil microorganisms to biodegrade organic pollutants in the soil by supplying air or oxygen under pressure into the saturated zone. This technique is very similar to bioventing (Gautam et al. 2022). However, in biosparging technique, the air is injected at the saturated zone, and therefore this technique is more applicable for reducing the concentration of hydrocarbons dissolved in groundwater or adsorbed to soil particles below the groundwater table (Yadav et al. 2021).

Bioaugmentation refers to the addition of exogenous microorganisms such as bacteria, yeasts or filamentous fungi into the soil to degrade specific soil and groundwater pollutants. Bioaugmentation can be divided into two main approaches: bioaugmentation by adding indigenous microorganisms; and bioaugmentation by adding nonindigenous microorganisms. Choice of bioaugmentation approaches primarily depends on the type and concentration of pollutants and soil physical and chemical properties (da Silva and Alvarez 2010).

Phytoremediation is a relatively recently developed remediation technique that reduces or stabilizes heavy metals and some other soil pollutants using plants. It can also be defined as plants' ability to remove, stabilize, degrade or immobilize toxic substances in the environment. Phytoremediation is one of the most widespread remediation techniques for removing heavy metals from polluted soils. It is an efficient, eco-friendly and non-invasive in situ remediation technique which doesn't require any special equipment for implementation, and therefore it is well accepted by the public. Besides, the operating cost of phytoremediation is much lower than that of other remediation techniques (Sarwar et al. 2017). However, phytoremediation of heavy metal polluted soils has some shortcomings. Among those are the incomplete metal removal i.e. relatively low remediation efficiency, slow processing time and environmental risks associated with disposal of biomass polluted by heavy metal(s). Namely, inappropriate disposal of heavy metal contaminated biomass can lead to leaching of heavy metals, causing secondary environmental pollution. Furthermore, phytoremediation techniques are usually used to remove or degrade only one pollutant in the soil, although the soils can be polluted by many different pollutants. In this case, the phytoremediation efficiency is limited. Also, the ecological perturbation caused by introduction of hyperaccumulators i.e. plants that can accumulate high concentration of heavy metal(s) in their above-ground parts from other environmental sites is a further issue that cannot be disregarded (Che-Castaldo and Inouye 2015).

This study investigates the role of plants in heavy metals remediation as well as tools and equipment required for phytoremediation. The study also investigates potential ways of removing heavy metals from polluted soil using different types of phytoremediation including phytoextraction, phytostabilization, phytofiltration, phytovolatilization and phytodegradation. Furthermore, the present study provides insights into the advantages and limitations of any type of phytoremediation used for the remediation of heavy metal polluted soils.

6.2 Factors Affecting Phytoremediation Efficiency

The main factors that affect the efficiency of phytoremediation in heavy metal polluted soil are shown in Fig. 6.1.

The characteristics of plants species is the key factor that affect the phytoremediation efficiency. Different plant species have different abilities to remove, degrade, or detoxify toxic heavy metal in the soil, and therefore, the appropriate selection of the plant species is crucial for effective phytoremediation.

Another crucial factor affecting phytoremediation is the bioavailability of heavy metals in the polluted soil. The bioavailability of heavy metals in soil is generally governed by their chemical speciation, which is deeply influenced by the soil properties including pH, texture, organic matter, redox potential, root zone activity and microbial activity (Nedjimi 2021).

Soil pH directly influences the heavy metal bioavailability in the soil. At low pH, heavy metals tend to be found as free ions in soil solution due to the higher competition between H^+ and free metal ions for binding on soil surface and therefore are more bioavailable, whereas at high pH heavy metals tend to form insoluble heavy

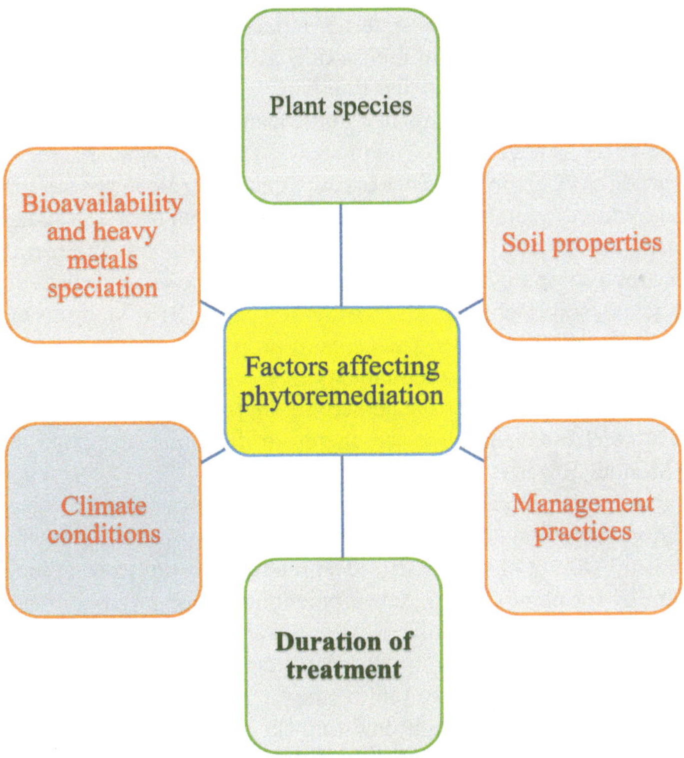

Fig. 6.1 Factors affecting phytoremediation of heavy metal(s) polluted soil

metal hydroxides and carbonates, resulting in lower bioavailability (Olaniran et al. 2013). Furthermore, heavy metal speciation in soil also significantly influence their bioavailability. The most mobile and thus more bioavailable heavy metals are Cd, Mo and Zn, while the least mobile are Pb, Ni and Cr (Prasad and Freitas 2003).

The texture as well as physical properties of soils also greatly affect the bioavailability of heavy metals in the soil. Soils high in clay exhibit higher heavy metal adsorption capacities owing to large surface area and permanent or variable charges, thus reducing the bioavailability of heavy metals in soils. Contrastingly, sandy soils usually exhibit low heavy metal holding capacity, which can increase heavy metals bioavailability and the leachability of heavy metals to lower soil layers (Angelaki et al. 2022).

Soil organic matter has a great influence on holding of heavy metals in the soil. Components of soil organic matter usually have negatively charged sites on the surface area that help retain positively charged heavy metal ions in the soil. In this point of view, increasing the level of organic matter in the soil helps to reduce the mobility of heavy metal ions in the soil and thus their availability to plants (Aigberua 2018).

Redox conditions is another factor that influences the heavy metal speciation and their bioavailability. Heavy metals such as As, Fe and Mn are more bioavailable and toxic in their lower oxidation state i.e. reduced form, whereas some of them are more bioavailable and toxic in their oxidized form e.g. Cr (Seraj and Rahman 2018). However, redox conditions and thus bioavailability of heavy metals are not uniform in soils as they are continually influenced by pH, organic matter content, concentration of oxygen and microbial processes (Sazykin et al. 2023).

Root zone activity i.e. root exudates play a significant role in phytoremediation of soil polluted by heavy metals. Root exudates exert effects on heavy metal bioavailability mainly by modifying the soil properties in the rhizosphere. In general, exudates i.e. organic carbon compounds released from roots into the soil increases the mobility of heavy metals in soil (Debela et al. 2010; Wang et al. 2016; Mahrous et al. 2019). However, the changes in the soil rhizosphere properties caused by root exudates significantly differ according to the plant species, indicating the choice of the plant species creates a variation at root zone activity that results in an increase/decrease in the heavy metal bioavailability and therefore determine the phytoremediation efficiency (Montiel-Rozas et al. 2016).

Microbial activities have a potent impact on phytoremediation. Microbes can enhance phytoremediation efficiency by increasing or decreasing heavy metal availability in the soil (Ma et al. 2016). Besides, microbes can release compounds that can be beneficial for plant growth, development and immunity and thus help them survive in heavy metal polluted soils (Alves et al. 2022).

Climate conditions, such as precipitation and temperature, can affect plant growth and consequently plants' ability to remove heavy metals from polluted soil (Priya et al. 2023). For instance, the high environment temperature strongly affects the transpiration rate of plants during phytoremediation treatment, which in turn, can greatly influence the uptake as well as translocation and accumulation of heavy metals in plant (Dodgen et al. 2015).

Appropriate management practices, such as applying organic amendments (i.e. compost) and microbial biostimulant applications (i.e. plant growth-promoting bacteria) can also improve phytoremediation efficiency (Visconti et al. 2023). These approaches improve phytoremediation efficiency mainly by stimulating the uptake and accumulation of heavy metals in plants and increasing the plant tolerance level to heavy metals in polluted soils.

6.3 Phytoremediation Types

There are five basic types of phytoremediation, each with a different mechanism for remediating heavy metal-polluted soils: (1) phytoextraction, (2) phytostabilization, (3) phytovolatization, (4) phytodegradation and (5) phytofiltration.

6.3.1 Phytoextraction

Phytoextraction or phytoaccumulation refers to the use of plants to take up heavy metals or other pollutants from soil through their uptake and accumulation into the aboveground biomass (Yan et al. 2020). It is clean, nature-friendly and cost-effective biological remediation approach and therefore is one of the most widely accepted and widespread strategies for removal heavy metal(s) from polluted soils (Petelka et al. 2019).

The general concept for the implementation of phytoextraction includes the appropriate selection of plant species, their cultivation on polluted soils, the harvesting of heavy metal-contaminated biomass and treating it to reduce its volume and mass. A key factor for a successful implementation of phytoextraction is the appropriate selection of plant species (Suman et al. 2018). In order to be suitable for phytoextraction purposes, plant species should have the following attributes: (i) ability to avoid the adverse effects of heavy metals in the polluted soil, (ii) high capacity to absorb heavy metals from soil and translocate them into the above-ground parts of plant (iii) high biomass production, fast growth and extensive root system, (iv) easy harvesting (v) good adaptation to different environmental conditions, (vi) high tolerance to pathogens and pests and (vii) non-edibility for animals (Seth 2012; Ali et al. 2013).

Verbruggen et al. (2009) reported that plants can be regarded as a suitable for phytoextraction, if they can concentrate at least 100 mg/kg of Cd, 1000 mg/kg of Cr, As, Co, Cu, Ni, Pb, Sb or 10,000 mg/kg of Mn or Zn (dry weight) in their aboveground parts when grown in heavy metal-rich environments. These plants are called heavy metal hyperaccumulator plants i.e. heavy metal hyperaccumulators. Currently, more than 450 plant species have been identified as heavy metal hyperaccumulators and most belong to the family Brassicaceae (Balafrej et al. 2020).

The ability of plant to absorb and accumulate heavy metals from polluted soil is strongly dependent on many plant factors including plant seasonality, genotype and their morphological, anatomical and physiological characteristics (Mondol et al. 2011). Yoon et al. (2006) reported that indigenous native heavy metal-tolerant plants should be preferred as phytoextraction tools because they have the ability to adapt to changes in local environmental conditions and could easily grow, develop and reproduce under the environmental extremes of that area compared to non-native plants.

However, regardless of plant genotype, it is important to note that the key requirement for removing heavy metals from soils by phytoextraction is the heavy metal(s) bioavailability. Namely, it is impossible to remove heavy metals from soils using any plants if the heavy metals in the soil exist as insoluble form which is not available for plant uptake. Accordingly, heavy metal speciation and the resulting bioavailability rather than total heavy metal concentration determines the environmental risk derived by heavy metals. In this light, clearly and explicitly incorporating heavy metal bioavailability into the risk assessment can certainly help to improve the phytoremediation management practices and thus phytoremediation efficiency.

In general, phytoremediation is a very promising, low-cost and nature-friendly approach for the remediation of heavy metal pollutes soils; however, this treatment is limited to the soil surface area and takes a long time to achieve success (Skuza et al. 2022).

6.3.2 Phytostabilization

Phytostabilization refers to the use of plants to immobilize heavy metal ions in the soil and prevent further dispersal. After phytostabilization process, heavy metals become less mobile and less bioavailable, thus preventing their migration into the ecosystem and reducing potential environmental risks of heavy metals (Gerhardt et al. 2017).

Immobilization of heavy metals by plants can be achieved through various mechanisms, including precipitation/complexation of heavy metals within the rhizosphere and accumulation of heavy metals in root tissues. Plant species suitable for phytostabilization accumulate heavy metals in the roots rather than in the above-ground parts of plant, thus avoiding the disadvantages associated with phytoextraction such as the disposal of polluted plant material (Peco et al. 2021).

Native plants that exhibit high tolerance to heavy metal polluted soils and produce a sizable root mass that can accumulate large quantities of heavy metal ions inside the root can be regarded as candidates for stabilization of potentially toxic heavy metals in the soil. Suitable plants for phytostabilization purposes should also have a low rate of translocation of heavy metals from roots to the above-ground parts of the plant (Burges et al. 2018).

Many scientific studies have shown that some plants such as *Acanthus ilicifolius* L., *Agrostis capillaris* L., *Brassica juncea* (L.) Czern., *Chenopodium album* L., *Juncus effusus* L., *Alternanthera philoxeroides* (Mart.) Griseb., *Artemisia princeps*

Pamp., *Bidens pilosa* L., *Cynodon dactylon* (L.) Pers., *Festuca rubra* L. *Arundo donax* L., *Digitaria sanguinalis* (L.) Scop., and *Erigeron canadensis* L. have great potential for the phytostabilization of heavy-metal polluted soils (Shackira and Puthur 2017; Nandillon et al. 2019; Cristaldi et al. 2020).

In general, phytostabilization is a phytoremediation approach that does not involve the removal of heavy metals from soil, they are still retained in the soil but in less mobile and available form. In this light, phytostabilization facilitates the preservation of soil health. Furthermore, phytostabilization is considerably less expensive than other remediation approaches including phytoextraction because it does not require the removal of heavy metal polluted biomass. In addition, phytostabilization can prevent soil erosion and reduce the spread of heavy metals from polluted areas and therefore is well accepted by public.

6.3.3 Phytovolatilization

Phytovolatilization refers to the uptake of pollutants by plant roots and their release into the atmosphere in a less toxic volatile form. Volatile compounds may be volatilized from stems or leaves through plant transpiration process or from soil as a result of plant root activities (Limmer and Burken 2016). Phytovolatilization is usually used for the remediation of soil polluted by organic pollutants and some heavy metals such as Hg, Se and As (Mahar et al. 2016; Muthusaravanan et al. 2018).

Plants that have both high evapotranspiration rate and the ability to transform pollutants into less toxic volatile forms before releasing them into the environment are considered good candidates for phytovolatilization.

Many scientific studies have suggested that some plants such as *Brassica napus* L. and *Brassica juncea* (L.) Czern. could be successful in the phytovolatilization of Se from soil (Bañuelos et al. 1993; Ghosh and Singh 2005). Bizily et al. (2000) also reported that *Arabidopsis thaliana* (L.) Heynh. has the ability to absorb Hg from soil, transform them into relatively less toxic volatile forms, and subsequently release them into the atmosphere through transpiration process.

The main advantage of phytovolatilization as compared to phytoextraction is that potential toxic metals As, Hg and Se can be removed from the polluted soil and dispersed as volatile compounds into the atmosphere, without any need for plant harvesting and subsequent disposal of biomass polluted by heavy metals. However, phytovolatilization does not remove the heavy metals completely, they still exist in the environment. Moreover, they may be redeposited to the pedosphere by precipitation. Thus, prior evaluation of possible environmental risk linked to the phytovolatilization is certainly required before its application in the field.

6.3.4 Phytodegradation

Phytodegradation refers to the breakdown of pollutants into non-toxic or less toxic substances by the action of enzymes released from the roots (rhyzodegradation) or by metabolic activities within plant tissues after the roots take up pollutants from the soil (direct phytodegradation). Phytodegradation is very similar to phytovolatilization since both remediation approaches aim to transform pollutants into non-toxic or less toxic substance; however, in the phytodegradation the transformed compounds are accumulated within plant tissues, whereas in the phytovolatilization they are released into the atmosphere in a volatile form.

Among plant species demonstrating a high phytodegradation potential are *Blumea malcolmii* Hook., *Elodea canadensis* Michx., *Datura innoxia* Mill., *Ipomoea carnea* Jacq. and *Phragmites australis* (Cav.) Trin. ex Steud. (Kagalkar et al. 2011; Jha et al. 2016; He et al. 2017; Singh et al. 2019).

Phytodegradation, also known as phytotransformation, can be applied to remediate soils polluted mainly by organic compounds. The successful implementation of phytodegradation implies that the transformed compounds have become non-toxic or significantly less toxic than the initial compounds in the biotransformation reaction (Casellas et al. 2013).

The main drawback of phytodegradation is its relatively low efficiency since this remediation approach usually requires more than one growing season to be efficient. Besides, transform compounds that accumulate within plant tissues pose a potential threat to human health via the food chain.

6.3.5 Phytofiltration

Phytofiltration, also known as rhyzofiltration, refers to the use of plants to remove pollutants from water and liquid waste, mainly through their root system. This technique is mainly used for industrial wastewater treatment primarily burdened by heavy metals (Hanus-Fajerska and Koźmińska 2016).

Plants with fibrous and rapid-growth root system and elevated tolerance to certain types of pollutants are well-suited for phytofiltration (Kanwar et al. 2020). Plants used for phytofiltration purposes are initially grown hydroponically and later transferred to polluted water after they have developed an extensive root area.

Unfortunately, only a few plant species can efficiently remove toxic heavy metals through phytofiltration. Some examples of plant species exploited in phytofiltration of heavy metals are *Eichhornia crassipes* (Mart.) Solms, *Pistia stratiotes* L. and *Eleocharis acicularis* (L.) Roem. and Schult. (Agunbiade et al. 2009; Sakakibara et al. 2011; Guarino et al. 2020). In this light, the selection of suitable plant material is the most important step to improve the phytofiltration strategy in the future.

In general, phytofiltration is a relatively inexpensive and environmentally friendly in-situ remediation techniques. This technique is exceptionally effective in implementations where large volumes of water with low heavy metal concentrations are involved.

6.4 Potential of Plant Species for Heavy Metal Remediation

Phytoremediation of heavy metal polluted soils primarily depends on the ability of plants to absorb, accumulate, transform, volatilize or stabilize heavy metals. Basically, there are two defense approaches adopted by plants to confront heavy metal toxicity in the polluted environment: avoidance and tolerance.

The avoidance approach refers to the ability of plants to reduce the uptake of heavy metals from the soil or to prevent the movement of heavy metal ions from the roots to the aerial parts of plants. This approach include several mechanisms such as (i) immobilization of metals by mycorrhizal association, (ii) vacuolar sequestration of heavy metals in root cell and (iii) complexation of heavy metals by organic acids exuded from plant roots. Some root exudates also increase the pH of the rhizosphere, leading to precipitation of heavy metals, thus limiting their bioavailability and consequently toxicity. Embedding the heavy metals in the root cell wall and heavy metal exclusion mechanism to prevent heavy metal uptake and root-to-shoot transport are also important mechanisms of heavy metal avoidance (Hall 2002; Memon and Schröder 2009).

The tolerance approach refers to the ability of plants to accumulate high concentrations of heavy metals in their aboveground parts without severe damage in plant growth and physiological processes (Jacobs et al. 2018). This is achieved through different mechanisms including heavy metal inactivation, intracellular complexation or chelation and heavy metal sequestration in various intracellular compartments in leaf cells (e.g., vacuole) (Memon et al. 2001; Pasricha et al. 2021).

In general, plant species have developed and adopted different mechanisms for their survival in the heavy metal-polluted soil, which can be used for phytoremediation purposes. Plants adopting avoidance mechanisms to cope with heavy metal toxicity are termed as excluders. These plants are good candidates for immobilization of heavy metal in the polluted soils. A list of some plant species that show a high ability to stabilize/immobilize potential toxic heavy metals in soil is given in Table 6.1.

Plants that can reduce heavy metal toxicity by accumulating large amounts of heavy metal ions in their above-ground parts in places far from the area of metabolic activity are termed hyperaccumulators. This adaptation allows plants to successfully remove heavy metals from polluted soil through phytoextraction. A list of some plant species that could be used for phytoextraction is given in Table 6.2.

Some plant species have developed and adopted other approaches to cope with heavy metal toxicity such as transformation of toxic heavy metals into non-toxic

Table 6.1 List of some plants that could be used for phytostabilization purposes

Plant species	Family	Heavy metal(s)	References
Acantus ilicifolius L.	Poaceae	Cd	Shackira and Puthur (2017)
Agrostis vulgaris With.	Poaceae	Pb	Nandillon et al. (2019)
Athyrium wardii (Hook.) Makino	Athyriaceae	Pb, Cd	Zhao et al. (2016)
Atriplex halimus L.	Amaranthaceae	Cd	Nedjimi and Daoud (2009)
Brassica juncea (L.) Czern.	Brassicaceae	Pb	Meyers et al. (2008)
Dactylis glomerata L.	Poaceae	Cd, Pb, Zn	Visconti et al. (2020)
Populus deltoides L.	Salicaceae	As	Hussain et al. (2017)
Dahlia pinnata Cav.	Asteraceae	As	Raza et al. (2019)
Ricinus communis L.	Euphorbiaceae	Pb	Kiran and Prasad (2021)
Typha latifolia L.	Typhaceae	Co, Cd, Cr, Ni	Varun et al. (2011)
Salix purpurea L.	Salicaceae	As	Simiele et al. (2020)
Lupinus albus L.	Fabaceae	Cd	Vázquez et al. (2006)
Quercus ilex L.	Fagaceae	Cd	Domínguez et al. (2009)
Lolium perenne L.	Poaceae	Cu	Santibáñez et al. (2008)
Jatropha curcas L.	Euphorbiaceae	Al, Zn, Cu	Wu et al. (2011)

or less toxic substances through phytovolatilization or phytodegradation; however, their implementation for phytoremediation purposes is not widespread.

6.5 Improving Phytoremediation Efficiency in Heavy Metal-Polluted Soils

Phytoremediation is a nature-friendly technique that offers a cost-effective solution to remediate soil polluted with toxic heavy metals. However, the field application of different phytoremediation approaches is still limited, mainly because of its low efficiency (Oh et al. 2013). The low efficiency of phytoremediation can be overcome by using several approaches including (i) microbial assisted phytoremediation, (ii) plant genetic engineering and (iii) application of various soil amendments (Kafle et al. 2022).

Numerous studies have shown that microbial-assisted phytoremediation is a very-promising strategy to improve phytoremediation efficiency (Glick 2010; Gerhardt et al. 2017). Moreover, plant–microbe interactions are pivotal for plants to tolerate heavy metal toxicity, and therefore understanding of this intricate relationship is invaluable for the reclamation of heavy metal polluted soils (Alves et al. 2022).

Microorganisms can tolerate heavy metal toxicity through various mechanisms including precipitation, chemical adsorption and complexation. As a consequence

Table 6.2 List of some plants that could be used for phytoextraction purposes

Plant species	Heavy metal	References
Arabis paniculata Franch., *Brassica nigra* (L.) K. Koch, *Brassica juncea* (L.) Czern., *Xanthium strumarium* L., *Sedum alfredii* Hance, *Thlaspi rotundifolium* (L.) Gaudin, *Thlaspi caerulescens* J. Presl and C. Presl, *Pelargonium* × *hortorum* L. H. Bailey, *Euphorbia cheiradenia* Boiss. and Hohen., *Chenopodium album* L., *Helianthus annuus* L., *Deschampsia cespitosa* (L.) P. Beauv., *Medicago sativa* L., *Panicum virgatum* L.	Pb	Lombi et al. (2000), Salido et al. (2003), Tang et al. (2009), Xie et al. (2009), Krämer (2010), Midhat et al. (2019), Ning et al. (2019), Hart et al. (2022)
Atriplex halimus L., *Arabis gemmifera* (Matsum.) Makino, *Brassica juncea* (L.) Czern., *Pelargonium* × *hortorum* L. H. Bailey, *Thlaspi caerulescens* J. Presl and C. Presl, *Nicotiana tabacum* L., *Arabis* paniculata Franch., *Xanthium strumarium* L., *Sedum alfredii* Hance, *Thlaspi praecox* Wulfen, *Azolla pinnata* R. Br., *Eleocharis acicularis* (L.) Roem. and Schult., *Phytolacca americana* L., *Deschampsia cespitosa* (L.) P. Beauv., *Salsola kali* L., *Prosopis laevigata* (Willd.) M. C. Johnst., *Arabidopsis halleri* (L.) O'Kane and Al-Shehbaz	Cd	Escarre et al. (2000), Cosio et al. (2004), Boucher et al. (2005), Wieshammer et al. (2007), Rai (2008), Meyer et al. (2009), Nedjimi and Daoud (2009), Tang et al. (2009), Xie et al. (2009), Kaur et al. (2018)
Alyssum bertolonii Desv., *Alyssum lesbiacum* (Candargy) Rech. f., *Alyssum murale* Waldst. and Kit., *Alyssum montanum* L., *Alyssum corsicum* Duby, *Alyssum chalcidicum* Janka, *Alyssum serpyllifolium* Desf., *Berkheya coddii* Roessler, *Xanthium strumarium* L., *Pycnandra acuminata* (Pierre ex Baill.) Swenson and Munzinger, *Isatis pinnatiloba* P. H. Davis, *Sebertia acuminata* Pierre ex Baill. *Phyllanthus serpentinus* S. Moore	Ni	Kerkeb and Krämer (2003), Becerra Castro et al. (2009), Mengoni et al. (2009), Altinözlü et al. (2012), Centofanti et al. (2012), Álvarez-López et al. (2016), Broadhurst and Chaney (2016), Jaffré et al. (2018), Mesjasz-Przybyłowicz and Przybyłowicz (2020), Risse et al. (2023)
Pteris vittata L., *Pteris biaurita* L., *Pteris cretica* L., *Pteris quadriaurita* Retz., *Pteris ryukyuensis* Tagawa, *Pteridium aquilinum* (L.) Kuhn, *Ludwigia erecta* (L.) H. Hara, *Corrigiola telephiifolia* Pourr	As	Zhao et al. (2002), Salido et al. (2003), Caille et al. (2004), Srivastava et al. (2006), Onyia et al. (2021)
Arabidopsis halleri (L.) O'Kane and Al-Shehbaz, *Arabis paniculata* Franch., *Arabis gemmifera* (Matsum.) Makino, *Thlaspi caerulescens* J. Presl and C. Presl, *Thlaspi ochroleucum* Boiss. and Heldr., *Thlaspi praecox* Wulfen, *Dichapetalum sumatranum* (Miq.) Boerl., *Noccaea caerulescens* (J. Presl and C. Presl) F. K. Mey., *Deschampsia cespitosa* (L.) P. Beauv., *Cynodon dactylon* (L.) Pers., *Sedum alfredii* Hance, *Justicia procumbens* L.	Zn	Cosio et al. (2004), Peer et al. (2006), Wieshammer et al. (2007), Meyer et al. (2009), Tang et al. (2009), Xie et al. (2009), Lu et al. (2013), van der Ent et al. (2013), Ciarkowska et al. (2016), Reeves et al. (2018), Balafrej et al. (2020)

(continued)

Table 6.2 (continued)

Plant species	Heavy metal	References
Leersia hexandra Swartz, *Phragmites australis* (Cav.) Trin. ex Steud., *Jatropha curcas* L., *Spartina argentinensis* Parodi, *Nopalea cochenillifera* (L.) Salm-Dyck, *Pennisetum sinese* Roxb., *Typha angustifolia* L.	Cr	Zhang et al. (2007), Bareen and Khilji (2008), Mangkoedihardjo et al. (2008), Redondo-Gómez et al. (2011), Adki et al. (2013), Chen et al. (2020)
Elsholtzia splendens Nakai ex F. Maek., *Andryala integrifolia* L., *Commelina communis* L., *Alternanthera bettzickiana* (Regel) G. Nicholson, *Allium polyanthum* Schult. and Schult. f., *Rumex acetosella L.*, *Conyza albida* Willd. ex Spreng, *Phytolacca americana* L., *Plantago lanceolata* L., *Chenopodium album* L., *Ricinus communis* L.	Cu	Jiang et al. (2004), Lange et al. (2017), Asensio et al. (2018), Campillo-Cora et al. (2019) Khalid et al. (2020)
Berkheya coddii Roessler, *Alternanthera bettzickiana* (Regel) G. Nicholson, *Haumaniastrum robertii* (Robyns) P. A. Duvign. and Plancke	Co	Morrison et al. (1979), Malik et al. (2000), Keeling et al. (2003), Lange et al. (2017), Khalid et al. (2020)
Astragalus bisulcatus (Hook.) A. Gray, *Atriplex confertifolia* (Torr. and Frém.) S. Watson, *Stanleya pinnata* (Pursh) Britton	Se	Galeas et al. (2007), Sura-de Jong et al. (2015), White (2016)
Azolla pinnata R. Br., *Euphorbia granulata* Forssk	Hg	Parvez (2022), Rai (2008)

of those processes, they can be responsible for increasing or decreasing heavy metal bioavailability, which can be used to enhance phytoremediation efficiency. However, improving phytoremediation efficiency through plant–microbe associations is highly variable depending on the microbial strain, type of pollutants and the environment conditions. Therefore, the characterization of heavy metal tolerant microbial strains, and the understanding of their mechanisms to enhance heavy metal tolerance are crucial factors for improving microbial assisted phytoremediation efficiency.

Phytoremediation efficiency can also be improved by using plant genetic engineering. It refers to all the techniques that allow direct transfer one or more of specific genes between either closely or distantly related plant species to optimize the performance or plant or to obtain the desired agronomic traits. From a phytoremediation point of view, the transfer of any gene responsible for uptake, translocation, as well as sequestration of heavy metals into candidate plants can potentially improve their phytoremediation properties. Numerous studies have demonstrated the immense potential of plant genetic engineering for enhanced phytoremediation efficiency (Fasani et al. 2018; Kafle et al. 2022). However, plant genetic engineering as a relatively new approach is mainly used in research but without a doubt it provides promising opportunities in the field of phytoremediation.

The use of soil amendments such as biochar (Paz-Ferreiro et al. 2014) humic substances (Zhang et al. 2023), ethylene diamine tetraacetic acid (EDTA) (Shahid et al. 2014) and plant growth regulators (Bartucca et al. 2022) can also improve phytoremediation of soil polluted with heavy metals. These amendments can alter heavy metal speciation, increase or decrease their bioavailability by changing redox status and pH value of the soil, thus enhancing the plant efficiency in phytoremediation processes (Alvarenga et al. 2009; Burges et al. 2018). Besides, the application of beneficial organic or inorganic amendments to soil promotes the plant growth and soil fertility, thereby also increasing the effectiveness of phytoremediation.

6.6 Conclusions

Phytoremediation, as a practical and environmentally friendly remediation technique, has excellent potential to remediate heavy metal polluted soils. However, its successful application is usually limited due to slow and incomplete remediation by plants. Therefore, improving plant performance is a crucial factor for developing highly effective phytoremediation. In this light, a better understanding of the mechanisms of uptake, transport, and detoxification of heavy metals in plants can help overcome the existing limitations of phytoremediation.

Modification of plant genotypes via genetic engineering is a very promising approach to improve plant phytoremediation efficiency. The efficiency of phytoremediation can also be improved through using organic or inorganic amendments, chelating and acidifying agents, using microbial assisted phytoremediation and applying plant growth regulators. Besides, the combination of different phytoremediation approaches could offer new opportunities to improve the effectiveness of phytoremediation.

References

Adki VS, Jadhav JP, Bapat VA (2013) Nopalea cochenillifera, a potential chromium (VI) hyperaccumulator plant. Environ Sci Pollut Res 20:1173–1180. https://doi.org/10.1007/s11356-012-1125-4

Agunbiade FO, Olu-Owolabi BI, Adebowale KO (2009) Phytoremediation potential of *Eichornia crassipes* in metal-contaminated coastal water. Bioresour Technol 100(19):4521–4526. https://doi.org/10.1016/j.biortech.2009.04.011

Aigberua AO (2018) Soil organic matter: application for heavy metal source diagnostics in sediments of the Middleton River, Bayelsa State, Nigeria. MOJ Toxicol 1(4):410–415. https://doi.org/10.15406/mojt.2018.04.00138

Ali H, Khan E, Sajad MA (2013) Phytoremediation of heavy metals-concepts and applications. Chemosphere 91:869–881. https://doi.org/10.1016/j.chemosphere.2013.01.075

Alori ET, Gabasawa AI, Elenwo CE, Agbeyegbe OO (2022) Bioremediation techniques as affected by limiting factors in soil environment. Front Soil Sci 2:937186. https://doi.org/10.3389/fsoil.2022.937186

Alsafran M, Saleem MH, Al Jabri H, Rizvan M, Usman H (2023) Principles and applicability of integrated remediation strategies for heavy metal removal/recovery from contaminated environments. J Plant Growth Regul 42:3419–3440. https://doi.org/10.1007/s00344-022-10803-1

Altinözlü H, Karagöz A, Polat T, Ünver İ (2012) Nickel hyperaccumulation by natural plants in Turkish serpentine soils. Turk J Bot 36(3):6. https://doi.org/10.3906/bot-1101-10

Alvarenga P, Gonçalves A, Fernandes R, De Varennes A, Vallini G, Duarte E, Cunha-Queda AC (2009) Organic residues as immobilizing agents in aided phytostabilization: (I) effects on soil chemical characteristics. Chemosphere 74:1292–1300. https://doi.org/10.1016/j.chemosphere.2008.11.063

Álvarez-López A, Prieto-Fernández Á, Cabello-Conejo MI, Kidd PS (2016) Organic amendments for improving biomass production and metal yield of Ni-hyperaccumulating plants. Sci Total Environ 548:370–379. https://doi.org/10.1016/j.scitotenv.2015.12.147

Alves ARA, Yin Q, Oliveira RS, Silva EF, Novo LAB (2022) Plant growth-promoting bacteria in phytoremediation of metal-polluted soils: current knowledge and future directions. Sci Total Environ 838:156435. https://doi.org/10.1016/j.scitotenv.2022.156435

Anekwe IMS, Isa YM (2021) Comparative evaluation of wastewater and bioventing system for the treatment of acid mine drainage contaminated soils. Water-Energy Nexus 4:134–140. https://doi.org/10.1016/j.wen.2021.08.001

Angelaki A, Dionysidis A, Sihag P, Golia EE (2022) Assessment of contamination management caused by copper and zinc cations leaching and their impact on the hydraulic properties of a sandy and a loamy clay soil. Land 11(2):290. https://doi.org/10.3390/land11020290

Asensio V, Flórido FG, Ruiz F, Perlatti F, Otero X, Ferreira T (2018) Screening of native tropical trees for phytoremediation in copper-polluted soils. Int J Phytoremed 20:1456–1463. https://doi.org/10.1080/15226514.2018.1501341

Azubuike CC, Chikere CB, Okpokwasili GC (2016) Bioremediation techniques—classification based on site of application: principles, advantages, limitations and prospects. World J Microbiol Biotechnol 32:180. https://doi.org/10.1007/s11274-016-2137-x

Babalola OA, Adesodun JK, Olasantan FO, Adekunle AF (2012) Responses of some soil biological, chemical and physical properties to short-term compost amendment. Int J Soil Sci 7:28–38. https://doi.org/10.3923/ijss.2012.28.38

Bala S, Garg D, Thirumalesh BV, Sharma M, Sridhar K, Inbaraj BS, Tripathi M (2022) Recent strategies for bioremediation of emerging pollutants: a review for a green and sustainable environment. Toxics 10(8):484. https://doi.org/10.3390/toxics10080484

Balafrej H, Bogusz D, Triqui ZA, Guedira A, Bendaou N, Smouni A, Fahr M (2020) Zinc hyperaccumulation in plants: a review. Plants 9(5):562. https://doi.org/10.3390/plants9050562

Banuelos G, Cardon G, Mackey B, Ben–Asher J, Wu L, Beuselinck P, Akohoue S, Zambrzuski S (1993) Boron and selenium removal in boron–laden soils by four sprinkler irrigated plant species. J Environ Qual 22:786–792. https://doi.org/10.2134/jeq1993.00472425002200040021x

Bareen F, Khilji S (2008) Bioaccumulation of metals from tannery sludge by *Typha angustifolia* L. Afr J Biotechnol 7:3314–3320

Bartucca ML, Cerri M, Del Buono D, Forni C (2022) Use of biostimulants as a new approach for the improvement of phytoremediation performance—a review. Plants 11:1946. https://doi.org/10.3390/plants11151946

Becerra Castro C, Monterroso C, Garcia Leston M, Prieto Fernandez A, Acea M, Kidd P (2009) Rhizosphere microbial densities and trace metal tolerance of the nickel hyperaccumulator *Alyssum serpyllifolium* subsp. lusitanicum. Int J Phytoremed 11:525–541. https://doi.org/10.1080/15226510902717549

Beretta G, Mastorgio AF, Pedrali L, Saponaro S, Sezenna E (2018) Support tool for identifying in situ remediation technology for sites contaminated by hexavalent chromium. Water 10:1344. https://doi.org/10.3390/w10101344

Bizily SP, Rugh CL, Meagher RB (2000) Phytodetoxification of hazardous organomercurials by genetically engineered plants. Nat Biotechnol 2:213–217. https://doi.org/10.1038/72678

Boucher U, Balabane M, Lamy I, Cambier P (2005) Decomposition in soil microcosms of leaves of the metallophyte *Arabidopsis halleri*: effect of leaf-associated heavy metals on biodegradation. Environ Pollut 135:187–194. https://doi.org/10.1016/j.envpol.2004.10.020

Broadhurst CL, Chaney RL (2016) Growth and metal accumulation of an *Alyssum murale* nickel hyperaccumulator ecotype co-cropped with *Alyssum montanum* and perennial ryegrass in serpentine soil. Front Plant Sci 7:451. https://doi.org/10.3389/fpls.2016.00451

Burges A, Alkorta I, Epelde L, Garbisu C (2018) From phytoremediation of soil contaminants to phytomanagement of ecosystem services in metal contaminated sites. Int J Phytoremed 20:384–397. https://doi.org/10.1080/15226514.2017.1365340

Caille N, Swanwick S, Zhao FJ, McGrath SP (2004) Arsenic hyperaccumulation by *Pteris vittata* from arsenic contaminated soils and the effect of liming and phosphate fertilisation. Environ Pollut 132(1):113–120. https://doi.org/10.1016/j.envpol.2004.03.018

Campillo-Cora C, Fernández-Calviño D, Pérez-Rodríguez P, Fernández-Sanjurjo MJ, Núñez-Delgado A, Álvarez-Rodríguez E, Arias-Estévez M, Nóvoa-Muñoz JC (2019) Copper and zinc in rhizospheric soil of wild plants growing in long-term acid vineyard soils. Insights on availability and metal remediation. Sci Total Environ 672:389–399. https://doi.org/10.1016/j.scitotenv.2019.03.301

Casellas C, Chiron S, Fenet H, Gomez E (2013) Environmental transformation of organic substances in the context of aquatic ecotoxicology. In: Férard JF, Blaise C (eds) Encyclopedia of aquatic ecotoxicology. Springer, Dordrecht. https://doi.org/10.1007/978-94-007-5704-2_44

Centofanti T, Siebecker MG, Chaney RL, Davis AP, Sparks DL (2012) Hyperaccumulation of nickel by *Alyssum corsicum* is related to solubility of Ni mineral species. Plant Soil 359:71–83. https://doi.org/10.1007/s11104-012-1176-9

Chaudhary DK, Kim J (2019) New insights into bioremediation strategies for oil-contaminated soil in cold environments. Int Biodeterior Biodegrad 148:58–72. https://doi.org/10.1016/j.ibiod.2019.05.001

Che-Castaldo JP, Inouye DW (2015) Interspecific competition between a non-native metal-hyperaccumulating plant (*Noccaea caerulescens*, Brassicaceae) and a native congener across a soil-metal gradient. Aust J Bot 63:141. https://doi.org/10.1071/BT15045

Chemlal R, Abdi N, Lounici H, Drouiche N, Pauss A, Mameri N (2013) Modeling and qualitative study of diesel biodegradation using biopile process in sandy soil. Int Biodeterior Biodegrad 78:43–48. https://doi.org/10.1016/j.ibiod.2012.12.014

Chen X, Tong J, Su Y, Xiao L (2020) Pennisetum sinese: a potential phytoremediation plant for chromium deletion from soil. Sustainability 12(9):3651. https://doi.org/10.3390/su12093651

Ciarkowska K, Gargiulo L, Mele G (2016) Natural restoration of soils on mine heaps with similar technogenic parent material: a case study of long-term soil evolution in Silesian-Krakow Upland Poland. Geoderma 261:141–150. https://doi.org/10.1016/J.GEODERMA.2015.07.018

Cosio C, Martinoia E, Keller C (2004) Hyperaccumulaton of cadmium and zinc in *Thlaspi caerulescens* and *Arabidopsis halleri* at leaf cellular level. Plant Physiol 134(2):716–725. https://doi.org/10.1104/pp.103.031948

Crawford RL (2006) Bioremediation. In: Dworkin M, Falkow S, Rosenberg E, Schleifer KH, Stackebrandt E (eds) The prokaryotes. Springer, New York. https://doi.org/10.1007/0-387-30741-9_26

Cristaldi A, Oliveri Conti G, Cosentino SL, Mauromicale G, Copat C, Grasso A, Zuccarello P, Fiore M, Restuccia C, Ferrante M (2020) Phytoremediation potential of *Arundo donax* (Giant Reed) in contaminated soil by heavy metals. Environ Res 185:109427. https://doi.org/10.1016/j.envres.2020.109427

da Silva MLB, Alvarez PJJ (2010) Bioaugmentation. In: Timmis KN (eds) Handbook of hydrocarbon and lipid microbiology. Springer, Berlin, Heidelberg. https://doi.org/10.1007/978-3-540-77587-4_356

Debela F, Arocena JM, Thring RW, Whitcombe T (2010) Organic acid-induced release of lead from pyromorphite and its relevance to reclamation of Pb-contaminated soils. Chemosphere 80:450–456. https://doi.org/10.1016/j.chemosphere.2010.04.025

Dodgen LK, Ueda A, Wu X, Parker DR, Gan J (2015) Effect of transpiration on plant accumulation and translocation of PPCP/EDCs. Environ Pollut 198:144–153. https://doi.org/10.1016/j.envpol.2015.01.002

Domínguez MT, Madrid F, Marañón T, Murillo JM (2009) Cadmium availability in soil and retention in oak roots: potential for phytostabilization. Chemosphere 76:480–486. https://doi.org/10.1016/j.chemosphere.2009.03.026

Escarre J, Lefebvre C, Gruber W (2000) Zinc and cadmium hyperaccumulation by Thlaspi caerulescens from metalliferous and nonmetalliferous sites in the Mediterranean area: implications for phytoremediation. New Phytol 145(3):429–437. https://doi.org/10.1046/j.1469-8137.2000.00599.x

Fasani E, Manara A, Martini F, Furini A, DalCorso G (2018) The potential of genetic engineering of plants for the remediation of soils contaminated with heavy metals. Plant Cell Environ 41:1201–1232. https://doi.org/10.1111/pce.12963

Galeas ML, Zhang LH, Freeman JL, Wegner M, Pilon-Smits EAH (2007) Seasonal fluctuations of selenium and sulfur accumulation in selenium hyperaccumulators and related nonaccumulators. New Phytol 173:517–525. https://doi.org/10.1111/j.1469-8137.2006.01943.x

Gautam K, Gaur P, Sharma P (2022) Bioremediation of radioactive contaminants/radioactive metals. In: Tripathi M, Singh DS (eds) Bioremediation: challenges and advancements. Bentham Science Publishers, Singapore. https://doi.org/10.2174/9789815036039122010006

Gerhardt KE, Gerwing PD, Greenberg BM (2017) Opinion: taking phytoremediation from proven technology to accepted practice. Plant Sci 256:170–185. https://doi.org/10.1016/j.plantsci.2016.11.016

Ghosh M, Singh SP (2005) A review on phytoremediation of heavy metals and utilization of its bioroducts. Asian J Energy Environ 6:18. https://doi.org/10.15666/aeer/0301_001018

Glick BR (2010) Using soil bacteria to facilitate phytoremediation. Biotechnol Adv 28:367–374. https://doi.org/10.1016/j.biotechadv.2010.02.001

Guarino F, Miranda A, Castiglione S, Cicatelli A (2020) Arsenic phytovolatilization and epigenetic modifications in Arundo Donax L. Assisted by a PGPR consortium. Chemosphere 251:126310. https://doi.org/10.1016/j.chemosphere.2020.126310

Hall J (2002) Cellular mechanisms for heavy metal detoxification and tolerance. J Exp Bot 53:1–11. https://doi.org/10.1093/jexbot/53.366.1

Hanus-Fajerska EJ, Koźmińska A (2016) The possibilities of water purification using phytofiltration methods: a review of recent progress. Bio Technologia 97(4):315–322. https://doi.org/10.5114/bta.2016.64549

Hart G, Gilly A, Koether M, McElroy T, Greipsson S (2022) Phytoextraction of lead (Pb) contaminated soil by switchgrass (*Panicum virgatum* L): impact of BAP and NTA applications. Front Energy Res 10:1032404. https://doi.org/10.3389/fenrg.2022.1032404

He Y, Langenhoff AAM, Sutton NB, Rijnaarts HHM, Blokland MH, Chen F, Huber C, Schröder P (2017) Metabolism of ibuprofen by *Phragmites australis*: uptake and phytodegradation. Environ Sci Technol 51(8):4576–4584. https://doi.org/10.1021/acs.est.7b00458

Hejna M, Onelli E, Moscatelli A, Bellotto M, Cristiani C, Stroppa N, Rossi L (2021) Heavy-metal phytoremediation from livestock wastewater and exploitation of exhausted biomass. Int J Environ Res Public Health 18(5):2239. https://doi.org/10.3390/ijerph18052239

Hussain S, Akram M, Abbas G, Murtaza B, Shahid M, Shah NS, Bibi I, Niazi NK (2017) Arsenic tolerance and phytoremediation potential of *Conocarpus erectus* L. and *Populus deltoides* L. Int J Phytoremed 19:985–991. https://doi.org/10.1080/15226514.2017.1303815

Jacobs A, Drouet T, Noret N (2018) Field evaluation of cultural cycles for improved cadmium and zinc phytoextraction with *Noccaea caerulescens*. Plant Soil 430:381–394. https://doi.org/10.1007/s11104-018-3734-2

Jaffré T, Reeves RD, Baker AJM, Schat H, van der Ent A (2018) The discovery of nickel hyperaccumulation in the New Caledonian tree *Pycnandra acuminata* 40 years on: an introduction to a virtual issue. New Phytol 218(2):397–400. https://doi.org/10.1111/nph.15105

Jha P, Jobby R, Desai NS (2016) Remediation of textile azo dye acid red 114 by hairy roots of *Ipomoea carnea* Jacq. and assessment of degraded dye toxicity with human keratinocyte cell line. J Hazard Mater 311:158–167. https://doi.org/10.1016/j.jhazmat.2016.02.058

Jiang LY, Yang X, He Z (2004) Growth response and phytoextraction of copper at different levels in soils by *Elsholtzia splendens*. Chemosphere 55:1179–1187. https://doi.org/10.1016/j.chemosphere.2004.01.026

Kafle A, Timilsina A, Gautam A, Adhikari K, Bhattarai A, Aryal N (2022) Phytoremediation: mechanisms, plant selection and enhancement by natural and synthetic agents. Environ Adv 8:100203. https://doi.org/10.1016/j.envadv.2022.100203

Kagalkar AN, Jadhav MU, Bapat VA, Govindwar SP (2011) Phytodegradation of the triphenylmethane dye Malachite Green mediated by cell suspension cultures of *Blumea malcolmii* Hook. Bioresour Technol 102:10312–10318. https://doi.org/10.1016/j.biortech.2011.08.101

Kanwar VS, Sharma A, Srivastav AL, Rani L (2020) Phytoremediation of toxic metals present in soil and water environment: a critical review. Environ Sci Pollut Res 27:44835–44860. https://doi.org/10.1007/s11356-020-10713-3

Kaur P, Bali S, Sharma A, Vig AP, Bhardwaj R (2018) Role of earthworms in phytoremediation of cadmium (Cd) by modulating the antioxidative potential of Brassica juncea L. Appl Soil Ecol 124:306–316. https://doi.org/10.1016/j.apsoil.2017.11.017

Keeling SM, Stewart RB, Anderson CWN, Robinson BH (2003) Nickel and cobalt phytoextraction by the hyperaccumulator *Berkheya coddii*: implications for polymetallic phytomining and phytoremediation. Int J Phytoremed 5:235–244. https://doi.org/10.1080/713779223

Kerkeb L, Krämer U (2003) The role of free histidine in xylem loading of nickel in Alyssum lesbiacum and Brassica juncea. Plant Physiol 131:716–724. https://doi.org/10.1104/pp102.010686

Khalid A, Farid M, Zubair M, Rizwan M, Iftikhar U, Ishaq HK, Farid S, Latif U, Hina K, Ali S (2020) Efficacy of *Alternanthera bettzickiana* to remediate copper and cobalt contaminated soil physiological and biochemical alterations. Int J Environ Res 14:243–255. https://doi.org/10.1007/s41742-020-00251-8

Kiran BR, Prasad MNV (2021) Assisted phytostabilization of Pb–spiked soils amended with charcoal and banana compost and vegetated with *Ricinus communis* L. (Castor bean). Environ Geochem Health 43:1507–1521. https://doi.org/10.1007/s10653-021-00825-1

Krämer U (2010) Metal hyperaccumulation in plants. Annu Rev Plant Biol 61(1):517–534. https://doi.org/10.1146/annurev-arplant-042809-112156

Lange B, van der Ent A, Baker AJ, Echevarria G, Mahy G, Malaisse F, Meerts P, Pourret O, Verbruggen N, Faucon MP (2017) Copper and cobalt accumulation in plants: a critical assessment of the current state of knowledge. New Phytol 2017(2):537–551. https://doi.org/10.1111/nph.14175

Limmer M, Burken J (2016) Phytovolatilization of organic contaminants. Environ Sci Technol 50(13):6632–6643. https://doi.org/10.1021/acs.est.5b04113

Liu L, Li W, Song W, Guo M (2018) Remediation techniques for heavy metal-contaminated soils: principles and applicability. Sci Total Environ 633:206–219. https://doi.org/10.1016/j.scitotenv.2018.03.161

Lombi E, Zhao FJ, Dunham SJ, McGrath SP (2000) Cadmium accumulation in populations of *Thlaspi caerulescens* and *Thlaspi goesingense*. New Phytol 145:11–20. https://doi.org/10.1046/j.1469-8137.2000.00560.x

Lu L, Tian S, Zhang J, Yang X, Labavitch JM, Webb SM, Latimer M, Brown PH (2013) Efficient xylem transport and phloem remobilization of Zn in the hyperaccumulator plant species *Sedum alfredii*. New Phytol 198(3):721–731. https://doi.org/10.1111/nph.12168

Ma Y, Oliveira RS, Freitas H, Zhang C (2016) Biochemical and molecular mechanisms of plant-microbe-metal interactions: relevance for phytoremediation. Front Plant Sci 7:918. https://doi.org/10.3389/fpls.2016.00918

Mahar A, Wang P, Ali A, Awasthi MK, Lahori AH, Wang Q, Li R, Zhang Z (2016) Challenges and opportunities in the phytoremediation of heavy metals contaminated soils: a review. Ecotox Environ Safe 126:111–121. https://doi.org/10.1016/j.ecoenv.2015.12.023

Mahrous NN, Columbus MP, Southam G, Macfie SM (2019) Changes in microbial community structure and increased metal bioavailability in a metal-contaminated soil and in the rhizosphere of corn (Zea mays). Rhizosphere 11:100169. https://doi.org/10.1016/j.rhisph.2019.100169

Malik M, Chaney RL, Brewer EP, Li YM, Angle JS (2000) Phytoextraction of soil cobalt using hyperaccumulator plants. Int J Phytoremed 2:319–329. https://doi.org/10.1080/15226510008500041

Mangkoedihardjo S, Ratnawati R, Alfianti N (2008) Phytoremediation of hexavalent chromium polluted soil using Pterocarpus indicus and *Jatropha curcas* L. World Appl Sci J 2008(4):338–342

Memon AR, Schröder P (2009) Implications of metal accumulation mechanisms to phytoremediation. Environ Sci Pollut R 16:162–175. https://doi.org/10.1007/s11356-008-0079-z

Memon AR, Aktoprakligil D, Zdemur A, Vertii A (2001) Heavy metal accumulation and detoxification mechanisms in plants. Turk J Bot 25(3):111–121. https://journals.tubitak.gov.tr/botany/vol25/iss3/1

Mengoni A, Pini F, Huang LN, Shu WS, Bazzicalupo M (2009) Plant-by-plant variations of bacterial communities associated with leaves of the nickel hyperaccumulator *Alyssum bertolonii* Desv. Microb Ecol 58:660–667. https://doi.org/10.1007/s00248-009-9537-5

Mesjasz-Przybyłowicz J, Przybyłowicz WJ (2020) Ecophysiology of nickel hyperaccumulating plants from South Africa—from ultramafic soil and mycorrhiza to plants and insects. Metallomics 12(7):1018–1035. https://doi.org/10.1039/c9mt00282k

Meyers D, Auchterlonie GJ, Webb RI, Wood B (2008) Uptake and localisation of lead in the root system of *Brassica juncea*. Environ Pollut 153:323–332. https://doi.org/10.1016/j.envpol.2007.08.029

Meyer CL, Vitalis R, Saumitou-Laprade P, Castric V (2009) Genomic pattern of adaptive divergence in Arabidopsis halleri, a model species for tolerance to heavy metal. Mol Eco 18:2050–2062. https://doi.org/10.1111/j.1365-294x.2009.04159.x

Midhat L, Ouazzani N, Hejjaj A, Ouhammou A, Mandi L (2019) Accumulation of heavy metals in metallophytes from three mining sites (Southern Centre Morocco) and evaluation of their phytoremediation potential. Ecotoxicol Environ Saf 169:150–160. https://doi.org/10.1016/j.ecoenv.2018.11.009

Mitra S, Chakraborty AJ, Tareq AM, Emran TB, Nainu F, Khusro A, Idris AM, Khandaker MU, Osman H, Alhumaydhi FA, Simal-Gandara J (2022) Impact of heavy metals on the environment and human health: novel therapeutic insights to counter the toxicity. J King Saud Univ Sci 34(3):101865. https://doi.org/10.1016/j.jksus.2022.101865

Mondol MN, Chamon AS, Faiz B, Elahi SF (2011) Seasonal variation of heavy metal concentrations in water and plant samples around Tejgaon industrial area of Bangladesh. J Bangladesh Acad Sci 35(1):19–41. https://doi.org/10.3329/jbas.v35i1.7968

Montiel-Rozas MM, Madejón E, Madejón P (2016) Effect of heavy metals and organic matter on root exudates (low molecular weight organic acids) of herbaceous species: an assessment in sand and soil conditions under different levels of contamination. Environ Pollut 216:273–281. https://doi.org/10.1016/j.envpol.2016.05.080

Morrison RS, Brooks RR, Reeves RD, Malaisse F (1979) Copper and cobalt uptake by metallophytes from Zaïre. Plant Soil 53:535–539. https://doi.org/10.1007/BF02140724

Muthusaravanan S, Sivarajasekar N, Vivek JS, Paramasivan T, Naushad M, Prakashmaran J, Gayathri V, Al-Duaij OK (2018) Phytoremediation of heavy metals: mechanisms, methods and enhancements. Environ Chem Lett 16:1339–1359. https://doi.org/10.1007/s10311-018-0762-3

Nandillon R, Lebrun M, Miard F, Gaillard M, Sabatier S, Morabito D, Bourgerie S (2019) Contrasted tolerance of Agrostis capillaris metallicolous and non-metallicolous ecotypes in the context of a mining technosol amended by biochar, compost and iron sulfate. Environ Geochem Health 43:1457–1475. https://doi.org/10.1007/s10653-019-00447-8

Nedjimi B (2021) Phytoremediation: a sustainable environmental technology for heavy metals decontamination. SN Appl Sci 3:286. https://doi.org/10.1007/s42452-021-04301-4

Nedjimi B, Daoud Y (2009) Cadmium accumulation in Atriplex halimus subsp. schweinfurthii and its influence on growth, proline, root hydraulic conductivity and nutrient uptake. Flora 204:316–324. https://doi.org/10.1016/j.flora.2008.03.004

Ning Y, Liu N, Song Y, Luo J, Li T (2019) Enhancement of phytoextraction of Pb by compounded activation agent derived from fruit residue. Int J Phytoremed 21:1449–1456. https://doi.org/10.1080/15226514.2019.1633266

Nnaji ND, Onyeaka H, Miri T, Ugwa C (2023) Bioaccumulation for heavy metal removal: a review. SN Appl Sci 5:125. https://doi.org/10.1007/s42452-023-05351-6

Oh K, Li T, Cheng H, Hu X, He C, Yan L, Shinichi Y (2013) Development of profitable phytoremediation of contaminated soils with biofuel crops. J Environ Prot 4:58–64. https://doi.org/10.4236/jep.2013.44A008

Olaniran AO, Balgobind A, Pillay B (2013) Bioavailability of heavy metals in soil: impact on microbial biodegradation of organic compounds and possible improvement strategies. Int J Mol Sci 14(5):10197–10228. https://doi.org/10.3390/ijms140510197

Onyia PC, Ozoko DC, Ifediegwu SI (2021) Phytoremediation of arsenic-contaminated soils by arsenic hyperaccumulating plants in selected areas of Enugu State, Southeastern, Nigeria. Geol Ecol Landsc 5(4):308–319. https://doi.org/10.1080/24749508.2020.1809058

Parvez M (2022) Heavy metals levels and associated health risk assessment of *Euphorbia granulata* Forssk. Environ Sci Pollut Res 29:1998–2008. https://doi.org/10.1007/s11356-021-15569-9

Pasricha S, Mathur V, Garg A, Lenka S, Verma K, Agarwal S (2021) Molecular mechanisms underlying heavy metal uptake, translocation and tolerance in hyperaccumulators-an analysis: heavy metal tolerance in hyperaccumulators. Environ Chall 4:100197. https://doi.org/10.1016/j.envc.2021.100197

Paz-Ferreiro J, Lu H, Fu S, Méndez A, Gascó G (2014) Use of phytoremediation and biochar to remediate heavy metal polluted soils: a review. Solid Earth 5:65–75. https://doi.org/10.5194/se-5-65-2014

Peco JD, Higueras P, Campos JA, Esbrí JM, Moreno MM, Battaglia-Brunet F, Sandalio LM (2021) Abandoned mine lands reclamation by plant remediation technologies. Sustainability 13:6555. https://doi.org/10.3390/su13126555

Peer WA, Mahmoudian M, Freeman JL, Lahner B, Richards EL, Reeves RD, Murphy AS, Salt DE (2006) Assessment of plants from the Brassicaceae family as genetic models for the study of nickel and zinc hyperaccumulation. New Phytol 172:248–260. https://doi.org/10.1111/j.1469-8137.2006.01820.x

Petelka J, Abraham J, Bockreis A, Deikumah JP, Zerbe S (2019) Soil heavy metal(loid) pollution and phytoremediation potential. Water Air Soil Pollut 230:267. https://doi.org/10.1007/s11270-019-4317-4

Prasad MNV, Freitas H (2003) Metal hyperaccumulation in plants—biodiversity prospecting for phytoremediation technology. Electron J Biotechnol 6:285321. https://doi.org/10.2225/vol6-issue3-fulltext-6

Priya AK, Muruganandam M, Ali SS, Kornaros M (2023) Clean-up of heavy metals from contaminated soil by phytoremediation: a multidisciplinary and eco-friendly approach. Toxics 11(5):422. https://doi.org/10.3390/toxics11050422

Rai PK (2008) Phytoremediation of Hg and Cd from industrial effluents using an aquatic free floating macrophyte Azolla pinnata. Int J Phytoremed 10:430–439. https://doi.org/10.1080/15226510802100606

Raza A, Khan AHA, Nawaz I, Qu Z, Yousaf S, Ali MA, Sayal AU, Iqbal M (2019) Evaluation of arsenic-induced stress in Dahlia pinnata Cav.: morphological and physiological response. Soil Sediment Contam 28:716–728. https://doi.org/10.1080/15320383.2019.1657380

Redondo-Gómez S, Mateos-Naranjo E, Vecino-Bueno I, Feldman SR (2011) Accumulation and tolerance characteristics of chromium in a cordgrass chromium-hyperaccumulator Spartina argentinensis. J Hazard Mater 185:862–869. https://doi.org/10.1016/j.jhazmat.2010.09.101

Reeves RD, Baker AJ, Jaffré T, Erskine PD, Echevarria G, van der Ent A (2018) A global database for plants that hyperaccumulate metal and metalloid trace elements. New Phytol 218:407–411. https://doi.org/10.1111/nph.14907

Risse SBL, Puschenreiter M, Tognacchini A (2023) Rhizosphere processes by the nickel hyperaccumulator Odontarrhena chalcidica suggest Ni mobilization. Plant Soil. https://doi.org/10.1007/s11104-023-06161-w

Sakakibara M, Ohmori Y, Ha NTH, Sano S, Sera K (2011) Phytoremediation of heavy metal-contaminated water and sediment by *Eleocharis acicularis*. Clean: Soil Air Water 39:735–741. https://doi.org/10.1002/clen.201000488

Salido AL, Hasty KL, Lim JM, Butcher DJ (2003) Phytoremediation of arsenic and lead in contaminated soil using Chinese brake ferns (*Pteris vittata*) and Indian mustard (*Brassica juncea*). Int J Phytoremed 5(2):89–103. https://doi.org/10.1080/713610173

Santibáñez C, Verdugo C, Ginocchio R (2008) Phytostabilization of copper mine tailings with biosolids: implications for metal uptake and productivity of *Lolium perenne*. Sci Total Environ 395:1–10. https://doi.org/10.1016/j.scitotenv.2007.12.033

Sarwar N, Imran M, Shaheen MR, Ishaque W, Kamran MA, Matloob A, Rehim A, Hussain A (2017) Phytoremediation strategies for soils contaminated with heavy metals: modifications and future perspectives. Chemosphere 171:710–721. https://doi.org/10.1016/j.chemosphere.2016.12.116

Sazykin I, Khmelevtsova L, Azhogina T, Sazykina M (2023) Heavy metals influence on the bacterial community of soils: a review. Agriculture 13(3):653. https://doi.org/10.3390/agriculture13030653

Seraj F, Rahman T (2018) Heavy metals, metalloids, their toxic effect and living systems. Am J Plant Sci 9:2626–2643. https://doi.org/10.4236/ajps.2018.913191

Seth CS (2012) A review on mechanisms of plant tolerance and role of transgenic plants in environmental clean-up. Bot Rev 78:32–62. https://doi.org/10.1007/s12229-011-9092-x

Shackira AM, Puthur JT (2017) Enhanced phytostabilization of cadmium by a halophyte *Acanthus ilicifolius* L. Int J Phytoremed 19:319–326. https://doi.org/10.1080/15226514.2016

Shahid M, Austruy A, Echevarria G, Arshad M, Sanaullah M, Aslam M, Nadeem M, Nasim W, Dumat C (2014) EDTA-enhanced phytoremediation of heavy metals: a review. Soil Sediment Contam Int J 23:389–416. https://doi.org/10.1080/15320383.2014.831029

Sharma I (2020) Bioremediation techniques for polluted environment: concept, advantages, limitations, and prospects. In: Murillo-Tovar MA, Saldarriaga-Norena HA, Saeid A (eds) Trace metals in the environment: new approaches and recent advances. IntechOpen, London. https://doi.org/10.5772/intechopen.90453

Shrivastava NG (2022) Bioremediation of solid waste management. In: Baskar C, Ramakrishna S, Baskar S, Sharma R, Chinnappan A, Sehrawat R (eds) Handbook of solid waste management. Springer, Singapore. https://doi.org/10.1007/978-981-16-4230-2_48

Simiele M, Lebrun M, Miard F, Trupiano D, Poupart P, Forestier O, Scippa GS, Bourgerie S, Morabito D (2020) Assisted phytoremediation of a former mine soil using biochar and iron sulphate: effects on As soil immobilization and accumulation in three Salicaceae species. Sci Total Environ 710:136203. https://doi.org/10.1016/j.scitotenv.2019.136203

Singh V, Pandey B, Suthar S (2019) Phytotoxicity and degradation of antibiotic ofloxacin in duckweed (*Spirodela polyrhiza*) system. Ecotoxicol Environ Saf 179:88–95. https://doi.org/10.1016/j.ecoenv.2019.04.018

Skuza L, Szućko-Kociuba I, Filip E, Bożek I (2022) Natural molecular mechanisms of plant hyperaccumulation and hypertolerance towards heavy metals. Int J Mol Sci 23(16):9335. https://doi.org/10.3390/ijms23169335

Srivastava M, Ma LQ, Santos JA (2006) Three new arsenic hyperaccumulating ferns. Sci Total Environ 364:24–31. https://doi.org/10.1016/j.scitotenv.2005.11.002

Suman J, Uhlik O, Viktorova J, Macek T (2018) Phytoextraction of heavy metals: a promising tool for clean-up of polluted environment? Front Plant Sci 9:1476. https://doi.org/10.3389/fpls.2018.01476

Sura-de Jong M, Reynolds RJ, Richterova K, Musilova L, Staicu LC, Chocholata I, Cappa JJ, Taghavi S, van der Lelie D, Frantik T, Dolinova I, Strejcek M, Cochran AT, Lovecka P, Pilon-Smits EA (2015) Selenium hyperaccumulators harbor a diverse endophytic bacterial community characterized by high selenium resistance and plant growth promoting properties. Front Plant Sci 6:113. https://doi.org/10.3389/fpls.2015.00113

Tang YT, Qiu RL, Zeng XW, Ying RR, Yu FM, Zhou XY (2009) Lead, zinc, cadmium hyperaccumulation and growth stimulation in *Arabis paniculata* Franch. Environ Exp Bot 66:126–134. https://doi.org/10.1016/j.envexpbot.2008.12.016

van der Ent A, Baker AJ, Reeves RD, Pollard AJ, Schat H (2013) Hyperaccumulators of metal and metalloid trace elements: facts and fiction. Plant Soil 362:319–334. https://doi.org/10.1007/s11104-012-1287-3

Varun M, Souza RD, Pratas J, Paul MS (2011) Evaluation of phytostabilization, a green technology to remove heavy metals from industrial sludge using *Typha latifolia* L. Biotechnol Bioinf Bioeng 1:137–145

Vázquez S, Agha R, Granado A, Sarro MJ, Esteban E, Peñalosa JM, Carpena RO (2006) Use of white lupin plant for phyto-stabilization of Cd and as polluted acid soil. Water Air Soil Pollut 177:349–365. https://doi.org/10.1007/s11270-006-9178-y

Verbruggen N, Hermans C, Schat H (2009) Mechanisms to cope with arsenic or cadmium excess in plants. Curr Opin Plant Biol 12:364–372. https://doi.org/10.1016/j.pbi.2009.05.001

Visconti D, Álvarez-Robles MJ, Fiorentino N, Fagnano M, Clemente R (2020) Use of *Brassica juncea* and *Dactylis glomerata* for the phytostabilization of mine soils amended with compost or biochar. Chemosphere 260:127661. https://doi.org/10.1016/j.chemosphere.2020.127661

Visconti D, Ventorino V, Fagnano M, Woo SL (2023) Compost and microbial biostimulant applications improve plant growth and soil biological fertility of a grass-based phytostabilization system. Environ Geochem Health 45:787–807. https://doi.org/10.1007/s10653-022-01235-7

Wang ST, Liu HL, Liu W, Zuo QQ (2016) Effect of low-molecular-weight organic acids on nano-hydroxyapatite adsorption of cadmium and lead. J Biomater Tissue Eng 6:433–439. https://doi.org/10.1166/jbt.2016.1465

White PJ (2016) Selenium accumulation by plants. Ann Bot 117(2):217–235. https://doi.org/10.1093/aob/mcv180

Wieshammer G, Unterbrunner R, García TB, Zivkovic MF, Puschenreiter M, Wenzel WW (2007) Phytoextraction of Cd and Zn from agricultural soils by Salix ssp. and intercropping of *Salix caprea* and *Arabidopsis halleri*. Plant Soil 298:255–264. https://doi.org/10.1007/s11104-007-9363-9

Wu Q, Wang S, Thangavel P, Li Q, Zheng H, Bai J, Qiu R (2011) Phytostabilization potential of *Jatropha curcas* L. in polymetallic acid mine tailings. Int J Phytoremed 13:788–804. https://doi.org/10.1080/15226514.2010.525562

Xie HL, Jiang RF, Zhang FS, McGrath SP, Zhao FJ (2009) Effect of nitrogen form on the rhizosphere dynamics and uptake of cadmium and zinc by the hyperaccumulator *Thlaspi caerulescens*. Plant Soil 318:205–215. https://doi.org/10.1007/s11104-008-9830-y

Yadav M, Singh G, Jadeja RN (2021) Bioremediation of organic pollutants: a sustainable green approach. In: Mishra VK, Kumar A (eds) Sustainable environmental clean-up. Elsevier. https://doi.org/10.1016/b978-0-12-823828-8.00006-2

Yan A, Wang Y, Tan SN, Mohd Yusof ML, Ghosh S, Chen Z (2020) Phytoremediation: a promising approach for revegetation of heavy metal-polluted land. Front Plant Sci 11:359. https://doi.org/10.3389/fpls.2020.00359

Yoon J, Cao X, Zhou Q, Ma LQ (2006) Accumulation of Pb, Cu, and Zn in native plants growing on a contaminated Florida site. Sci Total Environ 368(2–3):456–464. https://doi.org/10.1016/j.scitotenv.2006.01.016

Zhang L, Lan S, Dou Q, Hao S, Wang Y, Wang X, Zhang R, Peng Y, Yang J (2023) Metagenomic insights into responses of microbial population and key functional genes to fulvic acid during partial nitritation. J Environ Sci 124:952–962. https://doi.org/10.1016/j.jes.2022.03.003

Zhang XH, Liu J, Huang HT, Chen J, Zhu YN, Wang DQ (2007) Chromium accumulation by the hyperaccumulator plant *Leersia hexandra* Swartz. Chemosphere 67:1138–1143. https://doi.org/10.1016/j.chemosphere.2006.11.014

Zhao FJ, Dunham SJ, McGrath SP (2002) Arsenic hyperaccumulation by different fern species. New Phytol 156:27–31. https://doi.org/10.1046/j.1469-8137.2002.00493.x

Zhao L, Li T, Zhang X, Chen GD, Zheng ZC, Yu HY (2016) Pb uptake and phytostabilization potential of the mining ecotype of *Athyrium wardii* (Hook.) grown in Pb contaminated soil. Clean (Weinh) 44:1184–1190. https://doi.org/10.1002/clen.201400870

Chapter 7
Heavy Metal Stress in Plants: Causes, Impact and Effective Management

Susmita Shukla, Subhankar Das, Sparsh Phutela, Ayushi Triathi, Chanchal Kumari, and Shiv Kant Shukla

Abstract Heavy metal toxicity is a persistent and significant havoc to plant system. It has been a subject of serious environmental and biological concern pertaining to be an arena of intense research. Plants are sessile organisms and presence of heavy metal even in the rudimentary concentrations can possess detrimental impacts to the physiological functioning of the plant system. There is a plethora of anthropogenic reasons for the release and accumulation of heavy metals in or near the plant system. Commercial enterprises, industrialization and Urbanization have led to the wide distribution of heavy metals such as Mn, Cu, Ni, Fe, Co, Cd, Zn, Hg and As etc. The geological and biological redistribution of these metal ions causes and interference with plant functioning and manifests in different ways. The current piece of work will provide a bigger conglomerate for heavy metal toxicity and plant stress. It will discuss the causes of aggregation, underlying mechanisms of impact and strategies incorporated for management of heavy metal toxicity. The heavy metal toxicity impacts the plant stress via dysregulating photosynthesis, cell wall functioning, altering the synthesis of plant secondary metabolites, sexual reproduction, ageing, mineral uptake, and enzymatic impairments. These physiological deteriorations have catastrophic implication on agricultural process as the accumulation in plant tissues makes the yield unfit for consumption and shrinks the overall yield. The recent physiological management intricacies are approached via nutrition, soil test and gauging, remediation, supplementing with organic material adjoined with the techniques such as Crop rotation and intercropping, addition of chelating agents, and selecting tolerant plant species. There is also a dire need to educate masses about the hazards of heavy metal toxicity, thus public awareness is also a suitable strategy as preventive measure.

S. Shukla (✉) · S. Phutela · A. Triathi · C. Kumari
Applied Plant Biotechnology Research Lab, Centre for Plant and Environment Biotechnology, Amity Institute of Biotechnology, Amity University, Noida, UP, India
e-mail: sshukla3@amity.edu

S. Das
Biotechnology Unit, Mangalore University, Mangalore, Karnataka, India

S. K. Shukla
Kaushalya—The Skill University, Ahmedabad, Gujarat, India

N. Kumar (ed.), *Heavy Metal Toxicity*, Environmental Science and Engineering,
https://doi.org/10.1007/978-3-031-56642-4_7

Keywords Heavy metal toxicity · Plant stress · ROS · Heavy metal toxicity and management strategies

7.1 Introduction

7.1.1 Heavy Metals and It's Impact on Ecosystem

The term "heavy metals" refers to metallic elements that possess a higher density in comparison to water. The problem of heavy metal stress has emerged as a significant global concern in many terrestrial and aquatic ecosystems. In the present era, the widespread industrialization has been observed to have adverse impacts on soil quality and crop yield due to the accumulation of heavy metals (Shahid et al. 2015). The sources of heavy metals in the environment have been reported to include geogenic, industrial, agricultural, pharmaceutical, residential effluents, and atmospheric sources (He et al. 2005), However this pollution is highly prevalent in specific locations such as mining sites, foundries and smelters, as well as other industrial operations that are primarily focused on metal-based activities. Although some heavy metals for example Zn, Cu, Mo, Mn, Co, and Ni have crucial role in biological processes and developmental pathways (Salla et al. 2011; Shahid et al. 2015) but these metals along with toxic heavy metals can cause some deleterious effect even at low level of exposure. There exist numerous anthropogenic causes that contribute to the emission and build-up of heavy metals in close proximity to plant systems. The process of urbanization and industrialization has resulted in the extensive distribution of several heavy metals, including mercury (Hg), manganese (Mn), copper (Cu), cobalt (Co), cadmium (Cd), nickel (Ni), zinc (Zn), and arsenic (As), iron (Fe), among other elements (Aydinalp and Marinova 2009). The geological and biological redistribution of heavy metal ions results in their presence within the plant environment, leading to disruptions in various aspects of plant functioning through numerous mechanisms. The effects of heavy metal toxicity on plant systems can be observed through a range of symptoms, varying from evident physical harm to more nuanced physiological alterations. Potential consequences of this phenomenon encompass inhibited growth, chlorosis (manifested as leaf yellowing), irregularities in root development, diminished photosynthetic function, modified nutrient absorption, and perhaps, molecular-level harm to cellular structures (Öztürk et al. 2015). In addition, the presence of heavy metal contamination has the ability to disrupt the intricate equilibrium of crucial nutrient elements within plants, hence potentially causing imbalances in nutrient levels and exacerbating their general physiological well-being. Plants possess a range of molecular and physiological systems that enable them to mitigate the effects of heavy metal stress. These strategies include intricate biochemical and genomic processes. Certain mechanisms inside the biological system are inherent components of the homeostatic process, hence being considered constitutive. Plants employ many techniques to restrict the absorption of heavy metals from the soil. One potential mechanism by which plants can regulate the influx of hazardous metals into their root systems is through the modulation of permeability in the membranes of root cells. Furthermore, several plant species possess the ability to actively extrude heavy metals from their root cells, therefore diminishing their internal concentration.

Heavy metals are pivotal constitutes of the biochemical and physiological processes in both the plants and animals systems. Heavy metals are cofactors for several metabolic enzymes and contribute to the significant oxidation–reduction reactions catalysing various metabolites (WHO 1996). The concentration of trace metals in soil plays a crucial role in determining the potential inhibitory or stimulating effects of these metals on plant growth, as highlighted by Abolghassem et al. (2018). The issue of heavy metal toxicity continues to pose a persistent and substantial risk to the overall well-being of plant ecosystems, garnering significant interest from academics in the fields of environmental science and biology. The effects of heavy metal toxicity on plant stress represents a continuously developing field of research, bearing significant consequences for the overall health and vitality of plants, which, by nature, are organisms that lack mobility. Even at low concentrations, heavy metals can have adverse impacts on the physiological processes that are crucial for the survival and growth of plants. The phenomenon of heavy metal stress leads to the inactivation or denaturation of certain crucial enzymes and proteins, while also disrupting the substitution events involving critical metal ions inside biomolecules. The aforementioned reaction disrupts the structural integrity of cellular membranes, leading to modifications in fundamental metabolic processes in plants, including photosynthesis, respiration, and homeostasis (Hossain et al. 2012). Given the aforementioned obstacles, current research endeavours are focused on gaining a deeper comprehension of the mechanisms underlying heavy metal toxicity in plants and devising efficient approaches to alleviate its detrimental consequences. The management of plant stress is crucial for maintaining the well-being and productivity of plant systems when exposed to heavy metal contamination. This involves implementing many strategies, including phytoremediation, genetic engineering, and soil management methods. This research endeavour seeks to enhance our overall comprehension of plant stress and the sustainable management of plant ecosystems in an evolving global context by providing a thorough analysis of heavy metal toxicity, including its causes and potential treatments.

7.2 Heavy Metal Toxicity and Causes of Aggregation

Heavy metal pollutants can be categorized into two fundamental classes i.e., Anthropogenic and natural sources. In comparison to anthropogenic vertical, which encompasses commercial enterprises, agriculture, mining, and effluents, natural sources include sedimentary rocks, volcanic eruptions, soil formation, and rock weathering. However, the usage of pollution indicators is a beneficial way to distinguish between anthropogenic and geogenic soil pollutants. Despite the usage of advance research

Table 7.1 Range of heavy metal concentrations (ppm) in igneous and sedimentary rocks

Toxicants	Granite igneous	Sandstone	Black shales	Basaltic igneous	Granite igneous
Zn	6–30	<1–31	7–150	2–18	6–30
Cu	5–140	2–41	34–1500	48–240	5–140
Pb	4–30	–	20–200	30–160	4–30
Cd	0.003–0.18	–	<0.3–8.4	0.006–0.6	0.003–0.18

and strategies, it is difficult to identify the source of apportionment. The various sources of heavy metals are exceedingly variable, consisting of aerosol particle deposition, heavy steel-containing raindrops, and agrochemicals. The data presented largely part focused on the identical investigated heavy metals from various sources like—Cd, Pb, Cu, and Zn. Natural reasons for the heavy metal aggregation is majorly due to igneous and sedimentary rocks. Table 7.1 lists the ranges of heavy metal concentrations (ppm) measured both the and igneous and sedimentary rocks. Researches have unveiled that the concentrations of distinct heavy metal components vary with one to another type of rock and also within the regions of a single type of rock. Rocks are situated in the native environment which along with the type of rock can be a determining factors in the type and concentration of heavy metals present in them. The rocks are weathered to form soil thus along with the category of river sediments the process of soil formation is also a pivotal reason in the build-up of heavy metals (Alengebawy et al. 2021).

Plethora of anthropogenic sources such as industries, agriculture, mining, and wastewater are the bigger evils for the discharge of Heavy metals pollutants in the environment. These factors are a major cause of aggravation of heavy metals pollutants leading to polluting ecosystem as whole and decreasing the resilience of ecosystem. This increases the chances of reaching a tipping point in an ecosystem (Cannon et al. 1978). The anthropogenic factors such as smelting releases Cu, Zn, and As. The use of insecticides, which helps to release As in the soil system disturbing the soil horizons. The factors as burning fossil fuels for combustive needs of transportations produces Hg; and vehicle exhaust, which helps to release Pb. Along with the regular human activities like agricultural production, manufacturing, and industrial processes are also an all-time treat by disturbing the self-sustainable nature of any ecosystem and biome (Fig. 7.1).

7.3 Impact of Heavy Metal on Plant Stress

Heavy metals are present in soil, exhibiting significant variations in terms of their dissolvability and bioavailability in the plant tissues. The influence of soil physicochemical parameters on the geochemical behaviour of heavy metals, their absorption by plants via sap transportation and pertaining effect on the yield from the crop

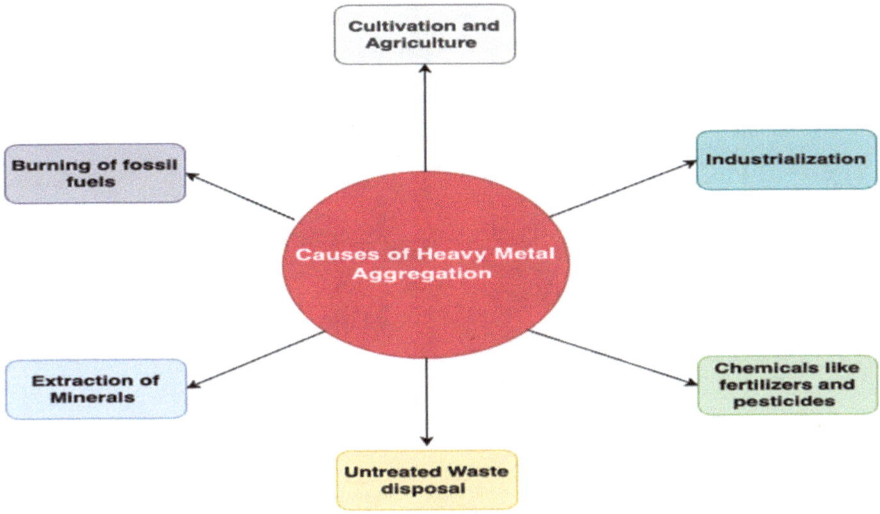

Fig. 7.1 Anthropogenic sources of heavy metals

production has been widely recognized (Shahid et al. 2015). Physicochemical properties of soil are influenced by a wide range of geochemical behaviour of heavy metals in soil, impacting their transduction by plants and pertaining impact on crop productivity. The presence of heavy metals in the environment has been found to have a negative impact on crop yield. This is due to the detrimental impacts they impose on several important physiological processes in plants, such as, the accumulation and remobilization of seed reserves for germination, vegetative plant growth, and photosynthetic fixation. Heavy metal toxicity at the cellular level has been observed to result in a decrease in crop productivity. This is mostly attributed to the generation of reactive oxygen species, which disrupt the redox balance and induce oxidative stress. Plants exhibit multiple defence strategies to cope with heavy metal stress, enabling them to effectively mitigate heavy metal toxicity and sustain their productivity. These mechanisms encompass diminished uptake of heavy metals by plants, sequestration of heavy metals within vacuoles, binding of heavy metals by phytochelatins, and activation of diverse antioxidant systems.

Heavy metals have been observed to have an impact on several cellular organelles and components inside biological systems. These include the cell membrane, mitochondria, lysosomes, endoplasmic reticulum, nuclei, as well as certain enzymes that play a role in metabolism, detoxification, and damage repair (Wang and Shi 2001). Previous studies have demonstrated that metal ions have the ability to engage with cellular constituents, including DNA and nuclear proteins. This interaction has been observed to result in detrimental effects such as DNA damage and alterations in molecular structure. These consequences have the potential to influence critical cellular processes such as cell cycle regulation as well as the development of cancerous cell (Fig. 7.2).

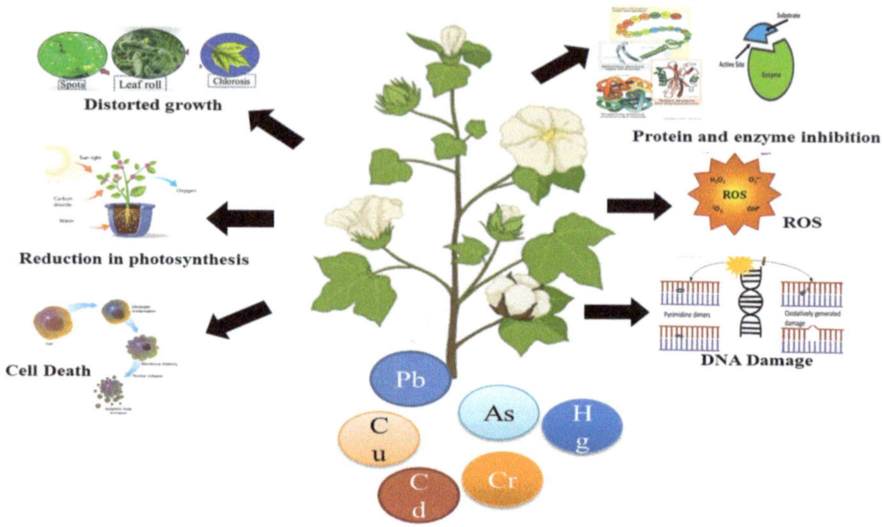

Fig. 7.2 Adverse effects of heavy metals on plants

7.3.1 Heavy Metal Toxicity and Inhibition of Photosynthesis

The elevated concentration of metals has a detrimental impact on plant metabolism, specifically in relation to the process of photosynthesis (Aggarwal et al. 2012). The adverse effects of metals on the process of photosynthesis mostly stem from the alterations in the structure and function of the photosynthetic apparatus in plants generated by metal exposure. Metal ions have the ability to hinder the functioning of many enzymes involved in photosynthesis, as well as the manufacture of chlorophyll. Metals have been observed to adversely affect the integrity of the chloroplast membrane system and disrupt the normal functioning of photosynthetic electron transport systems. Metals primarily induce toxicity in plants through the impairment of chloroplasts and disruption of photosynthetic processes. The interaction of metal ions with photosynthetic enzymes and chloroplast membranes leads to the inhibition of photosynthesis (Aggarwal et al. 2012). The accumulation of heavy metals in leaves of higher plants has an indirect impact on photosynthesis, as it disrupts the functioning of stomata. Consequently, this disruption impacts both photosynthesis and transpiration rates in a comprehensive manner. The reduction of photosynthetic pigments by heavy metals has an indirect impact on photosynthesis (Table 7.2). Consequently, the frequent utilization of non-destructive techniques and the ease of measuring photosynthetic pigments make them a valuable tool for assessing stress levels for regulatory purposes (Aggarwal et al. 2012).

While it is true that higher plants have the ability to accumulate heavy metals in their roots, it should be noted that certain detrimental elements are still able to penetrate the chlorophyll-containing cells. The process of photosynthesis exhibits a high degree of sensitivity towards heavy metals. This has been demonstrated by many

Table 7.2 Reduction in photosynthesis of different plant species by heavy metals

Heavy metals	Plant species got impacted	References
Cd	Lettuce	Gao et al. (2020)
	Wheat	Yotsova et al. (2020)
	White mustard	Mazur et al. (2016)
	Peanut	Shi and Cai (2008)
	Acer saccharinum L.	Lamoreaux and Chaney (1978)
	Lycopersicum esculentum Mill	Baszyńki et al. (1980)
Cr	*Spirodela polyrhiza*	Appenroth et al. (2003)
	Rice	Tripathi et al. (2012)
	Peanut	Zeid (2001)

studies conducted by Bertrand and Poirier (2005). The influence of heavy metals on chloroplasts, particularly triggered by Cd, has been found this leads to suppression of photosynthesis due to sub-lethal doses of Cd^{2+} in a range of taxa, from cyanobacteria to higher plants (Shukla et al. 2003).

The photosynthetic light processes, specifically Photosystem II (PS2), have been found to be particularly impacted by heavy metals (Tůmová and Sofrová 2002). The lack of organisation in the structure of chloroplasts results in the deactivation of the oxygen-evolving centres located on the donor side of photosystem II. This disorganisation also hampers the efficient transport of electrons. According to Voigt and Nagel (2002), in order to address the shortcoming of the PS II, it is necessary to increase the photo-fluence rates. Meanwhile, it is worth noting that the susceptibility of PS II to photoinhibition may be heightened as a result of the decline in anti-oxidative defence caused by metal pollution (Pätsikkä et al. 2002).

PS II abnormalities have been observed in many photosynthetic species that are contaminated with metals by other authors as well. The replacement of manganese (Mn) with zinc (Zn) at the photolysis site has been found to impede the emission of oxygen Mercury (Hg) also caused a reduction in the coefficients of photochemical and non-photochemical quenching. It was reported that a decrease in photochemical quenching (qP) and excitation capture efficiency of open PS2 (Fv'/Fm') in rice bean treated with Mn, accompanied by an increase in non-photochemical quenching (qN).

The decrease in photophosphorylation rates in the presence of Cd can be attributed to the malfunctioning of PS II, rather than a direct inhibition of the ATP. The impact of metal pollution on carbon metabolism in non metal-tolerant plants has been documented by many (Voigt and Nagel 2002), specifically in relation to the inhibition of photosynthetic dark reactions. The reduction in CO_2 absorption might occur due to a decrease in the number of stomata in plants that are exposed to Cd pollution (Baryla et al. 2001). According to a decrease in the demand for ATP and NADPH within the Calvin cycle leads to the down-regulation of PS2 photochemistry.

7.3.2 Heavy Metal Toxicity and Altered Water Uptake and Transport

The root is the first plant organ to come in contact with heavy metals, resulting in decreased elongation of the primary root, impaired secondary growth, or reduction in the formation of root hair. As a result, water absorption decreases in root tissues and the plant faces osmotic stress in Aroca et al. (2012). Heavy metals also influence water delivery to the shoot due to transpiration inhibition, reduction in stomatal density, and decrease in the leaf size of plants.

7.3.3 Heavy Metal Toxicity Stunted Growth and Reduced Yield

The existence of cadmium and copper in a plant's body competes with other minerals in apoplast and root vacuoles and thus decreases their transport among plant organs. The increasing concentration of chromium in paddy crops has shown a reduction in micronutrients (Mn, Cu, Zn, Fe) and macronutrients (N, P, K) in plants, ultimately leading to changes in growth, yield, and nutrient content of plants (Sundaramoorthy et al. 2010). In Arabidopsis thaliana cell death, decreased root growth, and stimulated callose deposition due to Chromium accumulation (Eleftheriou et al. 2015).

7.3.4 Heavy Metal Toxicity and Impact on Cell Wall Formation

One of the reactions happening inside the cell mass of plants when they are presented to outside pressure is the fast age of responsive oxygen species (ROS) because of NOX movement (Raggi et al. 2015). The development of superoxide revolutionaries (O–) thusly prompts the formation of hydrogen peroxide (H_2O_2) through the activity of superoxide dismutase (Grass). In the cell wall, class III peroxidases work with the oxidation of different substrates through the peroxidative cycle, bringing about the cross-connecting of the phone wall and the discontinuance of development. By and by, peroxidases can likewise work with cell wall unwinding (in this way advancing extension) through the hydroxylic cycle, in which the Fenton response produces hydroxyl extremists. These hydroxyl extremists start the non-enzymatic breakdown of cell wall polysaccharides.

7.3.5 Heavy Metal Toxicity and Decreased Pollen Tube Formation

To investigate the impact of Ni^{2+} on the processes involved in the initiation and maintenance of polar cell growth, we employed a well-established model system, specifically the germination of angiosperm pollen grains. In a liquid medium, tobacco pollen grains form elongated tubes with growth occurring exclusively at the tip. Ni^{2+} did not interfere with the formation of pollen tube initiators but hindered their subsequent growth, exhibiting an IC50 value of 550 μM. At a concentration of 1 mM, Ni^{2+} completely halted polar growth, although all pollen grains remained viable, with only a slight impact on respiration and no increase in the production of reactive oxygen species (ROS).

When Ni^{2+} was introduced after the germination had begun, it had a dual impact on tube development. Some tubes displayed notable extra-long growth, appearing to expand rapidly, while the growth of many other tubes was impaired. Our investigation into potential sites of Ni^{2+} influence revealed that it could affect structures both in the cell wall and within the cytoplasm, as confirmed by specific staining techniques. Ni^{2+} disrupted the proper distribution of transport vesicles at the tips of these tubes and notably decreased the relative calcium content in the aperture region of pollen grains, as determined through X-ray microanalysis. These factors are considered crucial for normal polar cell growth. Additionally, Ni^{2+} induced the deposition of callose in the tips of the tube initiators and in the pollen tubes that had ceased their growth. It is conceivable that the disruption of calcium balance induced by Ni^{2+} can lead to disturbances in vesicle transport and abnormal callose deposition, consequently blocking polar growth (Sawidis et al. 2018).

7.3.6 Heavy Metal Toxicity and Accumulation in Plant Tissues

Heavy metals including cadmium (Cd), lead (Pb), nickel (Ni), Zinc (Zn), or chromium (Cr) comprise primary abiotic stress leading to severe phytotoxicity. The root is the primary plant organ to come in contact with these hazardous metals, leading to interference in cell division, elongation, and proper functioning of root cells. This leads to formation of defective root architecture (Yadav et al. 2021). Heavy metals like, Cr(VI) and cadmium affect cell elongation in roots by reducing cyclin-dependent kinase expression leading to a defective mitosis process (Lešková et al. 2020). Firstly, the root tip comes in contact with metals, and exposure of the Arabidopsis root tip to Nickel (Ni) containing media showed root length inhibition significantly. Nickel toxicity leads to auxin sensitivity by reducing PIN-FORMED protein (PIN) levels in roots. Additionally, excess Ni increases the accumulation of reactive oxygen species and disorganizes microtubule integrity and orientation, ultimately affecting auxin shootward distribution.

The hyperaccumulator plants relocate heavy metal absorbed from roots to the shoots via the xylem with the help of different classes of proteins such as heavy metal-transporting ATPases, cation diffusion facilitator (CDF) family proteins, zinc–iron permease (ZIP) family proteins, and MATE (Multidrug and Toxin Efflux) protein family (Williams and Pittman 2010). The study has confirmed, various negative effects of metals on stem epidermal and cortical cells. For example, in *Pisum sativum* L. presence of Pb resulted in a ruptured epidermis, irregular cell division, and abnormal enlargement of cortical cells in the plant stem (Chaudhari et al. 2016; Tripathi et al. 2017).

7.3.7 *Heavy Metal Toxicity and Enzymatic Stress*

Heavy metal stress affects both chlorophyll biosynthesis enzyme porphobilinogen (PBG), δ-aminolevulinic acid (δ-ALA) and uroporphyrinogen III (Urogen III), chlorophyll degrading enzyme chlorophyllase (Chlase) and Mg-dechelatase (MDCase). This affects photosynthesis, accumulation of bio-assimilate, plant growth, and yield negatively (Yang et al. 2020). The effect of zinc toxicity, on the decrease in RUBISCO activity has been observed in Phaseolus vulgaris plant by replacing Mg^{+2} cofactor with Zn^{+2}, ultimately affecting photosynthesis (Monnet et al. 2001). Recently, the effect of cadmium (Cd) and copper (Cu) was observed on the perennial legume Pigeon pea with nutritional and medicinal values at the metabolite level. The reduction in heavy metal toxicity inflicted high levels of oxidative stresses in plants by increasing lipid peroxidation, hydrogen peroxide (H_2O_2), and electrolyte leakage. To combat this stress expression of antioxidant enzymes such as superoxide dismutase (SOD), peroxidase (POD), and catalase (CAT) was increased. In contrast, enzymes related to phenolic and flavonoid biosynthesis were reduced, leading to a decrease in the medicinal activity of plants (Hayat et al. 2021).

7.4 Impact on Soil Health Under Heavy Metal Stress

Soil health is the primary factor in determining the nutritional content of the food material which is further consumed by humans and other living organisms. Natural and anthropogenic activities such as energy-intensive industries, sewage water, and vehicular pollution are sources of heavy metal contamination in soil, which in turn is getting absorbed by plants. Accumulation of heavy metals in soil adversely affects soil microbiota and the interaction of microbe with soil. Beneficial soil microbes such as nematodes, insects, invertebrates such as earthworms, and mammals all get affected (Rai et al. 2019). The soil microbial community plays a primary role in the fundamental processes in the mineralization of organic matter degradation and allows the recycling of nutrients. Heavily metal-contaminated sites showed a significant decrease in the ratio of the microbial biomass carbon to oxidizable carbon content (Friedlova 2010).

The soil serves as a multifaceted system situated at the juncture of the atmosphere, lithosphere, hydrosphere, and biosphere. Soils have an enormous impact on determining the economic standing of nations, making them a significant form of natural capital. This is primarily due to its pivotal function in food production and its contribution to sustainability by supporting essential social and ecological functions. Soils play a crucial role in facilitating plant development and sustenance by supplying essential components such as micronutrients and macronutrients (e.g., nitrogen, phosphorus, sulphur, potassium, calcium, and magnesium), as well as oxygen and carbon that provide structural support in soils, allowing plants to grow and thrive, resulting in crop production that caters to human needs for food, fuel, and fibre. But because of climate change and anthropogenic factors, soil is severely degrading, putting access to healthy food at risk for generations to come. One form of pollution that poses a significant environmental challenge is the presence of heavy metals, which is closely linked to the progress of agriculture and industrial activities. The phenomenon has been seen to have adverse impacts on plant life and poses a potential risk to human well-being due to the accumulation of heavy metals in crops (Pereira et al. 2018; Jogawat et al. 2021; Lehmann et al. 2020). Heavy metals and metalloids have polluted the soils of 5 million locations around the globe, spanning 20 million hectares of land. It is further projected that heavy metal pollution costs the global economy more than $10 billion annually (Liu et al. 2018).

There are around 70 metallic chemical elements that are categorised as heavy metals, with quantities in the earth's crust ranging from less than 0.1% to less than 0.01%. Since heavy metals are non-biodegradable in nature, they cannot be easily eradicated from the environment. Certain heavy metals have been identified as immobile, meaning they lack the ability to displace themselves from the location where they have gathered. Conversely, there are heavy metals classified as mobile due to their capacity to be absorbed by the root system of plants through processes such as diffusion, endocytosis, or metal transporters. It is worth noting that heavy metals are also considered essential components for plant growth. However, in cases where these heavy metals are present in inadequate concentrations, plants have been seen to exhibit deficiency symptoms that are often characterised by the discoloration or browning of leaves and overall growth suppression. Further, these heavy metals are constituents of metalloproteins, cofactors in enzyme catalysis, and in several other cellular activities. On the contrary, when present at concentrations higher than ideal, these heavy metals can exhibit phytotoxic effects, leading to leaf chlorosis, stunted plant growth, etc. On the other hand, some non-essential heavy metals, such as mercury, arsenic, lead, cadmium, etc., are phytotoxic and disruptive even at minimal concentrations and hence induce detrimental effects on the normal growth process of plants.

The primary factors that govern the adsorption of heavy metals and their dispersion between soil and water are the type of soil, metal speciation, concentrations of different metals, pH of the soil, solid-to-solution mass ratio, and contact duration.

Heavy metals are a significant class of contaminants also found in landfill leachate. Further, the presence of such leachate has the potential to result in widespread contamination that has a substantial influence on the chemical characteristics of soil. This, in turn, disrupts the equilibrium of nutrients, leading to adverse effects on both the productivity and growth of crops. The process of acidification has a detrimental effect on soil fertility as it leads to the depletion of essential nutrients for plant growth. For instance, the process of smelting sulphur dioxide minerals results in the production of SO_2. If this SO_2 is not captured and treated, it is released into the atmosphere, contributing to the phenomenon of acid deposition. Although it has been observed that the acidification process is reversible in nature when a subsequent drop in sulphur emissions occurs, leading to an increase in soil pH. However it is imperative to acknowledge that the length of the recovery period might exhibit substantial variation and is dependent upon the individual attributes of the soil (Beinabaj et al. 2023).

As per the study conducted by Wang et al. (2006), the availability of metals to plant roots in the soil is greatly influenced by the pH and diminishes as the pH of the soil increases. In addition, it should be noted that soil colloidal particles exhibit substantial interface and specific surface areas, thus playing a pivotal role in the control of trace element and heavy metal concentrations within natural soil environments. Therefore, the existence of soil particles with a substantial specific surface area might potentially decrease the soluble concentration of some metals within the contaminated soil.

It is important to acknowledge that in the majority of plants, the roots serve as the initial organ that comes into contact with the soil, hence potentially encountering heavy metals. As the heavy metal ions engage in competition with cations that are important nutrients for the purpose of binding and absorption on the surface of the root, This implies that soil has a substantial role in the manifestation of phytotoxic effects caused by heavy metals, particularly through the roots, which ultimately hinder the overall growth performance of the affected plants (Singh et al. 2016). However, both the underground and aerial surfaces of plants have the capacity to absorb heavy metals, which can have can exert either a direct or indirect influence on the overall health of the plant (Dutta et al. 2018). Additionally, it has been noticed that fine soil particles have a higher affinity for heavy metals than coarse soil particles (Mishra et al. 2017). The dispersal of small particles originating from the top layer of soil as a result of different environmental variables, such as rain and wind, is a significant problem in the context of heavy metal contamination (Liu et al. 2017).

Another aspect that must be taken into consideration is the presence of heavy metal-contaminated soil, which has the potential to adversely affect the microbial community of the soil. Soil microorganisms have a crucial role in the process of soil organic matter breakdown as well as in the promotion and maintenance of plant development under heavy metal stress. Consequently, a reduction in microbial variety or abundance might have detrimental consequences for the ability of plants to absorb nutrients from the soil. Nevertheless, soil microorganisms have been seen to actively participate in the process of phytoremediation in soils that are polluted with heavy metals. It has been previously mentioned that soil microbes, whether existing free-living or in symbiosis, inside the rhizosphere of plants cultivated in metal-contaminated soils have the potential to augment plant biomass output and facilitate the phytoremediation process (Xie et al. 2016; Chu 2018).

7.5 Strategies in Managing Heavy Metal Toxicity

Soil pollution by heavy metals has thc potential to negatively impact the functioning of natural ecosystems and ultimately pose an imminent danger to human health through their incorporation into the food chain. Heavy metals are recognised as potent carcinogens capable of inducing cancer. Heavy metals are among the most potent carcinogens, and studies have shown that prolonged exposure to elevated concentration of heavy metals can result in enduring intellectual and developmental impairments, such as difficulties with reading and learning disabilities, behavioural issues, hearing loss, attention deficits, disruptions in the development of visual and motor functions, etc.

Several remediation strategies have been developed to effectively manage, purify, or rehabilitate soils that have been polluted with heavy metals. In order to mitigate the adverse effects of metal toxicity, there are three primary approaches: physical (which includes calcination, excavation, landfilling, and leaching), chemical (such as precipitation, soil washing, reverse osmosis, flocculation, nanofiltration, chemical reduction, ion exchange, and evaporation), and biological (which includes phytoremediation or microbial-mediated remediation) processes. Among these options, the physicochemical strategies are known for their speed but are hindered by their high cost, technical intricacy, and further generation of secondary pollutants. The biological strategy has been discovered to be cost-effective and ecologically sustainable. Biologically derived remediation solutions, such as phytoremediation, have been predicted to be much more cost-effective, ranging from four to a thousand times less expensive per unit volume compared to existing non-biological technologies. In recent times, there has been a notable emphasis on the utilisation of plants (phytoremediation process) as well as the bacterial and fungal microbiome (microbial remediation) within the realm of bioremediation of heavy metals from soil, which shows possibilities within the biological approach. Research on the utilisation of nanomaterials for the extraction of heavy metals has gained significant attention. This is due to the potential of nanomaterials as effective adsorbents and catalysts in the field of environmental remediation. Their notable chemical reactivity, expansive adsorption surface, and capacity for low-temperature modification contribute to their growing relevance in this area. Despite the extensive research conducted on nanomaterials in the domain of heavy metal remediation from soil. In the subsequent discussion, we will address the remediation techniques and various management methods that could potentially be implemented to effectively manage soil pollution caused by heavy metals (Rizvi et al. 2022; Yu et al. 2021).

7.5.1 Remediation of Heavy Metal-Contaminated Soil

The proliferation of industrialization, natural causes like weathering of rocks, volcanic eruptions, and the exponential increase in global population have led to a significant and pressing concern over the pollution of the environment and its subsequent impact on food security. The primary concerns associated with heavy metal pollution are its challenging detectability as well as its persistent and irreversible nature. The presence of heavy metal pollution has a detrimental impact on the quality of water resources, the atmosphere, and food crops. Consequently, this presents a substantial risk to the health and well-being of living species, since these pollutants have the potential to accumulate within the food chain. Hence, in instances where heavy metals have permeated the soil, the utilisation of remediation procedures becomes crucial. These techniques are expected to enhance soil quality while simultaneously achieving their primary objectives of reducing metal concentrations and enhancing phytoavailability (Li et al. 2019; Morkunas et al. 2018; Lwin et al. 2018). The following section provides a brief summary of the biological and nanotechnology-based approaches utilised in the remediation of soil polluted with heavy metals.

7.5.1.1 Microbial Assisted Heavy Metal-Contaminated Soil Remediation

Soil microorganisms play a vital role as active decomposers within the soil ecosystem, hence contributing to the maintenance of ecosystem stability. Soil microbial variety is indicative of the extensive microbial population inhabiting the soil and serves as a significant factor in preserving soil fertility and promoting the health of plants. Furthermore, the elevated accumulation of metals within the soil–plant systems induces selection pressure, leading to the emergence of resistance mechanisms within the microbiome against these hazardous metals. Microorganisms commonly encounter several heavy metals when residing in their native habitats. Hence, the microorganisms exhibit a natural inclination to withstand these metallic elements, so showcasing their viability for application in the restoration of soils contaminated with heavy metals. The capacity of microorganisms, such as arbuscular mycorrhizal fungi, bacteria, and fungi, to remediate soil contaminated with toxic heavy metals has been documented in scientific literature. The interaction that occurs between heavy metals and microbial communities can be categorised into various fractions, including extra-cellular, exo-cellular, and intra-cellular interactions. Within these categories, microbial cells have the capacity to perform several actions, such as mobilising, immobilising, transforming, precipitating, accumulating, coordinating, exchanging, and absorbing metals. Additionally, microbial communities can also form complexes with heavy metals. Arbuscular mycorrhizal fungi, which are well recognised as significant soil microorganisms, have been observed to possess the ability to store heavy metals inside their mycelia. Furthermore, it has been revealed that arbuscular mycorrhizal (AM) fungi employ cellular pathways to metabolise heavy metals and mitigate their cytotoxic effects. Thus, the accumulation of these heavy metals serves the purpose of impeding the transport of heavy metals to plant cells, hence mitigating the harmful effects associated with such metals (Gao et al. 2022; Passarini et al. 2022).

Fungal strains possess the advantageous characteristic of exhibiting cell wall material that has exceptional metal-binding capabilities. Furthermore. There have been several instances. Numerous fungal species, including *Rhizopus arrhizus*, *Penicillium spinulum*, and *Aspergillus niger*, have undergone comprehensive investigation on their capacity for biosorption of heavy metals. However, the underlying mechanism of this process appears to vary according to the specific species involved. In a research investigation, a strain of Aspergillus flavus known as KRP1 was shown to possess the ability to withstand mercury (II) contamination. The fungus strain exhibited the capacity to eliminate 97.50 and 98.73% of mercury (II) from a solution containing 10 mg/L of the contaminant in both agitated and static conditions, respectively, when grown in PDB medium (Kurniati et al. 2014; Iram et al. 2013). In another study whereby the *T. harzianum* and *K. phaffii*, two fungal species isolated from soil, exhibited heightened resistance to escalating levels of heavy metals like Cadmium, Chromium, and lead (Liaquat et al. 2020).

The pigments seen in different strains of fungi and bacteria are also hypothesised to provide cellular protection against heavy metal toxicity. The physicochemical characteristics of melanins, along with their prevalence in the microbial realm, render these compounds and the microorganisms harbouring them valuable for cellular protection and as potential agents for heavy metal remediation. It is suggested that some fungal laccases have a role in the synthesis of diverse pigments. Accordingly, one potential role of laccase may include the copper-induced synthesis of melanin. According to research conducted, the fungal strain *T. pubescens* exhibited a dark brown coloration in its mycelia when cultivated in the presence of copper (Cu). The stimulation of extracellular laccase synthesis in this fungal strain can be attributed to the addition of Cu(II) to a glucose-based growth medium. In addition Moreover, it appears that the basidiomycetes inhabiting wood also demonstrate the phenomenon of biosorption. The ability of this fungus to bind heavy metals is attributed to the composition of its cell walls, which consist of polysaccharides, peptides, and pigments. Hence, the use of fungal and bacterial pigments holds promise for the development of an environmentally friendly remediation approach to address the issue of heavy metal contamination in soil (Cao et al. 2020).

Bacteria and other microorganisms engage in a range of methods to remediate metals found in polluted soil. These mechanisms encompass cell surface adsorption, enzyme-mediated transformation, precipitation, metabolic intake, biomineralization, bioaccumulation, and microbial-mediated oxidation or reduction of metal species, etc. The bacterial strains have the capability to encounter toxic heavy metals throughout the remediation process. It has been shown that these bacteria have developed alterations in their cell wall, membrane, envelope, or S-layer as a means to eliminate these metals. In this particular mechanism, bacteria employ a defensive strategy to safeguard their vital cellular constituents that are susceptible to the detrimental effects of metallic elements. In this particular mechanism, bacteria employ a defensive strategy to safeguard their vital cellular constituents that are susceptible to the deleterious effects of metallic elements. Metal adsorption mediated by bacteria occurs within the functional groups of bacterial cells, which serve as binding sites. In addition, several bacterial species possess a cell wall or outer cell envelope that has the ability to passively adsorb significant quantities of dissolved metals, often by charge-mediated attraction (Devi et al. 2022). Biosorption represents an alternative approach employed for the purpose of heavy metal elimination. It comprises a series of processes that encompass adsorption, covalent binding, Vander Waal's forces, ion exchange, chelating, and precipitation reactions and explain the movement of ions from the solution phase to the solid phase. Furthermore, the bacterial surfaces facilitate the adsorption of soluble metal ions by engaging in interactions with different functional groups, including carboxylate, amine, amide, imidazole, phosphate, thioether, hydroxyl, and other similar groups present in cell wall biopolymers (Ayangbenro and Babalola 2017).

In an investigation conducted by Kalaimurugan et al. in 2019, it was stated that Pseudomonas fluorescens and Bacillus safensis bacterial strains have been discovered to have significant chromium degradation capabilities, achieving rates of up to 84% and 72%, respectively (Jiang et al. 2008). Another study has reported the isolation of the bacterial strain Enterobacter cloacae B1 from soil contaminated with pollutants in Ghaziabad, India. The bacterial strain has demonstrated a notable capacity for bioaccumulation, with an exceptionally high percentage of accumulation seen for copper (95.25%), followed by cadmium (64.17%) and nickel (36.77%) (Banerjee et al. 2015).

7.5.1.2 Implementation of Nanotechnology in Heavy Metal-Contaminated Soil Remediation

The discipline of nanotechnology has emerged as a significant area of study in bioremediation due to its capacity to fabricate materials with specific properties, such as a high surface area to volume ratio distinct from those of bulk materials. Nanotechnology refers to the production process of particles that are smaller than 100 nm in size. The synthesis of nanoparticles may be achieved by three main approaches, including biological, physical, and chemical methods. However, in terms of environmental sustainability, there is a preference for the biological-mediated production of nanoparticles (Das et al. 2018, 2019). In the realm of biogenic synthesis, which encompasses many organisms such as plants, fungi, actinobacteria, algae, bacteria, etc., there exists a diverse array of processes that need further investigation.

Various strains of bacteria, algae, and Plant extracts have been shown to be capable of synthesising metallic nanoparticles through both intracellular and extracellular processes (Das et al.; Verma and Bharadvaja 2022). Moreover, the nanoparticle possesses many functional groups, including $-COOH$, $-NH_2$, $-OH$ etc. Studies have demonstrated that modifying the surface of the nanoparticle through physical or chemical activation can result in improved removal of heavy metals. As nanotechnology is utilised in the production of biosorbents, the resulting nanobiomaterials have exhibited notable characteristics such as a large surface area for reaction, enhanced heavy metal adsorption abilities, unique optical and magnetic properties and the ability to serve as immobilisation carriers (Verma et al. 2021). Nanoremediation is a novel remediation approach that is predicated on the utilisation of nanoparticles. Nanoremediation employs the inherent properties of nanoparticles, such as their mobility, reactivity (specifically as catalysis), and adsorption capacities, which have been effectively utilised in the process of eliminating heavy metals in soil and various other environments. A recent study has documented the capacity of highly reactive nano-scale biogenic magnetite (BnM), which is produced by the Fe(III)-reducing bacteria Geobacter sulfurreducens, to effectively reduce Cr(VI) in alkaline solutions (Ganie et al. 2021; Mansoor et al. 2021; Abed Hussein et al. 2022).

Biochar is a substance that is derived through the process of pyrolysis, which involves the thermal decomposition of biomass obtained from various sources such as agricultural residues, sewage, and animal byproducts. In addition, biochar possesses the ability to effectively trap contaminants within soil environments. In addition to its aforementioned features, biochar possesses characteristics that contribute to its efficacy in enhancing soil quality. Through its application, biochar not only aids in the reduction of soil organic matter but also facilitates the sequestration and mitigation of carbon dioxide emissions from the soil. Furthermore, biochar exerts a positive influence on the physical and chemical attributes of the soil, fostering improved growth rates and increased yields. The conversion of biochar into nano biochar can be achieved by many transformative procedures. The pores of nanobiochar may be categorised into three groups based on their size: (1) macrospores, which have a diameter above 50 nm; (2) mesopores, which have a diameter ranging from 2 to 50 nm; and (3) microspores, which have a diameter less than 2 nm. Micrometre porosity, in a broad sense, serves the function of facilitating the ingress of oxygen-containing functional groups within the biochar matrix, hence providing a conducive environment for microorganisms to thrive and safeguard themselves from unfavourable metal-contaminated soil Based on the findings of research, arsenic can be effectively removed using iron oxide-permeated mesoporous rice husk nanobiochars. In a separate investigation, the synthesis of nanoB from goat dung was conducted, followed by its utilisation in conjunction with CuO nanoparticles to exert an impact on soil microorganisms, nutritional composition, and wheat yield. Additional studies have indicated that the combination of nanobiochar with copper oxide nanoparticles, a vital element required for seed and chlorophyll synthesis, resulted in an enhanced absorption of copper in wheat plants. In addition, the combined application of nanobichar and copper nanoparticles had an advantageous impact on both microbial biomass and the availability of micronutrients (nitrogen, phosphorus, and potassium) in the soil (Elbehiry et al. 2022; Rashid et al. 2023). Hence, nanotechnology has emerged as a promising avenue for the rehabilitation of soil contaminated with heavy metals.

7.5.1.3 Phytoremediation Mediated Heavy Metal-Contaminated Soil Remediation

Phytoremediation is an established method that employs the use of live plants to facilitate the cleanup of metallic contaminants present in soil. It is characterised by its environmentally conscious and sustainable approach. Nevertheless, this strategy is not without its limitations, one of which being its rather slow rate. Furthermore, the frequency and intensity of this phenomenon are often contingent upon seasonal variations and geographical locations, making it particularly susceptible to the impacts of climatic fluctuations. Furthermore, plants possess the ability to function as hyperaccumulators, enabling them to accumulate potentially phytotoxic substances at concentrations exceeding those observed in non-accumulators by more than 100-fold. The plants that accumulate metals possess robust systems for sequestering metals and may exhibit higher internal demands for certain metals. Certain species have the ability to extract metals from soil fractions that have lower solubility, a characteristic that distinguishes them from non-hyperaccumulating species (Rashid et al. 2023). The concept of tolerance to heavy metals in plants may be described as the capacity to endure and thrive in a soil environment that is hazardous to other plant species. This ability is often observed through various genetic, biochemical, or molecular alterations in response to the conditions in the environment. Additionally, the use of perennial crops in soils polluted with heavy metals with the intention of phytoremediation circumvents the issue of competing land usage with food crops. In a study conducted by Barbosa et al., the researchers assessed the metal tolerance capability of two plant species, namely Giant reed (*Arundo donax*) and *Miscanthus* spp., in order to address the issue of heavy metal contamination in soil. And further, heavy metal contamination does not have any impact on the quality of the biomass. Metal-accumulating plants, while possessing potential as viable candidates for phytoremediation, typically exhibit limited annual biomass production, rendering them unsuitable for extensive soil detoxification on a broad scale. Therefore, to implement field-scale phytoremediation, it is recommended to utilise high-biomass crops that have the ability to collect adequate amounts of metals in their shoots. This approach is advantageous due to the possibility of a significant increase in above-ground biomass output, which can effectively offset the decreased concentration of potentially harmful elements in plant tissues. In an investigation, a greenhouse experiment was undertaken to assess the efficacy of *Rhazya stricta* as a phytoremediator for heavy metals in contaminated soil. The quick development of the plant and subsequent studies have provided more insights into the translocation factor, which follows the order Zn > Cu > Cd > Pb, with Zn exhibiting the highest value and Pb the lowest.

7.5.2 Strategic Management of Heavy Metal-Contaminated Soil

Heavy metals, which form naturally occurring elements that reside within the Earth's crust and often exhibit low natural quantities in soil. As a result of various anthropogenic activities, heavy metals get discharged into the environment. Different countries face significant variations in their understanding of and approaches to addressing the issue of heavy metal pollution in soil.

7.5.2.1 pH Adjustment and Management of Heavy Metal Toxicity

The byproduct of pyrolyzing biomass is biochar, which can be added to contaminated soil to stabilise metal levels. Although little is known about how biochar influences metal mobility in soils. The study concentrated on three areas to determine the main processes as the underlying reasons (i) whether the transport within the biochar particles creates a kinetic limitation; (ii) how the mineral phases in the biochar change over time; and (iii) how the pH of the soil is affected by biochar. In order to evaluate the sorption–desorption relationships for lead (Pb), zinc (Zn), copper (Cu), cadmium (Cd), and nickel (Ni) in wood-derived biochar. The sorption results were compared with extraction tests conducted using biochar.

Despite having comparable specific surface areas, results showed that the size of the biochar particles affected the rate at which metals were sorbent. This implied a constraint pertaining to the dispersion of metals within the particles of biochar. According to sorption isotherms, the order of metal sorption to biochar was Pb > Cu > Cd > Zn > Ni, and the sorption was partially reversible. The primary explanation for this observation was the metals' precipitation or adsorption onto particular mineral phases in the biochar.

Metals in both contaminated soils were successfully immobilised by biochar, though this wasn't predicted purely by the sorption isotherms. Metals in both contaminated soils were successfully immobilised by biochar, though this wasn't predicted purely by the sorption isotherms. The biochar-induced increase in soil pH that resulted in increased metal retention on soil particles appeared to have an impact on this immobilisation. Therefore, both the processes that raise soil pH and the diffusion within biochar particles appear to have an impact on the short-term effects of biochar on contaminated soils (Rees et al. 2014).

7.5.2.2 Addition of Organic Matter to Manage Heavy Metal Toxicity

It's an effective practise to immobilise pollutants in the soil. In this regard, the soil's organic matter content—in particular, its fulvic and humic acids—is advantageous. This is due to their potent adsorption powers against a wide range of pollutants, including heavy metals, which can cause them to immobilise and protect food and groundwater from contamination. Human activity-modified soils frequently lack organic matter, or humic substances, which diminishes the natural protective capacity of the soil. Restoring the protective function of the soil and thereby lowering the risks to the environment and public health in areas affected by human activity may be accomplished by the rejuvenation of humic substances through the humification of organic matter from diverse sources added to the soil. The effectiveness of organic materials such as those derived from brown coal, brown coal-derived preparations, and farmyard manure was studied for their ability to reduce the uptake of heavy metals by plants and their mobility within groundwater by immobilizing them in the soil. A conceptual framework was created to illustrate the soil's function as a protective barrier against the absorption of heavy metals by plants and their movement into groundwater. Various organic additives aimed at diminishing the immobilization and, consequently, the accessibility of heavy metals in soils were assessed using field stone containers. The presence of the tested heavy metals in plants was at its lowest when a brown coal-derived preparation was utilized as an organic supplement, as evidenced by the lowest bio-accumulation index (BI) values: 0.24 for Cd (in winter wheat grain), 0.021 for Pb (in winter wheat straw), and 0.48 for Zn (in spinach) (Kwiatkowska-Malina 2018).

7.5.2.3 Selection of Stress Tolerant Plant Species

Abiotic stress factors, such as heavy metals, are dangerous environmental pollutants that are present in soil and can significantly affect plant development, metabolism, overall crop yield, and sustainability. Plants that are exposed to elevated concentrations of heavy metal stress undergo oxidative stress, which can lead to diverse modifications at the molecular, biochemical, morpho-physiological, and plant level. It is essential for plants to evolve specialised defence mechanisms in heavily heavy metal-contaminated soils in order to offset the negative effects of these pollutants and support healthy growth and development. Through a process known as phytoremediation, plants have developed efficient ways to either absorb and store heavy metals in various sections of their structure or to change them into non-toxic forms. Genetic engineering, plant stress physiology, transgenics, and other recent bioscience advances have made it possible to identify and characterise compounds, transcription factors, gene products, and external agents that are essential to signalling pathways and stress-related proteins involved in heavy metal detoxification and tolerance. The problem of improving plant resistance to heavy metal stress using designed techniques is still not well studied, despite recent achievements. The possibility of developing genetically altered crop plants that are more resistant to heavy metal stress may open the door to the development of plant tolerance through engineering, which might result in the creation of better crop types that can survive risks brought on by heavy metals. This strategy is a potentially effective way to manage plant stress and improve the economic yield and sustainability of crop plants in agriculture (Wani et al. 2018).

7.5.2.4 Addition of Chelating Agents

The prolong pace of phytoremediation due to limited metal availability in the soils, making it a slow natural process. The interface between solid particles and solution is the site of the main reactions governing the bioavailability of heavy metals in soil. The elements like carbonates, clay minerals, organic matter, soil microbes, iron and manganese oxides, and organic matter all have an impact on the bioavailability and transport of metals. Lead (e.g., Pb) and other metals are mostly found in residual form in soils with higher clay content, cation exchange capacity (CEC), organic carbon (OC), biological activity, and crystalline and amorphous forms of iron and aluminium. It is respectively followed by Mn-Fe-associated and organic matter-associated forms. It is reported that a higher percentage of these metals are in carbonate-associated form in soils with higher sand content, pH, and equivalent calcium carbonate (EEC). It is important to remember that different mechanisms can alter these distributions depending on whether soil microorganisms are present or absent. The use of chelating agents in soil is advised to improve the bioavailability of heavy metals and the effectiveness of phytoremediation. Chelating agents improve the availability of nutrients in the soil and increase the bioavailability of metals, which enhances the effectiveness of phytoremediation. Chelating agents fall into two primary categories: natural (such as siderophores, organic acids, and manures) and synthetic (such as polycarboxylates and hydroxycarboxylates). The most effective chelating agent for boosting metal availability in soil and facilitating their translocation to plant parts above ground is synthetic chelate EDTA, which is widely mentioned in literature. Nevertheless, EDTA might not be as effective when used to increase metal concentration as it is when it comes to biomass production, which is crucial for effective phytoremediation. This is because both the toxicity of EDTA by itself and the elevated toxicity of heavy metals that EDTA causes can prevent plants from producing biomass. During plant growth, it might be wise to apply EDTA to the soil at various times and in lower concentrations.

Natural chelating agents are a significant improvement in biomass production, but they might not be as good as EDTA at increasing metal concentration. Electrokinetics is another method used to boost the availability of heavy metals and their movement through the soil in the direction of plant roots. The use of chelating agents in conjunction with electrokinetics has been studied recently, and the results have demonstrated a notable increase in plant uptake of metals. Because the electric field increases the permeability of biomembranes and facilitates the exchange of metals across the disrupted membrane structure, plants that are negatively charged are able to absorb more cations through their roots. When chelating agents simultaneously supply more bioavailable metals to plant roots, promising the better results. Thus, the use of chelating agents in conjunction with electric fields significantly improves phytoremediation (Sinegani et al. 2015).

7.5.3 Monitoring and Assessing Heavy Metal Contamination in Soil

Numerous studies have revealed an increasing worldwide occurrence of sites that are polluted with high concentrations of heavy metals. Hence, it has become imperative to develop effective methodologies for accurately determining the content of heavy metals in soil. The quantification of heavy metal concentrations in soils is feasible; however, its accurate estimation relies on the use of extensive sampling and the utilisation of physical or traditional analytical methods. Nevertheless, the process of systematic sampling and subsequent laboratory examination is characterised by its time-consuming nature, labour-intensive requirements, and significant financial costs. Moreover, the implementation of techniques such as machine learning and remote sensing is crucial for the mapping of vegetated zones and mining sites in order to monitor heavy metal contamination. The dispersion and diffusion of heavy metals throughout mining regions pose a significant threat to the ecosystem due to the potential for abandoned tailings to serve as a prominent contributor to environmental contamination. Heavy metals present in mine tailings have the potential to be transferred by many mechanisms, such as wind dispersion, surface water flow (which includes acid mine drainage), and groundwater leaching. Remote sensing has been employed in studies examining the dispersion of heavy metals as a prompt technique of initial analysis, utilising high-resolution imaging spectrometer (hyperspectral) data. It has also shown promise as a tool for field and imaging spectroscopy in the detection of heavy-metal pollution in mining sites by identifying minerals. Therefore, it is important to enhance the investigation of the distribution and source of heavy metal contaminants in soil. This will subsequently facilitate the assessment of risks and inform decision-making processes pertaining to remediation strategies.

7.5.4 Public Health Risk Assessment and Awareness

Various food sources, including vegetables, meat, cereals, fruits, and fish, are essential components of the human diet since they contribute significantly to the intake of protein, vitamins, carbohydrates, calcium, iron, and other vital micronutrients. The human body necessitates these nutrients in order to fulfil its nutritional requirements. The issue of heavy metal contamination has emerged as a notable worry due to its potential to infiltrate the food chain and then be absorbed into our bodies through our consumption of food. As Plants that are cultivated in soils contaminated with heavy metals have the inherent ability to accumulate and retain these elements inside their tissues. For instance, the element Cd has the highest tendency to transmit into the food chain due to its higher mobility in both soil and plants compared to other heavy metal contaminants. Excessive consumption of cadmium (Cd) has been found to have detrimental effects on many organs, including the liver, lungs, kidneys, and bones. A build-up of Cd in humans is associated with an increased likelihood of developing lung adenocarcinoma, cardiovascular disease, low immunity, etc. Likewise, arsenic, known for its detrimental effects on human health even in minuscule quantities, poses a particular risk to pregnant women and young children and exhibits heightened susceptibility to the poisonous properties of heavy metals (Liang et al. 2019; Bhatia et al. 2015).

Periodic testing and monitoring of soil, plant, and water quality in agricultural areas are essential for the timely adoption of preventive measures aimed at mitigating the entry of metals from soil to plants. Consequently, this strategy is essential for the purpose of mitigating any health risks for consumers. Furthermore, it is necessary to establish regulations pertaining to the release and disposal of various types of garbage, including domestic waste, e-waste, industrial waste, and mining waste. In addition, there is the possibility of promptly implementing a comprehensive system for the collection and recycling of waste materials, as well as the restoration of land contaminated with heavy metals through the use of various remediation techniques.

Additionally, the implementation of thorough monitoring and effective removal techniques for heavy metals in irrigation systems, specifically targeting metal-contaminated wastewater, river water, and groundwater, As the use of water that is polluted with heavy metals, there is a possibility that metals will accumulate in agricultural soils.

In addition to the general awareness among consumers regarding food quality and their understanding of the adverse consequences associated with heavy metal contamination. The dissemination of knowledge to farmers regarding their active involvement in the remediation of heavy metal contaminants, along with the regular monitoring of soil quality in their cultivated fields, holds equal significance. Hence, it is imperative to establish comprehensive methods, regulations, and an extensive awareness programme to effectively monitor and mitigate the dissemination of heavy metals into the food consumed by individuals (Fig. 7.3).

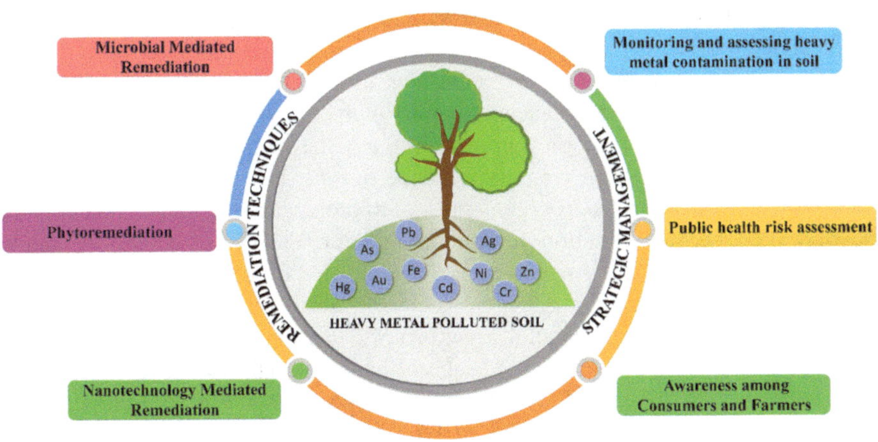

Fig. 7.3 Comprehensive method of monitoring and mitigation of heavy metals

7.6 Conclusion

The impact of heavy metal toxicity due to its natural and anthropogenic causes of aggregation has significant effect on environment, soil quality and plant health. Heavy metals aggregation is due to anthropogenic factors includes activities like agriculture, mining, industrial production. While the natural sources of heavy metal production include factors like volcanic eruptions and sedimentary rocks. There are numerous ways in which the heavy metal stress impacts the plant stress. Heavy metals can disrupt various physiological processes in plants, reducing crop productivity. These perturbations include effects on seed germination, growth, photosynthesis, and cellular processes. Plants have evolved defence mechanisms to cope with heavy metal stress, including absorption of heavy metals and activation of antioxidant systems Heavy metals as well can inhibit photosynthesis, affect plant metabolism and reduce yields. Interference with the chloroplast and photosynthetic electron transport systems can reduce the rate of photosynthesis. In addition, heavy metals can interfere with water absorption and transport, stunting plant growth and reducing yields. The accumulation of heavy metals in plant roots can affect the nutritional status of plants and overall health, ultimately affecting humans and consuming animals Plants accumulate heavy metals in their roots and shoots, and causes phytotoxicity. Heavy metal toxicity can also cause enzymatic stress, affecting enzymes involved in chlorophyll biosynthesis and degradation. This can reduce photosynthesis and plant growth. The impact of heavy metal stress extends to soil health, as contaminated soils can disrupt soil microbial and nutrient cycles. Heavy metals can affect beneficial soil bacteria and cause soil nutrient imbalances, affecting plant growth and crop productivity.

Heavy metallic toxicity is a complex problem with a long way-attaining results for the surroundings, plants, and soil health. Efforts to mitigate heavy metallic pollutants and its results on ecosystems are critical to make certain a sustainable and healthful environment for future generations. In end, soil pollutants via heavy metals poses a large threat to herbal ecosystems and human fitness via the infection of the meals chain. Heavy metals are robust carcinogens and can lead to numerous developmental and highbrow impairments, making their remediation crucial. Several remediation strategies were advanced, together with bodily, chemical, and biological methods. Among those, biological approaches, together with microbial remediation and phytoremediation, have won prominence due to their cost-effectiveness and ecological sustainability. Microorganisms like micro-organism and fungi have been shown to play a critical function in remediating heavy steel-infected soil via mechanisms like adsorption and precipitation. They can also form complexes with heavy metals, therefore reducing their dangerous effects. Additionally, the usage of fungal and bacterial pigments, like melanin, gives promise in mitigating heavy metallic toxicity. Nanotechnology, especially the usage of nanomaterials, has shown extremely good potential in heavy metal remediation. Nanoparticles can adsorb and catalyze the removal of heavy metals from soil, making them precious tools in environmental remediation. Biochar, derived from pyrolysis, can also be used to trap contaminants in the soil and enhance soil excellent. Phytoremediation, the usage of plants to smooth up heavy metallic-contaminated soil, is an environmentally conscious approach, but it has obstacles in phrases of velocity and effectiveness. The desire of stress-tolerant plant species and the genetic engineering of plants for heavy metallic tolerance can enhance the phytoremediation procedure. Strategic control of heavy steel-contaminated soil includes pH adjustment, the addition of organic count, and the usage of chelating sellers to improve steel bioavailability and translocation. Monitoring and assessing heavy metallic infection in soil are vital for identifying polluted web sites and informing remediation techniques. Remote sensing and system getting to know strategies can be valuable tools for assessing and tracking heavy metal infection in soil. The remediation of heavy steel-infected soil is a complex and multifaceted method that requires a aggregate of bodily, chemical, and biological techniques. The desire of remediation method relies upon at the specific website situations and the goals of the remediation attempt. Effective management of heavy metallic-contaminated soil is crucial to protect the surroundings and human fitness. The presence of heavy metals in our soil and meals deliver represents a significant hazard to each human health and the surroundings. A numerous array of food sources, vital for our dietary needs, are at risk of heavy steel contamination, probably main to extreme health consequences. To counter this problem, a multifaceted strategy is imperative. Regular and vigilant tracking of soil, plant life, and water first-class in agricultural regions is critical to right away stumble on and address heavy metallic contamination. Simultaneously, regulatory frameworks ought to be hooked up to control waste disposal and promote comprehensive waste management, which includes recycling and land restoration via remediation techniques. Efficient water control, with a focus on getting rid of heavy metals from irrigation structures, is important to save you their accumulation in agricultural soils. Consumer recognition about meals high-quality and education for farmers on lively participation in soil tracking and remediation are indispensable components of the answer. Addressing the fundamental hassle of heavy metal infection calls for a collective commitment to responsible business practices and minimizing the discharge of those pollution into the environment. Moreover, embracing revolutionary technologies, consisting of machine gaining knowledge of and faraway sensing, can offer invaluable equipment for tracking and decreasing heavy metal presence in soil. In this holistic approach, we are able to aspire to make sure a safer and greater sustainable meals supply while protective both human well-being and the environment.

References

Abed Hussein B, Mahdi AB, Emad Izzat S, Acwin Dwijendra NK, Romero Parra RM, Barboza Arenas LA, Mustafa YF, Yasin G, Thaeer Hammid A, Kianfar E (2022) Production, structural properties nano biochar and effects nano biochar in soil: a review. Egypt J Chem 65(12):607–618

Abolghassem E, Ding Y, Farzad M, Xie Y (2018) Antioxidant response of bamboo (Indocalamus latifolius) as affected by heavy metal stress. J Elementol 23(1):341–352

Aggarwal A, Sharma I, Tripathi BN, Munjal AK, Baunthiyal M, Sharma V (2012) Metal toxicity and photosynthesis. In: Itoh S, Mohanty P, Guruprasad KN (eds) Photosynthesis: overviews on recent progress and future perspectives. IK International Publishing House (Pvt.) Limited, New Delhi, pp 229–236

Alengebawy A, Abdelkhalek ST, Qureshi SR, Wang MQ (2021) Heavy metals and pesticides toxicity in agricultural soil and plants: ecological risks and human health implications. Toxics 9(3):42

Appenroth KJ, Keresztes A, Sárvári É, Jaglarz A, Fischer W (2003) Multiple effects of chromate on Spirodela polyrhiza: electron microscopy and biochemical investigations. Plant Biol 5(03):315–323

Aroca R, Porcel R, Ruiz-Lozano JM (2012) Regulation of root water uptake under abiotic stress conditions. J Exp Bot 63(1):43–57

Ayangbenro AS, Babalola OO (2017) A new strategy for heavy metal polluted environments: a review of microbial biosorbents. Int J Environ Res Public Health 14(1):94

Aydinalp C, Marinova S (2009) The effects of heavy metals on seed germination and plant growth on alfalfa plant (Medicago sativa). Bulgarian J Agr Sci 15(4):347–350

Banerjee G, Pandey S, Ray AK, Kumar R (2015) Bioremediation of heavy metals by a novel bacterial strain Enterobacter cloacae and its antioxidant enzyme activity, flocculant production, and protein expression in presence of lead, cadmium, and nickel. Water Air Soil Pollut 226:1–9

Baryla A, Carrier P, Franck F, Coulomb C, Sahut C, Havaux M (2001) Leaf chlorosis in oilseed rape plants (Brassica napus) grown on cadmium-polluted soil: causes and consequences for photosynthesis and growth. Planta 212:696–709

Baszyńki T, Wajda L, Krol M, Wolińska D, Krupa Z, Tukendorf A (1980) Photosynthetic activities of cadmium-treated tomato plants. Physiologia Plantarum 48(3):365–370

Beinabaj SMH, Heydariyan H, Aleii HM, Hosseinzadeh A (2023) Concentration of heavy metals in leachate, soil, and plants in Tehran's landfill: investigation of the effect of landfill age on the intensity of pollution. Heliyon 9(1)

Bertrand M, Poirier I (2005) Photosynthetic organisms and excess of metals. Photosynthetica 43(3):345–353

Bhatia A, Singh S, Kumar A (2015) Heavy metal contamination of soil, irrigation water and vegetables in peri-urban agricultural areas and markets of Delhi. Water Environ Res 87(11):2027–2034

Cannon HL, Connally GG, Epstein JB, Parker JG, Thornton I, Wixson G (1978) Rocks: geological sources of most trace elements. In: Report to the workshop at south seas plantation Captiva Island, FL, US. Geochem Environ 3:17–31

Cao X, Alabresm A, Chen YP, Decho AW, Lead J (2020) Improved metal remediation using a combined bacterial and nanoscience approach. Sci Total Environ 704:135378

Chaudhari J, Patel K, Patel V (2016) Exploring the toxic effects of Pb and Ni on stem anatomy of Pisum Sativum L. Int J Chem Environ Biol Sci (IJCEBS) 4(1):28–32

Chu D (2018) Effects of heavy metals on soil microbial community. In: IOP conference series: earth and environmental science, vol 113. IOP Publishing, p 012009

Das S, Kalita MC, Shukla S (2018) Rapid biosynthesis of silver nanoparticles using leaf extract of *Brassica olearacea* var. gongylodes and their antimicrobial activity against bacteria. World J Pharm Pharm Sci 7(8):1135–1145

Das S, Gupta V, Kalyani MI, Kalita MC, Shukla S (2019) Biological synthesis and characterization of silver nanoparticles using stem extract of *Langenaria siceraria* and their antibacterial activity against *Escherichia coli* and *Staphylococcus aureus*. Biomedicine 39(4):580–586

Devi R, Kaur T, Kour D, Hricovec M, Mohan R, Yadav N, Rai PK, Rai AK, Yadav A, Kumar M, Yadav AN (2022) Microbes-mediated alleviation of heavy metal stress in crops: current research and future challenges. J Appl Biol Biotechnol 10(2):25–37

Dutta S, Mitra M, Agarwal P, Mahapatra K, De S, Sett U, Roy S (2018) Oxidative and genotoxic damages in plants in response to heavy metal stress and maintenance of genome stability. Plant Signal Behav 13(8):e1460048

Elbehiry F, Darweesh M, Al-Anany FS, Khalifa AM, Almashad AA, El-Ramady H, El-Banna A, Rajput VD, Jatav HS, Elbasiouny H (2022) Using biochar and nanobiochar of water hyacinth and black tea waste in metals removal from aqueous solutions. Sustainability 14(16):10118

Eleftheriou EP, Adamakis IDS, Panteris E, Fatsiou M (2015) Chromium-induced ultrastructural changes and oxidative stress in roots of Arabidopsis thaliana. Int J Mol Sci 16(7):15852–15871

Friedlova M (2010) The influence of heavy metals on soil biological and chemical properties. Soil Water Res 5(1):21–27

Ganie AS, Bano S, Khan N, Sultana S, Rehman Z, Rahman MM, Sabir S, Coulon F, Khan MZ (2021) Nanoremediation technologies for sustainable remediation of contaminated environments: recent advances and challenges. Chemosphere 275:130065

Gao M, Chang X, Yang Y, Song Z (2020) Foliar graphene oxide treatment increases photosynthetic capacity and reduces oxidative stress in cadmium-stressed lettuce. Plant Physiol Biochem 154:287–294

Gao T, Wang X, Liu Y, Wang H, Zuo M, He Y, Li H, Li G, Li C, Li X, Li X (2022) Characteristics and diversity of microbial communities in lead–zinc tailings under heavy metal stress in north-west China. Lett Appl Microbiol 74(2):277–287

Hayat K, Khan A, Bibi F, Murad W, Fu Y, Batiha GES, Al-Harrasi A (2021) Effect of cadmium and copper exposure on growth, physio-chemicals and medicinal properties of Cajanus cajan L.(Pigeon Pea). Metabolites 11(11):769

He ZL, Yang XE, Stoffella PJ (2005) Trace elements in agroecosystems and impacts on the environment. J Trace Elem Med Biol 19(2–3):125–140

Hossain MA, Ngo HH, Guo WS, Setiadi T (2012) Adsorption and desorption of copper (II) ions onto garden grass. Biores Technol 121:386–395

Iram S, Perveen K, Shuja N, Waqar K, Akhtar I, Ahmad I (2013) Tolerance potential of different Species of Aspergillus as bioremediation tool—comparative analysis. J Microbiol Res 1(1):1–8

Jiang CY, Sheng XF, Qian M, Wang QY (2008) Isolation and characterization of a heavy metal-resistant Burkholderia sp. from heavy metal-contaminated paddy field soil and its potential in promoting plant growth and heavy metal accumulation in metal-polluted soil. Chemosphere 72(2):157–164

Jogawat A, Yadav B, Chhaya, Narayan OP (2021) Metal transporters in organelles and their roles in heavy metal transportation and sequestration mechanisms in plants. Physiol Plant 173(1):259–275

Kurniati E, Arfarita N, Imai T, Higuchi T, Kanno A, Yamamoto K, Sekine M (2014) Potential bioremediation of mercury-contaminated substrate using filamentous fungi isolated from forest soil. J Environ Sci 26(6):1223–1231

Kwiatkowska-Malina J (2018) Functions of organic matter in polluted soils: the effect of organic amendments on phytoavailability of heavy metals. Appl Soil Ecol 123:542–545

Lamoreaux RJ, Chaney WR (1978) The effect of cadmium on net photosynthesis, transpiration, and dark respiration of excised silver maple leaves. Physiol Plant 43(3):231–236

Lehmann J, Bossio DA, Kögel-Knabner I, Rillig MC (2020) The concept and future prospects of soil health. Nat Rev Earth Environ 1(10):544–553

Lešková A, Zvarík M, Araya T, Giehl RF (2020) Nickel toxicity targets cell wall-related processes and PIN2-mediated auxin transport to inhibit root elongation and gravitropic responses in Arabidopsis. Plant Cell Physiol 61(3):519–535

Li C, Zhou K, Qin W, Tian C, Qi M, Yan X, Han W (2019) A review on heavy metals contamination in soil: effects, sources, and remediation techniques. Soil Sedim Contam: Int J 28(4):380–394

Liang G, Gong W, Li B, Zuo J, Pan L, Liu X (2019) Analysis of heavy metals in foodstuffs and an assessment of the health risks to the general public via consumption in Beijing, China. Int J Environ Res Public Health 16(6):909

Liaquat F, Munis MFH, Haroon U, Arif S, Saqib S, Zaman W, Khan AR, Shi J, Che S, Liu Q (2020) Evaluation of metal tolerance of fungal strains isolated from contaminated mining soil of Nanjing, China. Biology 9(12):469

Liu G, Wang J, Xue W, Zhao J, Wang J, Liu X (2017) Effect of the size of variable charge soil particles on cadmium accumulation and adsorption. J Soils Sedim 17:2810–2821
Liu L, Li W, Song W, Guo M (2018) Remediation techniques for heavy metal-contaminated soils: principles and applicability. Sci Total Environ 633:206–219
Lwin CS, Seo BH, Kim HU, Owens G, Kim KR (2018) Application of soil amendments to contaminated soils for heavy metal immobilization and improved soil quality—a critical review. Soil Sci Plant Nutr 64(2):156–167
Mansoor S, Kour N, Manhas S, Zahid S, Wani OA, Sharma V, Wijaya L, Alyemeni MN, Alsahli AA, El-Serehy HA, Paray BA (2021) Biochar as a tool for effective management of drought and heavy metal toxicity. Chemosphere 271:129458
Mazur R, Sadowska M, Kowalewska Ł, Abratowska A, Kalaji HM, Mostowska A, Krasnodębska-Ostręga B (2016) Overlapping toxic effect of long term thallium exposure on white mustard (Sinapis alba L.) photosynthetic activity. BMC Plant Biol 16:1–17
Mishra J, Singh R, Arora NK (2017) Alleviation of heavy metal stress in plants and remediation of soil by rhizosphere microorganisms. Front Microbiol 8:1706
Monnet F, Vaillant N, Vernay P, Coudret A, Sallanon H, Hitmi A (2001) Relationship between PSII activity, CO_2 fixation, and Zn, Mn and Mg contents of Lolium perenne under zinc stress. J Plant Physiol 158(9):1137–1144
Morkunas I, Woźniak A, Mai VC, Rucińska-Sobkowiak R, Jeandet P (2018) The role of heavy metals in plant response to biotic stress. Molecules 23(9):2320
Öztürk M, Ashraf M, Aksoy A, Ahmad MSA, Hakeem KR (eds) (2015) Plants, pollutants and remediation. Springer, Berlin/Heidelberg, pp 251–268
Passarini MRZ, Ottoni JR, Costa PEDS, Hissa DC, Falcão RM, Melo VMM, Balbino VQ, Mendonça LAR, Lima MGDS, Coutinho HDM, Verde LCL (2022) Fungal community diversity of heavy metal contaminated soils revealed by metagenomics. Arch Microbiol 204(5):255
Pätsikkä E, Kairavuo M, Šeršen F, Aro EM, Tyystjärvi E (2002) Excess copper predisposes photosystem II to photoinhibition in vivo by outcompeting iron and causing decrease in leaf chlorophyll. Plant Physiol 129(3):1359–1367
Pereira P, Bogunovic I, Muñoz-Rojas M, Brevik EC (2018) Soil ecosystem services, sustainability, valuation and management. Curr Opin Environ Sci Health 5:7–13
Raggi S, Ferrarini A, Delledonne M, Dunand C, Ranocha P, De Lorenzo G, Ferrari S (2015) The Arabidopsis class III peroxidase AtPRX71 negatively regulates growth under physiological conditions and in response to cell wall damage. Plant Physiol 169(4):2513–2525
Rai PK, Lee SS, Zhang M, Tsang YF, Kim KH (2019) Heavy metals in food crops: health risks, fate, mechanisms, and management. Environ Int 125:365–385
Rashid MI, Shah GA, Sadiq M, Amin NU, Ali AM, Ondrasek G, Shahzad K (2023) Nanobiochar and copper oxide nanoparticles mixture synergistically increases soil nutrient availability and improves wheat production. Plants 12(6):1312
Rees F, Simonnot MO, Morel JL (2014) Short-term effects of biochar on soil heavy metal mobility are controlled by intra-particle diffusion and soil pH increase. Eur J Soil Sci 65(1):149–161
Rizvi A, Ahmed B, Khan MS, Rajput VD, Umar S, Minkina T, Lee J (2022) Maize associated bacterial microbiome linked mitigation of heavy metal stress: a multidimensional detoxification approach. Environ Exp Bot 200:104911
Salla V, Hardaway CJ, Sneddon J (2011) Preliminary investigation of Spartina alterniflora for phytoextraction of selected heavy metals in soils from Southwest Louisiana. Microchem J 97:207–212. https://doi.org/10.1016/j.microc.2010.09.005
Sawidis T, Baycu G, Cevahir–Öz G, Weryszko-Chmielewska E (2018) Effect of mercury on pollen germination and tube growth in Lilium longiflorum. Protoplasma 255(819–828):3
Shahid M, Khalid S, Abbas G, Shahid N, Nadeem M, Sabir M, Dumat C (2015) Heavy metal stress and crop productivity. In: Crop production and global environmental issues, pp 1–25
Shi GR, Cai QS (2008) Photosynthetic and anatomic responses of peanut leaves to cadmium stress. Photosynthetica 46:627–630

Shukla UC, Singh J, Joshi PC, Kakkar P (2003) Effect of bioaccumulation of cadmium on biomass productivity, essential trace elements, chlorophyll biosynthesis, and macromolecules of wheat seedlings. Biol Trace Elem Res 92:257–273

Sinegani AAS, Tahmasbian I, Sinegani MS (2015) Chelating agents and heavy metal phytoextraction. In: Heavy metal contamination of soils: monitoring and remediation, pp 367–393

Singh S, Parihar P, Singh R, Singh VP, Prasad SM (2016) Heavy metal tolerance in plants: role of transcriptomics, proteomics, metabolomics, and ionomics. Front Plant Sci 6:1143

Sundaramoorthy P, Chidambaram A, Ganesh KS, Unnikannan P, Baskaran L (2010) Chromium stress in paddy:(i) nutrient status of paddy under chromium stress;(ii) phytoremediation of chromium by aquatic and terrestrial weeds. CR Biol 333(8):597–607

Tripathi DK, Singh VP, Kumar D, Chauhan DK (2012) Rice seedlings under cadmium stress: effect of silicon on growth, cadmium uptake, oxidative stress, antioxidant capacity and root and leaf structures. Chem Ecol 28(3):281–291

Tripathi P, Singh PC, Mishra A, Srivastava S, Chauhan R, Awasthi S, Nautiyal CS (2017) Arsenic tolerant Trichoderma sp. reduces arsenic induced stress in chickpea (Cicer arietinum). Environ Pollut 223:137–145

Tůmová E, Sofrová D (2002) Response of intact cyanobacterial cells and their photosynthetic apparatus to Cd^{2+} ion treatment. Photosynthetica 40:103–108

Verma A, Bharadvaja N (2022) Plant-mediated synthesis and characterization of silver and copper oxide nanoparticles: antibacterial and heavy metal removal activity. J Cluster Sci 33(4):1697–1712

Verma S, Bhatt P, Verma A, Mudila H, Prasher P, Rene ER (2021) Microbial technologies for heavy metal remediation: effect of process conditions and current practices. Clean Technol Environ Policy 1–23

Voigt J, Nagel K (2002) The donor side of photosystem II is impaired in a Cd2+-tolerant mutant strain of the unicellular green alga Chlamydomonas reinhardtii. J Plant Physiol 159:941–950

Wang AS, Angle JS, Chaney RL, Delorme TA, Reeves RD (2006) Soil pH effects on uptake of Cd and Zn by Thlaspi caerulescens. Plant Soil 281:325–337

Wang S, Shi X (2001) Molecular mechanisms of metal toxicity and carcinogenesis. Mol Cell Biochem 222:3–9

Wani W, Masoodi KZ, Zaid A, Wani SH, Shah F, Meena VS, Mosa KA (2018) Engineering plants for heavy metal stress tolerance. Rendiconti Lincei. Scienze Fisiche e Naturali 29:709–723

WHO/FAO/IAEA (1996) Trace elements in human nutrition and health. World Health Organization, Geneva, Switzerland

Williams LE, Pittman JK (2010) Dissecting pathways involved in manganese homeostasis and stress in higher plant cells. Cell Biol Metals Nutr 95–117

Xie Y, Fan J, Zhu W, Amombo E, Lou Y, Chen L, Fu J (2016) Effect of heavy metals pollution on soil microbial diversity and bermudagrass genetic variation. Front Plant Sci 7:755

Yadav V, Arif N, Kováč J, Singh VP, Tripathi DK, Chauhan DK, Vaculík M (2021) Structural modifications of plant organs and tissues by metals and metalloids in the environment: a review. Plant Physiol Biochem 159:100–112

Yang Y, Zhang L, Huang X, Zhou Y, Quan Q, Li Y, Zhu X (2020) Response of photosynthesis to different concentrations of heavy metals in Davidia involucrata. PLoS ONE 15(3):e0228563

Yotsova E, Dobrikova A, Stefanov M, Misheva S, Bardáčová M, Matušíková I, Apostolova E (2020) Effects of cadmium on two wheat cultivars depending on different nitrogen supply. Plant Physiol Biochem 155:789–799

Yu G, Wang X, Liu J, Jiang P, You S, Ding N, Guo Q, Lin F (2021) Applications of nanomaterials for heavy metal removal from water and soil: a review. Sustainability 13(2):713

Zeid IM (2001) Responses of Phaseolus vulgaris chromium and cobalt treatments. Biol Plant 44:111–115

Chapter 8
An Overview on the Current State of Heavy Metal Pollution and Recent Remediation Approaches

Sadia Sikder, Mohammad Toha, and Md. Mostafizur Rahman

Abstract Heavy metal pollution has become a significant concern in recent decades due to its persistent nature, carcinogenic properties, toxicity, and bioaccumulation. Heavy metals have been detected in numerous kinds of fruits, vegetables, milk products, and natural oils in recent years, indicating the severe contamination of our food chain systems. Nevertheless, due to the acute and chronic nature of heavy metals' effects, this excessive concentration can negatively impact on human health and on ecosystems. Various remediation technologies and/or approaches have already been invented in the global industrial and scientific arena to reduce heavy metal pollution. Most conventional treatment systems have some limitations due to their high mechanism cost and the generation of toxic sludge. Thus, the present world is more focused on eco-friendly treatment systems. As a result, bacteria, fungi, diatoms, and various plants are used to remove heavy metals from the environmental matrices. Therefore, this chapter aims to discuss the sources, fates, transport, and recent heavy metal pollution remediation techniques. This chapter also summarizes the present state of heavy metal pollution in terms of global scale and regulatory frameworks.

Keywords Eco-friendly · Effects · Heavy metal · Pollution · Remediation

S. Sikder · M. Toha · Md. M. Rahman (✉)
Department of Environmental Science, Bangladesh University of Professionals, Mirpur, Cantonment, Dhaka 1216, Bangladesh
e-mail: rahmanmm@juniv.edu

Md. M. Rahman
Laboratory of Environmental Health and Ecotoxicology, Department of Environmental Sciences, Jahangirnagar University, Savar, Dhaka 1342, Bangladesh

S. Sikder · M. Toha · Md. M. Rahman
Department of Environmental Sciences, Jahangirnagar University, Savar, Dhaka 1342, Bangladesh

N. Kumar (ed.), *Heavy Metal Toxicity*, Environmental Science and Engineering,
https://doi.org/10.1007/978-3-031-56642-4_8

8.1 Introduction

Since the third industrial revolution, a tremendous amount of mineral resources have been used globally due to urbanization, population growth, industrialization, and modernization, where varieties of emerging pollutants, nanoparticles, organometallic compounds, and radioactive isotopes have emerged rapidly, making our environment more polluted and attractive as a global issue (Ahirvar et al. 2023; Sodhi et al. 2022). Moreover, the United States Environmental Protection Agency (USEPA) and the US Agency for Toxic Substances and Diseases Registry (ATSDR) have listed heavy metals (HMs) as emerging contaminants due to their eco-toxicity, non-biodegradability, persistent nature, and potent carcinogens (Vardhan et al. 2019). These HMs are generally composed of metals and metalloids with a great density of 5 g/cm^3 (Roy et al. 2022). Currently, cadmium (Cd), zinc (Zn), arsenic (As), copper (Co), chromium (Cr), cobalt (Co), nickel (Ni), molybdenum (Mo), mercury (Hg), and lead (Pb) are considered significant types of HMs worldwide (Li et al. 2022). Although some HMs are essential for plant growth and human health, their excessive use has dramatically changed their standard levels. It is hindering the development of the environment and human civilization. The scary thing is that HMs can create human health and environmental problems in both high (Fe, Cu, Co) and low (Cr, Cd, Pb, As) concentrations (Kumar et al. 2017; Ahirvar et al. 2023). However, man-made (industrial discharge, mining, smelting, automobiles, sewage sludge, roadways, domestic activities, and agricultural activities) and natural (bedrock erosion and rock weathering) are the two pathways of HMs in our environment where man-made is recognized as a significant source (Li et al. 2022; Rafique et al. 2022). In 2013, Ni and Pb production were 1 million metric tons (Mt) and 5 MT, respectively (Vardhan et al. 2019). Currently, the global production of Cr and Pb is 37.5 Mt and 11.3 Mt, respectively. Moreover, Cr and Pb production rates increased by 514% and 232% over the last fifty years, respectively (Li et al. 2022). After that, these HMs enter the water and soil environment due to surface run-off and industrial discharge. Consequently, massive amounts of HMs have already been found in all continents. Asian countries like Bangladesh, Pakistan, India, China, and Thailand have investigated to find the above permissible levels of HMs concentration (Zhou et al. 2020; Khalid et al. 2021). In Bangladesh, Ni, Pb, and Hg concentrations were found to be 0.81, 0.86, and 0.24 mg/L in water samples (Sarker et al. 2022). Notably, 75–96% of European seas face high levels of heavy metal pollution (HMP) (Sodhi et al. 2022). Cr, Zn, Pb, Cd, and Cr have been observed at excessive levels in the soil samples of India, the North East of Iran, Pakistan, Bangladesh, China, and other Asian European nations (Zhou et al. 2020). In Pakistan, the amount of HMs found in soil samples was Mn (137.3) > Cr (6.62) > Fe (968), and the order of the elements was Cr (17.89) > Ni (9.24) > Zn (2.31) > Pb (0.84) > Cu (0.29) > Cd (0.186). However, over the last two decades, HMs have also been found significantly in the road dust samples of our globe (Roy et al. 2022). The concentration (mg/kg) of Pb, Zn, and Cu were found in Dhaka, Bangladesh (67.60, 144.2, 98.9), in Delhi, India (1899, 370.5, 1016), in Baoji, NW China (408.40, 715.10, 123.17), in Coventry, UK (47.1, 385.70, 226.40), in Angola (315.30, 316.60, 41.78), in Canada (39.05, 112.50, 65.84), and in Sydney, Australia (487, 523, 164), respectively (Rahman et al. 2021; Shahab et al. 2020; Roy et al. 2022). For the eight elements, including Cd, Co, Cr, Cu, Mn, Ni, Pb, and Zn, the average total of all ecological risk factors suggested a very high risk; Cd was the primary ecological risk factor, and Cu was rated second. Cu was shown to be the most significant risk factor in a study evaluating the contamination status of soil obtained from a dumpsite in central Thailand (Zhou et al. 2020). These states indicate the rapid production of HMs and the miserable situation of our environment.

HMs can move from one trophic level to another, so their effects are widely distributed in ecosystems. Heavy metal buildup will pollute soils, having a long-lasting, quite harmful effect. These harmful substances enter the water, ingest cultivated plants, assimilate into plant organs, enter the food chain, and eventually endanger people. Three possible routes of exposure to HMP in soil are ingestion, inhalation, and dermal contact through various carriers such as food, drink, air, and skin. It has been investigated if long-term direct exposure to heavy metal-contaminated soil or consumption of agricultural products, including rice, fruit, and vegetables, causes human health concerns (Amer et al. 2019; Shaheen et al. 2016). As HMs can interfere with ionic homeostasis and the activity of enzymes, they can imbalance plants germination processes, photosynthesis, water balance, essential digestion, and multiplication. Moreover, it can create an oxidative stress-metabolic imbalance and reduce the biomass of plants. The prospective ecological risk indices were Cd > Cu > Co > Zn > Pb > Cr > Mn in that order (Aziz et al. 2023; Pandiyan et al. 2021). As HMs have already entered the food web, they can damage human nervous activities, blood circulation systems, respiratory systems, and other organs (Vardhan et al. 2019). For example, the world has already seen the destructive effects of Minamata disease.

In these circumstances, a proper regulatory approach and modern treatment systems are necessary to control the distribution of HMP. Otherwise, the world will face the destructive effects of emerging diseases like Minamata disease again. Various physical, chemical, and biological treatment systems are present to remedy heavy metal pollution. Chemical remediation contains chemical precipitation, coagulation, and electrolysis (Sodhi et al. 2022). Ion-exchange resins and membrane separation are included in the physical remediation process. Due to extra chemicals, energy consumption, and secondary toxic pollutants, these two approaches (physical and chemical) have already lost their attention globally (Mahapatra et al. 2019; Ahirvar et al. 2023). Hence, the world is moving toward biological treatment and adsorption material-based systems because they are eco-friendly, effective, and less costly. In addition, various bacteria, zeolite-activated carbon, carbon–metal oxide nanomaterials, and ligand-based materials are used in these two approaches (Chen et al. 2022).

Therefore, this chapter aims to give a detailed overview of HMP sources, fate, transport, and effects. This chapter will also cover HMP detection methods on a global scale. Lastly, this chapter will focus on the standards and guidelines of HM and the recent remediation techniques.

8.2 Sources of Heavy Metal Pollution (HMP)

Due to the prevalence and effects (direct and indirect) of heavy metals (HMs) at all levels of the environment, researchers worldwide have been interested in knowing about their sources for a long time. The sources of heavy metals are depicted in Fig. 8.1.

In the environmental matrix, HMs can have various forms, like oxides, sulfides, hydroxides, phosphates, carbonates, and silicates (Vardhan et al. 2019). They can also create and present an organic matrix with other elements in low concentrations. Consequently, they are recognized as trace elements. However, natural and anthropogenic activities are two sources of HMs. Specifically, anthropogenic activities are considered a significant source of HMs (Briffa et al. 2020). A descriptive discussion of these two sources is given below.

8.2.1 Natural Sources

Heavy metals (HMs) can enter our environment in natural processes, such as weathering, volcanic eruptions, seismic activities, natural forest fires, and decomposition (Rafique et al. 2022; Pachana et al. 2010). Weathering is an onsite process involving physical, chemical, and biological activities to break down the rock and minerals. In physical processes, heat, pressure, and water directly influence the breakdown of minerals (composed of heavy metals). On the contrary, the chemicals of rocks react with heat, pressure, water, and ice in a chemical process, creating a favourable situation to release HMs from the rocks to the nearby environment (Rezapour et al. 2022). However, these processes simultaneously occur in a specific system where physical abrasion of rock or minerals can enhance the chemical reactions and the release of HMs into the environment. In biological processes, microorganisms use minerals and rocks to survive (Zaynab et al. 2022; Briffa et al. 2020).

Moreover, plant roots can disintegrate the rocks. As a result, HMs containing rocks and minerals easily break down and generate various types of HMs. Due to the volcanic eruption, HMs can enter the earth's surface. The ash of volcanic eruptions can also contain a variety of HMs. After that, these HMs can be transported into our surrounding environment through rainfall, wind, and surface runoff (Vardhan et al. 2019). During the earthquake, underground geology gets fractured and oxidized. Then, when heavy rainfall or underground leakage occurs, this water will react with underground fractured components and release HMs (Kuhn et al. 2022). However, heavy rainfall, inadequate soil cover, and decaying vegetation are some natural processes that can release heavy metals into our surrounding environment.

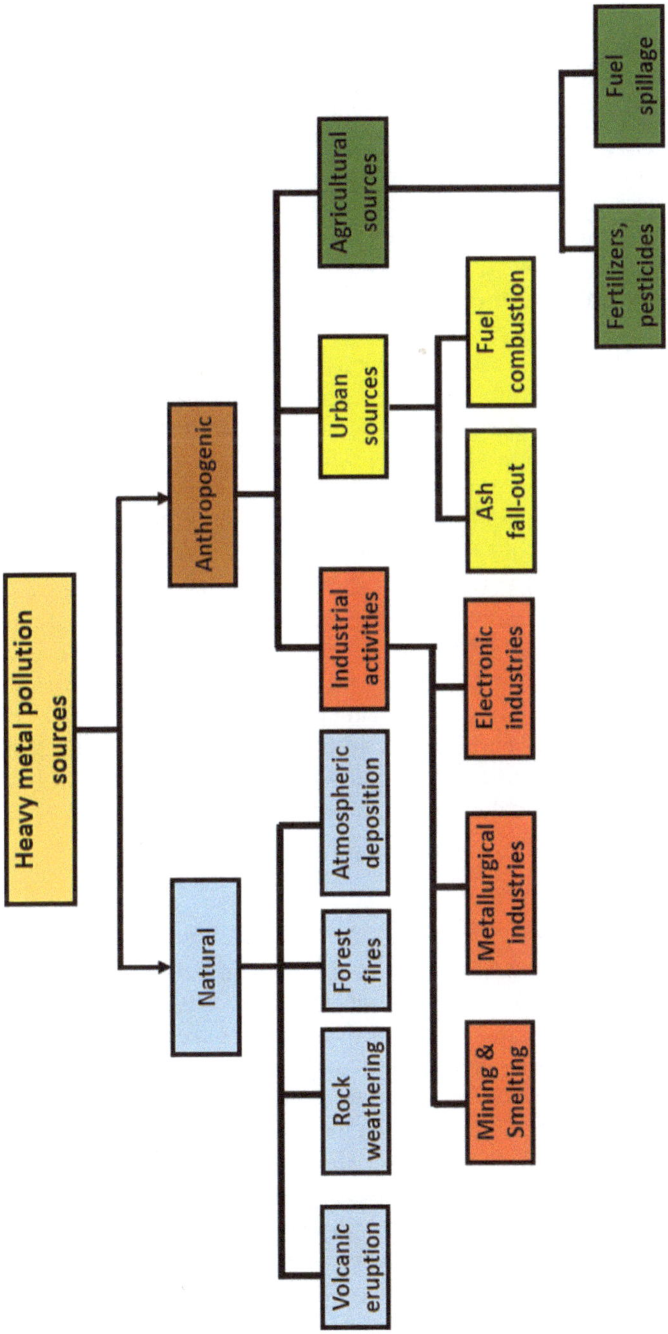

Fig. 8.1 Natural and anthropogenic sources of heavy metal pollution

8.2.2 *Anthropogenic Sources*

In this modern era, population growth, socio-economic activities, and the release of heavy metals (HMs) have a great connection. Various socio-economic activities such as mining, industrial, agricultural, manufacturing, municipal effluents, metallurgical, municipal runoff, and weathering of tires are mainly responsible for generating and distributing HMs in the surrounding environment (Khalid et al. 2021). The description of these anthropogenic sources is given below.

8.2.2.1 Agricultural Activities

Various organic and inorganic soil amendments, fertilizers, and pesticides are utilized in the agricultural sector to fulfil the growing food production requirements. Generally, pesticides control the growth of fungi, insects, and herbicides. Heavy metals (HMs) like Hg, As, and Pb are found in pesticides of phenylmercuric acetate and lead arsenate (Vardhan et al. 2019; Li et al. 2022). Hence, excessive pesticide use of these HMs can contaminate agricultural soil and water. Bio-solids can contain both soil growth nutrients and HMs of Pb, Ni, Cd, Cr, Cu and Zn that are detrimental at high concentrations (Pachana et al. 2010; Masindi et al. 2021). Rocks are used in the production of various fertilizers, especially phosphate-based fertilizers. Seventeen types of HMs have already been found in nitrogen, phosphate, and potassium fertilizers, where Cd, As, and Pb are the most common (Ahirvar et al. 2023; Sodhi et al. 2022). Agricultural sludge also contains a variety of HMs. As plants can efficiently uptake these HMs, excessive use of fertilizers, pesticides, and soil amendments should be immediately stopped. Moreover, proper monitoring systems and sustainable organic agricultural activities should be promoted.

8.2.2.2 Industrial Activities

The contribution of mining to the blooming of the world economy and the development of different countries is undeniable. Unfortunately, the rapid activities of mining, poor mining procedures, and a lack of resources have already led to a variety of heavy metals (HMs) in our environment. In the mining process, tailings (mining waste) and ore rocks create oxidizing conditions that can enhance the weathering of rocks (Pachana et al. 2010). Some sulphide bearing minerals: Pyrite (FeS_2), covellite (CuS), chalcopyrite ($CuFeS_2$), cinnabar (HgS), arsenopyrite (FeAsS), sphalerite (ZnS), galena (PbS), and molybdenite (MoS_2) are lead to the production of varieties of HMs (Li et al. 2022; Masindi et al. 2021). Coal and petroleum are utilized during the production of energy. Coals' byproducts are composed of heavy amounts of As, Cd, Co, Cr, Cu, Hg, Ni, Pb, Sc, and Ag (Masindi et al. 2021). Hence, this energy-producing activity is also responsible for environmental heavy metal pollution (HMP). After mining activities, specific HMs are going through a metallurgical process for purification. Leaching, solvent extraction, and digestion activities are involved in this metallurgical process (Masindi et al. 2021). Finally, a large amount of residues is generated from this process that contain HMs of Zn, Hg, Cu, and Ni. However, traffic, wearing of tires, weathering of pavement, abrasion of vehicles, and paint elements are some sources of HMP in urban road dust (Vardhan et al. 2019). As excessive industrial activities are the major causes of HMP, it is necessary to control this. Moreover, the use of renewable energy technologies and modern remediation technologies can reduce this pollution to a great extent.

8.3 Fate and Transport of Heavy Metals (HMs)

Environmental compartments exhibit a perpetual state of dynamism. In this natural process, if any physical changes occurred, it would directly or indirectly influence the changes of land, living things, and atmospheric activities. For example, chemical pollutants are created by industry but will circulate, contaminate, and accumulate in various environmental compartments. This circulation and contamination process of air–water-soil-biota are known as contaminant pathways of nature. The fate and transport of heavy metals are represented in Fig. 8.2.

Due to industrial activities, coal combustion, road transportation, vehicle exhaust, weathering of pavements, and atmospheric depositions, wide varieties of heavy metal pollutants are deposited on the land surface (Ahirvar et al. 2023). Recent studies have also found Cu, Hg, Pb, Cd, and Zn types of heavy metals in urban deposited road dust. These heavy metal pollutants do not remain stationary for extended periods and exhibit mobility (Briffa et al. 2020). In this situation, pollutant mobility depends on the reactions between pollutants' properties (concentration level, polarity, and volatilization) and environmental factors (temperature, wind speed, water flow, organic and inorganic pollutants) (Pachana et al. 2010). At this moment, some heavy metal pollutants are dispersed into the air, and most of them are transported to nearby aquatic bodies by stormwater and run-off. Hence, the primary routes through which chemicals enter and diffuse into the environment are air and water. Heavy metals (HMs) bind with organic matter after entering the aquatic body, using their positive ions, known as organometallic compounds (Vardhan et al. 2019). Consequently, they are found in sediments, animals, and plants. However, sinks and remobilization directly influence the fate and transport of heavy metals. The description of these processes is given below.

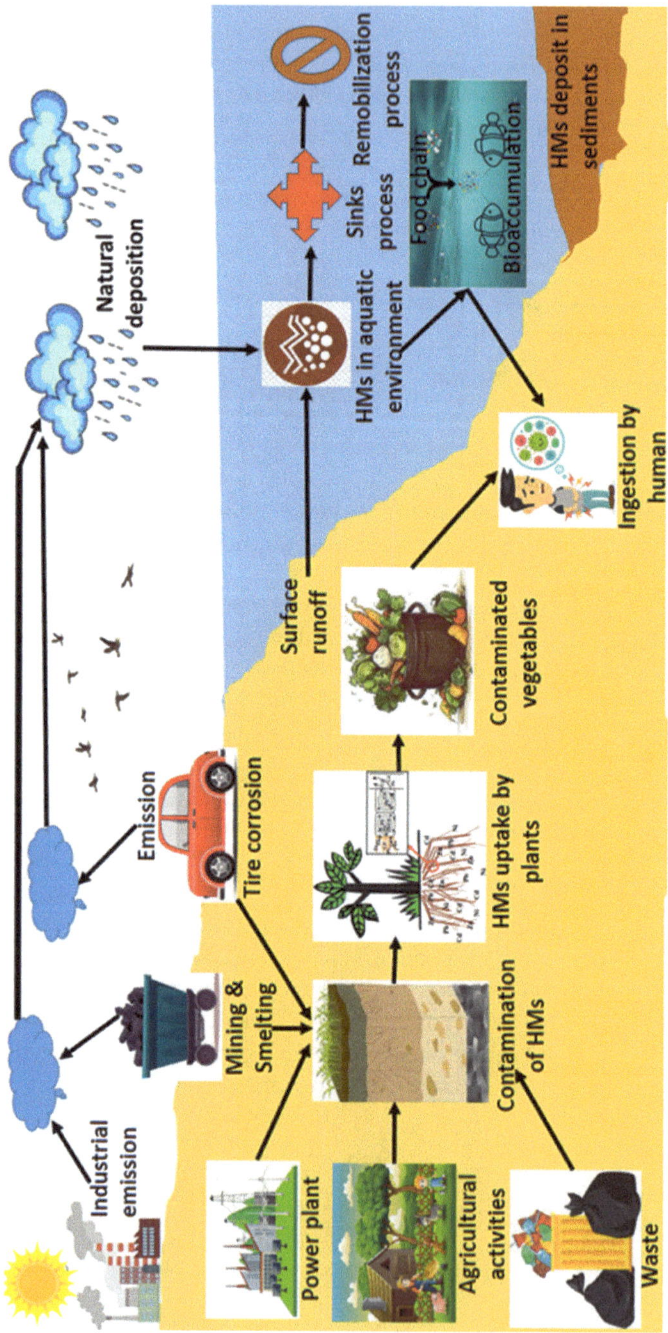

Fig. 8.2 Fate and transport of heavy metals in various environmental compartments

8.3.1 *Procedure of Sinks*

8.3.1.1 Adsorption and Co-precipitation

Initially, heavy metals (HMs) adsorb on particulate matter in aquatic environments. As HMs are positively charged, they adsorb with negatively charged sorptive properties such as aluminium hydroxides ($AlOH^-$), silicon hydroxides ($SiOH^-$), manganese hydroxides ($MnOH^-$), and ferrous hydroxides ($FeOH^-$). Moreover, they react with the organic substances of carboxyl and phenolic OH^- groups (Pachana et al. 2010; Briffa et al. 2020). After that, under oxidizing conditions, these hydrous and hydroxide ions, Fe and Mn, work as a sink of HMs (Pachana et al. 2010). On the other hand, under reducing conditions, sorbed HMs are rapidly mobilized.

8.3.1.2 Precipitation

The solubility product (Ksp) equilibrium constant is fixed for each metal. Ksp is utilized to describe a dynamic equilibrium of soild substance and its aqueous solution. Precipitation of the compound will occur if the Ksp of a particular compound is exceeded (Briffa et al. 2020; Vardhan et al. 2019). Hence, precipitation will occur due to the exceeded solubility product, which is determined by comparing the product of ion concentrations with the Ksp value (Pachana et al. 2010). Thus, sediments large surface area helps heavy metals (HMs) to act as sorbents for the dissolved.

8.3.1.3 Cooperation in Biological Activity

Since metals contain non-biodegradable properties, they cannot decompose into less harmful components. Metals like phosphorus, sodium, calcium, potassium, chlorine, and sulfur are generally used for ionic balance. Moreover, they are integral parts of amino acids and nucleic acids. On the other hand, Hg, Pb, Cd, and As are bioaccumulated in organisms and affect them. However, Zn and Fe act as essential elements of enzymes and carry blood, respectively (Pachana et al. 2010).

8.3.2 *Procedure of Remobilization*

Heavy metals (HMs) are remobilized from sediments for various reasons, leading to detrimental impacts on the whole food chain system. In the estuarine environment, various alkali metal group cations (Na^+, Mg^{2+}, Ca^{2+}) increase the salt concentration level (Pachana et al. 2010). After that, they will compete with the sediments' adsorbed materials and release them into water. Finally, the desorption or dissolution process helps them to remobilize again in aquatic environments (Pachana et al.

2010). The large concentration of Fe and Mn in the depth of (10–15) cm reduces the redox potential value. As a result, HMs are easily remobilized from sediments into the nearby water (Vardhan et al. 2019; Pachana et al. 2010). Lowering pH increases the H^+ concentrations in the aquatic environment; these ions compete with the adsorbed metals of sediments. Lastly, HMs are released into water compartments. Organic complexing agents such as nitrilotriacetic acid (NTA) create highly soluble organometallic complexes with HMs. After that, these complexing agents accelerate the release of HMs from sediments (Briffa et al. 2020; Pachana et al. 2010).

The global community has witnessed significant levels of heavy metal contamination across various environmental spheres. Research investigations have demonstrated that industrialized nations have implemented various measures to address this pollution issue. Regrettably, developing countries continue to lag behind developed countries regarding heavy metal research and essential remediation methods. Hence, understanding the fate and transport of HMs would be beneficial to assess the environmental monitoring of this pollution for developing nations.

8.4 Common Heavy Metals and Their Effects

Environmental literature often refers to metals and metalloids as "heavy metals" because they pollute, poison, and harm organisms. Density, relative atomic mass, and atomic number have been used to define the word (Ali and Khan 2018).

Heavy metal pollution is a major environmental issue nowadays. Rapid industrialization increases production, which raises metal concentrations. Due to improper disposal and dumping of industrial waste directly into water and land bodies, this causes severe contamination (Dixit et al. 2015). Heavy metals, the earth's crust's main elements, are persistent environmental contaminants that biodegrade slowly but can be transformed into non-toxic forms (Mishra et al. 2018).

A total of 35 metals are significant owing to residential or occupational exposure. These metals include 23 heavy metals. Antimony, arsenic, bismuth, cadmium, cerium, chromium, cobalt, copper, gallium, gold, iron, lead, manganese, mercury, nickel, platinum, silver, tellurium, thallium, tin, uranium, vanadium, and zinc are of concern. These heavy metals are found in nature and food. Jaishankar et al. (2014) state that minimal levels of these compounds are necessary for optimal health, while excessive amounts can be poisonous or harmful.

Metallic components are natural. Their existence is unique since they are difficult to remove from the ecosystem. Metals are common occupational and environmental hazards. Due to their extensive use in many locations, these toxic compounds are being studied for their effects on human health. Using various metals in industrial and daily uses has raised environmental worries about metal contamination (Mahurpawar 2015).

Scientific research on heavy metal toxicity is needed to understand its harmful mechanisms better and discover practical ways to reduce its health risks. Heavy metal wastewater discharge has increased, especially in developing nations. This increase is due to the rapid rise of metal plating, mining, fertilizer, tannery, battery, paper, and pesticide sectors. Heavy metals accumulate in organisms and are non-biodegradable. Many heavy metal ions are toxic and cancerous. Heavy metals like cadmium, copper, and zinc, especially dissolved, cause pollution and hazards. Due to their toxicity and mobility, heavy metals can harm people and undermine environmental benefits (Khalef et al. 2022). Some of these common heavy metals along with their effects on human health, plants, organisms and environment are given in Table 8.1.

8.5 Detection Methods of Heavy Metal

Heavy metals present a significant risk to biological and natural ecosystems, even when present in low environmental concentrations. The practical and accurate detection of heavy metals concealed in the environment is a crucial requirement for pollution assessment and prevention, as it serves as a potent tracer for monitoring the effects of human activities. In recent years, there has been a notable advancement in heavy metal detection technologies, with a shift towards automated, intelligent, comprehensive, and precise rapid detection methods. Among these methods, the most commonly employed techniques are inductively coupled plasma mass spectrometry/atomic emission spectrometer (ICP-MS/AES), atomic absorption spectrometry (AAS), atomic fluorescence spectrometry (AFS), and laser-induced breakdown spectrometry (LIBS) (Jin et al. 2020; Malik et al. 2019). Some of the commonly used detection techniques, along with the detected heavy metals, have been given in Tables 8.2 and 8.3.

8.6 Global Scale of Heavy Metals (HMs) Pollution

Heavy metals (HMs) are more than permissible limits worldwide. Road dust is considered one of the significant pathways of heavy metals from the earth's surface to other mediums as they are deposited here resulting from various anthropogenic sources and transported through surface run-off. The highest concentration level of HMs in road dust has been found in the Asian continent (Roy et al. 2022). Moreover, Asian countries like China, India, Bangladesh, and Pakistan showed a higher prevalence of HMs than others due to their rapid economic development and population growth (Roy et al. 2022; Rahman et al. 2021). Excess levels of HMs have also been found in various European nations since 1991. The global scale of heavy metals pollution is shown in Table 8.1. In Nigeria, heavy metal concentration ranges were found with a massive range (Zn = 2.7–85.1, Cu = 1.3–49.9, Ni = 2.3–22.1, Cd = 2.1–5.7, and Cr = 1.9–61.5) mg/kg (Roy et al. 2021). Other African nations, Ghana, Angola, and Egypt also showed the same trends of heavy metal pollution. It has also been observed in South America, America, and Australia (Table 8.4).

Table 8.1 Common heavy metals and their impacts on human health and environment, plants and organisms

Heavy metal	Impact on human health		Impact on environment, plants and organisms
	Toxicity	Health effects	
Arsenic (As)	Neurotoxicity, carcinogenicity, genotoxicity, reproductive and developmental toxicity, skin toxicity	Clinicopathological conditions, developmental anomalies, neurobehavioral disorders, cardiovascular disorders, and auditory disorders, along with anemia, leukopenia, eosinophilia, and carcinoma, dermal, ocular, and hepatic impairments, as well as potential carcinogenicity, impact renal function and the central nervous system	Arsenic can induce chlorosis, stunted growth, and impaired photosynthesis in plants. Plants can die from nutritional deficits and water and nutrient uptake disruptions caused by arsenic. Herbivores can also be poisoned by arsenic in plant tissue
			Arsenic kills, inhibits growth, reproduction, and behavior
			Arsenic-contaminated environments have few species and fewer individuals
Cadmium (Cd)	Neurotoxicity, nephrotoxicity, carcinogenicity, hepatotoxicity, immunological toxicity, cardiovascular toxicity, reproductive and developmental toxicity	The potential adverse health outcomes associated with the aforementioned conditions include renal impairment, acute pulmonary manifestations, renal carcinoma, nephropathy, prostate dysfunction, skeletal problems, respiratory ailments, and renal pathologies	Cadmium bioaccumulates in vertebrates, especially in kidneys and livers, as well as invertebrates, algae, and plants
			Birds and animals suffer mostly from renal disease
			Cd inhibits photosynthesis, chlorophyll production, and stomatal closure
Copper (Cu)	Neurotoxicity, hepatotoxicity	Brain and kidney damage, elevated levels result in liver cirrhosis, chronic anemia, stomach and intestine irritation	Copper poisoning in aquatic habitats can harm fish, invertebrates, plants, and amphibians. Acute toxicity can kill organisms; chronic toxicity can reduce survival, reproduction, and growth
			Farm animals absorb harmful copper quantities from polluted soils. Copper poisoning mostly affects sheep due to its low quantities

(continued)

Table 8.1 (continued)

Heavy metal	Impact on human health		Impact on environment, plants and organisms
	Toxicity	Health effects	
Lead (Pb)	Neurotoxicity, nephrotoxicity, carcinogenicity, hepatotoxicity, genotoxicity, reproductive and developmental toxicity, immunological toxicity, cardiovascular toxicity	Renal impairment, cognitive decline, neuropathological conditions, hepatic and renal dysfunction, diminished fertility, reduced cognitive abilities, and impaired memory function	Pb causes biodiversity loss, community composition alterations, plant and animal development and reproduction delays, and vertebrate neurological consequences
Chromium (Cr)	Nephrotoxicity, carcinogenicity, hepatotoxicity immunological toxicity, cardiovascular toxicity, skin toxicity, genotoxicity	Nasal irritation, hear impairment, lung carcinoma, skin diseases, anemia, cancers	Cr reduces crop growth, yield, vegetable and grain quality by affecting plant metabolism In addition, Cr can cause genotoxicity in certain plant species
Manganese (Mn)	Neurotoxicity	Central and peripheral neuropathies	Chronic exposures disrupt the insect brain's dopaminergic signaling pathway, affecting insect behavior, particularly honey bee foraging
			Manganese can damage the lungs, liver, arteries, blood pressure, and brains of animal foetuses. Animal experiments have revealed that extreme manganese toxicity can cause tumors

(continued)

Table 8.1 (continued)

Heavy metal	Impact on human health		Impact on environment, plants and organisms
	Toxicity	Health effects	
Nickel (Ni)	Carcinogenicity, hepatotoxicity	Respiratory diseases, asthma, congenital malformations, cancers, itching, teratogenic, mutagenic, affects fertility and hair loss	Excess Ni can harm habitats for plants and animals. Ni also contributes to greenhouse gas emissions, habitat loss, and air, water, and soil contamination
Mercury (Hg)	Nephrotoxicity, carcinogenicity, immunological toxicity, cardiovascular toxicity, reproductive and developmental toxicity, skin toxicity	Some chemicals can harm various physiological systems. Lung and kidney damage may occur. Memory, sensory, and coordination problems can result from nervous system damage. These compounds can also harm mitochondria, lipids, collect neurotoxic chemicals, and destroy nervous system microtubules	Mercury released into the air settles in water or washes onto land. Certain microbes can convert it into methylmercury, which accumulates in fish, shellfish, and fish-eating animals. Methylmercury-contaminated fish and shellfish are the main source of mercury exposure in the US and internationally
Selenium (Se)	Neurotoxicity, nephrotoxicity	Decreased natural killer cell activity, hepatotoxicity, gastrointestinal issues, liver, kidney, spleen damage, and anxiety	Even small increases in waterborne concentrations can cause harmful effects like malformed embryos and reproductive failure in wildlife because aquatic organisms bioaccumulate it
			Selenium poisoning causes leaf stunting, chlorosis, and fading, disrupting chlorophyl combination
			Laboratory animals with acute poisoning developed garlicky breath odor, vomiting, dyspnea, tetanic spasms, and respiratory failure with histological alterations

(continued)

Table 8.1 (continued)

Heavy metal	Impact on human health		Impact on environment, plants and organisms
	Toxicity	Health effects	
Zinc (Zn)	Neurotoxicity	Dizziness, fatigue, vomiting, renal damage, and cramps	Per ton of zinc produced, 3 tons of CO_2 is released
			Living in zinc-contaminated streams causes fish to accumulate zinc. These fish absorb zinc, which biomagnifies and enters the food chain
			Zinc can also acidify water. Zinc disrupts soil activity by affecting microorganisms and earthworms. This may significantly inhibit organic matter degradation
Cobalt (Co)	Cardiovascular toxicity	Diarrhea, low blood pressure, and paralysis	Water and crops are polluted by cobalt. High cobalt concentrations have been associated to crop and worm deaths, which are essential for soil fertility
			Dietary cobalt poisons animals. Polycythemia, hyperglycemia, and pancreatic alpha-cell injury after cobalt treatment are common

References De and Roy (2023), Mitra et al. (2022), Prasad (2021), Awa and Hadibarata (2020), Briffa et al. (2020), Hocaoglu-Ozyigit et al. (2020), Sall et al. (2020), Malik et al. (2019), Bashir et al. (2018), Mishra et al. (2018), Azimi et al. (2017), Abosede (2017), Carolin et al. (2017), US EPA (2016), Bazrafshan et al. (2015), Mahurpawar (2015), US EPA (2014)

The concentration of heavy metal pollution in the aquatic body is summarized in Table 8.5. It has been found that HMs concentration levels were relatively low during the 1970s period. After that, the world observed the rapid growth of heavy metal concentration in the 1990s, 2000s, and 2010s. For example, in the 2010s, Dongting Lake in China noticed more than 2.2 times higher Pb concentration than the concentration level of the 1970s (Yang et al. 2008; Liu et al. 2015). Moreover, the Niger River in Africa experienced 1914 times greater concentration levels than in the 1970s, respectively (Wangboje and Ikhuabe 2015). However, according to a recent study, the concentrations of heavy metals exhibited higher levels in Africa, Asia, and South America. Comparatively lower levels were observed in Europe and North America (Zhou et al. 2020). Moreover, several studies have also found a significant amount of HMs in river sediments.

Table 8.2 Commonly used heavy metal detection methods along with source and detected heavy metals

Detection technique	Source	Heavy metal	References
Atomic absorption spectrometry (AAS)	Medical	Cd, Pb, Fe, Zn	Qualitest (2023)
	Pharmaceutical	Cr, Pb, As, Cd, Cu, Hg	
	Food	Pb, Cd, Cr, Cu, Fe, Mn, Ni, Sb, Sn, Zn	
	Water treatment	Cd, Cu, Fe, Pb, Zn	
	Geological/mining	Cu, Fe, Ni, Pb, Zn, etc.	
	Fodder/animal feed	Cu, Fe, Mn, Zn, Cd, Co, Cr, Pb	
Atomic fluorescence spectrometry (AFS)	Food	Cd, As, Se, Sb	Malik et al. (2019)
	Water	Pb, Hg	
	Mining	As	
X-ray fluorescence spectrometry	Aqueous solution	Cu, Pb, Hg	Malik et al. (2019)
	Coastal seawater	Fe, Ni, Mn, Cu, Zn, Pb	
	Fish tissue	Cr, Ni, Cu, Zn, Hg, Pb	
	Soil	As, Cr, Cu, Ni, Pb, V, Zn, Cd	
Graphitic furnace atomic absorption spectroscopy (GF-AAS)	Water	Co, Ni, Pb	Jin et al. (2020), Malik et al. (2019)
	Chinese tea	Pb, Cd, Cr, Cu, Ni	
	Rocks	Pt	
	Muscle samples of fish	Hg	
Inductively coupled plasma mass spectrometry (ICP-MS)	Water	Mn, Cr, Ni, Co, Ag, As, Hg, Cu, Cd, Pb, Ni, Zn	Sarker et al. (2021)
	Food	As, Cd, Pb, Cr, Mn, Ni, Cu, Zn, Fe	
Inductively coupled plasma optical emission spectrometry (ICP-OES)	Marine sediments	As, Co, Cr, Cu, Fe, Mn, Ni, Pb, and Zn	Sall et al. (2020)
	Wastewater	Pb, Cd, Ni, Cr, and Cu	
Laser-induced breakdown spectrometry (LIBS)	Solid phase samples	Cu, Pb, Cd, Cr, Hg	Jin et al. (2020)
	Water	Ni, Pb, Zn	
Nanotechnology based biosensors	Water	As, Pb, Hg, Cd, Co	Gong et al. (2021), Salek Maghsoudi et al. (2021)
	Food	Pb, Hg, Zn	

Table 8.3 Mostly used analytical techniques in heavy metal detection in various food stuffs

Detection method	Detected heavy metals	Sample (food)	Country
ICP-MS	As, Cd, Pb, Sb	Cocoa beans and products	Malaysia
	Cu, Mn, Ni, Rb, Sr, Zn	Green coffee bean	Ethiopia
	Cr, Cu, Ni, As, Cd, Pb	Food	Bangladesh
AAS	Cr, Cd, Pb, Ni	Fruits, vegetables and cereals	Nigeria
FAAS	Fe, Co, Zn, Cd, Pb, Cu	Soft drink	Ghana
ICP-OES	Fe, Mn, Cd, Pb, Ni, Cr, Cu	Carbonated, flavored yogurt, juice drinks	Egypt
	As, Cd, Cr, Cu, Mn, Pb, Se, Zn	Vegetables	Congo
TXRF	Ti, Cr, Sb, As, Pb	Soft Drinks	Brazil

Source Deka et al. (2023)
FAAS Flame Atomic Absorption Spectrometry; *TXRF* Total Reflection X-Ray Fluorescence Spectroscopy

Table 8.4 The concentration (mg/kg) of heavy metals (HMs) in road dust worldwide

Countries	Cu	Ni	Cd	Pb	Mn	Fe	Zn	Cr	References
Dhaka, Bangladesh	59.30	–	–	59.60	–	27,350	189	–	Rahman et al. (2021)
Delhi, India	9.70	1.30	–	3.20	21.50	1220	27.2	5.70	Hama et al. (2021)
Guilin, China	149.4	15.2	–	123.8	1089	19,482	391.30	123	Shahab et al. (2020)
Kualalampur, Malaysia	67.32	57.345	0.70	31.05	–	–	73.973	27.07	Wahab et al. (2020)
Isfahan, Iran	182.2	66.63	–	393.3	–	–	707.19	–	Soltani et al. (2015)
Shiraz, Iran	49.81	39.40	0.31	36.80	245	16,300	160.90	31.60	Keshavarzi et al. (2015)
Kathmandu, Nepal	5.50	–	2.28	1.92	–	–	3.17	–	Shakya et al. (2019)
Nanjing, China	481.9	69	0.99	422.8	–	–	226.9	155.90	Liu et al. (2019)
Houston, USA	329	160	–	95	–	–	327	99	Fiala and Hwang (2021)
Colombia, South America	213.4	22.26	–	467.63	–	–	209.73	–	Trujillo-González et al. (2016)
Queensland, Australia	91.53	6.39	0.40	31.43	0.13	5.04	199.98	7.82	Gunawardana et al. (2012)
Castelo, Portugal	1816	110	–	276	1029	–	2162	296	Alves et al. (2020)
Ososbo, Nigeria	7.567	1.653	0.076	3.767	7.653	–	9.974	0.003	Taiwo et al. (2019)

'–' Stands for missing data

Table 8.5 The concentration of heavy metals (HMs) in river (µg/l) and sediments (mg/kg)

Rivers	Cu	Zn	Cd	Hg	Pb	Mn	Cr	Co	Fe	Ni	References
Buriganga, Bangladesh	62–712	–	2–19	–	43–3312	–	17–841	–	–	62–539	Islam et al. (2018a, b)
Brahmaputra, India	0.3–57.6	35.1–80.1	–	–	–	–	–	–	–	–	Borah et al. (2018)
Lee, England	0.66–55.1	1.04–46.7	0.01–0.44	0.01–1	0.21–13.3	–	–	–	–	–	Lundy et al. (2017)
River Nepal and China	1–43.63	4–93.36	7–0.59	–	6–36.17	–	9–123.05	3–19.59	–	42–52.65	Li et al. (2020)
Saigon, Vietnam	–	4.5–75	0.04–0.07	–	0.88–2.00	95.6–162.1	–	0.72–4.95	27.7–99	0.87–6.93	Nguyen et al. (2020)
Baram, Malaysia	30–410	60–250	–	–	11–123	77.45–805.44	–	4–8	–	0–40	Prabakaran et al. (2019)
Xiangjia, China	108.2	–	4.29	–	383.4	2283	99.61	21.74	–	54.32	Fang et al. (2019)
Houjing, Taiwan	2156	620	22.2	0.01–0.328	15.6–258	–	99.90	–	–	32.60	Vu et al. (2017)
Meghna, Bangladesh	27	400	18	–	–	500	20	9	250–2700	300	Bhuyan et al. (2017)
Feni, Bangladesh	–	–	–	0.71	6.47	37.85	35.28	31.02	–	33.27	Islam et al. (2018a, b)
Sangu, Bangladesh	2.23–31.40	22.8–80.83	–	–	12–33.42	–	9.52–31.4	–	–	15.6–79.7	Hossain et al. (2019)

'–' Stands for missing data

Several heavy metals pollution have also been found in soil compartments. Soil systems are contaminated due to both natural and anthropogenic sources. After that, plants uptake these HMs following various processes (Fig. 8.3). Consequently, the massive concentration level of HMs found in various vegetables and fruits worldwide indicates a tremendous human health concern as it is deadly (Shaheen et al. 2016; Amer et al. 2019). Table 8.6 showed the concentration of HMs in various fruits and vegetables.

Hence, the hazard of HMs has increased each year for various reasons. As HMs have already entered our food chain and have acute and chronic effects on plants and humans, we need to take immediate response on this issues.

8.7 Standards and Guidelines of Permissible Levels of Heavy Metal

Earth contains naturally occurring heavy metals. Human activities concentrate these compounds, which can be inhaled, ingested, or touched, accumulating in plant, animal, and human tissues. Cellular components may interfere. The WHO's top 10 compounds of public concern include cadmium, lead, mercury, and arsenic, according to Abosede (2017).

The World Health Organization (WHO), Environmental Protection Agency (EPA), and Food and Agricultural Organization (FAO) have set heavy metal limitations. Limits restrict and reduce heavy metal pollution. The WHO has classed heavy metals as hazardous air pollutants (HAPs). According to the WHO, Cadmium (Cd) and Manganese (Mn) recommended air values are 0.005 $\mu g/m^3$ and 0.15 $\mu g/m^3$, respectively. As, Ni, Cd, Cr, and Be are IARC Class 1 carcinogen (Kang et al. 2023). According to WHO air quality requirements and EPA regulations, Table 8.7 shows heavy metal thresholds and OSHA values (Morakinyo et al. 2021; Briffa et al. 2020). For drinking water quality, standards set by WHO, US EPA and Bangladesh Environmental Conservation Rules (ECR) have been added in Table 8.8. The permissible limit of heavy metals in soil, plants, sludge, vegetables and wastewater has been added in Table 8.9. The maximum limit of heavy metals in food set by the Codex Alimentarius Commission (commonly referred to as CODEX) has been added in Table 8.10.

8.8 Remediation Techniques

The utilization of hazardous metals in various industrial processes such as electroplating, painting, tanning, textiles and dyes, papermaking, and mining has witnessed a notable escalation, resulting in detrimental effects on our planet's biodiversity (Yadav et al. 2017). The unregulated discharge of metallic substances from these industrial

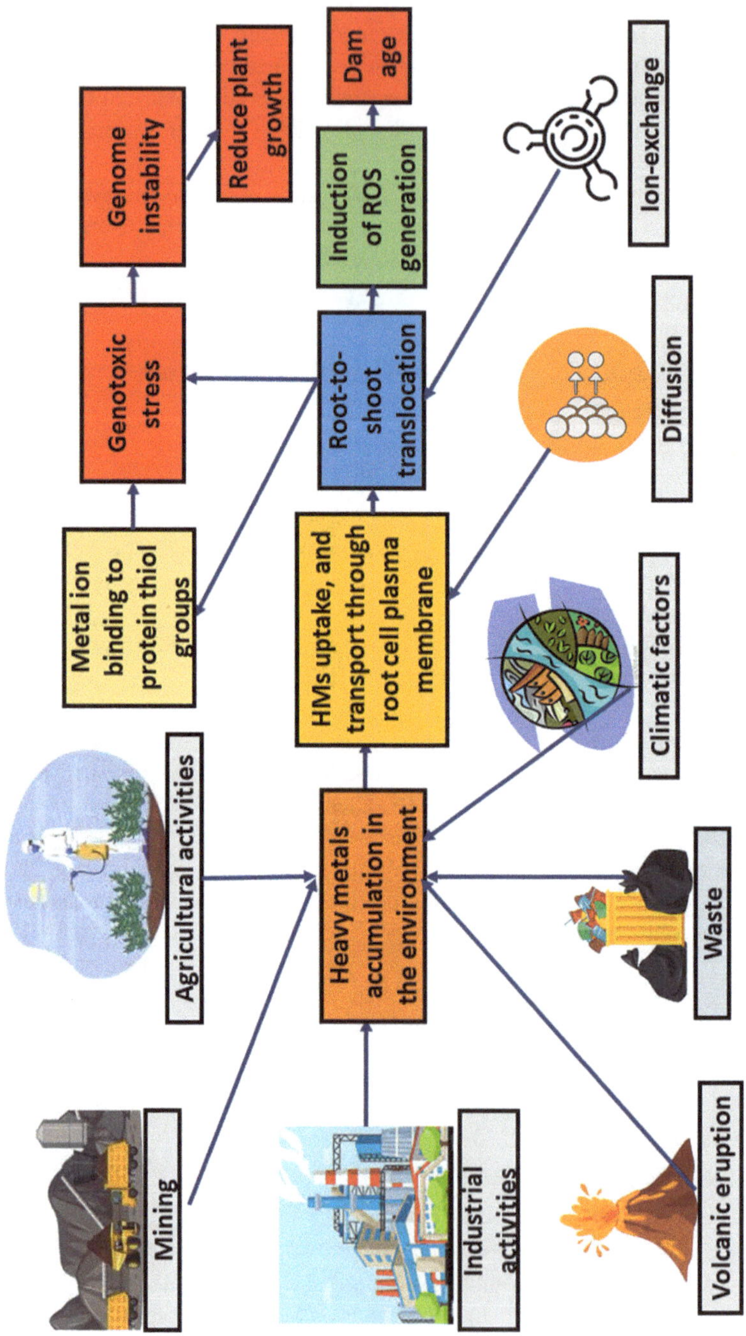

Fig. 8.3 Plants heavy metals uptake mechanism

Table 8.6 The concentration (mg/kg) of heavy metals (HMs) in various vegetables and fruits

Crops	Country	Cd	Pb	Cr	Mn	Ni	Cu	Zn	References
Mango	Bangladesh	0.005	0.642	0.893	6.06	0.293	7.891	0.604	Shaheen et al. (2016)
Potato	Bangladesh	0.013	0.007	0.528	6.928	0.643	4.3	3.019	Shaheen et al. (2016)
Banana	Bangladesh	–	0.003	0.317	10.74	0.037	0.946	0.235	Shaheen et al. (2016)
Radish	Egypt	–	–	–	–	–	0.83	–	Scutarașu and Trincă (2023)
Grapes	Egypt	–	–	1.06	–	0.3–1.78	0.72–3.52	–	Amer et al. (2019)
Lettuce	Romania	0.97–1.52	0.82–2.22	–	–	–	8.1–11.8	59.5–139.5	Dağcilar and Gezer (2021)
Tomato	Romania	0.1 μg/kg	–	–	(13–13.5) μg/kg	–	(32.5–46) μg/kg	–	Hegheduş-Mîndru et al. (2023)
Red potato	Romania	0.16–0.46	0.37–1.03	–	–	–	5.4–6.4	56.5–64	Dağcilar and Gezer (2021)
Onion	Romania	0.17–0.34	0.16–0.18	–	–	–	3.6–12.3	28–47	Dağcilar and Gezer (2021)
Carrot	Romania	0.27–0.69	0.54–0.94	–	–	–	6–9.4	51–75.5	Dağcilar and Gezer (2021)
Beans	Romania	0.04–0.05	0.08–0.52	–	–	–	7–11.1	58.5–91	Dağcilar and Gezer (2021)

(continued)

Table 8.6 (continued)

Crops	Country	Cd	Pb	Cr	Mn	Ni	Cu	Zn	References
Apples	Ukraine	–	–	–	–	0.06–0.38	0.36–2.2	–	Shaheen et al. (2016)
Apples	Kosovo	0.17–0.21	1.45–2.17	8.22–11.6	–	15.33–17.2	1.05–2.87	1.51–2.03	Scutarașu and Trincă (2023)

'–' Stands for missing data

Table 8.7 Permissible limit of concentration of heavy metals in ambient air

Heavy metal	Winter ($\mu g/m^3$)	Summer ($\mu g/m^3$)	Limit value ($\mu g/m^3$)	OSHA value (mg/m^3)
Arsenic (As)	0.035	0.07	0.006	0.01
Cadmium (Cd)	0.026	0.022	0.0002	0.005
Chromium (Cr)	0.354	0.309	0.012	0.005
Copper (Cu)	0.2	0.2	100	1
Iron (Fe)	4.3	3.4	10,000	5
Nickel (Ni)	0.067	0.061	0.00024	1
Lead (Pb)	0.5	0.5	0.5	0.5

Source Morakinyo et al. (2021), Briffa et al. (2020)

Table 8.8 Permissible limit of heavy metals in drinking water

Heavy metal	Permissible limit μg/L (US EPA)	Permissible limit μg/L (WHO)	Permissible limit mg/L (ECR, Bangladesh)
Arsenic (As)	10	10	0.05
Cadmium (Cd)	5	3	0.005
Chromium (Cr)	100	50	0.05
Copper (Cu)	1000	2000	1
Lead (Pb)	10	10	0.05
Iron (Fe)	300	300	0.3–1
Manganese (Mn)	50	400	0.1
Nickel (Ni)	20	70	0.1
Mercury (Hg)	2	6	0.001
Selenium (Se)	50	10	0.01
Zinc (Zn)	5000	3000	5
Cobalt (Co)	100	100	–

Source Hossain et al. (2021), Malik et al. (2019), Purushotham et al. (2013)
'–' Stands for missing data

activities, frequently exceeding authorized limits, presents a significant ecological hazard. Hence, it is imperative to eliminate or mitigate these pollutants to protect environmental integrity (Mishra et al. 2018). Conventional techniques for removing heavy metals typically encompass physical and chemical processes. However, contemporary approaches employ biological and physicochemical methods, such as adsorbent-based remediation.

Table 8.9 Permissible limit of heavy metals in soil, plants, vegetables, sludge and wastewater

Permissible limit	Heavy metal								References
	Cd	Cr	Cu	Pb	Ni	Zn	Fe	Hg	
Soil (mg/kg) according to WHO	0.8	100	36	85	35	50	–	–	Osmani et al. (2015)
Plants (mg/kg) according to WHO	0.02	1.3	10	2	10	0.60	–	–	
Vegetables (mg/kg) according to FAO/WHO	0.2	1.3	73.3	0.3	–	99.4	425.5	–	Mensah et al. (2009)
Sludge (mg/kg) according to US EPA	39	600	1500	300	420	2800	–	17	Agoro et al. (2020), US EPA (2020), DOE (2015)
Sludge (mg/kg) according to DOE, Bangladesh	10	900	800	900	200	2500	–	8	
Wastewater according to WHO (ppm)	0.003	0.05	–	0.01	0.02	–	–	0.001	Kinuthia et al. (2020)
Wastewater according to US EPA (ppm)	0.01	0.05	–	0.006	0.2	–	–	0.00003	
Wastewater according to WB (ppm)	0.1	0.5	–	0.1	0.5	–	–	0.01	

'–' Stands for missing data

8.8.1 *Heavy Metal Removal Techniques from Liquid Media (Water, Wastewater etc.)*

Heavy metal cations can be removed from liquid media using several methods. Figure 8.4. shows the different available methods for heavy metal removal from liquid media. The following will be briefly discussed. In many applications, chemical precipitation is the easiest way (Vareda et al. 2019). Adding a precipitation agent to effluent causes cations to react and precipitate insoluble substances. Lime is used to raise effluent pH, precipitating hydroxide ions. Chemical precipitation produces large amounts of sludges, requiring costly treatment and disposal. This method also raises wastewater salt levels, which violates disposal regulations (Vareda et al. 2019).

Using coagulation and precipitation agents simultaneously improves therapeutic efficacy. In addition to their traditional uses, coagulation and flocculation can remove

Table 8.10 Maximum limit of heavy metals in food according to CODEX (2019)

Heavy metal	Food	Max. limit mg/kg
Arsenic (As)	Rice, husked	0.35
	Rice, polished	0.2
	Salt, food grade	0.5
	Natural mineral water	0.01
	Edible fats and oils	0.1
Cadmium (Cd)	Wheat	0.2
	Rice, polished	0.4
	Salt, food grade	0.5
	Natural mineral water	0.003
	Cereal grains	0.1
	Pulses	0.1
	Legume vegetables	0.1
	Leafy vegetables	0.2
	Fruiting vegetables	0.05
	Brassica vegetables	0.05
	Bulb vegetables	0.05
	Root and tuber Vegetables	0.1
	Stalk and stem Vegetables	0.1
Mercury (Hg)	Salt, food grade	0.1
	Natural mineral water	0.001
Tin (Sn)	Canned foods	250
	Canned beverages	150
	Corned beef	50
	Cooked cured chopped meat	50
	Luncheon meat	50
Lead (Pb)	Salt, food grade	1
	Natural mineral water	0.01
	Milk	0.02
	Edible fats and oils	0.08
	Fish	0.3
	Meat and fat of poultry	0.1
	Cereal grains	0.2
	Fruit juices	0.03
	Canned vegetables	0.1
	Canned fruits	0.1

(continued)

Table 8.10 (continued)

Heavy metal	Food	Max. limit mg/kg
	Jams, jellies and marmalades	0.4
	Mango chutney	0.4
	Root and tuber Vegetables	0.1
	Pulses	0.1
	Fruits	0.1
	Legume vegetables	0.1
	Leafy vegetables	0.3
	Fruiting vegetables	0.05
	Brassica vegetables	0.1
	Bulb vegetables	0.1

Source CODEX (2019)

heavy metals and suspended particles. Coagulants like aluminum or ferrous sulfate make colloidal particles unstable, causing aggregation. Polyelectrolytes help particles aggregate in flocculation, forming more oversized agglomerates. Common flocculants include polyferric, aluminum, and polyacrylamide. Vareda et al. (2019) showed that sedimentation and filtering remove agglomerates.

Al-Saydeh et al. (2017) found that membrane filtering removes heavy metals from various sources. Ultrafiltration (UF) membranes have larger pores than hydrated cations and low-molecular-weight solutes. Surfactant micelles that have an affinity for cations or polymers that can form complexes with heavy metals are added to effluent to retain them. The membrane retains structures after this addition (Al-Saydeh et al. 2017). Nanofiltration (NF) may remove nickel, chromium, and arsenic. This work used membranes with an intrinsic charge, which caused steric (size exclusion) and electrical (Donnan exclusion) effects. These effects help the membrane selectively block charged solutes smaller than its pores (Al-Saydeh et al. 2017). This can also be done with reverse osmosis (RO). The semipermeable membrane used in this process excludes considerable dissolved contaminants under pressure. Vareda et al. (2019) found that this method retains most heavy metals. Ion exchange uses resins to reversibly exchange ions between the resin's solid and effluent's liquid phases. Cation-exchange resins must be carefully chosen to eliminate heavy metals. Selective resins make this method suitable for precious metal retrieval (Vareda et al. 2019).

Electrochemical methods can recover heavy metals from water sources, along with other methods. Metallurgy is often linked to them. Electrodialysis uses electric potential to exchange ionized species across a charged membrane (Al-Saydeh et al. 2017). According to Chowdhury et al. (2016), Vareda et al. (2016), Al-Saydeh et al. (2017), and Burakov et al. (2018), adsorption is a frequent method for heavy metal removal from liquids. The liquid adsorbate gathers on the solid adsorbent during

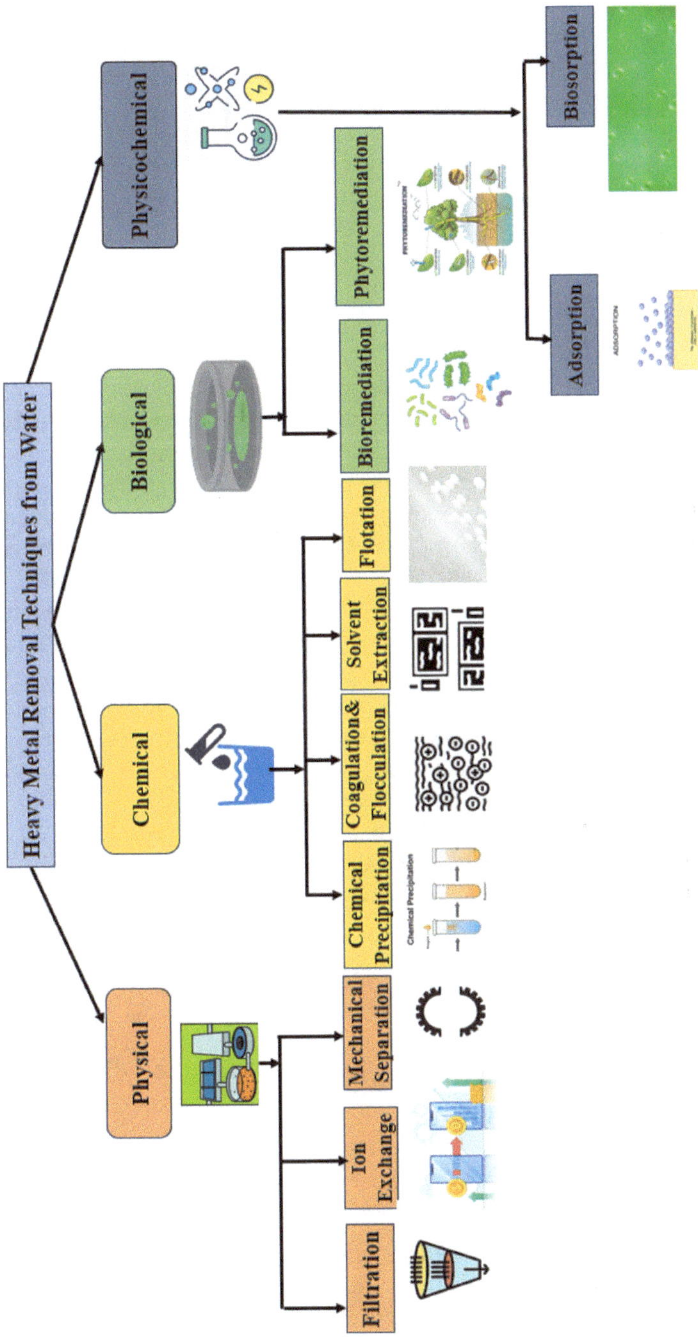

Fig. 8.4 Heavy metal removal methods from liquid media

this process. Adsorption reduces organic and inorganic contaminants in water and wastewater treatment to acceptable levels. Surface area or active sites, selectivity, and kinetics of the adsorbent affect process efficiency. Using the proper adsorbent can boost efficiency and speed up the process. Using cheap adsorbents can also make the process more viable. Multiple adsorbents (Lu and Astruc 2018) have improved adaptability, efficiency, design and operation simplicity, and possible selectivity for the technique. Table 8.11 shows the advantages and disadvantages of different heavy metal removal methods.

8.8.2 Removal Techniques of Heavy Metal from Soil

In recent decades, increased industrialization, human illiteracy, and avarice have made soil pollution with heavy metal ions a global issue. Heavy metal ions poison at low doses and are non-biodegradable. These toxins can build up in the body and cause chronic diseases like lung cancer, brain damage, respiratory issues, and renal impairment. Higher than permissible levels of metallic ions in the soil make it unfit for agricultural usage. Therefore, we must continuously monitor metal ion levels in soil and water ecosystems and apply more advanced technology to eradicate them. According to the literature, researchers extracted heavy metal ions from metal-contaminated soil samples using physical, chemical, and biological methods. These methods aimed to eliminate metal ions or convert them into safer, less harmful forms (Raffa et al. 2021). Figure 8.5 shows the heavy metal removal methods from the soil.

Over 20 million hectares of land worldwide are contaminated by heavy metals (loid) such as arsenic, cadmium, chromium, mercury, lead, cobalt, copper, nickel, zinc, and selenium. Current soil metal concentrations exceed geo-baseline or regulatory thresholds. Site heavy metal pollution can be remedied using in-situ and ex-situ methods. These methods include surface capping, encapsulation, landfilling, soil flushing, washing, electrokinetic extraction, stabilization, solidification, vitrification, phytoremediation, and bioremediation (Kumar et al. 2023). Table 8.12 shows the advantages and disadvantages of the heavy metal removal method from soil.

8.8.3 Recent and Mostly Used Heavy Metal Remediation Techniques

8.8.3.1 Bioremediation

Compared to standard treatment technologies, biological remediation using bacteria and plants is environmentally friendly, safe, economically viable, and sustainable. Conventional methods are expensive, need maintenance, and produce harmful byproducts or inefficiency. Instead, biological treatment methods offer cost-effective,

Table 8.11 Advantages and disadvantages of different heavy metal removal methods

Methods	Techniques	Advantages	Disadvantages
Physical	Filtration	• Small space requirement • High efficiency • Low pressure of operation (UF) • High rejection rate (RO)	• High operation cost • Membrane fouling • High pressure operation (RO) • High energy consumption (RO)
	Ion exchange	• No sludge production • High removal capacity • Fast kinetics • Can be selective	• Not all resins are adequate • Resins have to be regenerated • Treatment of regeneration solution • High cost
	Mechanical separation	• Can be applied to different metals	• Homogeneous • Distribution is the prerequisite for mechanical separation
	Electrokinetic remediation	• No introduction of other harmful ingredients • Environmentally friendly	• Takes a long time to remove heavy metals • Solubility is poor in solution • Increase the remediation time and energy consumption
Chemical	Chemical precipitation	• Low cost • Low energy requirements • Simple design/ operation • Safe operation	• Production of high volume of sludge • High cost for sludge disposal • High consumption of precipitant agent • Increasing difficulty in meeting disposal standards
	Coagulation and flocculation	• Reduced time to settle suspended solids • Improved sludge settling	• Sludge production • High cost for sludge disposal
	Electrochemical	• Selective separation • High efficiency • No sludge production	• High cost • High energy consumption
	Flotation	• It can purify wastewater and recover precious or rare metal ions • Simple, rapid, and cost efficient	• High reagent consumption • Environmental pollution

(continued)

Table 8.11 (continued)

Methods	Techniques	Advantages	Disadvantages
Biological	Bioremediation	• No side effects • Ecofriendly method • Natural approach	• Large amounts of waste can be generated • Requires specific environmental conditions to be effective • Can become expensive if the site is large
	Phytoremediation	• Economically feasible • Autotrophic system • Easy to handle • Low installation and maintenance cost • Eco-friendly • Prevents erosion and metal leaching • It can also improve soil fertility by releasing various organic matters to the soil	• Slow growth rate of plants • Global environmental obstacles, and the need for effective disposal methods for contaminated biomass • Improper disposal of contaminated biomass can lead to the leaching of heavy metals
Physicochemical	Adsorption	• Simple design and operation • Low initial cost • High efficiency • Low-cost adsorbents available • Can be selective	• Efficient adsorbents can be expensive • Loaded adsorbent needs disposal/regeneration
	Biosorption	• Low cost, • Limited use of chemicals, • No sludge generation • The regeneration of biosorbents	• Saturation of active sites of metal binding ligands • Reversible sorption of metals on biomass

References Sun et al. (2023), Sodhi et al. (2022), Liu et al. (2021), Lew (2020), Yan et al. (2020), Vareda et al. (2019), Shamim (2018)

user-friendly operating systems that do not produce secondary pollutants. Using microorganisms and plants, respectively, bioremediation and phytoremediation reduce the harmful effects of highly toxic contaminants by transforming them into less harmful and harmless forms.

Compared to expensive physicochemical methods for metal cleanup, bioremediation is feasible. Bioremediation uses natural microorganisms to convert toxic chemicals into less or non-toxic compounds. Indigenous microorganisms like bacteria, fungi, plants, and earthworms are used to safely eliminate or break down harmful pollutants and waste substances (Venkatesh et al. 2015). Polluted areas can be remedied by using indigenous microorganisms from the site or those from other sources.

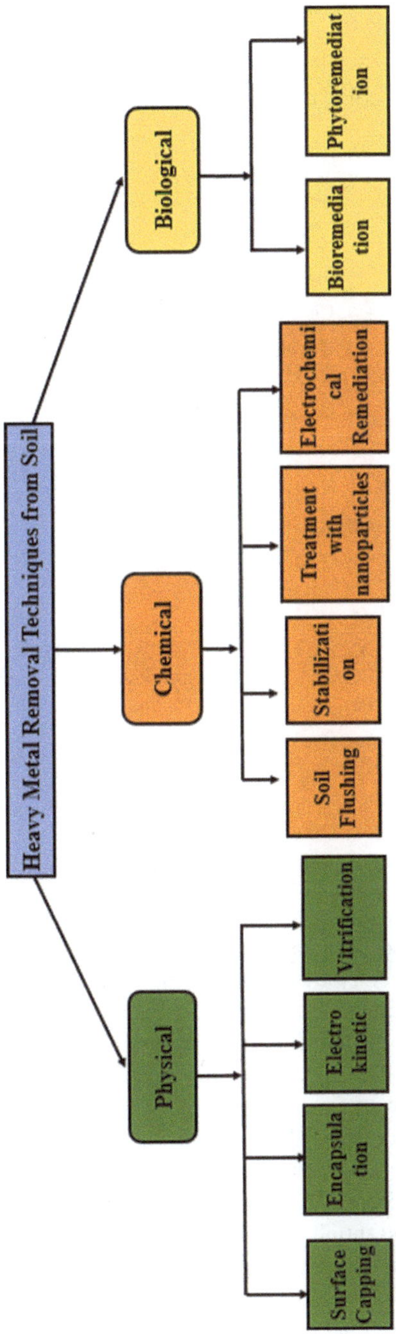

Fig. 8.5 Heavy metal removal methods from soil

Table 8.12 Removal methods of heavy metal from soil

Method	Techniques	Targeted heavy metals	Advantages	Disadvantages	Removal efficiency (%)
Physical	Soil replacement	Hg, Cd, Ni, Cu, Cr, Pb, Zn, As, Se	• Fast and simple process • High efficiency	• High cost	95
	Vitrification	Cu, Pb			
	Electrokinetic remediation	As, Pb, Cu, Ni, Zn			
Chemical	Soil flushing	Pb, Cd, Zn, Cu, Cr, Ni	• High efficiency • Simplicity • Rapidity	• Change the physiochemical properties of soil • High cost	90
	Electrochemical remediation	Cd, Cu, Hg, Ni, Zn, Pb			
	Soil washing	Cu, Zn, Cd, Pb			
	Stabilization	Zn, Cu, Cr, Ni, Co, Hg, Cd, As, Pb			
	Treatment with nanoparticle	Cd, Cr, Ni, Pb, Cu, Fe, Zn			
Biological	Phytoremediation	As, Zn, Cd, Pb, Cr, Hg, Fe	• Simple • Cost-effective • Eco friendly	• Slow remediation process • Only applicable for low contaminated sites	96
	Bioremediation	Cu, Pb, Ni, Cr,			

Source Kumar et al. (2023), Raffa et al. (2021), Liu et al. (2018), Khalid et al. (2017)

pH, temperature, texture and composition of the soil, nutrient inputs, and oxygen availability all have an impact on the bioremediation process. Bioremediation reduces the toxicity of inorganic pollutants like metals by microorganism adsorption, absorption, or accumulation. In situ and ex-situ bioremediation can remove pollutants. Microorganisms are used in bioremediation in situ to directly interact with dissolved and sorbed pollutants as substrates. This technology is compared to others for water and soil cleanup (Bharagava et al. 2017).

Fungi, bacteria, plants, and algae reduce pollutants. Bacteria are ubiquitous and essential to most biological processes that support life. Bacteria flourish even in toxic metals because metals are essential to their metabolism (Igiri et al. 2018). In higher concentrations, dangerous metals can affect osmotic equilibrium, oxidative phosphorylation, polypeptides, and nucleic acids, causing microorganism toxicity (Fashola et al. 2016). Bioremediation of heavy metals (HMs) may use HM-tolerant bacteria. Metal-stressed bacteria can remove heavy metals (Urra et al. 2019). Heterochromatin modifications (HMs) can only change states and withstand degradation. Microorganisms should turn hazardous chemicals into a less harmful and less bioavailable form. Heavy metal (HM) stress has created HM-resistant microorganisms. To endure HM stress, these bacteria have acquired hazardous metal tolerance mechanisms. These mechanisms include HM buildup in the cell wall, redox reactions, trapping (extracellular or intracellular), and cell membrane transfer (Jasmin et al. 2020). Plasmids mainly cause HM resistance. However, chromosomal components may also play a role (Sun et al. 2021). Biosorption passively removes heavy metals (HMs) from non-living biomass, while bioaccumulation involves bacterial cells aggregating on live cells. Bioremediation of heavy metals involves bacteria like Bacillus and Alcaligenes, fungi like Penicillium and Aspergillus, and plants like *Eichhornia* and *Pistia.* Table 8.13 shows the bioremediation of heavy metals using living organisms.

Table 8.13 Bioremediation of heavy metal using living organisms

Living organism involved in bioremediation	Type of living organism	Heavy metals removed
Bacillus sp.	Bacteria	Cu
Bacillus subtilis	Bacteria	Cd, Hg, Cr, Mn, Pb
Pseudomonas aeruginosa	Bacteria	Cd, Hg, Cr, Ni, Cu
Escherichia coli	Bacteria	Pb, Cd, Cr
Zoogloea ramigera	Bacteria	Cu
Exiguobacterium sp.	Bacteria	Cr
Nesterenkonia sp.	Bacteria	Cr
Burkholderia sp.	Bacteria	Cd
Alcaligenes sp.	Bacteria	Cu, Cr, Cd, Ni, Zn
Bacillus sp. strain	Bacteria	Cd
Penicillium notatum	Fungi	Cd
Aspergillus niger	Fungi	Ni, Cd, Cu, Pb, Co
Rhizopus arrhizus	Fungi	Cr, Cu, Ni, Zn, Pb, Fe
Saccharomyces cerevisiae	Fungi	Cu, Pb, Cd
Cladophora glomerata	Algae	Cr, Pb, Cu, Se
Chlorella vulgaris	Algae	Ni, Cu, Cr, Cd
Spirulina plantensis	Algae	Cr, Cd
Spirogyra spp. *Spirogyra insignis*	Algae	Cu, Pb, Cr, Zn
Laminaria japonica	Algae	Cr, Pb, Zn, Cd

Source Sodhi et al. (2022), Kumar et al. (2021)

8.8.3.2 Phytoremediation

Most conventional treatments are too expensive and worsen environmental issues. Most bioremediation methods that use green plants and their microbes to clean up contaminated soil and groundwater are called "phytoremediation". Since 1983, metal-accumulating plants have been used to remediate damaged soil and plants of heavy metals and other pollutants. The environmentally sustainable, visually pleasing, and economically efficient phytoremediation approach offers potential for underdeveloped nations. Depending on the approach, phytoremediation can involve phytoextraction, phytofiltration, phytostabilization, phytovolatilization, and phytodegradation. In phytoextraction, metal ions accumulate in aboveground plant parts, disposed of or burned to recover the metals. Plant roots or seedlings remove metals from watery waste via phytofiltration. Plant roots absorb contaminants from soil and store them in the rhizosphere during phytostabilization. Preventing leaching makes contaminants harmless. Foliage releases selenium (Se) and mercury (Hg) during phytovolatilization. Plants and microorganisms work together to degrade organic contaminants in phytodegradation (Mishra et al. 2018).

Microorganisms, speciation, and physicochemical properties affect heavy metal (HM) remediation (Abatenh et al. 2017). Decreased heavy metal (HM)-bacteria interaction may slow metal breakdown. Microbial biodegradation is hindered by heterogeneous metal and bacteria dispersion. Temperature, pH, and matrix type affect metal removal (Nedjimi 2021). As mentioned, HMs resist biodegradation. However, they can be made safer. However, the changed compounds are more persistent and hazardous than the original. Heavy metals (HMs) cleanup requires optimal circumstances, including specific biological activity and the right amounts of pollutants and nutrients. Phytoremediation uses plants and rhizospheric microorganisms to degrade environmental, water, and soil toxins (Nedjimi 2020). A cost-effective, ecologically friendly, and effective method is phytoremediation. Phytoremediation uses phytovolatilization, phytoextraction, phytodegradation, phytostabilization, and phytofiltration. After intake, HMs are translocated to plant structures and detoxified. Plant species, solubility, and speciation affect heavy metal (HM) dependence. Hyperaccumulator plants have high biomass and can gather heavy metals (HMs) in their shoots. Hyperaccumulator plants adapt to toxins by growing better in polluted soils. This trait depends on plants' ecological and physiological adaptation to heavy metal resistance. Current phytoremediation research focuses on plant growth-promoting microorganisms to remove heavy metals due to the constraints of conventional methods on a wide scale. HM decontamination can also be achieved by transgenic plants and phytohormone-assisted phytoremediation (Sodhi et al. 2022). Novel genes that improve heavy metal removal are added to transgenic plant genomes. Genetic engineering manipulates plant genomes by adding external genes. Genes that break down and sequester heavy metals (HMs) from other plants, fungi, and bacteria are introduced into plants. Nahar et al. (2017) observed that transgenic Nicotiana tabacum plants with the *Salicornia brachiata*-derived SbMT-2 gene resist harmful metals like Cd, Zn, and Cu and regulate ROS levels. Most transgenic plant research focuses on amplifying genes involved in metal-binding peptide pathways (Sodhi et al. 2022).

Table 8.14 Plant species and aquatic macrophytes used for phytoremediation of heavy metals

Plant species	English name	Heavy metals removed
Brassica juncea	Mustard plant	As, Cr, Ni
Pistia stratiotes	Water cabbage/water lettuce	Cd, Cr, Ni, Cu, Zn, Hg, Pb
Solanum nigrum	Black night shade	Cd
Thlaspi caerulescens	Alpine pennycress	Pb, Zn, Cd
Jatropha curcas	Physic nut	Pb, Ni, Cu, Cd
Zea mays	Corn	Zn, Cd, Pb
Salix fragilis	Crack willow	Pb, Cd, Cu, Zn
Populus trichocarpa	Black cotton wood	Cu, Pb, Cd, Zn
Aquatic macrophytes	English name	Heavy metals removed
Azolla pinnata	Feathered mosquito fern	Cd, Cr, Cu, Fe, Hg
Eichhornia crassipes	Water hyacinth	As, Cd, Cr, Ni, Cu, Zn, Se
Lemna minor	Common duckweed	Fe, Pb, Ni, Hg, As, Cu
Vallisneria spiralis	Eel grass	Cd, Cr, Ni
Nasturtium officinale	Watercress	Zn, Cu, Ni
Cyperus alternifolius	Umbrella sedge	Zn, Cu, Pb, Cd
Eleocharis acicularis	Needle spikerush	Pb, Ni, Cr, Cu, Zn, Mn

Source Mishra et al. (2018)

Table 8.14 shows the phytoremediation of heavy metals using plant species and aquatic macrophytes.

8.8.3.3 Nano-technological Innovations in Heavy Metal Removal

Water contamination with heavy metals is a result of growing urbanization and environmental damage. Human health and the health of other living things are at risk from contaminated water. The removal of heavy metals from contaminated water may be possible with the use of nano-technological advancements including layered double hydroxides, metal-oxide nanomaterials, and nano-polymer composites. Metal oxide nanoparticles have been shown by Mudzielwana et al. to be capable of extracting harmful metal ions from wastewater (Mudzielwana et al. 2019). According to reports, two semiconductors with potential impacts on the removal of heavy-metal contamination from water are titanium oxide (TiO_2) and zinc oxide (ZnO) nanoparticles (NPs) (Mudzielwana et al. 2019). According to reports, Cu, Ag, and Fe-induced NPs can be used to de-contaminate wastewater and remove harmful heavy metals (Mathur et al. 2022). Another inexpensive technique for reducing pollution in soil and water is nano-bioremediation. Fe_3O_4 NP application decreases the mobility of the heavy metal Cd, according to Sebastian et al. (2019).

"Nano-biosorbents" are a new method of removing heavy metals that combine nanotechnology and biosorption. Heavy metal absorption is facilitated by the carboxyl (–COOH) and hydroxyl (–OH) groups found in biosorbents (Sarubbo et al. 2022). For example, graphene quantum dots based on rice husk are a useful nano-biosorbent for the removal of Pb (III) and La (III) (Me et al. 2020). Heavy metals including zinc, copper, and nickel are removed by microbe-induced biosurfactants. Nano-bio-surfactants, also known as nanoparticle-capped bio-surfactants, are beneficial for heavy metal bioremediation. For instance, heavy metal removal can be accomplished using Rhamnolipid-capped Zn NP (derived from *Pseudomonas aeruginosa*) (Sarubbo et al. 2022; Me et al. 2020).

- **Nanoadsorbents based heavy metal removal**

Heavy metal removal uses nanoparticles. According to Wang et al. (2019), nanomaterials work by adsorbing heavy metals (HMs). Zeolite, metal oxide, and other nanoparticles remove heavy metals. Table 8.15 shows the different adsorbent types used in heavy metal removal. Recent years have seen widespread use of metal-oxide nanoparticles. Metallic nanoparticles are unstable and rarely used as adsorbents. They must be functionalized to make nanoparticle extraction from wastewater more accessible (Sun et al. 2016). Nano zero-valent iron (NZVI) nanoparticles (FeO nanoparticles) have essential properties. These include a higher surface area, stability, adsorption capability, and non-toxicity (Sun et al. 2016). Xiao et al. (2017) synthesized iron-based nanoparticles from $FeCl_3$ using Syzygium jambos (SJA) as a reducing agent. Their research showed that this approach eliminates chromate (Cr^{6+}). We also use bimetallic nanoparticles to eliminate metal ions. Wastewater purification uses metallic and metal-oxide nanoparticles (Sodhi et al. 2022). Due to their intrinsic magnetic characteristics, metal oxide nanoparticles are magnetic or non-magnetic. Nanoparticle metal ion removal depends on ionic radii, atomic mass, and electronegativity (Sodhi et al. 2022). The metal oxide surface can be modified with surfactants to increase nanoparticle adsorption. Recently developed nano adsorbents include carbon-based nanomaterials. Recent interest has focused on carbonaceous nanoparticles as nano adsorbents (Pham et al. 2017). In wastewater filtration, CNTs and graphene nano adsorbents are used. Carbon nanotubes (CNTs) are tiny, electrically conductive, cylindrical, hollow, and have a large surface area. Single-walled and multi-walled carbon nanotubes (CNTs) exist. High adsorption rates allow carbon nanotubes (CNTs) to remove Pb^{2+} and other heavy metals (HMs) (Gupta et al. 2021). Cd^{2+} ions were removed using MnO_2-oxidized multi-walled carbon nanotubes. Graphene is also a carbonaceous nanomaterial. Oxygen as a functional group makes graphene oxide hydrophilic. The material disperses in water due to its hydrophilicity. Gupta et al. (2021) says graphene oxide's large surface area helps purify wastewater. Adsorption of heavy metals (HMs) happens when metal ions are complex with graphene's oxide binding site (Sodhi et al. 2022).

Table 8.15 Adsorbent types used for heavy metal removal

Adsorbent	Heavy metals removed	Attributes
Activated carbon	Pb, Cu, Hg	Large surface area and high number of pores
Biosorbents	Pb, Cu, Cd, Zn, Ni, Cr	Eco-friendly, inexpensive, high efficiency, reuse
Carbon nanotubes	Pb, Cu, Cd, Zn	High surface area, good adsorption potential
Clays	Cu, Co, Pb, Zn, Cd, Cr, Ni, Mn	High cation exchange capacity, large surface area, chemical and mechanical stability and layered structure, high sorption capacity
Zeolites	Cd, Ni, Cr, Mn, Fe	High specific surface areas, high ion-exchange capacity, low prices
Metal oxides	Zn, Ni, Cu, Pb, Fe, Hg	Nanosized-metal oxides such as γ-Fe_2O_3, TiO_2, MnO_2 and γ-Al_2O_3 are environmentally benign and demonstrate amphoteric sorption behaviors because of specific affinity and high surface area for heavy metal adsorption from aqueous media

Source Ince and Kaplan İnce (2017)

8.8.3.4 Plant Phenolic Compounds for Heavy Metal Removal

Metal chelators essential to plant growth and development are phenolic compounds, a prominent class of secondary metabolites in plants (Deka et al., 2023). Hydroxyl (–OH) and carboxyl (–COOH) groups are present in phenolic compounds. Catechin, caffeic acid, gallic acid, ferulic acid, syringic acid, epi-catechin are a few examples of plant polyphenols. Phenols' carboxyl and hydroxyl groups have the ability to combine with heavy metals. Metal-polyphenol capping may be explained by the strong nucleophilicity of phenolic atomic rings. The synthesis of phenolic compounds in plants is enhanced by heavy metals, particularly essential elements like copper (Cu), iron (Fe), and zinc (Zn). Conversely, plants are at risk from lead, mercury, and arsenic (Deka et al. 2023).

Reactive oxygen species (ROS) are produced by plants contaminated with heavy metals, which changes their physiology. In comparison to ROS, plants with low polyphenol content or antioxidant activity are more susceptible. According to Deka et al. (2023), plant phenolic compounds offer protection against stress caused by heavy metals and other environmental variables.

8.8.3.5 Chemical Chelating Ligands for Heavy Metal Toxicity Removal

Coordination chemistry is used in chelation therapy, which has the potential to remove dangerous heavy metals. Metal compounds containing hazardous metals are produced by chelating ligands. Subsequent processes get rid of the body's metal complexes (Priya et al. 2022). For example, meso-2,3-dimercaptosuccinic acid removes metals, while 2,3-dimercaprol treats lead (Pb) and arsenic (As) poisoning (Priya et al. 2022).

Metal removal chelating ligands should be nontoxic, water-soluble, and cell membrane-penetrating (Velkova et al. 2018). Clinical chelation ligands include EDTA, Trientive, D-penicillamine, and Deferiorone (Velkova et al. 2018). Velkova et al. (2018) found that BAL (British anti-Lewisite) removes arsenic-based metal toxicity. BAL's toxicity was reduced in the 1960s by adding a thiol group. It was then transformed into DMSA.

Chelation therapy forms metal–ligand complexes to remove heavy metal toxicity. Thus, metal poisoning's adverse effects, such as heart attack and stroke, inflammation, and other health issues, are alleviated (Velkova et al. 2018).

8.8.3.6 Adsorption of Heavy Metal

Water and wastewater can effectively be free of trace metal ions through adsorption. An adsorbent is attached to dissolved metal ions from aqueous solutions through the process of adsorption. The adsorption of metal ions from water is controlled by both their characteristics and the adsorbent substance. These mechanisms are determined through metal ion-adsorbent interactions. The choice of adsorbent material is essential for successful adsorption separations (Lupa and Cocheci 2023).

Adsorbents are divided into two categories: synthetic and natural. Minerals, ores, charcoal, clays, and zeolites are examples of natural adsorbents. Synthetic adsorbents are made from sewage, agricultural, and industrial sludge (Hussain et al. 2021).

Adsorption is a low-cost and simple method of purifying water tainted with metal ions. The idea demonstrates the economic feasibility of industrial growth. This treatment method is adaptable due to the wide range of affordable adsorbent materials available. Regretfully, Lupa and Cocheci (2023) said that the existing procedure needs to be changed in order to be automated.

8.8.3.7 Microbial Culture Bio-absorbents for Bioremediation of Heavy Metal

Water pollution remediation via bio-absorption is a practical and economical approach (Deka et al. 2023). Microorganisms are efficiently used by bio-absorbents to remove heavy metals. The environment, the material being absorbed, and the heavy metal that needs to be eliminated all affect how effective microbial bio-absorbents are. Because of their ion exchange, physical absorption, complex formation, precipitation, and sensitivity to process variables such medium acidity, microorganisms such as bacteria, yeast, fungi, and algae can function as bio-absorbents. The removal of mercury (Hg), lead (Pb), arsenic (As), and cadmium (Cd), as well as precious metals like gold (Au), silver (Ag), and platinum (Pt), has been studied as a function of biosorbent concentration, sorption center density, and immobilization techniques (Deka et al. 2023). Biosorbents for cyanobacteria, yeast, gram-positive bacteria, and gram-negative bacteria that are tiny, elastic, low-density, and biocompatible are useful. *Bacillus subtillis* immobilizes Cd (III) ions, *Arthobacter* sp. extracts Cu (II), and *Bacillus drentensis* MG21831T biomass absorbs Pb and Cu, according to research by Seo et al. (2020).

Active cells are more expensive, take longer to sorb-desorb, and contain more nutrients than dead microbial biomass. Heavy metal removal from aqueous solutions depends on the microbial cell wall, which has functional groups including carbonyl, hydroxide, amino, sulfate, etc. (Deka et al. 2023).

8.9 Conclusion

The issue of heavy metal contamination is a matter of great concern in contemporary society since it poses detrimental effects on both fauna and flora. These chemicals possess carcinogenic effects, elicit congenital disabilities, and exhibit significant death rates. In order to effectively mitigate this issue, it is imperative to implement measures aimed at removing these deleterious metals from wastewater before their discharge into ecological systems.

The risk associated with hazardous materials has escalated due to human activities, releasing both treated and untreated heavy metals into the surrounding environment. The detrimental effects of heavy metals further worsen the circumstances. The growing public consciousness of environmental contamination has led to an upsurge in the emergence of cleanup approaches that are both environmentally helpful and long-lasting.

Several remediation techniques can be utilized to alleviate the pollution caused by hazardous materials and convert them into a harmless state. Nevertheless, achieving complete eradication of these entities is unachievable due to their inherent resistance to decomposition. Bioremediation presents several benefits, including cost reduction, improved environmental compatibility, and heightened social acceptance. The progress made in molecular biology and nanotechnology has enabled the emergence of innovative methodologies for detoxifying heavy metals.

The choice of methods for removing metal ions from water is contingent upon various aspects, including the level of competence, familiarity with a particular methodology, the composition of influential elements, the effectiveness of removal, and the operating circumstances. Prioritization of ecologically friendly technology is necessary to effectively mitigate the dangers associated with hazardous materials.

References

Abatenh E, Gizaw B, Tsegaye Z, Wassie M (2017) The role of microorganisms in bioremediation—a review. Open J Environ Biol 2(1):038–046

Abosede O (2017) Review on heavy metals contamination in the environment. Eur J Earth Environ 4(1). https://www.idpublications.org/wp-content/uploads/2017/01/Abstract-REVIEW-ON-HEAVY-METALS-CONTAMINATION-IN-THE-ENVIRONMENT.pdf

Agoro MA, Adeniji AO, Adefisoye MA, Okoh OO (2020) Heavy metals in wastewater and sewage sludge from selected municipal treatment plants in Eastern Cape Province, South Africa. Water 12(10):2746. https://doi.org/10.3390/w12102746

Ahirvar BP, Das P, Srivastava V, Kumar M (2023) Perspectives of heavy metal pollution indices for soil, sediment, and water pollution evaluation: an insight. Total Environ Res Themes 6:100039

Al-Saydeh SA, El-Naas MH, Zaidi SJ (2017) Copper removal from industrial wastewater: a comprehensive review. J Ind Eng Chem 56:35–44. https://doi.org/10.1016/j.jiec.2017.07.026

Ali H, Khan E (2018) What are heavy metals? Long-standing controversy over the scientific use of the term "heavy metals"—proposal of a comprehensive definition. Toxicol Environ Chem 100(1):6–19. https://doi.org/10.1080/02772248.2017.1413652

Alves CA, Vicente ED, Vicente AM, Rienda IC, Tomé M, Querol X, Amato F (2020) Loadings, chemical patterns and risks of inhalable road dust particles in an Atlantic city in the north of Portugal. Sci Total Environ 737:139596

Amer MM, Sabry BA, Marrez DA, Hathout AS, Fouzy AS (2019) Exposure assessment of heavy metal residues in some Egyptian fruits. Toxicol Rep 6:538–543

Awa SH, Hadibarata T (2020) Removal of heavy metals in contaminated soil by phytoremediation mechanism: a review. Water Air Soil Pollut 231(2). https://doi.org/10.1007/s11270-020-4426-0

Azimi A, Azari A, Rezakazemi M, Ansarpour M (2017) Removal of heavy metals from industrial wastewaters: a review. ChemBioEng Rev 4(1):37–59. https://doi.org/10.1002/cben.201600010

Aziz KHH, Mustafa FS, Omer KM, Hama S, Hamarawf RF, Rahman KO (2023) Heavy metal pollution in the aquatic environment: efficient and low-cost removal approaches to eliminate their toxicity: a review. RSC Adv 13(26):17595–17610

Bashir A, Malik LA, Ahad S, Manzoor T, Bhat MA, Dar GN, Pandith AH (2018) Removal of heavy metal ions from aqueous system by ion-exchange and biosorption methods. Environ Chem Lett 17(2):729–754. https://doi.org/10.1007/s10311-018-00828-y

Bazrafshan E, Mohammadi L, Ansari-Moghaddam A, Mahvi AH (2015) Heavy metals removal from aqueous environments by electrocoagulation process—a systematic review. J Environ Health Sci Eng 13(1). https://doi.org/10.1186/s40201-015-0233-8

Bharagava RN, Chowdhary P, Saxena G (2017) Bioremediation: an eco-sustainable green technology, its application and limitations. In: Environmental pollutants and their bioremediation approaches. CRC Press, Boca Raton

Bhuyan MS, Bakar MA, Akhtar A, Hossain MB, Ali MM, Islam MS (2017) Heavy metal contamination in surface water and sediment of the Meghna River, Bangladesh. Environ Nanotechnol Monit Manag 8(August 2016):273–279. https://doi.org/10.1016/j.enmm.2017.10.003

Borah R, Taki K, Gogoi A, Das P, Kumar M (2018) Contemporary distribution and impending mobility of arsenic, copper and zinc in a tropical (Brahmaputra) river bed sediments, Assam, India. Ecotoxicol Environ Saf 161 (336):769–776. https://doi.org/10.1016/j.ecoenv.2018.06.038

Briffa J, Sinagra E, Blundell R (2020) Heavy metal pollution in the environment and their toxicological effects on humans. Heliyon 6(9):e04691. https://doi.org/10.1016/j.heliyon.2020.e04691

Burakov AE, Galunin EV, Burakova IV, Kucherova AE, Agarwal S, Tkachev AG, Cai T, Wang L, Liu Y, Zhang S, Dong W, Chen H, Luo S (2018) Ag_3PO_4/Ti_3C_2 MXene interface materials as a Schottky catalyst with enhanced photocatalytic activities and anti-photocorrosion performance. Appl Catal B: Environ 239:545–554

Carolin CF, Kumar PS, Saravanan A, Joshiba GJ, Naushad M (2017) Efficient techniques for the removal of toxic heavy metals from aquatic environment: a review. J Environ Chem Eng 5(3):2782–2799. https://doi.org/10.1016/j.jece.2017.05.029

Chen L, Zhang X, Zhang M, Zhu Y, Zhuo R (2022) Removal of heavy-metal pollutants by white rot fungi: mechanisms, achievements, and perspectives. J Clean Prod 354:131681

Chowdhury S, Mazumder MAJ, Al-Attas O, Husain T (2016) Heavy metals in drinking water: occurrences, implications, and future needs in developing countries. Sci Total Environ 569–570:476–488. https://doi.org/10.1016/j.scitotenv.2016.06.166

Codex (2019) General standard for contaminants and toxins in food and feed. FAO. https://www.fao.org/fao-who-codexalimentarius/sh-proxy/en/?lnk=1&url=https%253A%252F%252Fworkspace.fao.org%252Fsites%252Fcodex%252FStandards%252FCXS%2B193-1995%252FCXS_193e.pdf

Dağcilar K, Gezer C (2021) Heavy metal residues in milk and dairy products produced in Northern Cyprus. Prog Nutr 23(1):e2021019

De A, Roy N (2023) Arsenic in environment—sources, implications and remedies. IntechOpen. www.intechopen.com. https://www.intechopen.com/online-first/1136350

Deka AK, Kumar KJ, Basumatary S (2023) Monitoring strategies for heavy metals in foods and beverages: limitations for human health risks. IntechOpen. www.intechopen.com. https://www.intechopen.com/chapters/86538

Dixit R, Wasiullah M, Pandiyan K, Singh BU, Sahu A, Shukla R, Singh BP, Rai JP, Sharma PK, Lade H, Paul D (2015) Bioremediation of heavy metals from soil and aquatic environment: an overview of principles and criteria of fundamental processes. Sustainability 7:2189–2212

DOE (2015) Bangladesh standards and guidelines for sludge management. https://doe.portal.gov.bd/sites/default/files/files/doe.portal.gov.bd/publications/2398e6c5_c300_472d_9a0c_0385522748f3/Bangladesh%20Standards%20and%20Guideline%20for%20sludge%20management-%20September%202016.pdf

Fang X, Peng B, Wang X, Song Z, Zhou D, Wang Q, Tan C (2019) Distribution, contamination and source identification of heavy metals in bed sediments from the lower reaches of the Xiangjiang River in Hunan province, China. Sci Total Environ 689:557–570. https://doi.org/10.1016/j.scitotenv.2019.06.330

Fashola MO, Ngole-Jeme VM, Babalola OO (2016) Heavy metal pollution from gold mines: environmental effects and bacterial strategies for resistance. Int J Environ Res Public Health 13(11):1047

Fiala M, Hwang HM (2021) Influence of highway pavement on metals in road dust: a case study in Houston, Texas. Water Air Soil Pollut 232(5):1–12

Gong Z, Chan HT, Chen Q, Chen H (2021) Application of nanotechnology in analysis and removal of heavy metals in food and water resources. Nanomaterials 11(7):1792. https://doi.org/10.3390/nano11071792

Gunawardana C, Goonetilleke A, Egodawatta P, Dawes L, Kokot S (2012) Source characterization of road dust based on chemical and mineralogical composition. Chemosphere 87:163–170

Gupta A, Sharma V, Sharma K, Kumar V, Choudhary S, Mankotia P, Mishra PK (2021) A review of adsorbents for heavy metal decontamination: growing approach to wastewater treatment. Materials (Basel) 14(16):4702

Hama S, Kumar P, Alam MS, Rooney DJ, Bloss WJ, Shi Z, Gupta SK (2021) Chemical source profiles of fine particles for five different sources in Delhi. Chemosphere 274:129913

Heghedűş-Mîndru G, Negrea P, Traşcă TI, Ştef DS, Cocan I, Heghedűş-Mîndru RC (2023) Food Intake of macro and trace elements from different fresh vegetables taken from Timisoara Market, Romania—chemometric analysis of the results. Foods 12(4):749

Hocaoglu-Ozyigit A, Bedriye N, Genc (2020) Cadmium in plants, humans and the environment. https://dergipark.org.tr/en/download/article-file/1245418

Hossain MB, Shanta TB, Ahmed ASS, Hossain MK, Semme SA (2019) Baseline study of heavy metal contamination in the Sangu River estuary, Chattogram, Bangladesh. Mar Pollut Bull 140(August 2018):255–261. https://doi.org/10.1016/j.marpolbul.2019.01.058

Hossain S, Anik AH, Khan R, Ahmed FU, Abu M, Khan A, Saha N, Idris AM, Shahid Alam M (2021) Public health vulnerability due to the exposure of dissolved metal(oid)s in tap water from a mega city (Dhaka, Bangladesh). Res Sq. https://doi.org/10.21203/rs.3.rs-849871/v1

Hussain A, Madan S, Madan R (2021) Removal of heavy metals from wastewater by adsorption. IntechOpen. www.intechopen.com. https://www.intechopen.com/chapters/75491

Igiri BE, Okoduwa SI, Idoko GO, Akabuogu EP, Adeyi AO, Ejiogu IK (2018) Toxicity and bioremediation of heavy metals contaminated ecosystem from tannery wastewater: a review. J Toxicol 2018

Ince M, Kaplan İnce O (2017) An overview of adsorption technique for heavy metal removal from water/wastewater: a critical review. Int J Pure Appl Sci 3(2):10–19. https://doi.org/10.29132/ijpas.358199

Islam MS, Hossain MB, Matin A, Islam Sarker MS (2018a) Assessment of heavy metal pollution, distribution and source apportionment in the sediment from Feni River estuary, Bangladesh. Chemosphere 202:25–32. https://doi.org/10.1016/j.chemosphere.2018.03.077

Islam MS, Proshad R, Ahmed S (2018b) Ecological risk of heavy metals in sediment of an urban river in Bangladesh. Hum Ecol Risk Assess 24:699–720. https://doi.org/10.1080/10807039.2017.1397499

Jaishankar M, Tseten T, Anbalagan N, Mathew BB, Beeregowda KN (2014) Toxicity, mechanism and health effects of some heavy metals. Interdiscip Toxicol 7(2):60–72. https://doi.org/10.2478/intox-2014-0009

Jasmin C, Anas A, Singh D, Purohit HJ, Gireeshkumar TR, Nair S (2020) Aberrations in the microbiome of cyanobacteria from a tropical estuary polluted by heavy metals. Mar Pollut Bull 160:111575

Jin M, Yuan H, Liu B, Peng J, Xu L, Yang D (2020) Review of the distribution and detection methods of heavy metals in the environment. Anal Methods 12(48):5747–5766. https://doi.org/10.1039/d0ay01577f

Kang S, Lee S, Kwoun J, Lee TJ, Jo YM (2023) Analysis of harmful heavy metals and carbonaceous components in urban school PM2.5. Aerosol Air Qual Res 23(2):220335–220335. https://doi.org/10.4209/aaqr.220335

Keshavarzi B, Tazarvi Z, Rajabzadeh MA, Najmeddin A (2015) Chemical speciation, human health risk assessment and pollution level of selected heavy metals in urban street dust of Shiraz, Iran. Atmos Environ 119:1–10

Khalef RN, Hassan AI, Saleh HM (2022) Heavy metal's environmental impact. IntechOpen. www.intechopen.com. https://www.intechopen.com/chapters/81083

Khalid N, Aqeel M, Noman A, Khan SM, Akhter N (2021) Interactions and effects of microplastics with heavy metals in aquatic and terrestrial environments. Environ Pollut 290:118104

Khalid S, Shahid M, Niazi NK, Murtaza B, Bibi I, Dumat C (2017) A comparison of technologies for remediation of heavy metal contaminated soils. J Geochem Explor 182:247–268. https://doi.org/10.1016/j.gexplo.2016.11.021

Kinuthia GK, Ngure V, Beti D, Lugalia R, Wangila A, Kamau L (2020) Levels of heavy metals in wastewater and soil samples from open drainage channels in Nairobi, Kenya: community health implication. Sci Rep 10(1):8434. https://doi.org/10.1038/s41598-020-65359-5

Kuhn R, Bryant IM, Jensch R, Bollmann J (2022) Applications of environmental nanotechnologies in remediation, wastewater treatment, drinking water treatment, and agriculture. Appl Nano 3(1):54–90

Kumar M, Gogoi A, Kumari D, Borah R, Das P, Mazumder P, Tyagi VK (2017) Review of perspective, problems, challenges, and future scenario of metal contamination in the urban environment. J Hazard Toxic Radioact Waste 21(4):04017007

Kumar M, Seth A, Singh AK, Rajput MS, Sikandar M (2021) Remediation strategies for heavy metals contaminated ecosystem: a review. Environ Sustain Indic 12:100155. https://doi.org/10.1016/j.indic.2021.100155

Kumar V, Rout C, Singh J, Saharan Y, Goyat R, Umar A, Akbar S, Baskoutas S (2023) A review on the clean-up technologies for heavy metal ions contaminated soil samples. Heliyon 9(5):e15472. https://doi.org/10.1016/j.heliyon.2023.e15472

Lew D (2020) Limitations of bioremediation—metal contaminated. Dr. Darrin Lew. https://www.drdarrinlew.us/metal-contaminated-2/limitations-of-bioremediation.html

Li C, Wang H, Liao X, Xiao R, Liu K, Bai J, He Q (2022) Heavy metal pollution in coastal wetlands: a systematic review of studies globally over the past three decades. J Hazard Mater 424:127312

Li M, Zhang Q, Sun X, Karki K, Zeng C, Pandey A, Rawat B, Zhang F (2020) Heavy metals in surface sediments in the trans-Himalayan Koshi River catchment: distribution, source identification and pollution assessment. Chemosphere. https://doi.org/10.1016/j.chemosphere.2019.125410

Liu E, Wang X, Liu H, Liang M, Zhu Y, Li Z (2019) Chemical speciation, pollution and ecological risk of toxic metals in readily washed of road dust in a megacity (Nanjing), China. Ecotoxicol Environ Saf 173:381–392

Liu J, Liang J, Yuan X, Zeng G, Yuan Y, Wu H, Li X (2015) An integrated model for assessing heavy metal exposure risk to migratory birds in wetland ecosystem: a case study in Dongting Lake Wetland, China. Chemosphere 135:14–19

Liu L, Li W, Song W, Guo M (2018) Remediation techniques for heavy metal-contaminated soils: principles and applicability. Sci Total Environ 633:206–219. https://doi.org/10.1016/j.scitotenv.2018.03.161

Liu Y, Serrano A, Villa-Gomez D (2021) Biological treatment of mine-impacted waters on the context of metal recovery. Future Effl Treat Plants 499–522. https://doi.org/10.1016/b978-0-12-822956-9.00026-x

Lu F, Astruc D (2018) Nanomaterials for removal of toxic elements from water. Coord Chem Rev 356:147–164. https://doi.org/10.1016/j.ccr.2017.11.003

Lundy L, Alves L, Revitt M, Wildeboer D (2017) Metalwater-sediment interactions and impacts on an urban ecosystem. Int J Environ Res Publ Health 14(7). https://doi.org/10.3390/ijerph14070722

Lupa L, Cocheci L (2023) Heavy metals removal from water and wastewater. IntechOpen. www.intechopen.com. https://www.intechopen.com/chapters/86109

Mahapatra B, Dhal NK, Dash AK, Panda BP, Panigrahi KCS, Pradhan A (2019) Perspective of mitigating atmospheric heavy metal pollution: using mosses as biomonitoring and indicator organism. Environ Sci Pollut Res 26:29620–29638

Mahurpawar M (2015) Effects of heavy metals on human health. Int J Res GRANTHAALAYAH 3(9SE):1–7. https://doi.org/10.29121/granthaalayah.v3.i9se.2015.3282

Malik LA, Bashir A, Qureashi A, Pandith AH (2019) Detection and removal of heavy metal ions: a review. Environ Chem Lett 17(4):1495–1521. https://doi.org/10.1007/s10311-019-00891-z

Masindi V, Mkhonza P, Tekere M (2021) Sources of heavy metals pollution. Remediat Heavy Metals 419–454

Mathur S, Singh D, Ranjan R (2022) Remediation of heavy metal(loid) contaminated soil through green nanotechnology. https://doi.org/10.3389/fsufs.2022.932424

Me M, Na F, Am A (2020) A novel nanobiosorbent of functionalized graphene quantum dots from rice husk with barium hydroxide for microwave enhanced removal of lead (II) and lanthanum (III). Biores Technol 298. https://doi.org/10.1016/j.biortech.2019.122514

Mensah E, Kyei-Baffour N, Ofori E, Obeng G (2009) Influence of human activities and land use on heavy metal concentrations in irrigated vegetables in Ghana and their health implications. Appropr Technol Environ Protect Dev World 9–14. https://doi.org/10.1007/978-1-4020-9139-1_2

Mishra S, Bharagava RN, More N, Yadav A, Zainith S, Mani S, Chowdhary P (2018) Heavy metal contamination: an alarming threat to environment and human health. Environ Biotechnol Sustain Future 103–125. https://doi.org/10.1007/978-981-10-7284-0_5

Mitra S, Chakraborty AJ, Tareq AM, Emran TB, Nainu F, Khusro A, Idris AM, Khandaker MU, Osman H, Alhumaydhi FA, Simal-Gandara J (2022) Impact of heavy metals on the environment

and human health: novel therapeutic insights to counter the toxicity. J King Saud Univ Sci 34(3):101865. https://doi.org/10.1016/j.jksus.2022.101865
Morakinyo OM, Mukhola MS, Mokgobu MI (2021) Health risk analysis of elemental components of an industrially emitted respirable particulate matter in an urban area. Int J Environ Res Public Health 18(7):3653. https://doi.org/10.3390/ijerph18073653
Mudzielwana R, Gitari MW, Ndungu P (2019) Uptake of As(V) from groundwater using Fe-Mn oxides modified kaolin clay: physicochemical characterization and adsorption data modeling. Water 11(6):1245. https://doi.org/10.3390/w11061245
Nahar N, Rahman A, Nawani NN, Ghosh S, Mandal A (2017) Phytoremediation of arsenic from the contaminated soil using transgenic tobacco plants expressing ACR2 gene of Arabidopsis thaliana. J Plant Physiol 218:121–126
Nedjimi B (2020) Germination characteristics of Peganum harmala L (Nitrariaceae) subjected to heavy metals: implications for the use in polluted dryland restoration. Int J Environ Sci Technol 17(4):2113–2122
Nedjimi B (2021) Phytoremediation: a sustainable environmental technology for heavy metals decontamination. SN Appl Sci 3(3):1–19
Nguyen BT, Do DD, Nguyen TX, Nguyen VN, Phuc Nguyen DT, Nguyen MH, Bach QV (2020) Seasonal, spatial variation, and pollution sources of heavy metals in the sediment of the Saigon River, Vietnam. Environ Pollut 256:113412. https://doi.org/10.1016/j.envpol.2019.113412
Osmani M, Bani A, Hoxha B (2015) Heavy metals and Ni phytoextractionin in the metallurgical area soils in Elbasan. Albanian J Agric Sci 14(4):414–419
Pachana K, Wattanakornsiri A, Nanuam J (2010) Heavy metal transport and fate in the environmental compartments. NU Sci J 7(1):1–11
Pandiyan J, Mahboob S, Govindarajan M, Al-Ghanim KA, Ahmed Z, Al-Mulhm N, Krishnappa K (2021) An assessment of level of heavy metals pollution in the water, sediment and aquatic organisms: a perspective of tackling environmental threats for food security. Saudi J Biol Sci 28(2):1218–1225
Pham TD, Do TT, Doan THY, Nguyen TAH, Mai TD, Kobayashi M, Adachi Y (2017) Adsorptive removal of ammonium ion from aqueous solution using surfactant modified alumina. Environ Chem 14(5):327–337
Prabakaran K, Nagarajan R, Eswaramoorthi S, Anandkumar A, Franco FM (2019) Environmental significance and geochemical speciation of trace elements in Lower Baram River sediments. Chemosphere 219:933–953. https://doi.org/10.1016/j.chemosphere.2018.11.158
Prasad S (2021) Chromium contamination and effect on environmental health and its remediation: a sustainable approaches. J Environ Manag 285:112174. https://doi.org/10.1016/j.jenvman.2021.112174
Priya AK, Gnanasekaran L, Dutta K, Rajendran S, Balakrishnan D, Soto-Moscoso M (2022) Biosorption of heavy metals by microorganisms: evaluation of different underlying mechanisms. Chemosphere 307:135957. https://doi.org/10.1016/j.chemosphere.2022.135957
Purushotham D, Rashid M, Lone MA, Rao AN, Ahmed S, Nagaiah E, Dar FA (2013) Environmental impact assessment of air and heavy metal concentration in groundwater of Maheshwaram watershed, Ranga Reddy district, Andhra Pradesh. J Geol Soc India 81(3):385–396. https://doi.org/10.1007/s12594-013-0049-z
Qualitest (2023) Atomic absorption spectrometer | Qualitest. www.worldoftest.com. https://www.worldoftest.com/atomic-absorption-spectrometer#:~:text=Atomic%20absorption%20spectroscopy%20(AAS)%20is
Raffa CM, Chiampo F, Shanthakumar S (2021) Remediation of metal/metalloid-polluted soils: a short review. Appl Sci 11(9):4134. https://doi.org/10.3390/app11094134
Rafique M, Hajra S, Tahir MB, Gillani SSA, Irshad M (2022) A review on sources of heavy metals, their toxicity and removal technique using physico-chemical processes from wastewater. Environ Sci Pollut Res 29(11):16772–16781

Rahman MS, Kumar P, Ullah M, Jolly YN, Akhter S, Kabir J, Begum BA, Salam A (2021) Elemental analysis in surface soil and dust of roadside academic institutions in Dhaka City, Bangladesh and their impact on human health. Environ Chem Ecotoxicol 3:197–208

Rezapour S, Siavash Moghaddam S, Nouri A, Khosravi Aqdam K (2022) Urbanization influences the distribution, enrichment, and ecological health risk of heavy metals in croplands. Sci Rep 12(1):1–16

Roy S, Gupta SK, Prakash J, Habib G, Kumar P (2022) A global perspective of the current state of heavy metal contamination in road dust. Environ Sci Pollut Res 1–22

Salek Maghsoudi A, Hassani S, Mirnia K, Abdollahi M (2021) Recent advances in nanotechnology-based biosensors development for detection of arsenic, lead, mercury, and cadmium. Int J Nanomed 16:803–832. https://doi.org/10.2147/ijn.s294417

Sall ML, Diaw AKD, Gningue-Sall D, Efremova Aaron S, Aaron J-J (2020) Toxic heavy metals: impact on the environment and human health, and treatment with conducting organic polymers, a review. Environ Sci Pollut Res 27(24):29927–29942. https://doi.org/10.1007/s11356-020-09354-3

Sarker A, Kim J-E, Islam ARMdT, Bilal M, Rakib MdRJ, Nandi R, Rahman MM, Islam T (2021) Heavy metals contamination and associated health risks in food webs—a review focuses on food safety and environmental sustainability in Bangladesh. Environ Sci Pollut Res. https://doi.org/10.1007/s11356-021-17153-7

Sarker A, Kim JE, Islam ARMT, Bilal M, Rakib MRJ, Nandi R, Islam T (2022) Heavy metals contamination and associated health risks in food webs—a review focuses on food safety and environmental sustainability in Bangladesh. Environ Sci Pollut Res 29(3):3230–3245

Sarubbo LA, da SilvaDurval IJB, Bezerra KGO, Ribeiro BG, Silva IA, Twigg MS, Banat IM MGC (2022) Biosurfactants: production, properties, applications, trends, and general perspectives. Biochem Eng J 181:108377. https://doi.org/10.1016/j.bej.2022.108377

Scutarașu EC, Trincă LC (2023) Heavy metals in foods and beverages: global situation, health risks and reduction methods. Foods 12(18):3340

Sebastian A, Nangia A, Prasad MNV (2019) Cadmium and sodium adsorption properties of magnetite nanoparticles synthesized from Hevea brasiliensis Muell. Arg. Bark: relevance in amelioration of metal stress in rice. J Hazard Mater 371:261–272. https://doi.org/10.1016/j.jhazmat.2019.03.021

Seo H, Lee M, Wang S (2020) Equilibrium and kinetic studies of the biosorption of dissolved metals on Bacillus drentensis immobilized in biocarrier beads. Environ Eng Res 18:45–53

Shahab A, Zhang H, Ullah H, Rashid A, Rad S, Li J, Xiao H (2020) Pollution characteristics and toxicity of potentially toxic elements in road dust of a tourist city, Guilin, China: ecological and health risk assessment☆. Environ Pollut 266:115419

Shaheen N, Irfan NM, Khan IN, Islam S, Islam MS, Ahmed MK (2016) Presence of heavy metals in fruits and vegetables: health risk implications in Bangladesh. Chemosphere 152:431–438

Shakya S, Baral S, Belbase P, Siddique MN, Samoh AN, Das B, Shrestha P, Shakya P (2019) Determination and contamination assessment of heavy metals in street dust from different types of land-use in Kathmandu District, Nepal. J Inst Sci Technol 24(1):6–18

Shamim S (2018) Biosorption of heavy metals. IntechOpen. www.intechopen.com. https://www.intechopen.com/chapters/58112

Sodhi KK, Mishra LC, Singh CK, Kumar M (2022) Perspective on the heavy metal pollution and recent remediation strategies. Curr Res Microb Sci 3:100166. https://doi.org/10.1016/j.crmicr.2022.100166

Soltani N, Keshavarzi B, Moore F, Tavakol T, Lahijanzadeh AR, Jaafarzadeh N, Kerman M (2015) Ecological and human health hazards of heavy metals and polycyclic aromatic hydrocarbons (PAHs) in road dust of Isfahan metropolis, Iran. Sci Total Environ 505:712–772

Sun MH, Huang SZ, Chen LH, Li Y, Yang XY, Yuan ZY, Su BL (2016) Applications of hierarchically structured porous materials from energy storage and conversion, catalysis, photocatalysis, adsorption, separation, and sensing to biomedicine. Chem Soc Rev 45(12):3479–3563

Sun Z, Zhao M, Chen L, Gong Z, Hu J, Ma D (2023) Electrokinetic remediation for the removal of heavy metals in soil: limitations, solutions and prospection. Sci Total Environ 165970–165970. https://doi.org/10.1016/j.scitotenv.2023.165970

Taiwo AM, Michael JO, Gbadebo AM, Oladoyinbo FO (2019) Pollution and health risk assessment of road dust from Osogbo metropolis, Osun State, Southwestern Nigeria. Hum Ecol Risk Assess Int J 1–16. https://doi.org/10.1080/10807039.2018.1563478

Trujillo-González JM, Torres-Mora MA, Keesstra S, Brevik EC, Jiménez-Ballesta R (2016) Heavy metal accumulation related to population density in road dust samples taken from urban sites under different land uses. Sci Total Environ 553:636–642

Urra J, Alkorta I, Mijangos I, Epelde L, Garbisu C (2019) Application of sewage sludge to agricultural soil increases the abundance of antibiotic resistance genes without altering the composition of prokaryotic communities. Sci Total Environ 647:1410–1420

US EPA (2014) Mercury emissions: the global context. US EPA. https://www.epa.gov/international-cooperation/mercury-emissions-global-context#:~:text=Mercury%20emitted%20into%20the%20air%20eventually%20settles%20into%20water%20or

US EPA (2016) Basic information about lead air pollution | US EPA. US EPA. https://www.epa.gov/lead-air-pollution/basic-information-about-lead-air-pollution

US EPA (2020) Regulatory determinations for pollutants in biosolids. US EPA. https://www.epa.gov/biosolids/regulatory-determinations-pollutants-biosolids

Vardhan KH, Kumar PS, Panda RC (2019) A review on heavy metal pollution, toxicity and remedial measures: current trends and future perspectives. J Mol Liq 290:111197

Vareda JP, Valente AJ, Durães L (2016) Heavy metals in Iberian soils: removal by current adsorbents/amendments and prospective for aerogels. Adv Colloid Interface Sci 237:28–42. https://doi.org/10.1016/j.cis.2016.08.009

Vareda JP, Valente AJM, Durães L (2019) Assessment of heavy metal pollution from anthropogenic activities and remediation strategies: a review. J Environ Manag 246:101–118. https://doi.org/10.1016/j.jenvman.2019.05.126

Velkova Z, Kirova G, Stoytcheva M, Kostadinova S, Todorova K, Gochev V (2018) Immobilized microbial biosorbents for heavy metals removal. Eng Life Sci 18(12):871–881. https://doi.org/10.1002/elsc.201800017

Venkatesh KR, More N, Kanoujia S (2015) Chromium accumulation in Eisenia fetida in modified vermi compost supplemented with tannery sludge. J Agroecol Nat Resour Manag 1(4):135–141

Vu CT, Lin C, Shern CC, Yeh G, Le VG, Tran HT (2017) Contamination, ecological risk and source apportionment of heavy metals in sediments and water of a contaminated river in Taiwan. Ecol Indicat. https://doi.org/10.1016/j.ecolind.2017.06.008

Wahab MIA, Razak WMAA, Sahani M, Khan MF (2020) Characteristics and health effect of heavy metals on non-exhaust road dusts in Kuala Lumpur. Sci Total Environ 703:135535

Wang L, Wang Y, Ma F, Tankpa V, Bai S, Guo X, Wang X (2019) Mechanisms and reutilization of modified biochar used for removal of heavy metals from wastewater: a review. Sci Total Environ 668:1298–1309

Wangboje OM, Ikhuabe AJ (2015) Heavy metal content in fish and water from River Niger at Agenebode, Edo State, Nigeria. Afr J Environ Sci Technol 9(3):210–217

Xiao Z, Zhang H, Xu Y, Yuan M, Jing X, Huang J, Sun D (2017) Ultra-efficient removal of chromium from aqueous medium by biogenic iron based nanoparticles. Sep Purif Technol 174:466–473

Yadav A, Chowdhary P, Kaithwas G, Bharagava RN (2017) Toxic metals in environment, threats on ecosystem and bioremediation approaches. In: Das S, Singh (eds) Handbook of metal-microbe interactions and bioremediation. CRC Press/Taylor & Francis Group, Boca Raton

Yan A, Wang Y, Tan SN, Mohd Yusof ML, Ghosh S, Chen Z (2020) Phytoremediation: a promising approach for revegetation of heavy metal-polluted land. Front Plant Sci 11(359). https://doi.org/10.3389/fpls.2020.00359

Yang ZF, Xia XQ, Yu T, Hou QY, Cao TN, Zhong J (2008) Distribution and fluxes of as and Cd trace metals in the Dongting Lake water system, Hunan Province, China. Geoscience 22:897e908 (in Chinese)

Zaynab M, Al-Yahyai R, Ameen A, Sharif Y, Ali L, Fatima M, Li S (2022) Health and environmental effects of heavy metals. J King Saud Univ Sci 34(1):101653

Zhou Q, Yang N, Li Y, Ren B, Ding X, Bian H, Yao X (2020) Total concentrations and sources of heavy metal pollution in global river and lake water bodies from 1972 to 2017. Glob Ecol Conserv 22:e00925

Chapter 9
Development of Nano-adsorbent for Heavy Metals Removal from Wastewater

Felix Sunday Nworie

Abstract The increasing industrialization and urbanization have led to a significant rise in heavy metal contamination in wastewater, posing a serious threat to environmental and human health. Conventional wastewater treatment methods often fall short in effectively removing these toxic heavy metals, necessitating the development of advanced and efficient technologies. Nano-adsorbents have emerged as promising candidates for heavy metal removal due to their exceptional adsorption capabilities and large surface area-to-volume ratio. This abstract presents a comprehensive overview of the recent advancements in nano-adsorbent technologies for the removal of heavy metals from wastewater. The synthesis methods and properties of various nanomaterials, such as nanoparticles, nanotubes, and nanofibers, which have shown excellent adsorption affinity towards heavy metal ions was discussed. Their unique surface characteristics, tunable functionalities, and enhanced reactivity make them attractive candidates for wastewater treatment. Furthermore, the mechanisms of heavy metal adsorption onto nano-adsorbents, including ion exchange, chemical bonding, and electrostatic interactions was discussed. Understanding these adsorption mechanisms is crucial for designing efficient and cost-effective nano-adsorbents tailored to specific heavy metal contaminants. The challenges associated with nano-adsorbent application, including issues of stability, regeneration, and potential release of nanoparticles into the environment was treated. Sustainable and environmentally friendly approaches for the synthesis and disposal of nano-adsorbents are explored to mitigate any adverse impacts on the ecosystem. Nano-adsorbent technologies are a viable and innovative solution for the removal of heavy metals from wastewater. By harnessing the unique properties of nanomaterials, the pressing issue of heavy metal pollution, contributing to cleaner and safer water resources for future generations can be addressed.

Keywords Nano-adsorbents · Wastewater · Heavy metals · Treatment technologies · Functionalization

F. S. Nworie (✉)
Department of Industrial Chemistry, Ebonyi State University, Abakaliki, Nigeria
e-mail: nworie.sunday@ebsu.edu.ng

N. Kumar (ed.), *Heavy Metal Toxicity*, Environmental Science and Engineering,
https://doi.org/10.1007/978-3-031-56642-4_9

9.1 Introduction

Nano-adsorbents, which are materials with nanoscale structures used to adsorb or remove substances from liquids or gases, gained significant attention in the field of research between 1980 and 1990 (Gusain et al. 2020). The late twentieth century marked the beginning of the nanotechnology revolution. Researchers started exploring the unique properties of nanomaterials, including their high surface area, which made them promising candidates for various applications, including adsorption. Carbon nanotubes were discovered in 1991 and are known to have significant implications for nano-adsorbent research (Backes and Hirsch 2010). These cylindrical carbon structures with nanoscale diameters and high aspect ratios showed exceptional adsorption capacities for gases and liquids. While initial research on CNTs began after 1990, their potential for adsorption became evident in the years following (Gupta and Saleh 2013). In the 1980s, zeolites and mesoporous materials were already known for their adsorption capabilities. Researchers were exploring ways to modify these materials at the nanoscale to enhance their adsorption efficiency and selectivity (Ajith et al. 2021). This era saw the development of synthetic routes to create zeolites and mesoporous materials with controlled pore sizes and surface chemistries. Researchers were working on modifying the surface properties of nanomaterials through functionalization, which involves attaching specific functional groups to enhance adsorption interactions (Akpomie and Conradie 2020). This allowed for tailoring nano-adsorbents to target specific contaminants in liquids and gases. The concept of combining different nanomaterials, like nanoparticles or nanofibers (nanocomposites), with adsorbent properties was gaining attention. These nanocomposites often exhibited synergistic effects, resulting in improved adsorption capacities and selectivities (Durán-Álvarez et al. 2016). Researchers explored the application of nano-adsorbents for environmental remediation and industrial processes which included the removal of heavy metals, organic pollutants, and other harmful substances from water and air (Dihom et al. 2022). As nano-adsorbent research advanced, the need for specialized characterization techniques arose. Techniques like transmission electron microscopy (TEM), scanning electron microscopy (SEM), and surface area analysis using methods like Brunauer–Emmett–Teller (BET) played a crucial role in understanding the structural and surface properties of nanomaterials (El-Sayed 2020). During this period, challenges associated with nano-adsorbent research included scalability of production, understanding the long-term stability of nanomaterials, and evaluating potential environmental and health impacts.

Notably, while some early groundwork was laid between 1980 and 1990, the explosion of nano-adsorbent research occurred in the following decades, driven by advances in nanotechnology and the growing recognition of nanomaterials' unique properties for various applications, including adsorption. The historical development of nano-adsorbents beyond 1990 has been marked by significant advancements in materials science and nanotechnology. An overview of the key developments in this field from the 1990s onwards is outlined with emphasis on the specific areas of interest of the researchers. In the early 1990s, the field of nanomaterials started to gain prominence as researchers began synthesizing and characterizing nanoscale materials like nanoparticles, nanotubes, and nanofibers with unique properties that made them suitable for adsorption applications (Aitken et al. 2006; Ajith et al. 2021; Ali et al. 2018). Nanoparticles, particularly those made of metal oxides like titanium dioxide (TiO_2) and iron(II)diiron(III) oxide (Fe_3O_4), were explored for their high surface area and adsorption capacity.

Carbon-based nanomaterials, such as carbon nanotubes (CNTs) and graphene, gained significant attention for their exceptional adsorption properties. The CNTs, in particular, were found to be efficient adsorbents for a wide range of contaminants due to their high surface area and tunable properties (Ali et al. 2020; Ren et al. 2011). In search for specificity and tunable properties, researchers started to functionalize nano-adsorbents by modifying their surfaces with various functional groups and chemicals (Li et al. 2011; Bhatnagar et al. 2013; Jana et al. 2021; Cheng et al. 2019). This allowed for better selectivity and enhanced adsorption capabilities, particularly for specific pollutants. Consequently, nanocomposites were highly researched on as nano-adsorbents were often combined with other matrix materials to create a new multifunctional hybrid nanocomposites with improved adsorption performance. This includes hybrid materials like CNTs decorated with metal nanoparticles which were implicated to exhibit enhanced catalytic and adsorption properties (Goda et al. 2019). Developments in analytical and characterization techniques, such as electron microscopy and spectroscopy, further enabled researchers to better understand the structure–property relationships of nano-adsorbents, leading to the design of more effective materials. Nano-adsorbents found applications in environmental remediation, including air purification systems, the removal of heavy metals, organic pollutants, and emerging contaminants like pharmaceuticals and personal care products from water sources (Jain et al. 2021; Tlili and Alkanhal 2019).

As research progressed, some nano-adsorbents and nanocomposite materials transitioned from the laboratory to commercial applications. Companies began producing and marketing nano-adsorbent products for various industries and hence, there is need for regulatory and safety concern. With the increased use of nano-adsorbents, concerns about their potential environmental and health impacts emerged. Researchers and regulatory bodies started studying and addressing these concerns to ensure the safe use of nano-adsorbents (Hurt et al. 2006; Fajardo et al. 2022). Research in the field of nano-adsorbents continues to advance, with a focus on improving adsorption efficiency, selectivity, and scalability. Sustainable and cost-effective synthesis methods are also a key area of investigation. Researchers have developed a wide range of nanomaterials with unique properties for adsorption (Enamul Hoque et al. 2017; Nworie et al. 2023). These nanomaterials offer high surface areas and tunable surface chemistries, making them highly effective adsorbents. Nano-adsorbents typically have extremely high surface areas due to their nanoscale size and unique structures. This increased surface area allows for greater adsorption capacity and efficiency compared to traditional bulk adsorbents.

Based on the trend in nano-adsorbent research, customization became inevitable and researchers can tailor nano-adsorbents to target specific contaminants or pollutants by modifying their surface properties. Functionalization and surface modifications can make nano-adsorbents selective for particular ions, molecules, or pollutants (Nworie et al. 2022; Liu et al. 2005; Zare et al. 2018). Nano-adsorbents often exhibit faster adsorption kinetics due to their small size and high surface area. This rapid adsorption is particularly advantageous in time-sensitive applications, such as wastewater treatment (Wilson et al. 2018; Kumari et al. 2019). As research in the field progressed, regeneration and reusability became necessary. Many nano-adsorbents can be easily regenerated and reused, reducing waste and operating costs. Regeneration methods often involve simple thermal or chemical treatments (Elella et al. 2022).

Currently, Scale-Up and Commercialization approaches have been incorporated in nano-adsorbent studies. Many nano-adsorbent materials developed in research laboratories have been scaled up and commercialized for practical applications, leading to a growing market for nanotechnology-based adsorption products (Elella et al. 2021).

Summary of the applications of nano-adsorbent as time progresses include; (i) Environmental Applications: Nano-adsorbents have found applications in treating contaminated water and air. They are used to remove heavy metals, organic pollutants, and emerging contaminants like pharmaceuticals and microplastics from water sources. (ii) Catalytic Properties: Some nano-adsorbents also possess catalytic properties, allowing them to simultaneously adsorb and degrade contaminants, providing a two-in-one solution for environmental cleanup. (iii) Industrial Processes: Nano-adsorbents are employed in various industrial processes, such as the removal of impurities from gases, purification of chemicals, and separation technologies. (iv) Healthcare and Biomedical Applications: Nano-adsorbents have been explored for drug delivery, diagnostic assays, and medical imaging due to their unique properties and biocompatibility (Jaspal and Malviya 2020; Hnamte and Pullikal 2022).

9.1.1 Nano-adsorbents

Nano-adsorbents
are a class of materials designed to remove pollutants, contaminants, or other substances from a fluid (gas or liquid) through the process of adsorption at the nanoscale level (Mudhoo and Sillanpää 2021; Tawade et al. 2023). They are composed of nanoscale particles, which can have a high surface area-to-volume ratio, making them highly efficient in adsorbing target molecules. Examples of nano-adsorbents include nanoparticles of various materials such as carbon-based nanomaterials (e.g., carbon nanotubes, graphene oxide), metal–organic frameworks (MOFs), zeolites, and various metal and metal oxide nanoparticles. Nano-adsorbents have gained significant attention in various fields, including environmental remediation, water purification, gas separation, and biomedical applications (Beyene and Ambaye 2019; Sengwa and Dhatarwal 2022; Fu et al. 2019). The unique properties of nano-adsorbents offer several advantages over conventional bulk-scale

adsorbents. The advantages include (i) High surface area due to their nanoscale size, which allows them to adsorb a large number of target molecules or contaminants (ii) Improved adsorption kinetics as it facilitates faster diffusion and interaction with target molecules, resulting in quicker adsorption rates (iii) Enhanced selectivity since they can be engineered and functionalized to exhibit selective adsorption of specific molecules, making them useful for the removal of particular contaminants (iv) Regenerability as some nano-adsorbents can be easily regenerated and reused multiple times without significant loss of adsorption capacity (v) Tailored properties as they can be synthesized with specific properties, such as controlled pore sizes, surface chemistry, and surface charge, to optimize their adsorption performance for particular applications (vi) Lower mass and volume requirements due to their high efficiency, a smaller amount of nano-adsorbents is typically needed compared to conventional adsorbents, leading to reduced material and transportation costs (Mercante et al. 2020; AboMarkeb et al. 2019, Cheng et al. 2019; Chandramouli et al. 2023).

9.1.2 Classification of Nano-adsorbents

Nano-adsorbents are nanomaterials that have a high surface area and can adsorb or remove substances from a solution. They are used in various applications, including wastewater treatment, environmental remediation, and purification processes. Nano-adsorbents can be classified based on several criteria:

Composition: (a) Carbon-Based Nano-adsorbents: These are nanomaterials primarily composed of carbon, such as graphene, carbon nanotubes (CNTs), and activated carbon nanoparticles. (b) Metal-Based Nano-adsorbents: These include nanomaterials containing metal nanoparticles like silver, gold, or iron oxide nanoparticles. (c) Polymeric Nano-adsorbents: These are nanomaterials based on polymers like chitosan nanoparticles, polymeric microgels, or dendrimers (Backes and Hirsch 2010).

Structure: (a) 2D Nano-adsorbents: Materials like graphene, graphene oxide, and graphene-based nanocomposites that have a two-dimensional structure. (b)3D Nano-adsorbents: Materials like metal–organic frameworks (MOFs), mesoporous silica nanoparticles, and dendrimers that have a three-dimensional structure with well-defined pores (El-sayed 2020).

Size: (a) Zero-Dimensional Nano-adsorbents: Nanoparticles with zero dimensions, such as quantum dots, which have a very small size and high surface area. (b) One-Dimensional Nano-adsorbents: Materials like carbon nanotubes (CNTs) that have a high aspect ratio and exhibit unique adsorption properties.

Surface Modification: (a) Functionalized Nano-adsorbents: Nano-adsorbents with modified surface chemistry through the addition of functional groups like –OH, $-NH_2$, or –COOH for enhanced adsorption capabilities. (b) Composite Nano-adsorbents: Nano-adsorbents that combine two or more different materials to harness their synergistic effects for improved adsorption performance.

Application-Specific Classification: (a) Water Treatment Nano-adsorbents: Nanomaterials designed specifically for the removal of pollutants, heavy metals, or contaminants from water sources. (b) Environmental Nano-adsorbents: Nano-adsorbents tailored for environmental remediation applications, such as soil or air purification (Jaspal and Malviya 2020).

Adsorbate Type: (a) Heavy Metal Adsorbents: Nano-adsorbents designed to adsorb heavy metals like lead, mercury, cadmium, and chromium. (b) Organic Pollutant Adsorbents: Nanomaterials with a high affinity for organic pollutants like dyes, pesticides, and pharmaceutical compounds (Ren et al. 2011).

Source of Nano-adsorbents: (a) Natural Nano-adsorbents: Nanomaterials derived from natural sources, such as clay nanoparticles, zeolites, or chitosan (b) Synthetic Nano-adsorbents: Nanomaterials synthesized through various chemical and physical methods (Mercante et al. 2020).

Mechanism of Adsorption: (a) Chemisorption Nano-adsorbents: Nanomaterials that form strong chemical bonds with the adsorbate. (b) Physisorption Nano-adsorbents: Nano-adsorbents that rely on weaker physical forces (Van Der Waals, electrostatic interactions) for adsorption (Ali et al. 2018). Therefore, understanding these classifications can help researchers and engineers choose the most suitable nano-adsorbent for their specific application, optimizing the adsorption process for effective pollutant removal or purification.

9.1.3 Types of Nano-adsorbent

There are several types of nano-adsorbents, each designed for specific applications. (1) Nanoparticles: Nanoparticles of various materials, such as metals (e.g., silver, gold), metal oxides (e.g., titanium dioxide, iron oxide), or carbon-based materials (e.g., carbon nanotubes, graphene), can be used as nano-adsorbents. They are often functionalized with specific groups to enhance adsorption properties. (2) Nanofibers: Nanofibers are ultrafine fibers with diameters in the nanometer range. Materials like polymers, ceramics, or carbon can be processed into nanofibers and used as adsorbents. Nanofibers have a high surface area and porosity, making them effective for filtration and adsorption. (3) Nanocomposites: Nanocomposites are hybrid materials composed of nanoparticles or nanofibers embedded in a matrix material. These materials combine the properties of the nanoparticles or nanofibers with the structural support of the matrix. For adsorption purposes, nanocomposites can be designed with enhanced adsorption capacity and selectivity. (4) Carbon Nanotubes: Carbon nanotubes (CNTs) are cylindrical carbon structures with nanoscale diameters. They have exceptional mechanical strength and a high surface area, making them excellent adsorbents for gases and organic compounds. Functionalized CNTs can further improve their adsorption capabilities. (5) Nanoporous Materials: Nanoporous materials, including metal–organic frameworks (MOFs) and zeolites, consist of regularly spaced nanoscale pores. These materials are highly tunable and can be tailored for specific adsorption applications, such as gas separation and purification.

(6) Nanocomposite Hydrogels: Nanocomposite hydrogels are three-dimensional networks of polymer chains with nanoscale fillers like nanoparticles or nanofibers. They can absorb and retain large amounts of water and are used for applications like wastewater treatment and drug delivery. (7) Biosorbents: Nano-adsorbents can also be designed using biological materials at the nanoscale, such as nanoscale enzymes, peptides, or DNA. These biosorbents can target specific molecules or ions for adsorption, making them useful for biomedical and environmental applications. (8) Magnetic Nano-adsorbents: Some nano-adsorbents are magnetic, allowing for easy separation from a liquid or gas mixture using an external magnetic field. These materials can be particularly useful in water treatment and pollutant removal. (9) Polymeric Nano-adsorbents: Nanoscale polymers or polymer-based nanoparticles can be tailored to adsorb specific contaminants. They are often used in water purification and environmental remediation (Ali et al. 2012; Sharma and Kumar 2018; Akpomie and Conradie 2020; Banerjee 2012). The choice of nano-adsorbent depends on the specific application and the type of substance that needs to be adsorbed. Researchers continue to develop and refine nano-adsorbents to address various environmental, industrial, and healthcare challenges.

9.1.4 Nano-adsorbent Selection

Nano-adsorbents play a crucial role in heavy metal removal from wastewater due to their unique properties, such as high surface area, tunable surface chemistry, and excellent adsorption capacities. Selecting the right nano-adsorbent is essential for efficient and cost-effective heavy metal removal. Factors to consider when choosing nano-adsorbents for heavy metal removal include; adsorption capacity, selectivity, chemical stability, regenerability, cost effectiveness, environmental impact, availability and scalability, compatibility, toxicity and regulatory compliance of the nano-adsorbent (Cheng et al. 2019; Chen and Li 2010). Commonly used nano-adsorbents for heavy metal removal include various types of nanoparticles, such as metal oxides (e.g., iron oxide, titanium dioxide), carbon-based nanomaterials (e.g., activated carbon, carbon nanotubes), and clay-based nanoparticles (e.g., montmorillonite, zeolites). It is worth noting that the performance of nano-adsorbents can vary depending on the specific heavy metal, wastewater composition, and other environmental factors. Therefore, laboratory-scale testing and optimization are typically carried out before implementing any nano-adsorbent-based treatment on a larger scale.

9.1.5 Synthesis of Nano-adsorbent

Nano-adsorbents are a class of materials with a high surface area and a nanoscale structure, which makes them highly effective in adsorbing pollutants, including heavy metals, from wastewater. The synthesis of nano-adsorbents for heavy metal removal involves various methods to create materials with specific properties that enhance their adsorption capacity and efficiency. It is important to note that the selection of the appropriate synthesis method depends on the specific heavy metal contaminants present in the wastewater and the desired adsorption properties of the nano-adsorbent material. Additionally, safety considerations should be taken into account when handling nanoparticles during synthesis and application to ensure there are no adverse environmental or health impacts. Many researchers have employed several synthetic techniques for nano-adsorbents used in heavy metal removal from wastewater and other aqueous media (Chowhury and Balasubramanian 2014; Deliyanni et al. 2009). The commonly used fabrication method includes;

1. Sol–Gel Method: The sol–gel process involves the conversion of a colloidal suspension (sol) into a gel by polymerization. In this method, metal precursors are dissolved in a liquid medium to form a sol, and then a gel is formed by controlled hydrolysis and condensation reactions (El-Geundi et al. 2016). The resulting gel can be further dried and calcined to produce nanostructured metal oxide adsorbents with high surface areas. The sol–gel method is a versatile and widely used technique for preparing nanomaterials, including nano-adsorbents. This method involves the formation of a sol, which is a colloidal suspension of nanoparticles, followed by gelation to form a three-dimensional network of interconnected nanoparticles (Dihom et al. 2022). This network can be used as an adsorbent for various applications, including environmental remediation and water purification. The sol–gel method use precursor chemicals which are the compounds that will form the nanoparticles in the final gel such as metal alkoxides (e.g., tetraethylorthosilicate (TEOS), for silica-based materials) or metal salts, organic solvent such as ethanol or methanol, catalyst; often an acid or base to initiate the hydrolysis and condensation reactions. The sol–gel method has been reported to follow a procedural pattern of hydrolysis, condensation, aging, calcinations (which could be optional). The first two steps are illustrated in Eqs. 9.1 and 9.2 respectively for a preparation involving silica-based materials.

$$\text{TEOS} + H_2O \rightarrow Si(OH)_4 + \text{EtOH (ethyl alcohol)} \tag{9.1}$$

$$Si(OH)_4 \rightarrow Si{-}O{-}Si + H_2O \tag{9.2}$$

The sol–gel method allows for precise control over the size, composition, and structure of the nanoparticles, making it a powerful technique for tailoring nano-adsorbents for specific applications. The choice of precursor and process parameters can be adjusted to achieve the desired properties and performance of the nano-adsorbent.

2. Co-precipitation Method: In this technique, metal precursors are simultaneously precipitated in the presence of a suitable precipitating agent, leading to the formation of nanoparticles with controlled composition and size. This method allows the incorporation of multiple metal ions to enhance the adsorption capacity for a range of heavy metal contaminants (Gul et al. 2021). The co-precipitation method is a commonly used technique for preparing nano-adsorbents, especially for the removal of various contaminants from aqueous solutions. It involves the simultaneous precipitation of metal or metal oxide nanoparticles along with the adsorbent material of interest. The co-precipitation method for nano-adsorbent preparation use precursor salts of the metal or metal oxide (e.g., metal chlorides, nitrates, or acetates), adsorbent material (e.g., nanoparticles, carbon-based materials, polymers), aqueous solution (typically deionized water and pH adjustment chemicals (e.g., NaOH or HCl). The co-precipitation method allows for the controlled synthesis of nano-adsorbents with specific properties and surface chemistries, making them suitable for various applications, including water purification, catalysis, and environmental remediation. Adjusting parameters like pH, temperature, and precursor concentrations can help tailor the nano-adsorbent's characteristics for specific adsorption tasks.

3. Hydrothermal/Solvothermal Method: Hydrothermal or solvothermal synthesis involves the use of high-temperature and high-pressure conditions to promote the controlled growth of nanomaterials (Nworie et al. 2023). This method allows the synthesis of a variety of metal-based nano-adsorbents with tunable properties. These methods involve the synthesis of nanoscale materials in a high-temperature, high-pressure, and controlled chemical environment. Nano-adsorbents prepared using these methods often have unique properties that make them suitable for various applications, including water purification and pollutant removal. The hydrothermal method is advantageous because it ensures controlled synthesis of nanomaterials with well-defined sizes and shapes, improved crystallinity and phase purity of the nanoparticles, enhanced adsorption properties due to the nanoscale structure.

The solvothermal method is similar to the hydrothermal method but uses organic solvents instead of water as the reaction medium. This method is particularly useful for synthesizing materials that are not stable in aqueous environments. The advantages of solvothermal method are that it allows the synthesis of nanomaterials that are not stable in water, provides control over the size, shape, and properties of nano-adsorbents and offers flexibility in choosing solvents and reaction conditions. Both hydrothermal and solvothermal methods can be tailored to produce nano-adsorbents with specific properties, such as high surface area, controlled porosity, and enhanced adsorption capacities. Researchers often optimize these methods to achieve the desired characteristics for various applications, including environmental remediation and water treatment (Mohammad et al. 2023).

4. Template-Assisted Method: Templates such as surfactant micelles, nanospheres, or porous materials can be used to guide nano-adsorbent synthesis (Chandramouli et al. 2023). A template-assisted method for preparing nano-adsorbents involves using a template or scaffold to guide the synthesis of nanoscale materials with specific properties for adsorption applications. This method is commonly used to create nano-adsorbents with a high surface area, tailored pore size distribution, and enhanced adsorption capacity. The template-assisted synthesis of nano-adsorbents involves the selection of Template, (common templates include silica nanoparticles, polymer beads, and surfactant micelles), preparation of Template (when using solid templates (e.g., silica nanoparticles), such can be dispersed in a solvent by sonication or stirring until a homogeneous suspension is obtained but if using micellar templates (e.g., surfactant micelles), the micellar solutions is prepared in a suitable solvent), synthesis of Nano-adsorbent by mixing the precursor materials (e.g., metal salts, organic ligands) with the template suspension or micellar solution, removal of Template through a suitable method (silica templates can be removed by etching with an acid (e.g., hydrochloric acid), polymer templates can be thermally decomposed or dissolved in a suitable solvent and surfactant micelles can be broken by changing the solvent conditions) (Nworie et al. 2022). The nano-adsorbent formed is always purified by centrifugation then washed and dried under vacuum or in an oven.

5. Green Synthesis: Green synthesis techniques involve the use of environmentally friendly agents or natural sources, such as plant extracts, microorganisms, or biomass, to synthesize nano-adsorbents (Tawade et al. 2023). These methods are gaining popularity due to their eco-friendliness and potential for large-scale production. Green synthesis of nano-adsorbents involves the preparation of nanoscale materials with adsorption properties using environmentally friendly and sustainable methods. These nano-adsorbents can be used for various applications, including the removal of pollutants from water and air. The preparation of the nano-adsorbents using green synthesis methods involves careful selection of green precursor which is environmentally friendly, often from renewable sources such as natural polymers, plant extracts, or biodegradable materials (cellulose, chitosan, starch, or various plant extracts rich in bioactive compounds), selection of the nanomaterial such as nanoparticles (e.g., metal nanoparticles), nanocomposites (e.g., polymer-nanoparticle hybrids), or carbon-based nanomaterials (e.g., graphene oxide). The synthesis Methods notably, green synthesis methods that minimize the use of toxic chemicals and reduce energy consumption such as biosynthesis which make use of microorganisms (bacteria, fungi) or plant extracts to reduce metal ions and create nanoparticles, microwave or ultrasound-assisted synthesis methods which reduce reaction times and energy consumption, making them more environmentally friendly, hydrothermal or solvothermal synthesis, green chemistry principles approach which adhere to green chemistry principles, such as minimizing waste, using renewable feedstocks, and designing safer chemicals and functionalization which modifies the surface properties of the nano-adsorbent to enhance its adsorption capacity achieved by attaching functional groups, adding ligands, or doping the nanomaterial with specific elements (Jana et al. 2021).

The adsorption Studies of the nano-adsorbent is conducted to evaluate the adsorption capacity and efficiency for the target pollutants or contaminants. This is usually followed by optimization to fine-tune the synthesis process and nano-adsorbent properties to achieve the best adsorption performance while maintaining green and sustainable principles and if successful, scale-up is done with much emphasis on safety and environmental impact assessment. Green synthesis of nano-adsorbents is an evolving field, and researchers are continually exploring new materials and methods to make the process more sustainable and environmentally friendly.

6. Electrospinning: Electrospinning is a technique that creates nanofibers by applying a high-voltage electric field to a polymer solution (Gupta and Saleh 2013). These nanofibers can be functionalized with metal nanoparticles or metal oxides to create effective nano-adsorbents for heavy metal removal. Electrospinning is a versatile technique used for preparing nanofibers and nanoparticles from various materials, including polymers, ceramics, and composites. While it's more commonly associated with nanofiber production, it can also be adapted for nanoparticle generation. Electrospinning method for preparing nano-adsorbent utilizes an electric field to draw and stretch a polymer solution or melt into ultrafine fibers or particles. The basic setup includes a high-voltage power supply, a syringe pump or spinneret for material delivery, and a grounded collector. The preparation process involves carefully choosing a polymer or material that can be dissolved or melted and processed into nanoparticles using electrospinning (polymers such as polyvinyl alcohol (PVA), polyethylene oxide (PEO), or biodegradable polymers like polylactic acid (PLA) or poly(lactic-co-glycolic acid) (PLGA). Thereafter, the solution or melt is prepared by dissolving the selected polymer or material in a suitable solvent (for polymer solutions) or heat it to a molten state (for polymer melts). In another way, the nanoparticles (e.g., metal nanoparticles, drug-loaded nanoparticles) can be introduced into the polymer solution or melt to incorporate them into the electrospun nanoparticles (Basha and Anand 2014; Biswas et al. 2017). Depending on the specific application of the nano-adsorbent, factors such as heat treatment, crosslinking, or surface modification can be fine-tuned to enhance the properties of the electrospun nanoparticles. Similarly, electrospinning parameters, such as solution/melt properties, voltage, flow rate, and collector distance, can be adjusted to control the size, morphology, and composition of the resulting nanoparticles.

7. Bottom-up Self-assembly: This approach involves the spontaneous organization of atoms or molecules into nanostructured materials with specific adsorption properties (Hurt et al. 2006; Crini 2005, 2006). It allows precise control over the material's architecture and surface chemistry. This approach relies on the spontaneous organization of nanoscale building blocks into a desired structure or morphology through weak intermolecular forces and chemical interactions. Bottom up assembly practically involves selection of nanoscale building blocks such as nanoparticles (e.g., metal, metal oxide, carbon), nanotubes, nanofibers, or nanosheets. The selection depends on the specific adsorption properties and intended application. This is always followed by surface modification which may involve functionalization with chemical groups or molecules that can selectively interact with target adsorbates (e.g., pollutants, ions, organic compounds). The selection of appropriate solvent or dispersion medium that is compatible with both the nanoscale building blocks and the target adsorbate. The solvent should facilitate the dispersion and assembly of the nanoscale materials. The self assembly is enhanced by mixing the surface-modified nanoscale materials with the solvent, creating a colloidal suspension or solution and consequently, the control of conditions (e.g., temperature, pH, concentration) to encourage self-assembly. The use of template or nucleating agent to guide the assembly process into a specific structure is necessary and highly advocated by researchers (Gupta et al. 2014).

9.1.6 Surface Functionalization of Nano-adsorbents

Surface functionalization of nano-adsorbents for heavy metal removal from wastewater is a crucial and effective approach to enhance their adsorption capacity and selectivity (Gul et al. 2021). Nano-adsorbents are nanoscale materials with a large surface area and high reactivity, making them excellent candidates for removing heavy metals from contaminated water. Functionalization involves modifying the surface of these nano-adsorbents by introducing specific chemical groups or ligands to target and attract heavy metal ions. After synthesizing the nano-adsorbents, their properties are often optimized by modifying factors such as surface functionalization, pH, temperature, and contact time to maximize heavy metal adsorption efficiency. The functionalization process can be achieved through various methods, and some of the commonly used techniques include:

1. Coating: Applying a thin layer of functional material onto the surface of the nano-adsorbent. This could be achieved through physical deposition, chemical deposition, or self-assembly processes. Coating of nano-adsorbents refers to the process of applying a thin layer of material onto the surface of nanoscale adsorbent particles (Bhatnagar et al. 2013). This coating can serve various purposes, including enhancing the adsorption capacity, improving selectivity, providing stability, or facilitating the targeted release of adsorbed substances. The choice of coating material and the coating method depend on the specific application and the properties desired for the nano-adsorbent.

The reasons for coating nano-adsorbents include enhanced adsorption capacity due to high surface area material for example; a silica or graphene oxide coating on magnetic nanoparticles can provide additional binding sites. Improved Selectivity: Coating can make nano-adsorbents more selective by modifying their surface chemistry. For instance, functional groups or ligands can be attached to the coating to target specific molecules. Stability and Durability: Coatings can protect the nano-adsorbent core from chemical degradation or aggregation, increasing its stability and lifespan. Controlled Release: Coatings can be designed to release adsorbed substances under certain conditions. For instance, pH-sensitive coatings may release adsorbates when the pH changes. The common coating materials are silica which is widely used coating material due to its inert nature and ease of functionalization. It can be applied through sol–gel processes or by the Stöber method (El-Geundi et al. 2016). Polymers like polyethylene glycol (PEG), chitosan, or polyvinyl alcohol (PVA) can provide stability and controlled release (Beyene and Ambaye 2019). They can be applied using techniques like layer-by-layer assembly or emulsion polymerization in the coating process. Noble metals like gold or metal oxides like titanium dioxide (TiO_2) can be used to enhance adsorption properties. These coatings can be deposited through various techniques like chemical reduction or physical vapor deposition (AboMarkeb et al. 2019). Coating nano-adsorbents with graphene or carbon nanotubes can improve their mechanical strength, electrical conductivity, and adsorption capacity. Adding functional groups like amino, carboxyl, or thiol groups to the coating can impart specific chemical interactions for selective adsorption. Biomolecules such as enzymes, antibodies, or DNA can be used as coatings for specific bioanalytical or biomedical applications (Tlili and Alkanhal 2019).

2. Chemical Modification: This involves introducing functional groups on the surface of the nano-adsorbent through chemical reactions (Ren et al. 2011). Common functional groups include carboxyl (–COOH), amino ($–NH_2$), hydroxyl (–OH), and thiol (–SH) groups, among others. Chemical modification of nano-adsorbents is a common technique used to enhance their adsorption capacity, selectivity, and stability for various applications, especially in the field of environmental remediation and wastewater treatment. Nano-adsorbents are typically composed of nanomaterials like nanoparticles, nanotubes, or nanofibers, which can be functionalized or modified through chemical reactions to improve their performance. Chemical modification can be achieved through surface functionalization, coating with polymers, ion exchange, silanization, immobilization of ligands, chemical precipitation, functional nanoparticles, crosslinking, surface engineering.

3. Doping: The process involves incorporating dopant atoms or ions into the nano-adsorbent's lattice structure to alter its properties and surface reactivity (Wilson et al. 2018). Doping of nano-adsorbents refers to the process of introducing foreign atoms or molecules, known as dopants, into the structure of nanomaterials that are used for adsorption purposes. This doping process is often carried out to enhance the adsorption properties of nano-adsorbents for specific applications. In doping of nano-adsorbents it is important to consider the purpose of doping, the type of nano-adsorbent, the dopant, methods of doping, effect of doping, application, characterization method and challenges emanating from optimization of variables. Generally, doping of nano-adsorbents is a versatile strategy to tailor their properties and improve their effectiveness in various adsorption applications.

4. Ion Exchange: Replacing existing ions on the nano-adsorbent's surface with specific metal-binding ions, which improve the affinity for heavy metal ions in the wastewater (Jaspal and Malviya 2020). Ion exchange is a widely used method for functionalizing nano-adsorbents, which are nanoparticles or nanomaterials with the ability to adsorb or capture specific ions or molecules from a solution.

5. Surface Oxidation/Reduction: This involves modifying the oxidation state of the nano-adsorbent's surface to enhance metal binding capacity (Li et al. 2011). Surface oxidation and reduction of nano-adsorbents are important processes that can significantly influence their performance in various applications, particularly in the field of environmental remediation and catalysis. These processes can alter the surface chemistry and reactivity of nano-adsorbents, ultimately affecting their ability to adsorb or catalyze specific reactions. Surface oxidation refers to the addition of oxygen or the removal of electrons from the surface of a nano-adsorbent material. It results in the formation of oxides or other oxygen-containing functional groups on the surface. Oxidation can introduce oxygen-containing functional groups such as hydroxyl (–OH), carbonyl (–C=O), or carboxyl (–COOH) groups on the surface. These functional groups can enhance the adsorption capacity of nano-adsorbents by providing more binding sites for target molecules. Oxidized surfaces are often more chemically reactive, making nano-adsorbents suitable for catalytic applications, in redox reactions and facilitate the removal of pollutants through oxidative processes. Surface oxidation can also enhance the stability of nano-adsorbents by preventing agglomeration or degradation over time.

Surface reduction is the process of adding electrons or removing oxygen from the surface of a nano-adsorbent (Nworie et al. 2022). It results in the reduction of metal ions or oxygen-containing functional groups. Surface reduction can occur through various mechanisms, including chemical reactions with reducing agents, exposure to reducing atmospheres, or electrochemical processes. Common reducing agents include hydrogen, sodium borohydride, and various organic molecules. In the case of metal-based nano-adsorbents, reduction can lead to the formation of metallic nanoparticles on the surface. Surface reduction can remove oxygen-containing functional groups, making the surface less polar and more hydrophobic. This can impact the adsorption selectivity and affinity for specific molecules. Reduced surfaces may facilitate electron transfer processes, making nano-adsorbents suitable for electrochemical applications or catalytic reactions involving redox reactions (Aitken et al. 2006). Advantages of surface functionalization of nano-adsorbent include increased surface area, enhanced selectivity, improved stability, tailored affinity and regeneration. The environmental and economic considerations of the functionalization process should be taken into account when designing nano-adsorbents for real-world applications in wastewater treatment.

9.1.7 Characterization of Nano-adsorbent

Characterization of a nano-adsorbent for heavy metal removal involves a series of tests and analyses to understand its physical, chemical, and structural properties. This process helps researchers and engineers to determine its efficiency, capacity, and suitability for removing heavy metals from various sources.

Surface Area Analysis: Determining the specific surface area of the nano-adsorbent using techniques like Brunauer–Emmett–Teller (BET) analysis provides valuable information about its adsorption capacity and potential (Biswas et al. 2017). Particle Size, Morphology and Pore Size Distribution: Scanning Electron Microscopy (SEM) and Transmission Electron Microscopy (TEM) can be used to visualize the nano-adsorbent's morphology, particle size, and shape. This helps understand its structure and porosity, which are essential for effective adsorption. Techniques like mercury intrusion porosimetry or gas adsorption (e.g., nitrogen) are used to assess the pore size distribution and pore volume of the nano-adsorbent (Dihom et al. 2022; Ali et al. 2020, 2022). This information is vital for understanding adsorption kinetics. X-ray Diffraction (XRD): The XRD analysis provides information about the crystalline structure of the nano-adsorbent. It helps identify the presence of specific crystalline phases and any changes that might occur during the adsorption process (Enamul Hoque et al. 2017). Fourier Transform Infrared Spectroscopy (FTIR): The

FTIR analysis is used to identify the functional groups present on the surface of the nano-adsorbent (Gul et al. 2021). This is crucial in understanding the chemical interactions responsible for heavy metal adsorption. Energy-Dispersive X-ray Spectroscopy (EDS): The EDS is often performed in conjunction with SEM and TEM to provide elemental analysis, confirming the presence of specific elements within the nano-adsorbent (Gul et al. 2021). Thermogravimetric Analysis (TGA): The TGA helps determine the thermal stability and weight loss behavior of the nano-adsorbent under different temperature conditions (Gul et al. 2021). Zeta Potential Measurement: Zeta potential analysis provides information about the surface charge of the nano-adsorbent (Gul et al. 2021). It indicates the electrostatic interactions between the nano-adsorbent and heavy metal ions. Adsorption isotherms and kinetics help to determine amount of heavy metal adsorbed as it relates concentration and the rate controlling step respectively (Nworie et al. 2023).

9.1.8 Applications of Nano-adsorbents

1. Environmental remediation: Nano-adsorbents are used to remove pollutants and contaminants from water and air, including heavy metals, organic compounds, dyes, and other harmful substances. Environmental remediation using nano-adsorbents is an innovative and promising approach to address various pollution and contamination issues in air, water, and soil (Chen and Li 2010). Nano-adsorbents are nanomaterials with high surface area and unique adsorption properties, making them highly efficient at removing contaminants from the environment. Some of the notable specific applications in environmental remediation include

(i) waste water treatment: Nano-adsorbents are used to remove heavy metals, organic pollutants, and dyes from wastewater. They can effectively adsorb contaminants like arsenic, lead, cadmium, and various organic compounds. (ii) Air Pollution Control and industrial purification processes: Nano-adsorbents can be incorporated into air filters to capture particulate matter, volatile organic compounds (VOCs), and other air pollutants They are especially useful for indoor air quality improvement and reducing emissions from industrial processes. They are employed in industrial settings to purify liquids and gases, such as in the food and beverage industry, pharmaceutical manufacturing, and wastewater treatment. (iii) Soil Remediation: Nano-adsorbents can immobilize or remove contaminants in soil, such as heavy metals and organic pollutants. They help in reducing the mobility and bioavailability of contaminants, preventing them from entering the food chain. (iv) Energy storage: Nano-adsorbents can be used in energy storage devices, like supercapacitors, due to their high surface area and capacitance properties. (v) Catalysis: Nano-adsorbents are used as catalyst supports to enhance reaction rates and efficiency in various catalytic processes (Gul et al. 2021; Barakat 2011).

9.2 Wastewater

Waste water refers to water that has been used and contaminated through various human activities and processes (Gupta et al. 2012). It contains a mixture of pollutants, including chemicals, biological contaminants, and other substances that make it unsuitable for immediate reuse or direct release into the environment. Managing wastewater is crucial for protecting human health, ecosystems, and water resources. Wastewater can originate from various sources, including residential, industrial, commercial, and agricultural activities. Common examples include sewage from households and businesses, industrial effluents, storm water runoff, and agricultural runoff. The composition of wastewater can vary widely depending on its source. It may contain organic matter, nutrients (such as nitrogen and phosphorus), pathogens (bacteria, viruses, etc.), heavy metals, chemicals, and suspended solids.

Improper disposal of untreated or poorly treated wastewater can have severe environmental consequences, including pollution of rivers, lakes, and oceans, as well as harm to aquatic life and ecosystems. Many countries and regions have established regulations and standards to control the discharge of wastewater into the environment, ensuring that it meets certain quality criteria to protect human health and the environment (Elella et al. 2022). Managing wastewater presents challenges, including population growth, urbanization, industrial expansion, and climate change, which can strain wastewater treatment infrastructure and water resources. Efficient wastewater management is essential for safeguarding public health, protecting the environment, and ensuring the sustainability of water resources. It involves a combination of infrastructure development, technological advancements, and regulatory measures to address the complex and dynamic nature of wastewater.

Wastewater treatment is the process of removing contaminants from wastewater to make it safe for disposal or reuse. Common treatment methods include primary, secondary, and tertiary treatment and typically involve physical, chemical, and biological processes to remove or neutralize pollutants (Pandey 2017). In primary treatment, large solids are removed from the wastewater through processes like screening and sedimentation. This step reduces the amount of solids in the wastewater. Secondary treatment involves biological processes in which microorganisms are used to break down organic matter and remove dissolved substances. This step further purifies the water. Tertiary treatment is an advanced stage that aims to remove remaining contaminants, including nutrients and pathogens, to a level where the treated water can be safely released into the environment or reused for non-potable purposes. Treated wastewater, known as reclaimed water or effluent, can be used for various purposes, such as irrigation, industrial processes, and even direct potable water reuse in some cases.

9.2.1 Heavy Metals in Waste Water

Heavy metals in wastewater are a significant environmental concern due to their toxicity, persistence, and potential to bioaccumulate in the food chain. Heavy metals are naturally occurring elements that possess a density higher than water, and they include elements like lead (Pb), cadmium (Cd), mercury (Hg), arsenic (As) and chromium (Cr) (Akpomie and Conradie 2020; Aziz et al. 2008). The primary sources of heavy metals in wastewater include (i) Industrial activities such as metal smelting, mining, electroplating, and manufacturing, can release heavy metals into wastewater through effluent discharges (ii) Urban Runoff: Rainwater washes heavy metals from roads, roofs, and other urban surfaces into storm water drains, which ultimately end up in water bodies or sewage treatment plants (iii) Agricultural Activities: The use of metal-based pesticides, fertilizers, and irrigation water containing heavy metals can introduce these contaminants into wastewater (iv) Domestic Sewage: Household products, including cleaning agents, personal care items, and pharmaceuticals, can contribute to heavy metal concentrations in wastewater (Chen and Li 2010).

9.2.2 Toxicity and Health Impacts of Heavy Metals

Heavy metals are toxic even at low concentrations and can pose serious health risks to humans and aquatic organisms. When wastewater containing heavy metals is discharged into natural water bodies, it can contaminate drinking water sources and disrupt aquatic ecosystems. Common health effects of heavy metal exposure include neurological disorders, kidney and liver damage, respiratory issues, developmental problems, and even cancer (Banerjee 2012).

9.2.3 Environmental Impact of Heavy Metals in Waste Water

Heavy metals can have severe environmental impacts. They may inhibit photosynthesis in aquatic plants, disrupt the reproductive cycles of aquatic organisms, and alter the behavior and physiology of aquatic species. In extreme cases, heavy metal pollution can lead to fish kills and decimate entire aquatic ecosystems (Akpomie and Conradie 2020).

9.2.4 Bioaccumulation and Biomagnification

One of the most concerning aspects of heavy metals in wastewater is their ability to bioaccumulate and biomagnify in the food chain. Bioaccumulation occurs when organisms absorb and retain heavy metals at a rate higher than they can eliminate them. As a result, the concentration of heavy metals increases as they move up the food chain, affecting higher-level predators the most.

Bioaccumulation and biomagnification of heavy metals in wastewater are important environmental concerns because they can have detrimental effects on ecosystems and human health. Bioaccumulation refers to the process by which organisms, such as aquatic plants and animals, accumulate heavy metals from their surrounding environment over time (Akpomie and Conradie 2020). In wastewater, heavy metals like lead, mercury, cadmium, and arsenic are often present due to industrial discharges, sewage, and runoff from urban areas. Aquatic organisms can absorb these heavy metals through various pathways, including direct uptake from water, sediment ingestion, and through their food. Once absorbed, the metals tend to accumulate in the tissues of these organisms. They can bind to proteins and other molecules, making them difficult to excrete. Biomagnification is a process in which the concentration of a substance, such as a heavy metal, increases at higher trophic levels in a food chain or web (Akpomie and Conradie 2020). This means that predators at the top of the food chain can have significantly higher levels of heavy metals in their bodies compared to organisms lower in the food chain. The biomagnification of heavy metals occurs because these substances are not easily metabolized or excreted by organisms. As a result, they accumulate over time as organisms consume other organisms containing these metals. For instance, small fish that have bioaccumulated heavy metals may be eaten by larger fish, and the heavy metals become concentrated in the larger fish. If humans consume these larger fish, they may be exposed to high levels of heavy metals. The consequences of bioaccumulation and biomagnification of heavy metals in wastewater are far-reaching and include (i) Heavy metals can harm aquatic ecosystems by disrupting the health of plants, animals, and microorganisms. (ii) Accumulated heavy metals can lead to reduced biodiversity, altered reproductive patterns, and even the death of sensitive species. (iii) Sediment contaminated with heavy metals can persist for a long time, continuing to release metals into the water (iv) Human Health Risks. Heavy metal exposure can lead to a range of health problems, including neurological disorders, kidney damage, cancer, and developmental issues, particularly in children and pregnant women (Nworie et al. 2022).

To mitigate the bioaccumulation and biomagnification of heavy metals in wastewater, it is essential to implement effective wastewater treatment processes to remove or reduce heavy metal concentrations before discharge into natural water bodies. Additionally, regulatory measures, pollution control, and public awareness are crucial in addressing this environmental issue and minimizing its impact on ecosystems and human health.

9.2.5 *Treatment and Regulation*

Effective wastewater treatment is crucial to removing heavy metals and mitigating their environmental impact. Common treatment methods include chemical precipitation, ion exchange, adsorption, and membrane filtration (Gupta et al. 2012). However, some conventional treatment processes may not be sufficient to remove all heavy metals from wastewater. Conventional treatment processes for heavy metal removal from wastewater typically involve physical, chemical, or biological methods. The choice of method depends on factors such as the type and concentration of heavy metals present, the volume of wastewater, and environmental regulations. The common conventional methods include

Chemical Precipitation: Chemical precipitation involves adding chemicals to the wastewater to form insoluble precipitates with heavy metals (Gupta et al. 2012; Rajasulochana and Preethy 2016). Commonly used chemicals include lime (calcium hydroxide), sodium hydroxide, and various metal salts such as ferric chloride or alum. The heavy metal ions bind with these chemicals and settle as solid particles, which can then be removed by sedimentation or filtration.

Coagulation-Flocculation: Coagulation and flocculation are often used in combination with chemical precipitation. Coagulants like aluminum sulfate (alum) or ferric chloride are added to destabilize and aggregate the heavy metal ions, and then flocculants like polymers are added to form larger flocs (Gupta et al. 2012).

Ion Exchange: Ion exchange involves passing the wastewater through a resin or bed of ion-exchange materials like zeolites or synthetic resins (Gupta et al. 2012). These materials exchange the heavy metal ions in the water with less harmful ions present on the resin. When the resin becomes saturated with heavy metals, it can be regenerated or replaced.

Adsorption: Adsorption is the process of adhering heavy metal ions to the surface of an adsorbent material (Nworie et al. 2019). Common adsorbents include activated carbon, clay minerals, and various natural materials like rice husk ash or coconut shell. The metal ions adhere to the surface of the adsorbent, allowing the clean water to pass through.

Membrane Filtration: Membrane filtration methods, such as ultrafiltration, nanofiltration, and reverse osmosis, can effectively remove heavy metals from wastewater by physically blocking the passage of metal ions through a semipermeable membrane (Gupta et al. 2012). These methods are especially useful for treating highly contaminated wastewater.

Electrochemical Precipitation: Electrochemical methods involve applying an electric current to wastewater containing heavy metal ions (Gupta et al. 2012). This causes the heavy metals to undergo reduction reactions, forming solid precipitates that can be separated from the water.

Sedimentation and Settling: In some cases, heavy metal removal can be achieved by allowing the wastewater to sit in a settling tank or basin (Gupta et al. 2012). Over time, the heavy metal particles settle to the bottom and can be removed as sludge.

Biological Treatment: Biological processes can be used in conjunction with other methods to enhance heavy metal removal (Gupta et al. 2012). Certain microorganisms have the ability to accumulate and precipitate heavy metals, a process known as biosorption. This method is often used in constructed wetlands or bioreactors.

Chemical Oxidation/Reduction: Some heavy metals can be transformed into less toxic forms through chemical oxidation or reduction processes (Gupta et al. 2012; Rajasulochana and Preethy 2016). For example, ferrous sulfate can be used to reduce hexavalent chromium (Cr(VI)) to trivalent chromium (Cr(III)), which is less toxic and easier to remove.

Conventional methods of heavy metals removal from wastewater have several disadvantages, which have led to the development and adoption of more advanced and sustainable treatment technologies. Some of the disadvantages of conventional methods include; Limited Metal Specificity, Sludge Generation, Chemical Usage, pH Dependency, Incomplete Removal, Energy Intensive, Regeneration and Disposal of Sorbents, Space Requirements, Maintenance and Monitoring, Environmental Impact, Long Treatment Times, Insufficient for Low Concentrations. In light of these disadvantages, there has been a growing interest in the development and adoption of more sustainable and efficient technologies for heavy metal removal from wastewater. These newer technologies aim to address some of the limitations associated with conventional methods while providing better performance and environmental benefits. Emerging technologies for heavy metals removal from wastewater continue to evolve as researchers and engineers seek more efficient and sustainable solutions. Notable types of emerging technologies for heavy metals removal include

Nanotechnology: Nanomaterials such as nanoparticles, nanotubes, and nanofibers have shown promise in adsorbing heavy metals from wastewater due to their large surface area and high reactivity (Ali et al. 2020). Materials like graphene oxide, carbon nanotubes, and various metal oxide nanoparticles can be used for this purpose. Biosorption: This involves the use of biological materials such as algae, bacteria, fungi, and agricultural waste products like biosorbents to adsorb and accumulate heavy metals. Bioengineered microorganisms can also be employed to enhance the biosorption capacity (Cheng et al. 2019). Emerging Electrochemical Methods: Electrochemical techniques like electrocoagulation, electroflotation, and electrodeposition are emerging as efficient methods for heavy metals removal (Gupta et al. 2012). They use electricity to induce chemical reactions that facilitate metal precipitation and removal. Advanced Ion Exchange Resins (Gusain et al. 2020): Advanced ion exchange resins with high selectivity for specific heavy metals are being developed. These resins can be more efficient and cost-effective than traditional ion exchange methods.

Modified Adsorbent Materials: There has been continuous work on developing new adsorbent materials, such as modified clays, zeolites, and activated carbon, with improved selectivity and capacity for heavy metals removal (Cheng et al. 2019; Ali et al. 2012, 2018). Advanced Membrane Filtration: Advances in membrane technology, including nanofiltration and reverse osmosis, allow for the removal of heavy metals from water through size-based separation. Selective membranes can target specific heavy metal ions. Magnetic Adsorbents: Magnetic materials like magnetic nanoparticles coated with ligands that have an affinity for heavy metals can be used for easy separation and recovery of heavy metal ions using magnetic fields (Gul et al. 2021). Hybrid Systems: Combining multiple technologies like adsorption, membrane filtration, and precipitation in a single treatment system can provide more effective heavy metal removal (Fajardo et al. 2022). Sonochemical and Ultrasonic Treatment: Ultrasonic and sonochemical methods can enhance the removal of heavy metals by creating cavitation bubbles that disrupt metal complexes, facilitating their separation and precipitation. Green Synthesis: Green synthesis methods involve the use of natural materials or plant extracts to synthesize nanoparticles or adsorbents with heavy metal removal capabilities. Artificial Intelligence (AI) and Machine Learning: Artificial intelligence and machine learning algorithms can be used to optimize treatment processes, predict heavy metal removal efficiencies, and enhance the overall performance of wastewater treatment systems. These methods are eco-friendly and sustainable.

The use of nanotechnology/nano-adsorbent as an emerging technology for heavy metal removal include presents several advantages which include: Enhanced Efficiency: Emerging technologies often exhibit higher removal efficiency compared to conventional methods. They can target specific heavy metals more effectively, resulting in cleaner water or soil. Selectivity: Many emerging technologies can selectively remove certain heavy metals while leaving others untouched. This selectivity reduces the need for complex and costly separation processes. Reduced Chemical Usage: Some emerging methods use fewer or less harmful chemicals, making them environmentally friendly and minimizing the generation of hazardous waste. Lower Energy Consumption: Advanced materials and processes can require less energy, reducing operational costs and the overall environmental footprint of heavy metal removal operations. Faster Remediation: Nano-adsorbents as an emerging technologies work more rapidly than conventional methods, allowing for quicker remediation of contaminated sites and minimizing the risk of further contamination spread. Scalability: Nano-adsorbents are scalable, making them suitable for both small-scale and large-scale applications. This adaptability allows for the efficient treatment of various contamination scenarios. Long-Term Stability: nano-adsorbents have greater long-term stability, ensuring the continued effectiveness of heavy metal removal over time. Minimal Residuals: they often produce fewer residual wastes or byproducts, reducing the need for additional disposal or treatment steps (Wilson et al. 2018). Flexibility: Innovative technologies can be customized to target specific heavy metal contaminants and adapt to diverse environmental conditions, offering a more tailored solution to contamination challenges. Remote Monitoring and Control: nanotechnology can incorporate automation and remote monitoring capabilities, allowing for real-time data analysis and control, which can enhance the overall efficiency and effectiveness of the remediation process. Sustainability: nanotechnology rely on sustainable materials or renewable energy sources, aligning with modern environmental and sustainability goals. Compliance with Regulations: Employing innovative methods that meet or exceed regulatory standards can help organizations avoid fines and legal repercussions related to heavy metal contamination. Reduced Overall Costs: synthesis and application of nano-adsorbents are facile and can lead to significant cost savings in the long run due to increased efficiency, reduced operational expenses, and minimized environmental liabilities.

9.2.6 Mechanism of Heavy Metal Removal

The removal of heavy metals from various environmental and industrial sources using nano-adsorbents involves several mechanisms that exploit the unique properties of nanomaterials. Heavy metal removal using nano-adsorbents can involve multiple mechanisms, including chelation, ion exchange, and complexation. These mechanisms depend on the specific properties of the nano-adsorbent and the nature of the heavy metal ions being removed. The mechanisms involved in heavy metal removal by nano-adsorbents include

1. Surface Adsorption: Nano-adsorbents have a high surface area-to-volume ratio, providing a large number of active sites for heavy metal ions to attach. Heavy metal ions in the solution are attracted to the surface of nanomaterials due to electrostatic interactions and chemical affinity. The surface of nano-adsorbents may possess functional groups or ligands that can form coordination bonds with heavy metal ions, enhancing adsorption capacity. The process of adsorption onto the nano-adsorbent can be represented by Eq. 9.3

$$M^{2+}(aq) + \text{Nano-adsorbent(s)} \rightarrow \text{Nano-adsorbent} - M^{2+}(s) \tag{9.3}$$

2. Ion Exchange: Some nanomaterials have exchangeable ions or functional groups on their surfaces that can replace heavy metal ions in the solution through ion exchange processes. This mechanism is particularly effective for nano-adsorbents like zeolites and layered double hydroxides. The nano-adsorbent might have surface functional groups that can exchange ions with the heavy metal ions in the solution. In this case, ion exchange is represented as Eq. 9.4

$$\text{NP–F–X} + Pb^{2+} \leftrightarrow \text{NP–F–Pb} + X^{-} \tag{9.4}$$

   where NP–F–X represents the nano-adsorbent with exchangeable ions (e.g., –COOH groups), and X^{-} represents the counter ion released during ion exchange.
3. Chemical Precipitation: Nano-adsorbents can promote the precipitation of heavy metal ions as insoluble compounds on their surfaces. This occurs when the nano-adsorbent provides nucleation sites for the formation of metal hydroxides, sulfides, or oxides, which are less soluble in water and can be easily separated.

The nano-adsorbent or nanoparticles are not explicitly shown in the equation but play a crucial role in facilitating the precipitation reaction. Equation 9.5 represents the precipitation of lead ions using sodium sulfide as the precipitating agent:

$$Pb^{2+}(aq) + S^{2+}(aq) \rightarrow PbS(s) \tag{9.5}$$

In this reaction, lead ions in the solution react with sulfide ions to form insoluble lead sulfide (PbS) precipitate.

4. Complexation: Nano-adsorbents with specific functional groups can form complexes with heavy metal ions, reducing their mobility and concentration in the solution. Chelating ligands on the nano-adsorbent's surface can bind with heavy metal ions to form stable complexes. Apart from chelation, the nano-adsorbent might form non-chelate complexes with heavy metal ions. This can be represented as Eq. 9.6

$$NP{-}F + Pb^{2+} \leftrightarrow NP{-}F{-}Pb \tag{9.6}$$

where NP–F–Pb represents a complex formed between the nano-adsorbent and the heavy metal ion.

5. Physical Adsorption: Nano-adsorbents may employ physical forces such as Van Der Waals interactions and hydrogen bonding to adsorb heavy metal ions. These weak interactions can contribute to the overall adsorption capacity of nano-adsorbents.
6. Electrostatic Attraction: Nano-adsorbents with charged surfaces can attract heavy metal ions with opposite charges through electrostatic interactions. This mechanism is particularly relevant for nanomaterials like nanoparticles and nanoclays. The electrostatic attraction between the charged nano-adsorbent surface (usually negatively charged) and the heavy metal ions (usually positively charged) can be described by Coulomb's law as shown in Eq. 9.7.

$$\mathrm{Fe} = k\frac{q1q2}{r^2} \tag{9.7}$$

where: Fe = is the electrostatic force of attraction, K = is Coulomb's constant, q1 and q2 are the charges of the nano-adsorbent surface and the heavy metal ions, respectively, R is the distance between the nano-adsorbent surface and the heavy metal ions. The attractive force increases as the charges on the nano-adsorbent surface and heavy metal ions increase, and it decreases as the distance between them increases.

7. Chelation Mechanism: The chelating ligand on the nano-adsorbent binds with the heavy metal ion, forming a stable chelate complex as shown in Eq. 9.8.

$$NP{-}F{-}L + Pb^{2+} \leftrightarrow NP{-}F{-}L{-}Pb \tag{9.8}$$

where NP–F–L–Pb represents the chelate complex between the nano-adsorbent, ligand, and lead ion. The specific mechanism of heavy metal removal may vary depending on factors such as the type of heavy metal ions present, the pH of the solution, the concentration of heavy metals, and the desired treatment efficiency.

9.2.7 Regeneration of Nano-adsorbent

The regeneration of a nano-adsorbent used for the adsorption of heavy metals involves the process of restoring the adsorbent's capacity to remove heavy metal ions from a solution after it has been saturated with metals (Nworie et al. 2023). This regeneration step is essential to extend the adsorbent's lifetime and to make it economically viable for long-term use. The specific regeneration method for a nano-adsorbent will depend on the type of adsorbent material and the nature of the heavy metal ions being adsorbed and include desorption with chemicals, thermal method, electrochemical, biological and ultrasonic regeneration.

9.3 Scaling up and Applying Nano-adsorbents

This is an important area of research and development, especially in the fields of environmental remediation, water purification, and industrial processes. The factors to consider include scalable and cost effective synthetic method, proper characterization method, optimization of adsorption parameters, reusability and regeneration, integration into existing system for real-world application, safety and environmental concerns, cost benefit analysis, regulatory compliance, collaboration and knowledge sharing between researchers, industries and policy makers and monitoring and performance assessment to ensure long term efficacy in real-world applications.

9.3.1 Challenges and Considerations

(i) Toxicity: The environmental and health impacts of nano-adsorbents themselves need to be thoroughly studied. (ii) Cost: production and synthesis of nano-adsorbents can be more expensive than traditional adsorbents, limiting their widespread application. The cost of production of nano-adsorbents is expected to decrease as technology advances. (iii) Stability: Ensuring the stability and long-term performance of nano-adsorbents in various environmental conditions is crucial. (iv) Nanoparticle release: There is a concern about the potential release of nanoparticles into the environment during the application of nano-adsorbents, which may have unintended consequences on ecosystems and human health. (v) Scale-up issues: The scalable production of nano-adsorbents for large-scale industrial applications may be challenging.

(vi) Regeneration efficiency: While some nano-adsorbents can be regenerated and reused, the efficiency of regeneration processes needs improvement.

9.3.2 Future Directions

Continued research is essential to optimize nano-adsorbent materials and manufacturing methods. Integration with other remediation techniques, such as photocatalysis and bioremediation, can enhance overall effectiveness. Regulations and guidelines regarding the use and disposal of nano-adsorbents need to be developed.

9.4 Conclusion

In conclusion, heavy metals in wastewater present a multifaceted environmental issue that requires concerted efforts from industries, governments, researchers, and individuals. Implementing effective treatment technologies, adhering to regulations, and promoting responsible waste management practices are vital steps toward mitigating the impact of heavy metals on water resources and safeguarding human and environmental health.

References

AboMarkeb A, Llimós-Turet J, Ferrer I, Blánquez P, Alonso A, Sánchez A, Moral-Vico J, Font X (2019) The use of magnetic iron oxide based nanoparticles to improve microalgae harvesting in real wastewater. Water Res 159:490–500

Aitken RJ, Chaudhry MQ, Boxall ABA, Hull M (2006) Manufacture and use of nanomaterials: current status in the UK and global trends. Occup Med (Chic Ill) 56(5):300–306

Ajith MP, Aswathi M, Priyadarshini E, Rajamani P (2021) Recent innovations of nanotechnology in water treatment: a comprehensive review. Bioresour Technol 342:126000

Akpomie KG, Conradie J (2020) Advances in application of cotton-based adsorbents for heavy metals trapping, surface modifications and future perspectives. Ecotoxicol Environ Saf 201:110825

Ali F, Khan SB, Kamal T, Alamry KA, Asiri AM (2018) Chitosan-titanium oxide fibers supported zero-valent nanoparticles: highly efficient and easily retrievable catalyst for the removal of organic pollutants. Sci Rep 81 8(1):1–18

Ali I, Asim M, Khan TA (2012) Low cost adsorbents for the removal of organic pollutants from wastewater. J Environ Manag 113:170–183

Ali ME, Hoque ME, Safdar Hossain SK, Biswas MC (2020) Nano-adsorbents for wastewater treatment: next generation biotechnological solution. Int J Environ Sci Technol 17(9):4095–4132

Ali SS, Al-Tohamy R, Sun J (2022) Performance of Meyerozyma caribbica as a novel manganese peroxidase-producing yeast inhabiting wood-feeding termite gut symbionts for azo dye decolorization and detoxification. Sci Total Environ 806:150665

Aziz HA, Adlan MN, Ariffin KS (2008) Heavy metals (Cd, Pb, Zn, Ni, Cu and Cr(III)) removal from water in Malaysia: post treatment by high quality limestone. Bioresour Technol 99(6):1578–1583
Backes C, Hirsch A (2010) Noncovalent functionalization of carbon nanotubes. Chem Nanocarbons 42(8):1–48
Banerjee K (2012) A novel agricultural waste adsorbent, watermelon shell for the removal of copper from aqueous solutions. Iran J Energy Environ. 3(2):143–156
Barakat MA (2011) New trends in removing heavy metals from industrial wastewater. Arab J Chem 4(4):361–377
Basha JS, Anand RB (2014) Performance, emission and combustion characteristics of a diesel engine using carbon nanotubes blended Jatropha methyl ester emulsions. Alexandria Eng J 53:259–273
Beyene HD, Ambaye TG (2019) Application of sustainable nanocomposites for water purification process. In: Inamuddin TS, Kumar Mishra R, Asiri A (eds) Sustainable polymer composites and nanocomposites, pp 387–412
Bhatnagar A, Hogland W, Marques M, Sillanpää M (2013) An overview of the modification methods of activated carbon for its water treatment applications. Chem Eng J 219:499–511
Biswas MC, Jeelani S, Rangari (2017) Influence of biobased silica/carbon hybrid nanoparticles on thermal and mechanical properties of biodegradable polymer films. Compos Commun 4:43–53
Chandramouli M, Basavanna V, Ningaiah S (2023) A comprehensive review on nontronite-chitosan based nanocomposites (a review). Russ J Gen Chem 93:1523–1534
Chen Y, Li F (2010) Kinetic study on removal of copper(II) using goethite and hematite nano-photocatalysts. J Colloid Interface Sci 347(2):277–281
Cheng SY, Show PL, Lau BF, Chang JS, Ling TC (2019) New prospects for modified algae in heavy metal adsorption. Trends Biotechnol 37(11):1255–1268
Chowdhury S, Balasubramanian R (2014) Recent advances in the use of graphene-family nano-adsorbents for removal of toxic pollutants from wastewater. Adv Colloid Interface Sci 204:35–56
Crini G (2005) Recent developments in polysaccharide-based materials used as adsorbents in wastewater treatment. Prog Polym Sci 30(1):38–70
Crini G (2006) Non-conventional low-cost adsorbents for dye removal: a review. Bioresour Technol 97(9):1061–1085
Deliyanni EA, Peleka EN, Matis KA (2009) Modeling the sorption of metal ions from aqueous solution by iron-based adsorbents. J Hazard Mater 172(2–3):550–558
Dihom HR, Al-Shaibani MM, Radin Mohamed RMS, Al-Gheethi AA, Sharma A, Bin Khamidun MH (2022) Photocatalytic degradation of disperse azo dyes in textile wastewater using green zinc oxide nanoparticles synthesized in plant extract: a critical review. J Water Process Eng 47:102705
Durán-Álvarez JC, Avella E, Ramírez-Zamora RM, Zanella R (2016) Photocatalytic degradation of ciprofloxacin using mono-(Au, Ag and Cu) and bi-(Au-Ag and Au-Cu) metallic nanoparticles supported on TiO_2 under UV-C and simulated sunlight. Catal Today 266:175–187
Elella MHA, Goda ES, Abdallah HM, Shalan AE, Gamal H, Yoon KR (2021) Innovative bactericidal adsorbents containing modified xanthan gum/montmorillonite nanocomposites for wastewater treatment. Int J Biol Macromol 167:1113–1125
Elella MHA, Aamer N, Mohamed Y, El Nazer HA, Mohamed RR (2022) Innovation of high-performance adsorbent based on modified gelatin for wastewater treatment. Polym Bull 2022:1–17
El-Geundi MS, Ismail HM, Attyia KME (2016) Activated clay as an adsorbent for cationic dyestuffs. 12(2):109–117
El-sayed MEA (2020) Nano-adsorbents for water and wastewater remediation. Sci Total Environ 739:139903
Enamul Hoque M, Maryanne Peiris A, Atiqure Rahman SM, Abdul Wahab M (2017) New generation antibacterial nanofibrous membrane for potential water filtration. Curr Anal Chem 13:567

Fajardo C, Martinez-Rodriguez G, Blasco J, Mancera JM, Thomas B, De Donato M (2022) Nanotechnology in aquaculture: applications, perspectives and regulatory challenges. Aquac Fish 7(2):185–200

Fu S, Sun Z, Huang P, Li Y, Hu N (2019) Some basic aspects of polymer nanocomposites: a critical review Nano Mater. Sci 1:2–30

Goda ES, Gab-Allah MA, Singu BS, Yoon KR (2019) Halloysite nanotubes based electrochemical sensors: a review. Microchem J 147:1083–1096

Gul A, Khaligh NG, Julkapli NM (2021) Surface modification of carbon-based nano-adsorbents for the advanced wastewater treatment. J Mol Struct 1235:130148

Gupta VK, Saleh TA (2013) Sorption of pollutants by porous carbon, carbon nanotubes and fullerene-an overview. Environ Sci Pollut Res 20(5):2828–2843

Gupta VK, Ali I, Saleh TA, Nayak A, Agarwal S (2012) Chemical treatment technologies for waste-water recycling—an overview. RSC Adv 2(16):6380–6388

Gupta VK, Nayak A, Agarwal S, Tyagi I (2014) Potential of activated carbon from waste rubber tire for the adsorption of phenolics: effect of pre-treatment conditions. J Colloid Interface Sci 417:420–430

Gusain R, Kumar N, Ray SS (2020) Recent advances in carbon nanomaterial-based adsorbents for water purification. Coord Chem Rev 405:213111

Hnamte M, Pulikkal AK (2022) Clay-polymer nanocomposites for water and wastewater treatment: a comprehensive review. Chemosphere 307:135869

Hurt RH, Monthioux M, Kane A (2006) Toxicology of carbon nanomaterials: status, trends, and perspectives on the special issue. Carbon N y. 44(6):1028–1033

Jain K, Patel AS, Pardhi VP, Flora SJS (2021) Nanotechnology in wastewater management: a new paradigm towards wastewater treatment. Molecules 26(6):580

Jana DK, Roy S, Bhattacharjee S, Dostal P, Roy S (2021) Saw dust-derived activated carbon in different impregnation ratios and its application in de-fluoridation of waste water using IT2FLC and RSM. Biomass Convers Biorefin 2021:1–21

Jaspal D, Malviya A (2020) Composites for wastewater purification: a review. Chemosphere 246:125788

Kumari P, Alam M, Siddiqi WA (2019) Usage of nanoparticles as adsorbents for waste water treatment: an emerging trend. Sustain Mater Technol 22:e00128

Li G, Zhao Z, Liu J, Jiang G (2011) Effective heavy metal removal from aqueous systems by thiol functionalized magnetic mesoporous silica. J Hazard Mater 192(1):277–283

Liu Y, Liang P, Guo L (2005) Nanometer titanium dioxide immobilized on silica gel as sorbent for preconcentration of metal ions prior to their determination by inductively coupled plasma atomic emission spectrometry. Talanta 68(1):25–30

Mercante LA, Andre RS, Schneider R, Mattoso LHC, Correa DS (2020) Free-standing SiO_2/TiO_2–MoS_2 composite nanofibrous membranes as nano-adsorbents for efficient Pb(II) removal. New J Chem 44:13030–13035

Mohammad SA, Kumar D, Banerjee S (2023) One-Dimensional polymeric nanocomposites: synthesis to emerging applications. CRC Press, p 21

Mudhoo A, Sillanpää M (2021) Magnetic nano-adsorbents for micropollutant removal in real water treatment: a review. Environ Chem Lett 19(6):4393–4413

Nworie FS, Nwabue FI, Oti W, Mbam E, Nwali BU (2019) Removal of methylene blue from aqueous solution using activated rice husk biochar: adsorption isotherms, kinetics and error analysis. J Chil Chem Soc 64(1):4365–4376

Nworie FS, Ike-Amadi AC, Agha-Akpu J, Nworu J (2022) Equilibrium, kinetic and mass transfer optimization studies of cationic dye adsorption onto chemically modified bio-sorbent derived from lignocellulosic waste. Acta Chemica Malaysia 6(2):70–77

Nworie FS, Nkoli M, Ike-Amadi AC, Ebunoha J (2023) Functionalized biochars for enhanced removal of heavy metals from aqueous solutions: mechanism and future industrial prospects. J Human Earth Future 3(3):377–395

Pandey SA (2017) Comprehensive review on recent developments in bentonite-based materials used as adsorbents for wastewater treatment. J Mol Liq 241:1091–1113

Rajasulochana P, Preethy V (2016) Comparison on efficiency of various techniques in treatment of waste and sewage water—a comprehensive review. Resour Technol 2(4):175–184

Ren X, Chen C, Nagatsu M, Wang X (2011) Carbon nanotubes as adsorbents in environmental pollution management: a review. Chem Eng J 170(2–3):395–410

Sengwa RJ, Dhatarwal P (2022) Toward multifunctionality of PEO/PMMA/MMT hybrid polymer nanocomposites: promising morphological, nanostructural, thermal, broadband dielectric, and optical properties. J Phys Chem Solids 166:110708

Sharma R, Kumar D (2018) Nano-adsorbents: an approach towards wastewater treatment. In: Mishra AK, Hussain CM (eds) Nanotechnology for sustainable water resources, 2nd edn. Scrivener Publishing LLC, Beverly, MA, pp 371–405

Tawade PV, Bhattacharjee S, Wasewar KL (2023) Nano-adsorbents for treatment of wastewater. In: Shah MP (eds) advanced application of nanotechnology to industrial wastewater. Springer, Singapore

Tlili I, Alkanhal TA (2019) Nanotechnology for water purification: electrospun nanofibrous membrane in water and wastewater treatment. J Water Reuse Desalin 9(3):232–248

Wilson R, George G, Jose AJ (2018) Polymer membranes reinforced with carbon-based nanomaterials for water purification. New Polym Nanocompos Environ Remediat 15:457–468

Zare EN, Motahari A, Sillanpää M (2018) Nano-adsorbents based on conducting polymer nanocomposites with main focus on polyaniline and its derivatives for removal of heavy metal ions/dyes: a review. Environ Res 162:173–195

Chapter 10
Potential Human Health Risks from Heavy Metals in Crayfish Shells Used as Food and Organic Fertilizer

Murat Topal, E. Işıl Arslan Topal, and Erdal Öbek

Abstract Potential health risks of using waste crayfish shells as food and organic fertilizer were evaluated to protect health and environment. The heavy metals investigated can remain unchanged in receiving environments for a long time and can have negative effects on health. Hazard index (HI) values for man and woman were >1 for Cr, Co, V, La, and Mn in crayfish shells. HI values for child were greater than 1 for Ni, Cr, Mn, Co, V, La, and Pb in crayfish shells. There was a non-carcinogenic human health risk because of Ni, Cr, Mn, Co, V, La, and Pb for child. Total maximum carcinogenic value was 8.84E−03 for Ni whereas minimum value was 1.32E−05 for Pb. The potential carcinogenic risks in terms of dietary (Ni, Cr, and Cd) were higher than permissible values (1×10^{-4} to 1×10^{-6}). A non-carcinogenic and carcinogenic potential health risk was determined in case of dietary intake of crayfish shells.

Keywords Cancer · Contaminant · Crawfish · Health risk · Shell

10.1 Introduction

Generated waste is globally rising. Therefore, it is becoming a significant environmental and health problem around the World (Peceño et al. 2021). Aquaculture becomes fundamental to supply global demand of protein (Fraga-Corral et al. 2022). Aquaculture industry produces some 6,000,000–8,000,000 metric tons of wastes annually in the world (Yan and Chen 2015; Peceño et al. 2021). The various wastes

M. Topal (✉)
Department of Chemistry and Chemical Processing Technologies, Tunceli Vocation School, Munzur University, Tunceli, Turkey
e-mail: murattopal@munzur.edu.tr

E. I. Arslan Topal
Department of Environmental Engineering, Faculty of Engineering, Fırat University, Elazığ, Turkey

E. Öbek
Department of Biongineering, Faculty of Engineering, Fırat University, Elazığ, Turkey

N. Kumar (ed.), *Heavy Metal Toxicity*, Environmental Science and Engineering,
https://doi.org/10.1007/978-3-031-56642-4_10

originating from aquaculture are used in diet of human. The crayfish shell can be processed into nutritious foods (Bueno-Solano et al. 2009; Chen et al. 2020). Bueno-Solano et al. (2009) successfully converted by products of shrimp into micronutrient by products rich in amino acid for recommendation in supplementations of diet of humans. There are various recipes which use crawfish shells. The crawfish shells are used to make broth. The shells and claws can be used to make a rich and flavorful crawfish stock to make various meals (URL 1; URL 2).

The various wastes originating from aquaculture are also used as organic fertilisers. The use of aquatic waste in organic fertiliser manufacture is one of the most common ways to utilize the aquatic waste (URL 3; Arvanitoyannis and Kassaveti 2008). Crustacean processing wastes, shrimp and crab wastes are used as fertilizers resulting in the consuming the wastes of crab and shrimp generated at processing ports (Costa 1977). Early usage of crustacean waste as fertilizer was discussed by Stevenson (1902). Horseshoe crabs were caught in Delaware Bay (America) in 1880 for solely using as fertilizer. Over 8 million pounds were caught. Prices ranged from \$4–8/ton fresh or \$15–25/ton dried and ground. It has been reported that horseshoe crabs were a good fertiliser for grains and fruits. Lobster cannery wastes were used to make fertilizer in New Brunswick, Maine, and Nova Scotia. The wastes of shrimp processing were used as fertilizers in large quantities in Louisiana and California. The shells that especially valued in strawberry and vegetable cultures, sold for about \$5 per ton. According to Turrentine (1913), over 250 tons of crab shells are used each year at a fertilizer mixing plant in Chesapeake Bay (Costa 1977). In Alaska, shell material from dungeness crab processing wastes was mixed with acid soils in the greenhouse (Laughlin et al. 1973; Costa 1977). Organic lobster shell meal used as a soil amendment is defined as an excellent dry organic source of nitrogen, phosphorous calcium, and magnesium (URL 4). Costa (1977) studied the determination of impacts of applications of shrimp and crab processing waste on soils. Wastes were affective as nitrogen and phosphorus sources for crops, and shrimp processing wastes were equivalent to conventional inorganic materials for the growth of orchardgrass. Ali et al. (1998) reported that seed yields from crab shells application as soil amendment were similar to the ones resulted with NPK fertilizer. Adiloğlu and Adiloğlu (2008) studied the determination of effect of powder of shell of crab on some element content of maize plant. They suggested crab shell powder as a new organic fertilizer and amendment for acid soils to increase the fertility and plant nutrition. In Japan, crab shells ground and lobster shells ground are used as organic fertilizer (Nawakawa 1990; Ali et al. 1998).

Heavy metals become more prevalent because societies become more industrialized (Nasab et al. 2022). The anthropogenic activities such as industry, waste management, and agriculture are key sources of heavy metals (Thanigaivel et al. 2024). These metals can be defined as high density metallic elements and exceedingly toxic even in very low magnitude (Naveed et al. 2023). Heavy metals have potential to impede growth, metabolism, and development in biological systems. Heavy metals are not biodegradable and remain in environment. Therefore, heavy metals are harmful to ecosystems and human (Thanigaivel et al. 2024). Due to the potential health hazard associated with some heavy metals, notice is being provided to their ways of entrance into the food chain of human directly or indirectly (Abdel-Haleem et al. 2001; Mahmoud et al. 2020). Heavy metals pose risk of toxicity even at low concentration. Kidney and brain are some of the organs having negative impacts (Thanigaivel et al. 2024).

Heavy metals researched in this chapter are lead (Pb), copper (Cu), cadmium (Cd), chromium (Cr), nickel (Ni), manganese (Mn), zinc (Zn), tungsten (W), vanadium (V), antimony (Sb), lanthanum (La), and cobalt (Co). General information about these heavy metals are as follows. Even low exposure to lead in air, water, and food results in causing toxicity (Wani et al. 2021). Lead is harmful to the body (Nasab et al. 2022). Besides cancer, lead causes diseases (e.g. neurological and cardiovascular) (Shrivastava et al. 2002; Lamas et al. 2016; Nasab et al. 2022). Cadmium is nonessential metal that is very toxic to livings (Kim et al. 2019). It is a carcinogen and it causes damages to various organs (Khan et al. 2018; Jafarzadeh et al. 2020; Eslami et al. 2022). Elevated copper level cause dizziness, nausea, diarrhea, and respiratory problems (Gujre et al. 2021). High chromium can cause liver problems (Eslami et al. 2022). Manganese is recognized as a toxicant with excessive manganese targeting developing central nervous system and contributing to neurological disorders although manganese is essential element (Yu et al. 2014; Horning et al. 2015; Bjørklund et al. 2017; ATSDR 2022; Tung et al. 2022). The health effects from chronic low dose exposure to nickel range from sensitization or exacerbation of existing dermatitis among sensitive human to impaired fetal growth (IARC 2006; ATSDR 2005; McDermott et al. 2015; St-Jean et al. 2018). Despite essential roles of zinc in humans, excess intake by human induce poor appetite, diarrhea, vomiting, headaches, and nausea. High zinc concentrations result in impaired immune response (Nriagu 2019; Natasha et al. 2022). Vanadium is potentially toxic to human beings (Liu et al. 2022). Vanadium is mainly ingested through foods (Bocio et al. 2015) and absorbed from gastrointestinal tract and transported in blood or transfer into tissues and organ (Zwolak 2014; Liu et al. 2022). High concentrations of antimony are toxic to ecosystem and human health via accumulations in food chain. Antimony is carcinogenic and poisonous (Bolan et al. 2022). The greater urinary antimony and tungsten concentrations were linked to high blood pressure by researchers (Shiue 2014; Nasab et al. 2022). High cobalt concentrations can be harmful to humans, resulting in liver damage, heart failure, and asthma (Yang et al. 2017; Abbas et al. 2021; Wu et al. 2022). Lanthanum is a non-essential element (Herrmann et al. 2016). Abundant La is in food, vegetable and water because of application of La in many areas. This situation increases exposures and intakes of La (Gwenzi et al. 2018; Han et al. 2022). Lanthanum entering the body through various ways accumulates in organs and tissues (Wang et al. 2017; Han et al. 2022).

Heavy metals accumulate in animals living in freshwater. Crayfish takes up the metals via gills from water and food chain from the sediment (Devesa et al. 2002; Suárez-Serrano et al. 2010; Nagarajan et al. 2019; Tan et al. 2021). The crayfish are macrobenthic organisms with constant contact with bottom sediment. Thus, the crayfish easily accumulate metals in tissues from surrounding environments (Kouba et al. 2010; Śmietana et al. 2020). Upon exposure of crayfish to metals, the highest accumulation occurs in the exoskeleton (Anderson et al. 1997). There are some studies reporting that exoskeleton of crayfish tends to exhibit the highest lead concentrations among tissues (Stinson and Eaton 1983; Knowlton et al. 1983; Rincon-Leon et al. 1988; Madigosky et al. 1991). Because of omnivory of them as well as necrophagia and life spans up to 6 years, the crayfish bioaccumulate heavy metals in shell (Peng et al. 2016).

Crayfish shells accounting for about percentage of 50–70 wt of crayfish (Zhu et al. 2017; Lv et al. 2020) are wasted from the traditional process of crayfish used as one of the common food sources for people (Chen et al. 2020). The usage of the waste shells in useful ways will solve the waste disposal problem also both supporting the sustainability and protection of the natural sources by preventing environmental pollution. However, the potential health risks of wastes that are used beneficially in human life should be investigated. In the literature there is not any research on health risk of heavy metals in crayfish shells used as food and organic fertilizer. The aim of our chapter is to fill this gap in scientific knowledge.

10.2 Material and Method

Natural crayfish shells used as material was collected from natural environment (Elazığ, Turkey) (Fig. 10.1). Crayfish shells were collected along the shore of the lake in the province Keban, Elazığ. The crayfish shell samples were brought to laboratory. The crayfish shells were dried and powdered. Samples were cold leached with HNO_3. As each sample leached in heating block of hot water bath, HCl, HNO_3, and DI H_2O were added to an aqua regia solution of equal parts. Samples were made up to volume with hydrochloric acid before filtration. Levels of heavy metals are determined by ICP/MS. QA/AC are in Table 10.1.

Non-carcinogenic risks from individual metals were expressed as hazard quotient (HQ) (Eq. 10.1). Sum of more than one HQ for multiple exposures can be expressed

Fig. 10.1 Crayfish

Table 10.1 QA/AC details for shell (STD CDV-1 is certified reference materials)

			Shell											
		Analyte	Cd	Cr	Cu	Pb	Mn	Ni	Zn	Co	Sb	V	La	W
		Unit	mg/kg	mg/kg	mg/kg	mg/kg	mg/kg	mg/kg	mg/kg	mg/kg	mg/kg	mg/kg	mg/kg	mg/kg
QC STD CDV-1	Std		0.03	13.6	8.32	0.75	386	5.9	22.8	1.86	0.03	5.0	2.04	<0.1
	Exp		0.04	12.1	8.61	1.00	385	6.4	23.3	2.0	0.03	8.7	2.50	–

as in Eq. 10.2 (Lian et al. 2019):

$$HQ = \frac{CDI}{RfD} \tag{10.1}$$

$$HQ = \sum_{k=1}^{n} \frac{CDIk}{RfDk} \tag{10.2}$$

where; RfD is reference dose (mg/kg day). CDI: daily exposure dose (mg/kg/day). CDI can be calculated according to USEPA (2001, 2011).

$$CDI_{dietary} = C_{shell} \times \frac{I_{intake} \times EF \times ED}{BW \times AT} \times 10^{-3} \tag{10.3}$$

$$CDI_{ing} = C_{shell} \times \frac{IngR \times EF \times ED}{BW \times AT} \times 10^{-6} \tag{10.4}$$

$$CDI_{Inh} = C_{shell} \times \frac{InhR \times EF \times ED}{BW \times AT \times PEF} \tag{10.5}$$

$$CDI_{dermal} = C_{shell} \times \frac{SA \times SL \times ABS \times EF \times ED}{BW \times AT} \times 10^{-6} \tag{10.6}$$

where; $CDI_{dietary}$, CDI_{ing}, CDI_{inh}, and CDI_{dermal}: dietary, ingestion, inhalation, and dermal (mg/kg day). Cshell is amount of metals in crayfish shell (mg/kg), IngR and InhR are ingestion and inhalation rates, I_{intake} is intake of crayfish shell (g/day), EF is exposure frequency and ED is exposure duration, AT is averaged time, SA is surface area and SL is adherence factor, PEF is particle emission factor, ABS is adsorption factor, BW is bodyweight. Hazard index (HI) and cancer risk (CR) were found as following (USEPA 1989):

$$HI = \sum HQ \tag{10.7}$$

$$CR = \sum_{k=1}^{n} LCDI_k \times SF_k \tag{10.8}$$

where; SF is slope factor (mg/kg day)$^{-1}$, SF in dermal risk is from extrapolated oral toxicity value (USEPA 2002), LCDI is chronic daily intake of carcinogens based on individual lifetime (mg/kg day).

Data were evaluated by IBM SPSS Statistics 22 (USA) (n = 48). A Pearson correlation test was performed for determination of whether there an association between heavy metals in crayfish shells.

10.3 Results and Discussion

The estimated daily exposure doses (EDs) of heavy metals are in Table 10.2. Total EDs for man, woman and child were Mn > Cr > Ni > Zn > V > Cu > La > Co > Pb > W > Sb > Cd. It was observed that highest heavy metals exposure for man, woman and child was via dietary and minimum heavy metals exposure for man, woman and child was by dermal. The dietary was followed by ingestion and inhalation. The highest exposure among child, woman, and man to heavy metals was child > woman > man. The exposure to pollutants is a major health concern especially among children. Heavy metals are widely available in environment. Therefore, exposure to them is inevitable in the susceptible child population (Tung et al. 2022). Adolescent and childhood exposures to lead, chromium, and zinc might be risk factors for development of obesity markers and cardiovascular diseases (Nasab et al. 2022).

HQ values of heavy metals for man, woman and child are in Table 10.3. HQs (dieatary) for man, woman, and child followed: Zn < Cd < Cu < W < Sb < Pb < Ni < V < Mn < Cr < Co < La. HQ values (dietary) for man and woman were above 10.0 for La. HQs (dietary) for child were >10.0 for La and Co. HQs (dietary) for V, Co, Mn and Cr were above 1 for woman and man. HQ values (dietary) for child were >10.0 for Cr. For child, HQ (dietary) for vanadium, lead, nickel, and manganese were >1. Possible effects from chronic low dose exposure to nickel and manganese may be neurodevelopmental effects in children (McDermott et al. 2015; Bjørklund et al. 2017; St-Jean et al. 2018). Highest HQ (dietary) was for child. HQ (dietary) risks were child > woman > man. As seen, lanthanum, cobalt, and vanadium are the common elements among man, woman, and child that have HQdie value >1. Lanthanum can cause cancer with human, and be threat to liver when accumulates (Vijayaraghavan et al. 2009). The negative effects of vanadium on immune system in children was reported (Liu et al. 2022). Children are particularly vulnerable to environmental exposure, as evidenced by higher urinary Co concentrations than adults (Cheyns et al. 2014; Garcia-Vazquez et al. 2021). Crayfish shells used as organic fertilizer can be accidentally or intentionally (some people willingly ingest soil and etc. materials) ingested by humans. HQingestion values for man, woman, and child were Zn < Cd < Cu < W < Sb < Pb < Ni < V < Mn < Cr < Co < La.

Table 10.2 (a) Estimated daily exposure doses of heavy metals (Cd, Cu, Ni, Zn, Cr, Mn, and Pb) (mg/kg day). (b) Estimated daily exposure doses of heavy metals (Co, Sb, V, La, and W) (mg/kg.day)

(a)

	Exposure	Cd	Cu	Ni	Zn	Cr	Mn	Pb
Man	Dietary	7.15 E−05	4.05 E−03	1.44 E−02	1.04 E−02	2.39 E−02	1.43 E−01	2.32E−03
	Dermal	5.09E−13	2.88E−11	1.03E−10	7.44E−11	1.70E−10	1.02E−09	1.65E−11
	Inhalation	7.44E−12	4.21E−10	1.50E−09	1.09E−09	2.49E−09	1.49E−08	2.41E−10
	Ingestion	3.16E−08	1.79E−06	6.39E−06	4.62E−06	1.06E−05	6.33E−05	1.02E−06
	Total	**7.15E−05**	**4.05E−03**	**1.44E−02**	**1.04E−02**	**2.39E−02**	**1.43E−01**	**2.32E−03**
Woman	Dietary	8.10E−05	4.58E−03	1.64E−02	1.18E−02	2.70E−02	1.62E−01	2.62E−03
	Dermal	5.07E−13	2.87E−11	1.02E−10	7.41E−11	1.69E−10	1.01E−09	1.64E−11
	Inhalation	8.43E−12	4.77E−10	1.70E−09	1.23E−09	2.82E−09	1.69E−08	2.73E−10
	Ingestion	3.58E−08	2.03E−06	7.24E−06	5.23E−06	1.20E−05	7.17E−05	1.16E−06
	Total	**8.10E−05**	**4.59E−03**	**1.64E−02**	**1.18E−02**	**2.71E−02**	**1.62E−01**	**2.63E−03**
Child	Dietary	1.81E−04	1.02E−02	3.65E−02	2.64E−02	6.03E−02	3.61E−01	5.85E−03
	Dermal	3.98E−13	2.25E−11	8.05E−11	5.82E−11	1.33E−10	7.97E−10	1.29E−11
	Inhalation	2.37E−11	1.34E−09	4.79E−09	3.47E−09	7.93E−09	4.75E−08	7.69E−10
	Ingestion	3.20E−07	1.81E−05	6.46E−05	4.67E−05	1.07E−04	6.39E−04	1.04E−05
	Total	**1.81E−04**	**1.02E−02**	**3.65E−02**	**2.64E−02**	**6.04E−02**	**3.62E−01**	**5.86E−03**

(b)

	Exposure	Co	Sb	V	La	W
Man	Dietary	2.44E−03	1.29E−04	8.58E−03	2.72E−03	1.43E−04
	Dermal	1.74E−11	9.17E−13	6.11E−11	1.94E−11	1.02E−12
	Inhalation	2.54E−10	1.34E−11	8.93E−10	2.83E−10	1.49E−11

(continued)

Table 10.2 (continued)

(b)

	Exposure	Co	Sb	V	La	W
	Ingestion	1.08E−06	5.69E−08	3.80E−06	1.20E−06	6.33E−08
	Total	**2.45E−03**	**1.29E−04**	**8.58E−03**	**2.72E−03**	**1.43E−04**
Woman	Dietary	2.77E−03	1.46E−04	9.72E−03	3.08E−03	1.62E−04
	Dermal	1.73E−11	9.13E−13	6.09E−11	1.93E−11	1.01E−12
	Inhalation	2.88E−10	1.52E−11	1.01E−09	3.20E−10	1.69E−11
	Ingestion	1.23E−06	6.45E−08	4.30E−06	1.36E−06	7.17E−08
	Total	**2.77E−03**	**1.46E−04**	**9.72E−03**	**3.08E−03**	**1.62E−04**
Child	Dietary	6.18E−03	3.25E−04	2.17E−02	6.86E−03	3.61E−04
	Dermal	1.36E−11	7.17E−13	4.78E−11	1.51E−11	7.97E−13
	Inhalation	8.12E−10	4.27E−11	2.85E−09	9.02E−10	4.75E−11
	Ingestion	1.09E−05	5.75E−07	3.84E−05	1.21E−05	6.39E−07
	Total	**6.19E−03**	**3.26E−04**	**2.17E−02**	**6.87E−03**	**3.62E−04**

When HQs of heavy metals for ingestion were evaluated, it was child > woman > man. Crayfish shells used as organic fertilizer can be inhalated by humans. HQs for man, woman, and child depend on inhalation were Zn < Cd < Cu < W < Sb < Pb < Ni < V < Cr < La < Co < Mn. When HQs of heavy metals were examined for inhalation, maximum HQs were for child. For inhalation, HQs were child > woman > man. People can come into contact with crayfish shells used as organic fertilizer through their skin. HQdermal exposure for man, woman, and child were Zn < Cu < Co < Ni < Pb < Cd < Sb < Mn < V < Cr. The HQs were man > woman > child for dermal.

According to the obtained data, maximum HI risk value in terms of man was 54.3 for La while minimum HI risk value was 0.03 for Zn. For man, HIs followed: Zn < Cd < Cu < W < Sb < Pb < Ni < V < Mn < Cr < Co < La. Maximum HI risk value in terms of woman was 61.6 for La whereas minimum HI risk was 0.04 for Zn. For woman, HIs followed: Zn < Cd < Cu < W < Sb < Pb < Ni < V < Mn < Cr < Co < La. Maximum HI risk value in child was 137.49 for La while minimum HI risk value was 0.09 for Zn. For child, HIs followed: Zn < Cd < Cu < W < Sb < Pb < Ni < V < Mn < Cr < Co < La. HI risk of heavy metals were child > woman > man. Maximum total hazard index (THI) value was 253.4 for La, and minimum THI was 0.16 for Zn. THIs followed: La > Co > Cr > Mn > V > Ni > Pb > Sb > W > Cu > Cd > Zn. THIs were above 1 for Cr, Mn, Pb, Ni, Sb, V, Co, La in crayfish shells. HI values higher than 1 indicate a non-carcinogenic risk. As a result, HI values for man and woman were greater than 1 for Cr, Co, V, La, and Mn in crayfish shells. HI values for child were greater than 1 for Ni, Cr, Mn, Co, V, La, and Pb in crayfish shells. There is a non-carcinogenic human health risk because of Ni, Cr, Mn, Co, V, La, and Pb for child.

The highest CR was calculated as 8.83 E−03 for dietary, lowest CR was 6.94 E−14 for dermal. CR values were dietary > ingestion > inhalation > dermal. The permissible level range for CR has been reported as 10^{-4} to 10^{-6} (Ferreira-Baptista and De Miguel 2005). Potential carcinogenic risk values in terms of dietary (Ni, Cr, and Cd) were higher than permissible values (1×10^{-4} to 1×10^{-6}).

The relationship between heavy metals in crayfish shells were shown in Table 10.4. Correlations are not significant ($p < 0.05$) (Zn–Cd, Zn–Ni, Cr–Cu, Cr–Zn, Mn–Cr, Pb–Cr, Co–Zn, Co–Cr, Sb–Cr, V–Cd, V–Ni, V–Co, and La–Cr). Correlations between Pb–Mn, Sb–Mn, and Sb–Pb were important ($r = 1.000$, $p = 0.01$). The correlation between La–Cu was important ($r = 1.000$, $p = 0.05$). r values were close to 1 in relationship between heavy metals in crayfish shells indicating strong relationship between them. Significance of statistically significant correlations may indicate that these metals have same input sources.

Table 10.3 (a) Non-carcinogenic risks (HQ) of heavy metals (Cd, Cu, Ni, Zn, Cr, Mn, and Pb). (b) Non-carcinogenic risks (HQ) of heavy metals (Co, Sb, V, La, and W) according to different exposure types

(a)

Elements	HQdietary			HQinhalation		
	Man	Woman	Child	Man	Woman	Child
Cd	7.15E−02	8.10E−02	1.81E−01	7.44E−09	8.43E−09	2.37E−08
Cu	1.01E−01	1.15E−01	2.56E−01	1.05E−08	1.19E−08	3.34E−08
Ni	7.22E−01	8.18E−01	1.82E+00	7.52E−08	8.52E−08	2.40E−07
Zn	3.48E−02	3.94E−02	8.79E−02	3.62E−09	4.10E−09	1.16E−08
Cr	7.96E+00	9.02E+00	2.01E+01	8.28E−07	9.39E−07	2.64E−06
Mn	3.11E+00	3.52E+00	7.85E+00	1.04E−03	1.18E−03	3.32E−03
Pb	6.62E−01	7.50E−01	1.67E+00	6.85E−08	7.76E−08	2.18E−07
Elements	HQIngestion			HQdermal		
	Man	Woman	Child	Man	Woman	Child
Cd	3.16E−05	3.58E−05	3.20E−04	5.09E−08	5.07E−08	3.98E−08
Cu	4.48E−05	5.07E−05	4.52E−04	2.40E−09	2.39E−09	1.88E−09
Ni	3.19E−04	3.62E−04	3.23E−03	1.91E−08	1.90E−08	1.49E−08
Zn	1.54E−05	1.74E−05	1.56E−04	1.24E−09	1.23E−09	9.69E−10
Cr	3.52E−03	3.99E−03	3.56E−02	2.84E−06	2.82E−06	2.22E−06
Mn	1.38E−03	1.56E−03	1.39E−02	5.54E−07	5.51E−07	4.33E−07
Pb	2.93E−04	3.32E−04	2.96E−03	3.14E−08	3.13E−08	2.46E−08

(b)

Heavy metals	HQdietary			HQinhalation		
	Man	Woman	Child	Man	Woman	Child
Co	8.15E+00	9.23E+00	2.06E+01	4.46E−05	5.05E−05	1.42E−04
Sb	3.22E−01	3.64E−01	8.13E−01	3.35E−08	3.79E−08	1.07E−07
V	1.70E+00	1.93E+00	4.30E+00	1.28E−07	1.45E−07	4.07E−07
La	5.43E+01	6.15E+01	1.37E+02	5.66E−06	6.41E−06	1.80E−05
W	1.79E−01	2.02E−01	4.51E−01	1.86E−08	2.11E−08	5.93E−08
Heavy metals	HQIngestion			HQdermal		
	Man	Woman	Child	Man	Woman	Child
Co	3.61E−03	4.09E−03	3.64E−02	3.11E−09	3.10E−09	2.43E−09
Sb	1.42E−04	1.61E−04	1.44E−03	1.15E−07	1.14E−07	8.96E−08
V	7.53E−04	8.53E−04	7.61E−03	8.73E−07	8.70E−07	6.83E−07
La	2.40E−02	2.72E−02	2.43E−01	–	–	–
W	7.91E−05	8.96E−05	7.99E−04	–	–	–

Table 10.4 Correlation coefficients among heavy metals in crayfish shell

		Cd	Cu	Ni	Zn	Cr	Mn	Pb	Co	Sb	V	La	W
Cd	Pearson correlation	1											
	Sig. (2-tailed)												
Cu	Pearson correlation	0. 474	1										
	Sig. (2-tailed)	0.685											
Ni	Pearson correlation	0.881	0.835	1									
	Sig. (2-tailed)	0.314	0.371										
Zn	Pearson correlation	−0.655	0.355	−0.219	1								
	Sig. (2-tailed)	0.546	0.769	0.860									
Cr	Pearson correlation	0.481	−0.543	0.009	−0.978	1							
	Sig. (2-tailed)	0.680	0.634	0.994	0.135								
Mn	Pearson correlation	0.000	0.880	0.473	0.756	−0.877	1						
	Sig. (2-tailed)	1.000	0.315	0.686	0.454	0.320							
Pb	Pearson correlation	0.000	0.880	0.473	0.756	−0.877	1.000**	1					
	Sig. (2-tailed)	1.000	0.315	0.686	0.454	0.320	0.000						
Co	Pearson correlation	0.866	0.851	1.000*	−0.189	−0.021	0.500	0.500	1				
	Sig. (2-tailed)	0.333	0.352	0.019	0.879	0.986	0.667	0.667					
Sb	Pearson correlation	0.000	0.880	0.473	0.756	−0.877	1.000**	1.000**	0.500	1			
	Sig. (2-tailed)	1.000	0.315	0.686	0.454	0.320	0.000	0.000	0.667				
V	Pearson correlation	−0.866	0.029	−0.526	0.945	−0.855	0.500	0.500	−0.500	0.500	1		
	Sig. (2-tailed)	0.333	0.981	0.647	0.212	0.347	0.667	0.667	0.667	0.667			
La	Pearson correlation	0.500	1.000*	0.850	0.327	−0.518	0.866	0.866	0.866	0.866	0.000	1	
	Sig. (2-tailed)	0.667	0.019	0.353	0.788	0.653	0.333	0.333	0.333	0.333	1.000		

(continued)

Table 10.4 (continued)

		Cd	Cu	Ni	Zn	Cr	Mn	Pb	Co	Sb	V	La	W
W	Pearson correlation	0.189	0.954	0.631	0.619	−0.770	0.982	0.982	0.655	0.982	0.327	0.945	1
	Sig. (2-tailed)	0.879	0.194	0.565	0.575	0.441	0.121	0.121	0.546	0.121	0.788	0.212	

* Correlation is significant at the 0.05 level (2-tailed)
** Correlation is significant at the 0.01 level (2-tailed)

10.4 Conclusions

In production of aquatic products for human consumption, shrimp, crab, and crayfish wastes total 6–8 million tons annually, causing unpleasant odors and other environmental problems (Xu et al. 2013; Bajaj et al. 2015; Mao et al. 2017; Dun et al. 2019). These wastes can be used for useful purposes as food and organic fertilizer. However, it should not be forgotten that these wastes may contain the human health risk that they may create due to the accumulation of heavy metals. Therefore, it is necessary to investigate potential health risks that may arise from these wastes that are used beneficially by people. As a result, there was potential non-carcinogenic health risk because of heavy metals (manganese, chromium, lead, nickel, cobalt, lanthanum, and vanadium). USEPA (2001) suggests permissible level of CR = 10^{-4} to 10^{-6}. Potential CR values in terms of dietary (cadmium, nickel, chromium) were greater than these levels.

References

Abbas MN, Al-Tameemi IM, Hasan MB, Al-Madhhachi A-ST (2021) Chemical removal of cobalt and lithium in contaminated soils using promoted white eggshells with different catalysts, South Africa. J Chem Eng 35:23–32

Abdel-Haleem A, Sroor A, El-Bahi S, Zohny E (2001) Heavy metals and rare earth elements in phosphate fertilizer components using instrumental neutron activation analysis. Appl Radiat Isotopes 55(2001):569–573

Adiloğlu A, Adiloğlu S (2008) Amendment of acid soils using crab shell powder. Asian J Chem 20(3):2156–2162

Ali M, Horiuchi T, Miyagawa S (1998) Effects of soil amendment with crab shell on the growth and nodulation of soybean plants (Glycine max Merr.). Plant Prod Sci 1(2):119–125

Anderson MB, Reddy P, Preslan JE, Fingerman M, Bollinger J, Jolibois L, Maheswarudu G, George WJ (1997) Metal accumulation in crayfish, Procambarus clarkii, exposed to a petroleum-contaminated Bayou in Louisiana. Ecotoxicol Environ Saf 37(3):267–272

Arvanitoyannis IS, Kassaveti A (2008) Fish industry waste: treatments, environmental impacts, current and potential uses. Int J Food Sci Technol 43:726–745

ATSDR (2005) Agency for toxic substances and disease registry, toxicological profile for nickel. U.S. Department of Health and Human Services, Atlanta, GA. https://www.atsdr.cdc.gov/toxprofiles/tp15.pdf

ATSDR (2022) https://wwwn.cdc.gov/TSP/ToxProfiles/ToxProfiles.aspx?id=102&tid=23

Bajaj M, Freiberg A, Winter J, Xu Y, Gallert C (2015) Pilot-scale chitin extraction from shrimp shell waste by deproteination and decalcification with bacterial enrichment cultures. Appl Microbiol Biotechnol 99:9835–9846

Bjørklund G, Chartrand MS, Aaseth J (2017) Manganese exposure and neurotoxic effects in children. Environ Res 155:380–384

Bocio A, Nadal M, Domingo JL (2015) Human exposure to metals through the diet in Tarragona, Spain-temporal trend. Biol Trace Elem Res 104:193–201

Bolan N, Kumar M, Singh E, Kumar A, Singh L, Kumar S, Keerthanan S, Hoang SA, El-Naggar A, Vithanage M, Sarkar B, Wijesekara H, Diyabalanage S, Sooriyakumar P, Vinu A, Wang H, Kirkham MB, Shaheen SM, Rinklebe J, Siddique KHM (2022) Antimony contamination and its risk management in complex environmental settings: a review. Environ Int 158:106908

Bueno-Solano C, López-Cervantes J, Campas-Baypoli ON, Lauterio-García R, Adan-Bante NP, Sánchez-Machado DI (2009) And biological characteristics of protein hydrolysates from fermented shrimp by-products. Food Chem 112:671–675

Chen S, Jiang S, Jiang H (2020) A review on conversion of crayfish-shell derivatives to functional materials and their environmental applications. J Bioresour Bioprod 5(4):238–247

Cheyns K, Nkulu BL, Kabamba Ngombe CL, Asosa JN, Haufroid V, De Putter T, Nawrot T, Muleka Kimpanga C, Luboya Numbi O, Kabyla Ilunga B, Nemery B, Smolders E (2014) Pathways of human exposure to cobalt in Katanga, a mining area of the D.R, Congo. Sci Total Environ 490:313–321

Costa RE (1977) The fertilizer value of shrimp and crab processing wastes. Master of Science, 167 pp

Devesa V, Súñer MA, Lai VWM, Granchinho SCR, Martínez JM, Vélez D, Cullen WR, Montoro R (2002) Determination of arsenic species in a freshwater crustacean Procambarus clarkii. Appl Organomet Chem 16:123–132

Dun Y, Li Y, Xu J, Hu Y, Zhang C, Liang Y, Zhao S (2019) Simultaneous fermentation and hydrolysis to extract chitin from crayfish shell waste. Int J Biol Macromol 123:420–426

Eslami H, Esmaeili A, Razaeian M, Salari M, Hosseini AN, Mobini M, Barani A (2022) Potentially toxic metal concentration, spatial distribution, and health risk assessment in drinking groundwater resources of southeast Iran. Geosci Front 13(1):101276

Ferreira-Baptista L, De Miguel EJ (2005) Geochemistry and risk assessment of street dust in Luanda, Angola: a tropical urban environment. Atmos Environ 39(25):4501–4512. ISSN 1352-2310. https://doi.org/10.1016/j.atmosenv.2005.03.026

Fraga-Corral M, Ronza P, Garcia-Oliveira P, Pereira AG, Losada AP, Prieto MA, Quiroga MI, Simal-Gandara J (2022) Aquaculture as a circular bio-economy model with Galicia as a study case: how to transform waste into revalorized by-products. Trends Food Sci Technol 119:23–35

Garcia-Vazquez E, Cartón P, Domínguez M, Rodríguez N, Bustillos A, Dopico E (2021) For a good selfie. Enhancing mobile phone recycling through simulated exposure to cobalt mining. Sustain Prod Consum 26:44–53

Gujre N, Rangan L, Mitra S (2021) Occurrence, geochemical fraction, ecological and health risk assessment of cadmium, copper and nickel in soils contaminated with municipal solid wastes. Chemosphere 271:129573

Gwenzi W, Mangori L, Danha C, Chaukura N, Dunjana N, Sanganyado E (2018) Sources, behaviour, and environmental and human health risks of high-technology rare earth elements as emerging contaminants. Sci Total Environ 636(2018):299–313

Han G-C, Jing H-M, Zhang W-J, Zhang N, Li Z-N, Zhang G-Y, Gao S, Ning J-Y, Li G-J (2022) Effects of lanthanum nitrate on behavioral disorder, neuronal damage and gene expression in different developmental stages of Caenorhabditis elegans. Toxicology 465:153012

Herrmann H, Nolde J, Berger S, Heise S (2016) Aquatic ecotoxicity of lanthanum—a review and an attempt to derive water and sediment quality criteria. Ecotoxicol Environ Saf 124:213–238

Horning KJ, Caito SW, Tipps KG, Bowman AB, Aschner M (2015) Manganese is essential for neuronal health. Annu Rev Nutr 35:71–108

IARC (2006) International Agency for Research on Cancer, IARC monographs on the evaluation of carcinogenic risks to humans-Volume 86: Cobalt in hard metals and cobalt sulfate, gallium arsenide, indium phosphide and vanadium pentoxide. WHO, Geneva

Jafarzadeh S, Fard RF, Ghorbani E, Saghafipour A, Moradi-Asl E, Ghafuri Y (2020) Potential risk assessment of heavy metals in the Aharchai River in northwestern Iran. Phys Chem Earth 115:102812

Khan SA, Khan H, Ishtiaq M, Saddique U, Muhammad S, Farooq M (2018) Shallow water proved higher levels of potentially harmful elements and human health risk along the Sadkal oil exploration and production. Desalin Water Treat 114(2018):146–154

Kim JJ, Kim YS, Kumar V (2019) Heavy metal toxicity: an update of chelating therapeutic strategies. J Trace Elem Med Biol 54(2019):226–231

Knowlton MF, Boyle TP, Jones JR (1983) Uptake of lead from aquatic sediment by submerged macrophytes and crayfish. Arch Environ Contam Toxicol 12:535–541

Kouba A, Buric M, Kozak P (2010) Bioaccumulation and effects of heavy metals in crayfish: a review. Water Air Soil Pollut 211(2010):5–16

Lamas GA, Navas-Acien A, Mark DB, Lee KL (2016) Heavy metals, cardiovascular disease, and the unexpected benefits of chelation therapy. J Am Coll Cardiol 67(20):2411–2418

Laughlin WM, Martin PF, Smith GR (1973) Processed crab waste: liming and fertilizer value on two Alaskan soils. Univ. of Alaska. Inst. Agric. Sci. Research Report 73-1

Lian M, Wang J, Sun L, Xu Z, Tang J, Yan J, Zeng X (2019) Profiles and potential health risks of heavy metals in soil and crops from the watershed of Xi River in Northeast China. Ecotoxicol Environ Saf 169:442–448

Liu J, Huang Y, Li H, Duan H (2022) Recent advances in removal techniques of vanadium from water: a comprehensive review. Chemosphere 287(Part 1):132021

Lv S, Xi C, Yang Q, Guo S, You L, Guo Y, Zhenga J (2020) Improvements on high-temperature stability, rheology, and stiffness of asphalt binder modified with waste crayfish shell powder. J Clean Prod 264:121745

Madigosky SR, Alvarez-Hemandez X, Glass J (1991) Lead, cadmium and aluminum accumulation in red swamp crayfish (Procambarus clarkii G.) collected from roadside drainage ditches in Louisiana. Arch Environ Contam Toxicol 20:253–258

Mahmoud ME, Fekry NA, Abdelfattah AM (2020) A novel nanobiosorbent of functionalized graphene quantum dots from rice husk with barium hydroxide for microwave enhanced removal of lead (II) and lanthanum (III). Biores Technol 298:122514

Mao X, Guo N, Sun J, Xue C (2017) Comprehensive utilization of shrimp waste based on biotechnological methods: a review. J Clean Prod 143:814–823

McDermott S, Salzberg DC, Anderson AP, Shaw T, Lead J (2015) Systematic review of chromium and nickel exposure during pregnancy and impact on child outcomes. J Toxicol Environ Health A 78(21–22):1348–1368

Nagarajan R, Anandkumar A, Hussain SM, Jonathan MP, Ramkumar M, Eswaramoorthi S, Saptoro A, Chua HB (2019) Chapter 12—Geochemical characterization of beach sediments of Miri, NW Borneo, SE Asia: implications on provenance, weathering intensity, and assessment of coastal environmental status. In: Ramkumar M, Arthur James R, Menier D, Kumaraswamy K (eds) Coastal zone management. Elsevier, pp 279–330

Nasab H, Rajabi S, Eghbalian M, Malakootian M, Hashemi M, Mahmoudi-Moghaddam H (2022) Association of As, Pb, Cr, and Zn urinary heavy metals levels with predictive indicators of cardiovascular disease and obesity in children and adolescents. Chemosphere 133664

Natasha N, Shahid M, Bibi I, Iqbal J, Khalid S, Murtaza B, Bakhat HF, Farooq ABU, Amjad M, Hammad HM, Niazi NK, Arshad M (2022) Zinc in soil-plant-human system: a data-analysis review. Sci Total Environ 808:152024

Naveed S, Oladoye PO, Alli YA (2023) Toxic heavy metals: a bibliographic review of risk assessment, toxicity, and phytoremediation technology. Sustain Chem Environ 2:100018

Nawakawa O (1990) Shells fertilizer. Nogyo Gijutsu Taikei 7:152.2–152.5 (Rural Culture Assoc, Tokyo)

Nriagu J (2019) Zinc toxicity in humans☆. In: Nriagu J (ed) Encyclopedia of environmental health, 2nd edn. Elsevier, Oxford, pp 500–508

Peceño B, Alonso-Fariñas B, Vilches LF, Leiva C (2021) Study of seashell waste recycling in fireproofing material: technical, environmental, and economic assessment. Sci Total Environ 790:148102

Peng Q, Nunes LM, Greenfield BK, Dang F, Zhong H (2016) Are Chinese consumers at risk due to exposure to metals in crayfish? A bioaccessibility-adjusted probabilistic risk assessment. Environ Int 88:261–268

Rincon-Leon P, Jerera-Cosano G, Pozo-Lora R (1988) Lead and cadmium concenuaiions in red crayfish (Procambarus clarkii G.) in the Guadalquivir River marshes (Spain). Arch Environ Contamin Toxicol 17:251–256

Shiue I (2014) Higher urinary heavy metal, phthalate, and arsenic but not parabens concentrations in people with high blood pressure, US NHANES, 2011–2012. Int J Environ Res Publ Health 11(6):5989–5999

Shrivastava R, Upreti RK, Seth PK, Chaturvedi UC (2002) Effects of chromium on the immune system. FEMS Immunol Med Microbiol 34(1):1–7

Śmietana N, Panicz P, Sobczak M, Nędzarek A, Śmietana P (2020) Variability of elements and nutritional value of spiny-cheek crayfish (Faxonius limosus, Rafinesque, 1817): variability of elements and nutritional value of F. limosus. J Food Compos Anal 94:103656

Stevenson CH (1902) Aquatic products in arts and industries. Fish oils, fats, and waxes. Fertilizers from aquatic products. In: Report US Comm fish and fisheries, pp 177–279

Stinson MD, Eaton DL (1983) Concentrations of lead, chromium, mercury and copper in crayfish (Pacifasticus leniusculus) obtained from a lake receiving urban runoff. Arch Environ Contam Toxicol 12:693–700

St-Jean A, Barguil Y, Dominique Y, Le Bot B, Ayotte P, Cordier S (2018) Nickel and associated metals in New Caledonia: exposure levels and their determinants. Environ Int 118:106–115

Suárez-Serrano A, Alcaraz C, Ibáñez C, Trobajo R, Barata C (2010) Procambarus clarkii as a bioindicator of heavy metal pollution sources in the lower Ebro River and Delta. Ecotoxicol Environ Saf 73(3):280–286

Tan Y, Peng B, Wu Y, Xiong L, Sun J, Peng G, Bai X (2021) Human health risk assessment of toxic heavy metal and metalloid intake via consumption of red swamp crayfish (Procambarus clarkii) from rice-crayfish co-culture fields in China. Food Control 128:108181

Thanigaivel S, Vinayagam S, Gnanasekaran L, Suresh R, Soto-Moscoso M, Che W-H (2024) Environmental fate of aquatic pollutants and their mitigation by phycoremediation for the clean and sustainable environment: a review. Environ Res 240(Part 1):117460

Tung PW, Burt A, Karagas M, Jackson BP, Punshon T, Lester B, Marsit CJ (2022) Association between placental toxic metal exposure and NICU Network Neurobehavioral Scales (NNNS) profiles in the Rhode Island Child Health Study (RICHS). Environ Res 204(Part A):111939

Turrentine JW (1913) The fish-scrap fertilizer industry of the Atlantic coast. Bull. U.S. Dept. Agric. no. 2

URL, 1. https://www.sweetdaddy-d.com/homemade-crawfish-stock/

URL, 2. https://blog.memeinge.com/crawfish-bisque/

URL, 3. http://www.fao.org/documents/show_cdr.asp?url_file/DOCREP/003/X9199E/X9199E00.HTM

URL, 4. https://www.compostwerks.com/organics/natural-fertilizers/product/37-crab-lobster-shell-meal

USEPA (1989) Risk assessment guidance for superfund human health evaluation manual (Part A), vol 1. Office of Emergency and Remedial Response, Washington, DC (1989) [EPA/540/1-89/002]

USEPA (2001) Supplemental guidance for developing soil screening levels for superfund sites. Office of Solid Waste and Emergency Response, Washington, DC [OSWER9355.4e24]

USEPA (2002) Supplemental guidance for developing soil screening levels for Superfund sites. Soild Waste and Emergency Response, Washington, DC [OSWER 9355.4-24]

USEPA (2011) United States Environmental Protection Agency. Exposure factors handbook. National Center for Environmental Assessment. Washington, DC (EPA/600/R-09/ 052F). https://cfpub.epa.gov/ncea/risk/recordisplay.cfm?deid=236252

Vijayaraghavan K, Mahadevan A, Joshi UM, Balasubramanian R (2009) An examination of the uptake of lanthanum from aqueous solution by crab shell particles. Chem Eng J 152(1):116–121

Wang YM, Yu Z, Zhao ZM, Jia L, Fang HQ, Zhang TF, Yuan X-Y, Shu Y-L, He J, Peng H, Li L-Z, Zhao J, Jia L-Z, Peng S-Q (2017) Subchronic toxicity study of yttrium nitrate by 90-day repeated oral exposure in rats. Regul Toxicol Pharmacol 90:116–125

Wani AL, Hammad GG, Shadab A, Afzal M (2021) Lead and zinc interactions—an influence of zinc over lead related toxic manifestations. J Trace Elem Med Biol 64:126702

Wu Y, Liu Q, Ma J, Zhao W, Chen H, Qu Y (2022) Antimony, beryllium, cobalt, and vanadium in urban park soils in Beijing: machine learning-based source identification and health risk-based soil environmental criteria. Environ Pollut 293:118554

Xu Y, Bajaj M, Schneider R, Grage SL, Ulrich AS, Winter J, Gallert C (2013) Optimisation of electrocatalytic dechlorination of 2,4-dichlorophenoxyacetic acid on a roughened silver–palladium cathode. Microb Cell Fact 12:90

Yan N, Chen X (2015) Sustainability: don't waste seafood waste. Nature 524(7564):155–157

Yang J, Teng JYG, Wu J, Chen H, Wang G, Song L, Yue W, Zuo R, Zhai Y (2017) Current status and associated human health risk of vanadium in soil in China. Chemosphcre 171:635–643

Yu XD, Zhang J, Yan CH, Shen XM (2014) Prenatal exposure to manganese at environment relevant level and neonatal neurobehavioral development. Environ Res 133:232–238

Zhu P, Gu Z, Hong S, Lian H (2017) One-pot production of chitin with high purity from lobster shells using choline chloride–malonic acid deep eutectic solvent. Carbohydr Polym 177:217–223

Zwolak I (2014) Vanadium carcinogenic, immunotoxic and neurotoxic effects: a review of in vitro studies. Toxicol Mech Methods 24:1–12

Chapter 11
Heavy Metals in *Cetraria islandica*: Health Risk Assessment

Murat Topal, E. Işıl Arslan Topal, Erdal Öbek, and Ali Aslan

Abstract In this chapter, potential human health risk of the heavy metals associated with usage of *Cetraria islandica* were evaluated. Heavy metals (lead, cadmium, manganese, chromium, copper, zinc, and nickel) were determined in *Cetraria islandica*. Highest and lowest heavy metal values were 87 ± 4.3 mg/kg for manganese (Mn) and 0.12 ± 0.01 mg/kg for cadmium (Cd). When potential health risk was assessed, the doses of exposure for man, woman, and child followed the order of bread > tea > dermal. Risk of health for bread and tea were child > woman > man. Also, risk of health for dermal was woman > child > man. Non-carcinogenic risk was woman > child > man for dermal way. Non-carcinogenic risks were child > woman > man for dietary. Highest total hazard index (THI) was 6.33 for chromium (Cr), lowest THI was 0.22 for zinc (Zn). THIs were zinc < copper < cadmium < nickel < lead < manganese < chromium. The THIs were higher than 1 for chromium, manganese and lead. Total highest carcinogenic value was 1.73E−03 for Ni and minimum was 1.19 E−05 for lead. Carcinogenic risks for bread (cadmium, nickel and chromium) were high. As a result, it was determined that *Cetraria islandica* may cause cancer risk if consumed as additive in bread. It was determined that if *Cetraria islandica* was consumed as tea and cream, there would be no risk of cancer and could not adversely affect human health.

Keywords *Cetraria islandica* · Exposure · Health risk · Lichen · Heavy metals

---

M. Topal (✉)
Department of Chemistry and Chemical Processing Technologies, Tunceli Vocation School, Munzur University, Tunceli, Turkey
e-mail: murattopal@munzur.edu.tr

E. I. Arslan Topal
Department of Environmental Engineering, Faculty of Engineering, Fırat University, Elazığ, Turkey

E. Öbek
Department of Biongineering, Faculty of Engineering, Fırat University, Elazığ, Turkey

A. Aslan
Department of Biology, Kyrgyz-Turkish Manas University, Bishkek, Kyrgyzstan

N. Kumar (ed.), *Heavy Metal Toxicity*, Environmental Science and Engineering,
https://doi.org/10.1007/978-3-031-56642-4_11

11.1 Introduction

Urbanization and industrialization are developing, causing growth in the population consuming more chemicals to improve life (Guan et al. 2021). Therefore, there is many stressors in the environment (Vlachokostas et al. 2014; Yao et al. 2021). Environmental stressors are highly diverse and include various exposures (e.g. nutritional, and chemical exposures). Continuous environmental exposures, such as toxic substances, raise concerns about possible effects on human health (Achillas et al. 2011; Vlachokostas et al. 2014). Heavy metals are among these substances. Term heavy metal denotes a potentially toxic source. The dictionary definition of heavy metal is an element with a specific gravity greater than 5 (Henry and Miles 2006). Recently, there has been increasing global public health and ecological concern associated with contamination of environment by the heavy metals (Tchounwou et al. 2012). The heavy metals exhibit characteristics of accumulating in the environment, displaying strong biological toxicity, and having slow decomposition rates (Mazurek et al. 2017; Zhou et al. 2024). Heavy metals copper (Cu), cadmium (Cd), chromium (Cr), nickel (Ni), zinc (Zn), and lead (Pb) are World Health Organization list of metals which have formidable concern. Anthropogenic activities (e.g. application of mineral phosphate fertilizers and coal combustion) have increased inputs of cadmium in the environment (Borovička et al. 2021). Therefore, it is a heavy metal of considerable occupational and environmental concern (Tchounwou et al. 2012). It shows high toxicity to human (Borovička et al. 2020). It is a carcinogenic metal with dubious distinction of being designated, carcinogenic to human, by WHO and the International Agency for Research on Cancer (Gujre et al. 2021). Main routes of exposure to it are via cigarette smoke or inhalation, and food ingestion. The foodstuffs that are rich in Cd can increase its concentrations in body of human. One of the important distribution route is circulatory system whereas blood vessels are considered to be main stream organs of Cd toxicity (Tchounwou et al. 2012). Prolonged exposure to heavy metals such as cadmium and lead at low doses may have unwanted effect on human health (Meli et al. 2018). Lead has many various applications (domestic, agricultural, and industrial) (Tchounwou et al. 2012). Lead exposure sources include domestic sources, industrial processes, food, smoking, and drinking water (Jaishankar et al. 2014). It is highly toxic and bioaccumulates and persistent in environment with serious health outcomes (UNEP 2019; WHO 2019; Mwai et al. 2023). Lead has devastating effects resulting in long-lasting intellectual disabilities among children (Nriagu et al. 1996; Tripathi et al. 2001; WHO 2019; Mwai et al. 2023). Chromium and nickel reach the environment by various natural processes and activities with anthropogenic origin. Chromium and nickel are used in different industrial processes (Kumar and Diwivedi 2021). Chromate production, tannery, stainless steel welding, metal processing, ferrechrome and chrome pigment production releases Cr (Tchounwou et al. 2012). Nickel is used in modern metallurgies in a broad variety of metallurgical processes, in production of nickel–cadmium batteries and as a catalyst in food and chemical industry (Genchi et al. 2020). Because of their toxic (mutagenic and carcinogenic) properties chromium and nickel become serious threat to fauna and flora of the environment and health of human. US Department of Health and Human Services categorized chromium and nickel as known to be human carcinogen (Kumar and Diwivedi 2021). Exposure to chromium can result in formation of ulsers persisting for months (Jaishankar et al. 2014). As a carcinogen and immunotoxic agent, nickel can cause respiratory tract cancer, contact dermatitis, asthma, lung fibrosis, and cardiovascular disease (Chen et al. 2017; Genchi et al. 2020). Copper is the 2nd most used non ferrous metal in industries (Boulamanti and Moya 2016; Izydorczyk et al. 2021). High level of copper can cause various diseases in human beings (Gujre et al. 2021). Copper poses potential threats to the human health such as dermatitis, mental retardation, liver and kidney dysfunction, poisoning, central nervous system damage, anemia, and bronchitis (Gadd and Griffiths 1978; Malamis et al. 2011; Jeremias et al. 2023). Zinc is widely used chemicals in industrial processes (US Geological Survey 2015; Chen et al. 2015). It can be toxic when present in excess (Van Eynde 2023). Exposure to environmentally relevant Zn concentration over a long period of time can induce cardiac effects in vivo (Pattanaik and Huggins 2020).

People have relied on wild plants for diets for hundreds of thousands of years. Many people continue to rely on wild-grown plants to meet at least some of daily nutritional needs of them (Turner et al. 2011). Furthermore, growing need for new food sources is of paramount importance at the global level (Redzic et al. 2010). Lichens are one of the wild harvested plant foods. They are one of the edible products (Turner et al. 2011). Unlike plants and mushrooms, lichens are not poisonous (except few species) and therefore are more suitable for human use. Lichens are an extraordinary source of dietetics, medicines, and nutrients (Redzic et al. 2010). However, toxic elements present in the environment penetrate into lichens. Lichens, which accumulate pollutants in the air, can pose serious health risks for consumers. Elements that accumulate in lichens can enter human body (Meli et al. 2018). It should also be kept in mind that many cultures have used lichens as embalming agents, preservatives, perfumes, cosmetics and dyes, as well as in alcohol production and food (Mitrović et al. 2011; Zhao et al. 2020).

The lichen *Cetraria islandica* from the Parmaliaceae family has been used in folk medicines through the ages. The *Cetraria islandica* (Lichen islandicus or Island moss) is used to treat irritation of throat and mouth, appetite loss, common cold, bronchitis, indigestion, dry cough, fevers, and susceptibility to infection (Meli et al. 2018). The throat pastilles and capsules are produced from Iceland moss. It is used as emergency food in famine and non famine years for bread as it is very nutritious and rich in minerals (Airasiksinen et al. 1986). It can be mixed with wheat, used as sweet jelly preparation, and used as flavoring in alcoholic beverages (Meli et al. 2018). Herbal tea of *Cetraria islandica* with a variety of medicinal uses is prepared (Xu et al. 2016). *Cetraria islandica* tea is commercially available. Furthermore, this lichen is a valued lichen in the industry of cosmetics. The good candidation of *Cetraria islandica* to be screened for UV-filters is reported by some researchers (Nybakken et al. 2004; Lohézic-Le Dévéhat et al. 2013). From Iceland moss, skin cream is produced. *Cetraria islandica* is directly applied to poorly healing wounds

(Meli et al. 2018) because of its properties against wounds and acne (Vitalini et al. 2015).

Currently, no information about potential health risk because of heavy metals associated with *Cetraria islandica* usage exists. Hence, the aim of this chapter is providing the assessment of the potential health risks that human may face from using/consuming *Cetraria islandica.* The novelty of this chapter is as follows; (i) Heavy metals were determined in *Cetraria islandica* (ii) It was investigated whether there are carcinogenic and non-carcinogenic health risks when *Cetraria islandica* is used for dietary and dermal.

11.2 Material and Method

Cetraria islandica was from Oltu-İnci (Erzurum, Turkey). *Cetraria islandica* was dried and powdered in the laboratory. Analysis procedure are briefly given: *Cetraria islandica* was cold leached with HNO_3. After cooling a modified Aqua Regia solution of equal parts concentrated HCl, HNO_3 and DI H_2O were added to samples for leaching in heating block of hot water bath. The samples were made up to volume with dilute HCl before filtration. ICP/MS was used for the analyses of PTEs (zinc, manganese, nickel, chromium, Cd, Cu, and lead). QA/AC was shown in Table 11.1. The health risk values were evaluated for the investigation of health risk of the heavy metals by *Cetraria islandica* usage.

Cetraria islandica is used as drink and food as well as cream in cosmetics. Heavy metals in *Cetraria islandica* may cause threat to health. Therefore, determination of the carcinogenic (CR) and non-carcinogenic (HQ) risks as a result of *Cetraria islandica* usage is important. The HQ risk is investigated by determination of heavy metals in *Cetraria islandica.* HQ is determined as following (Lian et al. 2019; Topal et al. 2021):

$$HQ = \frac{CDI}{RfD} \tag{11.1}$$

$$HQ = \sum_{k=1}^{n} \frac{CDIk}{RfDk} \tag{11.2}$$

Table 11.1 QA/AC details for *Cetraria islandica*

			Cetraria islandica						
		Analyte	Cd	Cr	Cu	Pb	Mn	Ni	Zn
		Unit	mg/kg	mg/kg	mg/kg	mg/kg	mg/kg	mg/kg	mg/kg
QC STD CDV-1	Standard		0.03	11.7	7.95	0.9	402	5.7	20.1
	Expected		0.04	12.1	8.61	1.0	385	6.4	23.3

where, RfD is reference dose (mg/kg day). Daily expsure dose (CDI) is determined as following (USEPA 2001, 2011):

$$CDI_{dietary} = C_{lichen} \times \frac{I_{intake} \times EF \times ED}{BW \times AT} \times 10^{-3} \quad (11.3)$$

$$CDI_{dermal} = C_{lichen} \times \frac{SA \times SL \times ABS \times EF \times ED}{BW \times AT} \times 10^{-6} \quad (11.4)$$

CDIdermal and CDIdietary: dermal and dietary (mg/kg day). Clichen: value in *Cetraria islandica* (mg/kg), ED and EF: exposure duration and frequency, AT and SA: averaged time and surface area, BW: body weight, SL and ABS: adherence and adsorption factor. Calculation of hazard index (HI) was as follows (USEPA 1989; Topal 2021):

$$HI = \sum HQ \quad (11.5)$$

Calculation of the cancer risk (CR) is as following (USEPA 1989):

$$CR = CDI \times SF \quad (11.6)$$

SF is slope factor (mg/kg day)$^{-1}$, slope factor for dermal was from value of extrapolated oral toxicity (USEPA 2002).

IBM SPSS Statistcs 21 (USA) is used for evaluation of data. The Pearson correlation tested for the indication of correlation between elements in *Cetraria islandica.*

11.3 Results and Discussion

Heavy metals in *Cetraria islandica* were determined. Highest and lowest heavy metal values were 87 ± 4.3 mg/kg for Mn and 0.12 ± 0.01 mg/kg for Cd. Copper, nickel, zinc, chromium and lead concentarations in *Cetraria islandica* were determined as 3.14 ± 0.16, 4.3 ± 0.21, 21.9 ± 1.1, 6.2 ± 0.3 and 3.17 ± 0.15 mg/kg, respectively. Heavy metals in *Cetraria islandica* followed order of cadmium < copper < lead < nickel < chromium < zinc < manganese. Relationship between heavy metals in *Cetraria islandica* were given in Table 11.2.

According to Table 11.2, correlations between lead-cadmium, cadmium-Ni, zinc-Cd, nickel-Cu, chromium-nickel, Mn–Ni, and Cr-zinc were negative. Pb-zinc is observed as positive and important ($r = 0.999$, $p = 0.05$). The correlations between Cr–Cd ($r = 0.996$) was lower than Pb–Zn, but it was important.

The estimated daily exposure doses (EDEs) of heavy metals were shown in Table 11.3.

Table 11.2 The relationship between toxic/harmful elements in *Cetraria islandica*

		Cd	Cu	Ni	Zn	Cr	Mn	Pb
Cd	Pearson correlation	1						
	Sig. (2-tailed)							
Cu	Pearson correlation	0.500	1					
	Sig. (2-tailed)	0.667						
Ni	Pearson correlation	−0.971	−0.277	1				
	Sig. (2-tailed)	0.154	0.821					
Zn	Pearson correlation	−0.116	0.802	0.351	1			
	Sig. (2-tailed)	0.926	0.407	0.772				
Cr	Pearson correlation	0.996	0.574	−0.946	−0.028	1		
	Sig. (2-tailed)	0.056	0.611	0.211	0.982			
Mn	Pearson correlation	0.756	0.945	−0.577	0.563	0.811	1	
	Sig. (2-tailed)	0.454	0.212	0.609	0.620	0.398		
Pb	Pearson correlation	−0.069	0.829	0.307	0.999*	0.019	0.601	1
	Sig. (2-tailed)	0.956	0.377	0.802	0.030	0.988	0.590	

* Correlation is significant at the 0.05 level (2-tailed)

Maximum total exposure dose (TED) for men was 5.72E−02 mg/kg day for manganese, minimum TED was 7.89E−05 mg/kg day for cadmium. Maximum TED for woman was 6.48E−02 mg/kg day for manganese and min. TED was 8.94E−05 mg/kg day for cadmium. Maximum TED for child was 1.45E−01 mg/kg day for manganese, minimum TED was 1.99E−04 mg/kg day for cadmium.

For men, highest EDEs for dermal, bread, and tea were 2.33E−07, 5.50E−02, 2.20E−03 mg/kg day for Mn, minimum EDEs were 7.59E−05, 3.04E−06 and 3.21E−10 mg/kg day for cadmium. Maximum EDEs for dermal, bread, tea were 2.64E−07, 6.24E−02, 2.49E−03 mg/kg day for Mn, minimum EDEs were 8.60E−05, 3.44E−06 and 3.64E−10 mg/kg day for cadmium, for woman. Maximum EDEs (dermal, bread, and tea) were 2.58E−07, 1.39E−01, 5.56E−03 mg/kg day for manganese, minimum EDEs were 1.92E−04, 7.67E−06, 3.55E−10 mg/kg day for cadmium, for child. The EDEs (bread, tea and dermal) were manganese > zinc > chromium > nickel > lead > copper > cadmium for man, woman and child. EDEs for man, woman and child were bread > tea > dermal. Exposure to the heavy metal bread and tea were child > woman > man. Exposure to the heavy metal for dermal was woman > child > man.

Non-carcinogenic risk (HQ) of heavy metal for man, woman, and child was shown in Table 11.4.

For bread, HQ maximum for man was 1.31E+00 for chromium, HQ minimum was 4.62E−02 for zinc. The HQ for man was chromium > manganese > lead > nickel > cadmium > copper > zinc. For HQs of woman, HQ maximum was 1.48E+00 for chromium and HQ minimum was 5.23E−02 for zinc. HQ was zinc < copper < cadmium < nickel < lead < manganese < chromium for woman. For child, HQ

Table 11.3 Estimated daily exposure doses of toxic/harmful elements (mg/kg day)

	Exposure	Cd	Cu	Ni	Zn	Cr	Mn	Pb
Men	Dietary (bread)	7.59 E−05	1.99 E−03	2.72 E−03	1.39 E−02	3.92 E−03	5.50 E−02	2.01 E−03
	Dietary (tea)	3.04 E−06	7.94 E−05	1.09 E−04	5.54 E−04	1.57 E−04	2.20 E−03	8.02 E−05
	Dermal	3.21 E−10	8.40 E−09	1.15 E−08	5.86 E−08	1.66 E−08	2.33E−07	8.49E−09
	Total	**7.89E−05**	**2.07E−03**	**2.83E−03**	**1.44E−02**	**4.08E−03**	**5.72E−02**	**2.09E−03**
Women	Dietary (bread)	8.60E−05	2.25E−03	3.08E−03	1.57E−02	4.44E−03	6.24E−02	2.27E−03
	Dietary (tea)	3.44E−06	9.00E−05	1.23E−04	6.28E−04	1.78E−04	2.49E−03	9.09E−05
	Dermal	3.64E−10	9.52E−09	1.30E−08	6.64E−08	1.88E−08	2.64E−07	9.61E−09
	Total	**8.94E−05**	**2.34E−03**	**3.20E−03**	**1.63E−02**	**4.62E−03**	**6.48E−02**	**2.36E−03**
Children	Dietary (bread)	1.92E−04	5.02E−03	6.87E−03	3.50E−02	9.91E−03	1.39E−01	5.07E−03
	Dietary (tea)	7.67E−06	2.01E−04	2.75E−04	1.40E−03	3.96E−04	5.56E−03	2.03E−04
	Dermal	3.55E−10	9.30E−09	1.27E−08	6.49E−08	1.84E−08	2.58E−07	9.39E−09
	Total	**1.99E−04**	**5.22E−03**	**7.15E−03**	**3.64E−02**	**1.03E−02**	**1.45E−01**	**5.27E−03**

Table 11.4 Non-carcinogenic risks (HQ) of toxic/harmful elements

Elements	HQdietary (bread)			HQdietary (tea)			HQ dermal		
	Men	Women	Children	Men	Women	Children	Men	Women	Children
Cd	7.59E−02	8.60E−02	1.92E−01	3.04E−03	3.44E−03	7.67E−03	3.21E−05	3.64E−05	3.55E−05
Cu	4.97E−02	5.63E−02	1.25E−01	1.99E−03	2.25E−03	5.02E−03	7.00E−07	7.94E−07	7.75E−07
Ni	1.36E−01	1.54E−01	3.44E−01	5.44E−03	6.16E−03	1.37E−02	2.13E−06	2.42E−06	2.36E−06
Zn	4.62E−02	5.23E−02	1.17E−01	1.85E−03	2.09E−03	4.67E−03	9.77E−07	1.11E−06	1.08E−06
Cr	1.31E+00	1.48E+00	3.30E+00	5.23E−02	5.92E−02	1.32E−01	2.77E−04	3.13E−04	3.06E−04
Mn	1.20E+00	1.36E+00	3.02E+00	4.79E−02	5.42E−02	1.21E−01	1.27E−04	1.43E−04	1.40E−04
Pb	5.73E−01	6.49E−01	1.45E+00	2.29E−02	2.60E−02	5.79E−02	1.62E−05	1.83E−05	1.79E−05

maximum was 3.30E+00 for chromium and HQ minimum was 1.17E−01 for zinc. The HQ was zinc < copper < cadmium < nickel < lead < manganese < chromium for child. The HQs for chromium and Mn were > 1 for woman, man, and child and HQ for lead was > 1 for child, showing adverse impact on health (Lian et al. 2019; Wei and Cen 2020). Highest HQ was for children when HQ values were evaluated for diet (bread). For bread, the HQ was C > W > M.

For tea, HQ maximum for man was 5.23E−02 for chromium, HQ minimum was 1.85E−03 for zinc. The HQ value for man was chromium > manganese > lead > nickel > cadmium > copper > zinc. For HQs of woman, HQ maximum was 5.92E−02 for chromium and HQ minimum was 2.09E−03 for zinc. HQ for woman was chromium > manganese > lead > nickel > cadmium > copper > zinc. HQ maximum was 1.32E−01 for chromium and HQ minimum was 4.67E−03 for zinc (for child). HQ value was zinc < copper < cadmium < nickel < lead < manganese < chromium for children. The HQs for the heavy metals were lower than 1 for woman, man, and child, showing no negative impacts on health, for tea. For tea, highest HQ was for child. HQ was C > W > M, for tea.

For dermal, HQ maximum for man was 2.77E−04 for chromium, HQ minimum was 7.00E−07 for Cu. HQ value was chromium > manganese > cadmium > lead > nickel > zinc > copper for man. For woman, maximum HQ was 3.13E−04 for chromium and minimum HQ was 7.94E−07 for Cu. HQ was chromium > manganese > cadmium > lead > nickel > zinc > copper for woman. For child, maximum HQ was 3.06E−04 for chromium and minimum HQ was 7.75E−07 for Cu. HQs were chromium > manganese > cadmium > lead > nickel > zinc > copper (for child). HQs were lower than 1 for woman, man, and child, showing no adverse impact on health of man, child, and woman, for dermal. Highest HQ was for woman when HQs were assessed for dermal. For dermal, HQ was woman > child > man.

HIs of the heavy metals for man, woman and child were given in Fig. 11.1.

The values of HI were <1 for cadmium, zinc, copper, and nickel. HIs for men and woman were >1 for chromium and manganese. Chromium and manganese values for men and woman were 1.36–1.24, and 1.54–1.41, respectively. Also, HI values for child was higher than 1 for chromium, manganese, and lead. HIs for child were 3.44 for chromium, 3.14 for manganese and 1.51 for lead. These values, when evaluated according to the USEPA, indicated a non-carcinogenic risk to humans (USEPA 2004).

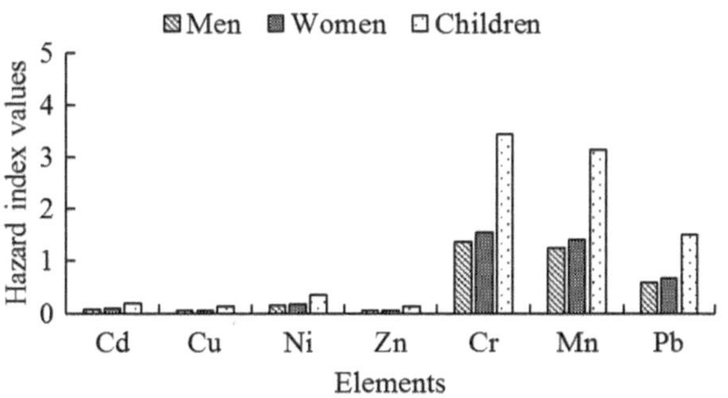

Fig. 11.1 Hazard index (HI) values of toxic/harmful elements

Table 11.5 Carcinogenic risk (CR) values of toxic/harmful elements

Elements	Dietary (bread)	Dietary (tea)	Dermal	Total CR
Cd	3.11E−04	1.24E−05	2.16E−08	3.24E−04
Ni	1.66E−03	6.65E−05	1.44E−07	1.73E−03
Cr	1.32E−03	5.27E−05	1.83E−07	1.37E−03
Pb	1.15E−05	4.58E−07	3.97E−11	1.19E−05

HIs were child > woman > man. However, highest THI was 6.33 for chromium, lowest THI was 0.22 for zinc. THIs were zinc < copper < cadmium < nickel < lead < manganese < chromium. THIs were higher than 1 for chromium, manganese and lead.

Carcinogenic risks (CRs) of the heavy metal by the consumption of *Cetraria islandica* were given in Table 11.5.

Risks of the heavy metal to be carcinogenic were calculated. Maximum CR (bread) was 1.66E−03 for Ni and minimum CR was 3.11E−04 for Cd. Maximum CR was 6.65E−05 for nickel and minimum CR was 4.58E−07 for lead for tea. Maximum CR for dermal was 1.83E−07 for chromium, minimum CR was 3.97E−11 for lead. Values of carcinogenic risk were dermal < tea < bread. Total highest and lowest carcinogenic values were 1.73E−03 for Ni and 1.19E−05 for Pb. The USEPA (2001) gives permissible level between 10–4 and 10–6. The values of carcinogenic risk for bread (cadmium, nickel, chromium) were high.

11.4 Conclusions

Cetraria islandica is collected by people in world, dried and used for different purposes. These are as following; (i) as additive in bread (ii) as tea and (iii) as cream for the treatment of some diseases. In this respect, determination of the element content of lichens is important. Since lichens are bioindicators of pollution, it is important to detect heavy metals in the lichens. Usage of lichen contaminated by the heavy metals causes potential health risk. Hence, determination of health risks of lichen with the heavy metal contents on human is important. Within the scope of this chapter, the following results were obtained;

(1) EDEs (bread, tea and dermal) were manganese > zinc > chromium > nickel > lead > copper > cadmium for man, woman and child. EDEs for man, woman and child were dermal < dietary (tea) < dietary (bread).

(2) HQs for Cr and Mn were higher than 1 for woman, man, and child. HQ for Pb was >1 for child. Hence, it was concluded that the consumption of *Cetraria islandica* as a bread may cause adverse effects on human.

(3) HQs (dermal and tea) for heavy metals were <1 for woman, man, and child. Hence, it was concluded that the consumption of *Cetraria islandica* as tea and the usage of it in dermal way will not cause adverse effects on human.

(4) HQs were man < woman < child for dietary (bread and tea). Unlike, HQ was woman > child > man for dermal.
(5) HIs for men and women were >1 for chromium and manganese. HIs for children was >1 for chromium, manganese, and lead. Therefore, it was determined to be a non-carcinogenic risk.
(6) In the present research, high values of CR for dietary (bread) (cadmium, nickel, and chromium) were obtained.

As a result, it was determined that *Cetraria islandica* may cause cancer risk if consumed as additive in bread. However, it was determined that if *Cetraria islandica* was consumed as tea and cream, there would be no risk of cancer and could not negatively affect human health.

References

Achillas C, Vlachokostas C, Moussiopoulos N, Banias G, Kafetzopoulos G, Karagiannidis A (2011) Social acceptance for the development of a waste-to-energy plant in an urban area: application for Thessaloniki, Greece. Resour Conserv Recycl 55:857–863
Airasiksinen MM, Peura P, Ala-Fossi-Solokangas L, Antere S, Lukkarinen J, Saikkonen M (1986) Toxicity of plant material used as emergency food during famines in Finland. J Ethnopharmacol 18:273–296
Borovička J, Ackerman L, Rejšek J (2020) Cadmium isotopic composition of biogenic certified reference materials determined by thermal ionization mass spectrometry with double spike correction. Talanta 18
Boulamanti A, Maya JA (2016) Production costs of the non-ferrous metals in the EU and other countries: copper and zinc. Res Pol 49:112–118
Chen B-C, Luo J, Hendryx M (2015) Zinc compound air releases from toxics release inventory facilities and cardiovascular disease mortality rates. Environ Res 142:96–103
Chen QY, Brocato J, Laulicht F, Costa M (2017) Mechanisms of nickel carcinogenesis. In: Mudipalli A, Zelikoff JT (eds) Essential and non-essential metals. Molecular and integrative toxicology. Springer International Publishing AG, New York, NY, pp 181–197
Gadd G, Griffiths A (1978) Microorganisms and heavy metal toxicity. Microb Ecol 4:303–317
Genchi G, Carocci A, Lauria G, Sinicropi MS, Catalano A (2020) Nickel: human health and environmental toxicology. Int J Environ Res Public Health 17(3):679
Guan G, Pan JH, Li Z (2021) Innovative utilization of molecular imprinting technology for selective adsorption and (photo)catalytic eradication of organic pollutants. Chemosphere 265:129077
Gujre N, Rangan L, Mitra S (2021) Occurrence, geochemical fraction, ecological and health risk assessment of cadmium, copper and nickel in soils contaminated with municipal solid wastes. Chemosphere 271:129573
Henry PR, Miles RD (2006) Heavy metals-vanadium in poultry. Ciencia Anim Brasileira 2(1):11–26
Izydorczyk GK, Mikula K, Skrzypczak D, Moustakas K, Witek-Krowiak A, Chojnacka K (2021) Potential environmental pollution from copper metallurgy and methods of management. Environ Res 197:111050
Jaishankar M, Tseten T, Anbalagan N, Mathew BB, Beeregowda KN (2014) Toxicity, mechanism and health effects of some heavy metals. Interdiscip Toxicol 7(2):60–72
Jeremias SD, Lin J-Y, Lourdes P, Dalida M, Lu M-C (2023) Abatement technologies for copper containing industrial wastewater effluents—a review. J Environ Chem Eng 11(2):109336

Kumar V, Dwivedi SK (2021) A review on accessible techniques for removal of hexavalent chromium and divalent nickel from industrial wastewater: recent research and future Outlook. J Clean Prod 295:126229

Lian M, Wang J, Sun L, Xu Z, Tang J, Yan J, Zeng X (2019) Profiles and potential health risks of heavy metals in soil and crops from the watershed of Xi River in Northeast China. Ecotoxicol Environ Saf 169:442–448

Lohézic-Le Dévéhat F, Legouin B, Couteau C, Boustie J, Coiffard L (2013) Lichenic extracts and metabolites as UV filters. J Photochem Photobiol B 120:17–28

Malamis S, Katsou E, Haralambous KJ (2011) Study of Ni(II), Cu(II), Pb(II), and Zn(II) removal using sludge and minerals followed by MF/UF. Water Air Soil Pollut 218:81–92

Mazurek R, Kowalska J, Gąsiorek M, Zadrożny P, Józefowska A, Zaleski T, Kępka W, Tymczuk M, Orłowska K (2017) Assessment of heavy metals contamination in surface layers of Roztocze National Park forest soils (SE Poland) by indices of pollution. Chemosphere 168:839–850

Meli MA, Desideri D, Cantaluppi C, Ceccotto F, Feduzi L, Roselli C (2018) Elemental and radiological characterization of commercial Cetraria islandica (L.) Acharius pharmaceutical and food supplementation products. Sci Total Environ 613–614:1566–1572. ISSN 0048-9697. https://doi.org/10.1016/j.scitotenv.2017.08.320

Mitrović T, Stamenković S, Cvetković V, Tošić S, Stanković M, Radojević I, Stefanovic O, Comic L, Dacic D, Curcic M, Marković S (2011) Antioxidant, antimicrobial and antiproliferative activities of five lichen species. Int J Mol Sci 12(8):5428–5448

Mwai L, Onyatta J, Were FH (2023) Lead content in automotive paints purchased at formal and informal outlets in Kenya. Heliyon 9(1):e12831

Nriagu JO, Blankson ML, Kwamena Ocran K (1996) Childhood lead poisoning in Africa: a growing public health problem. Sci Total Environ 181(2):93–100

Nybakken L, Solhaug K, Bilger W, Gauslaa Y (2004) The lichens Xanthoria elegans and Cetraria islandica maintain a high protection against UVB radiation in arctic habitats. Oecologia 140:211–216

Pattanaik S, Huggins FE (2020) Chemical nature of zinc in size fractionated particulate matter from residual oil combustion—a comparative study. Atmos Environ 221:117099

Redzic S, Barudanovic S, Pilipovic S (2010) Wild mushrooms and lichens used as human food for survival in war conditions; Podrinje-Zepa Region (Bosnia and Herzegovina, W. Balkan). Human Ecol Rev 17(2):175–187

Tchounwou PB, Yedjou CG, Patlolla AK, Sutton DJ (2012) Heavy metals toxicity and the environment, molecular, clinical and environmental. Toxicology 101:133–164

Topal M (2021) Investigation of the potential human health risk of toxic mercury determined in the grapevine exposed to mine gallery waters. J Food Sci Technol 58:1604–1610. https://doi.org/10.1007/s13197-020-04673-2

Topal M, Arslan Topal EI, Öbek E (2021) Evaluation of non-carcinogenic health risks of thallium in grapevine exposed to mine waters of an abandoned mining region in Turkey. Environ Dev Sustain. https://doi.org/10.1007/s10668-020-01125-x

Tripathi RM, Raghunath R, Kumar AV, Sastry VN, Sadasivan S (2001) Atmospheric and children's blood Lead as indicators of vehicular traffic and other emission sources in Mumbai, India. Sci Total Environ 267(1–3):101–108

Turner NJ, Łuczaj LJ, Migliorini P, Pieroni A, Dreon AL, Sacchetti LE, Paoletti MG (2011) Edible and tended wild plants, traditional ecological knowledge and agroecology. Crit Rev Plant Sci 30:198–225

UNEP (2019) United National environment programme, update on the global status of legal limits on lead in paint. https://www.unenvironment.org/resources/report/2019-update-global-status-legal-limits-lead-paint

USEPA (1989) Risk assessment guidance for superfund human health evaluation manual (Part A), vol 1. Office of emergency and remedial response, Washington, DC [EPA/540/1-89/002]

USEPA (2001) Supplemental guidance for developing soil screening levels for superfund sites. Office of Solid Waste and Emergency Response, Washington, DC [OSWER9355.4e24]

USEPA (2002) Supplemental guidance for developing soil screening levels for superfund sites. Soild Waste and Emergency Response, Washington, DC [OSWER 9355.4-24]

USEPA (2004) Risk assessment guidance for superfund. Volume I: Human health evaluation manual (Part E, Supplemental guidance for dermal risk assessment) Final OSWER 9285.7-02EP. July 2004

USEPA (2011) United States Environmental Protection Agency. Exposure factors handbook. National Center for Environmental Assessment. Washington, DC (EPA/600/R-09/052F). https://cfpub.epa.gov/ncea/risk/recordisplay.cfm?deid=236252

US Geological Survey (2015) Zinc statistics and information. http://minerals.usgs.gov/minerals/pubs/commodity/zinc/ (accessed 04.11.23)

Van Eynde E, Fendrich AN, Ballabio C, Panagos P (2023) Spatial assessment of topsoil zinc concentrations in Europe. Sci Total Environ 892:164512

Vitalini S, Puricelli C, Mikerezi I, Iriti M (2015) Plants, people and traditions: ethnobotanical survey in the Lombard Stelvio National Park and neighbouring areas (Central Alps, Italy). J Ethnopharmacol 173:435–458

Vlachokostas C, Banias G, Athanasiadis A, Achillas C, Akylas V, Moussiopoulos N (2014) Cense: a tool to assess combined exposure to environmental health stressors in urban areas. Environ Int 63:1–10

Wei J, Cen K (2020) Assessment of human health risk based on characteristics of potential toxic elements (PTEs) contents in foods sold in Beijing, China. Sci Total Environ 70310:134747

WHO (2019) World Health Organization, lead poisoning and health. https://www.who.int/news-room/fact-sheets/detail/lead-poisoning-and-health

Xu M, Heidmarsson S, Olafsdottir ES, Buonfiglio R, Kogej T, Omarsdottir S (2016) Secondary metabolites from cetrarioid lichens: chemotaxonomy, biological activities and pharmaceutical potential. Phytomedicine 23(5):441–459

Yao W, Zhang X, Gong Q (2021) The effect of exposure to the natural environment on stress reduction: a meta-analysis. Urban For Urban Green 57:126932

Zhao Y, Wang M, Xu B (2020) A comprehensive review on secondary metabolites and health-promoting effects of edible lichen. J Funct Foods 104283

Zhou H, Yue X, Chen Y, Liu Y (2024) Source-specific probabilistic contamination risk and health risk assessment of soil heavy metals in a typical ancient mining area. Sci Total Environ 906:167772

Chapter 12
Efficient Management of Heavy-Metal-Contaminated Soil and Water Using Sustainable and Greener Technologies

Adaobi Yvonne Onmonya, Sochi Otisi Anaga, and Chinedu Valentine Umeobika

Abstract This article discussed the issue of heavy metal contamination in soil and water, it highlights the risks to human health and the environment. It emphasizes the need for sustainable and greener technologies like phytoremediation and bioremediation to efficiently manage heavy metal contamination. The article also identified various techniques and methods used for remediation, such as physical, chemical, and nanotechnology-based solutions. This article gave a background of the problem and highlights the existing and recent green technologies being deployed for the efficient management of heavy-metal contaminated soil and water.

Keywords Green technologies · Heavy metal pollution · Bioremediation · Nanotechnology

12.1 Introduction

12.1.1 Overview of Heavy Metal Contamination in Soil and Water

The issue of heavy metal pollution is widespread across the globe, although the extent and levels of contamination vary depending on the location. The uncontrolled rise of industrial activities and urbanization due to a rapidly growing population has

A. Y. Onmonya (✉) · S. O. Anaga
Environmental Biotechnology and Bioconservation Department, National Biotechnology and Development Agency, Abuja, Nigeria
e-mail: onmonya.ya@nbrda.gov.ng; a.y.onmonya@gmail.com

C. V. Umeobika
Department of Civil Engineering, Federal Polytechnic Kaura Namoda, Kaura Namoda, Zamfara State, Nigeria

N. Kumar (ed.), *Heavy Metal Toxicity*, Environmental Science and Engineering,
https://doi.org/10.1007/978-3-031-56642-4_12

disrupted the natural balance of ecosystems. This has led to the pollution of air, soil, and water, endangering plants and animals. Ultimately, this hazardous situation puts our own lives at risk as we navigate through a dangerous interplay of pollutants and efforts to control them. The environmental presence of heavy metals presents a serious risk to human health and our surroundings well-being because of their elemental nature (Das et al. 2023). Heavy metals cannot undergo the well-being of our surroundings chemical breakdown, and their detoxification in the environment is primarily achieved either by stabilizing them in their original location or by removing them from the surrounding material (Mitra et al. 2022). Soil contamination can occur when heavy metals and metalloids build up in the soil due to various sources such as emissions from growing industrial areas, mine waste, leather factories, electronics manufacturing, electroplating, improper disposal of materials with high metal content, the use of leaded gasoline and paints, coal ash, accidental release of petrochemicals, and deposition of pollutants from the atmosphere (Gasparatos 2022).

The presence of heavy metals in agricultural soil reduces its economic worth due to the possibility of these dangerous substances being absorbed by the crops that are meant for human consumption. The reason for the existence of high levels of heavy metals in agricultural soil could be attributed to the prolonged application of fertilizers, animal waste, sewage sludge, pesticides, and fungicides, as well as the utilization of city wastewater for irrigation over an extended period. Heavy metal pollution can be caused by the prolonged usage of fertilizers, pesticides, and fungicides that contain heavy metals and also by utilizing municipal wastewater for irrigation purposes. Phytotoxic heavy metals such as cadmium, lead, cobalt, zinc, and chromium are harmful to plants at both low and high levels (El Ati-Hellal and Hellal 2021). These metals can be found in wastewater from various industries including mining, tanneries, electronics, electroplating, batteries, and petrochemicals, as well as textile mill products (Balali-Mood et al. 2021). Soil and water are the main areas where heavy metals released into the environment by human activities are collected. Unlike organic pollutants that are broken down by microbes into carbon dioxide, most metals do not decompose through microbial or chemical processes. As a result, their overall concentration in soil remains for a significant duration after they are first introduced. Changes in the forms of chemicals and how they are available for use by living organisms are possible. The existence of harmful metals in soil and water can greatly impede the breakdown of organic pollutants. The presence of heavy metals in soil and water may be dangerous to both humans and the environment. This can happen through consuming or touching contaminated soil, as well as through the food chain (Priya et al. 2023). The contamination can also affect the safety and marketability of food, as well as the availability of usable land for agriculture. This can lead to issues such as food insecurity, a shortage of clean water, and problems with land ownership.

Water pollution, combined with the scarcity of safe drinking water, has placed a significant strain on the natural surroundings. Approximately 40% of the global population is experiencing water shortage caused by factors such as climate change, quick urban development, increasing food demands, and excessive utilization of natural resources (He et al. 2021). In recent years, the increase in urbanization, industrialization, agricultural practices, geothermal water discharge, and the production of olive wastewater in olive-growing regions has led to a higher amount of polluted wastewater being released into the environment (Wang et al. 2015). Wastewater containing highly elevated levels of pollutants is extremely harmful to both aquatic ecosystems and human well-being. The reclamation of wastewater has become a primary choice for fulfilling the rising water needs in expanding industrial and agricultural fields. Untreated wastewater from both industrial and residential sources can contain various pollutants including pesticides, oils, dyes, phenol, cyanides, toxic organic compounds, phosphorous, suspended solids, and heavy metals. These toxic substances, including heavy metals, can readily accumulate in the surrounding environment. Industries involving metal processing, mining, geothermal energy plants, automobiles, paper production, pesticide manufacturing, tanning, dyeing, and plating are accountable for the worldwide pollution caused by heavy metals (Deblina et al. 2021). The majority of metals do not break down naturally and can move through various levels of the food chain, continuously building up in plants and animals, which can cause environmental issues. Research on toxins has shown that harmful metals can move from land and water into surface and groundwater (Ahmad et al. 2021). These heavy metals have serious health effects and are especially dangerous because they accumulate and increase in concentration throughout the food chain. This is because these metals cannot be broken down naturally. Getting rid of harmful contaminants is of utmost significance to reduce risks to both human health and the natural environment. Different methods like reverse osmosis, ion exchange, chemical precipitation, adsorption, solvent extraction and hybrid magnetic nanoparticles are used to eliminate heavy metals (Chang et al. 2020). However, these techniques typically result in high operational and maintenance expenses and are often harmful to the environment. The traditional methods used to treat heavy metals are typically expensive and take a long time. These treatment methods necessitate significant financial resources and ultimately face the challenge of how to dispose of the resulting sludge.

12.1.2 Importance of Efficient Management of Heavy-Metal Contaminated Soil and Water

Metal pollution negatively impacts biological systems due to its resistance to biodegradation, leading to harmful consequences. Toxic metals like Pb, Cd, As, Cr, and Hg can build up in organisms, leading to a range of illnesses and disorders, even when they are present in smaller amounts. Higher organisms may face various health risks due to the presence of heavy metals. They are also recognized for their impact on the growth of plants, ground covering, surface and underground water, as well as for harming soil microorganisms. The fact that heavy metals cannot be broken down chemically is widely recognized, thus requiring their physical removal or conversion into non-harmful substances (Hou et al. 2020). Arsenic is a harmful substance that is commonly present in the environment and known for its toxicity to humans and other living beings. This element is extremely poisonous, and the level of toxicity varies based on its specific type. The presence of arsenic can be influenced by various factors such as pH levels, redox conditions, the types of minerals in the surrounding environment, and microbial actions. These factors can also determine whether the arsenic is in an inorganic or organic form, as well as its state of oxidation. It is widely agreed that inorganic forms of arsenic, specifically arsenite [As^{3+}] and arsenate [As^{5+}], are the most common species found in most environments. However, there is a possibility of organic forms being present as well. Typically, inorganic arsenic compounds are considered more poisonous than organic forms, which are comparatively less toxic (Briffa et al. 2020). The toxicity of trivalent compounds (arsenites) is higher compared to pentavalent compounds (arsenates). According to reports, the solubility of As^{3+} in water is 4–10 times greater than that of As^{5+}. Nevertheless, the trivalent methylated forms of arsenic have shown to be more harmful compared to inorganic arsenic due to their increased effectiveness in causing DNA degradation. While As^{5+} is typically less toxic than As^{3+}, it is more thermodynamically stable. This stability allows As^{5+} to dominate under normal conditions and contribute to significant groundwater contamination. Arsenate, in its pentavalent state (As^{5+}), is also recognized as harmful and cancer-causing to humans.

Lead can be found in various forms in natural sources all around the world, and it has become one of the most widely and uniformly spread trace metals.

Lead from car exhaust, dust, and emissions from different industrial sources can contaminate soil, water, and plants. The presence of high levels of Pb^{2+} was discovered to be highly poisonous to humans. Due to its non-biodegradable and immobile nature, Pb^{2+} persists as a long-term source of exposure after contaminating the environmental media. The presence of lead in soil and water can occur due to various sources including industrial areas, the use of leaded fuels, old lead pipes in plumbing systems, and even past orchard sites where lead arsenate was once utilized. If no action is taken, soil lead levels will never go back to their original state (Srivastava et al. 2022). In nature, lead is recognized as harmful to plants, animals, and microorganisms. The presence of this pollutant in the surroundings is found in a form that cannot be dissolved, and the poisonous metals cause severe health issues for humans, such as impairing the brain and slowing down development.

Mercury is a naturally occurring metal that can react with oxygen, sulphur, chlorine, and other elements to form salts or inorganic compounds. It is a persistent environmental contaminant that can build up in fish, animals, and people. Mercury pollution of the environment is created by a number of sectors, such as petrochemicals, mining, and painting, as well as agricultural inputs such as fertilizer, fungicidal sprays, municipal wastewater, and wastewater from factories for crop irrigation (UNEP 2024). Mercury is present in a number of home items, such as bleach, acid, and caustic chemicals, as well as devices such as thermometers, barometric instruments, manometers, and medical instruments. Mercury has been shown to interfere with electron transport in mitochondria and chloroplasts, which may have an effect on oxidative metabolism and photosynthesis. The cumulative toxic effects of mercury and its derivatives can be hazardous to human health even at low concentrations

(Briffa et al. 2020). Traditional methods for remediating sediment contaminated with metals involve two main approaches: in situ treatment and relocation efforts.

12.1.3 Need for Sustainable and Greener Technologies for Management

There are currently various methods being utilized to remove heavy metal pollutants from the environment. However, many of these conventional treatment technologies used to eliminate contaminants from wastewater can be expensive to maintain, require significant manpower, take a long time to complete, harm the environment, and are often not very effective.

The use of chemical technologies results in the production of a significant amount of sludge, which leads to higher costs (Tawfik et al. 2022).

Additionally, both chemical and thermal methods are challenging and costly, and all of these techniques can also deteriorate the important elements present in the soil. Remedying soils contaminated with heavy metals can be done through two methods: managing the issue on-site or removing the soil and disposing of it in a designated landfill area. This disposal method only moves the issue of contamination to another location, along with the dangers of transporting contaminated soil and the movement of pollutants from the landfill to the surrounding environment. Soil washing is an alternative method to excavation and disposal of landfills, used to remove contaminated soil (Lianwen et al. 2018). This approach is quite expensive and generates waste materials that contain high levels of heavy metals, necessitating additional processing.

Concerns in recent times regarding the pollution of the environment have led to the creation of suitable technologies to evaluate the occurrence and ability of metals to move around in soil, water, and wastewater. Biological treatment processes refer to systems that employ microorganisms to break down organic pollutants found in soil and water. Phytoremediation, a highly researched approach in recent years, is a promising and affordable green technology that harnesses the natural ability of plants to absorb and store heavy metals (Bala et al. 2022). Phytoremediation is a more environmentally friendly and less intrusive approach compared to traditional methods of using physical and chemical technologies to remove heavy metals from soil and water. According to estimates, phytoremediation is a more cost-effective means of cleansing heavy metals, saving up to 60–80% compared to traditional methods (Conte et al. 2021). Furthermore, it does not require costly equipment or specialized staff. This can be applied to a wide variety of environments and pollutants to clean them up while causing minimal harm to the surrounding environment (Vijendra and Daverey 2020).

On the other hand, bioremediation is a process that involves the use of living organisms, such as bacteria, fungi, and algae, to cleanse soil and water that has been contaminated by heavy metals (Akar et al. 2005). These techniques eliminate harmful pollutants in the environment by converting, breaking down, and converting them into safer substances. This technology consists of three fundamental processes: bio-transformation, biodegradation, and mineralization (Bala et al. 2022). Bio-transformation converts harmful components into safer substances, biodegradation breaks down organic substances into smaller molecules, and mineralization completely breaks down organic materials into inorganic compounds like CO_2 or H_2O. Each of these processes can occur either at the site where the contamination is present (in situ) or outside of the contamination site (ex-situ). Additionally, they can take place in the presence of oxygen (aerobic) or the absence of oxygen (anaerobic). In situ bioremediation is the process of treating organic pollutants in their natural environment without moving them, while ex-situ bioremediation involves excavating and moving the pollutants from their original location to be treated elsewhere, making it easier to control the conditions for the bioremediation process (Mallavarapu et al. 2011). Although ex situ bioremediation is quicker than in situ bioremediation, it is more expensive and there is a higher risk of potential harm to the public due to the transportation of pollutants.

12.2 Sources and Types of Heavy Metal Contamination

12.2.1 Industrial Activities and Their Impact on Soil and Water Quality

The development of industries in a region not only takes away the natural beauty of the surroundings but also has negative consequences for the resident's well-being (Grillitsch and Hansen 2019). The production process involves the making of bullets, tubes, wire, metal mixtures, and colored coatings. Industries that do not properly treat their wastewater, such as battery manufacturing, coating, automotive, aeronautical and steel, printing, pigments, fuels, photographic materials, and explosive manufacturing, are significant contributor to lead pollution. The mining, smelting, refining, and electroplating sectors play a significant role in causing nickel contamination, as well as the casting of alloys. Industrial pollutants constitute a major source of soil and water contamination. Chemicals are emitted from industrial facilities in both liquid and solid states. The emission of arsenic fluorides and sulfur dioxide through industrial activities greatly contributes to the increase in soil acidity and hurts vegetation. Unintentional spills and leaks that occur while storing, transporting, or using substances are a source of soil and water pollution (Kshyanaprava and Das 2023). The extraction and processing of metal ores, along with the associated industries, have left behind a significant problem for many countries—a widespread presence of metal contaminants in the soil. While mining operations are ongoing, the heavier and larger particles known as tailings settle at the bottom of flotation cells (Wang et al. 2019). These tailings are then released directly into natural depressions such as onsite wetlands, causing an increase in their concentration levels. Extensive mining and smelting of lead and zinc ores have led to soil contamination, which presents a potential danger to the well-being of both humans and the environment. Many methods employed to restore these sites are time-consuming and costly, and there is no guarantee that they will successfully rejuvenate the soil's productivity. The risk that heavy metals in the soil pose to humans in terms of the environment is connected to their bioavailability. Ways in which assimilation can occur include consuming plants that have been grown in contaminated soil or directly consuming soil that is contaminated.

Different industries, like textile, tanning, petrochemicals, pesticides, and pharmaceuticals, create various materials that can have different compositions. These materials can be produced accidentally through oil spills or by using petroleum-based products. Although a few are discarded on land, they provide minimal advantages to agriculture or forestry. Furthermore, numerous substances can be dangerous due to their high levels of heavy metals such as chromium, lead, and zinc, or toxic organic compounds. These substances are rarely, if ever, used for any land applications. Some others have negligible amounts of nutrients for plants or do not possess any properties for improving soil conditions (Kandasamy et al. 2022).

Industrial agriculture contributes to soil pollution through various means. Throughout history, agriculture has been regarded as the initial significant impact made by humans on the soil. Several fertilizers consist of high levels of heavy metals that have the potential to disturb normal crop growth periods. Examples of these heavy metals include calcium, nitrate, and potassium chloride. To thrive and go through their entire life cycle, plants need to obtain both major nutrients (such as nitrogen, phosphorus, potassium, sulfur, calcium, and magnesium) as well as important minor nutrients (Rebecca 2022). Certain soils lack the necessary heavy metals, including Co, Cu, Fe, Mn, Mo, Ni, and Zn, which are crucial for promoting healthy plant growth. To compensate for this deficiency, crops can be given these metals either by adding them directly to the soil or by spraying them onto the leaves. Sometimes, cereal crops that are grown on soil lacking sufficient copper may be treated with copper by adding it to the soil. Similarly, manganese can also be provided for both cereal and root crops. To support the growth of crops, intensive farming systems often utilize substantial amounts of fertilizers, which are regularly applied to soils to ensure an adequate supply of nutrients such as nitrogen (N), phosphorus (P), and potassium (K). The elements provided by the compounds used contain small amounts of heavy metals, such as Cadmium (Cd) and Lead (Pb), as unwanted substances. Over time, the repeated use of these fertilizers can greatly raise the levels of these heavy metals in the soil (Beerling et al. 2018). The use of specific phosphatic fertilizers unintentionally introduces harmful substances such as Cd, as well as other potentially dangerous elements like F, Hg, and Pb, into the soil.

Many popular pesticides that were previously widely used in agriculture and horticulture contained high levels of metals. In recent times, approximately 10% of the insecticides and fungicides approved for use in the United Kingdom have been derived from compounds containing Cu, Hg, Mn, Pb, or Zn. Some pesticides that fall into this category include Bordeaux mixture (a combination of copper sulfate and lime) and copper oxychloride, both of which are used to control fungal diseases (Robin et al. 2021). For a long period, lead arsenate was employed in fruit orchards to manage certain parasitic insects. Arsenic-containing substances were commonly employed to manage cattle ticks and control pests in banana crops in New Zealand and Australia. Additionally, wood products were treated with mixtures of copper, chromium, and arsenic for preservation purposes. As a result, there are currently numerous abandoned areas where the levels of these elements in the soil far surpass normal levels. The potential for problems exists when contamination occurs, especially when sites are used for different agricultural or non-agricultural activities after redevelopment. In contrast to fertilizers, the utilization of these substances has been limited to specific areas or plant species.

The use of municipal wastewater and industrial effluent for crop irrigation is a common cause of mercury contamination in agricultural soil. Organic materials such as animal waste, compost, and sewage sludge are used as fertilizers on land, resulting in the build-up of toxic metals such as arsenic, cadmium, chromium, copper, lead, mercury, nickel, selenium, molybdenum, zinc, thallium, and antimony (Ajay Singh 2021). Biosolids, also known as sewage sludge, are mostly organic solids created through the treatment of wastewater and can be recycled beneficially. Biosolids have raised significant concern due to the possibility of contaminating soil with heavy metals, such as lead, nickel, cadmium, chromium, copper, and zinc. The levels of these metals in biosolids are influenced by the kind and intensity of industrial activity, as well as the methods used for treating the biosolids. Recent research conducted on certain soils in New Zealand that were treated with biosolids has revealed higher levels of Cd, Ni, and Zn in the liquid that drains from the soil.

The use of municipal and industrial wastewater, as well as related effluents, for land application has been a tradition for 400 years and is now widely practiced in various regions across the globe. Globally, around 20 million hectares of cultivable land are estimated to be watered using wastewater. Research indicates that in various Asian and African cities, around half of the vegetables provided to urban areas come from agricultural practices that utilize wastewater for irrigation purposes. Farmers, in general, do not prioritize or show concern for the positive or negative impacts on the environment. Their main focus is on maximizing their crop yields and financial gains. While the levels of metal in wastewater effluents are generally not high, continuous use of such water for irrigating land can lead to the gradual buildup of heavy metals in the soil over time. For example polluting gasses and fumes, vapor escaping through a pipe, or dusty products in storage. Such air sources spread a form of metallic particles known as fine and solids suspension (Joseph 2023). These metals become vapors at high temperatures and they contain arsenic, cadmium and lead. On exposure to air, these materials will form oxides which will undergo condensation to form small particles in the absence of a reducing environment. Big territory natural airflow drifts away stack emissions until wet, dry precipitation mechanisms take off them out of the gas stream. Fugitive emissions have much smaller scale because they are released relatively close to the ground level hence low concentration of the released gases in this case. Typically, pollutant concentrations are considerably lower in escapes than emissions from chimneys. However, at a particular site, the type and amount of metals that are likely to be leaked from any of the sources will depend with certain site's given conditions. Over an extended period, the settled dusts would comprise of all that existed in fire smoke and factory chimneys. Various types such as polluting substances like the number of different kinds of heavy metals and many others being dumped into the ecosystem compartments beginning with the era of industrial revolution. Abnormal concentrations of Cadmium, Lead and Zinc in plants and soils near smelters. One of the main causes of pollution in soils due to leading compounds is the leakage of Pb into the air during burnings of tetraethyl lead gasoline (Fasih et al. 2021). The number of lead particles found in urban soil, especially road side soil also goes up significantly. In addition, Zn and Cd are transported either by tires or lubricating oils to adjacent soil near the road.

12.2.2 Natural Sources of Heavy Metals

Heavy metals are naturally present in the soil environment at levels below 1000 mg kg^{-1} that are often considered nonhazardous due to their pedogenetic weathering processes of parent materials (Raymond and Felix 2011). In most rural and urban soils, concentrations of one or more of the heavy metals exceed the baseline levels, leading to increased risk to both ecological and anthropic systems. However, heavy metals that result from man-made cycle have fast rates because (i) the generation of heavy metals occurs faster through artificial cycles as opposed to natural cycles, (ii) they are transported from mines to different unplanned sites with a greater likelihood for contact, (iii) the concentration levels the metals are found in the waste products (Saikat et al. 2022).

12.2.3 Trace Elements in the Water

Trace elements are one of several water quality problems, a major challenge within these challenges.

Although water is normally not full of minerals, they are always present in water at some level. Water can sometimes contain harmful elements at high rates. Dissolved inorganic components of significance in natural waters can be categorized into three categories.

1. Major components: The components include calcium, magnesium, sodium, potassium, chloride, sulphate, sulphide, carbonate and bicarbonate.
2. Minor elements: Some of these components are iron, manganese, boron, fluoride, strontium, and nitrate, among others.
3. Trace elements: Trace elements like copper, cobalt, zinc, molybdenum, vanadium, nickel, arsenic, lead, cadmium, and chromium can be found in natural waters in a small quantity ranging from micrograms per liter (Franson et al.

2022). Definitions of major, minor, trace and some other water born solutes are not well defined. Therefore, most of the natural waters would be expected to include every element occurring naturally. Traces of these elements are essential for many life processes and may be toxic. With respect to this area, some heavy metals have positive and negative influences on the development of xenobiotic microbes.

12.2.4 Common Heavy Metals Found in Contaminated Soil and Water

Heavy metals such as cadmium, lead, mercury, arsenic, copper, nickel, cobalt and chromium are naturally found in the soils of environments according to reports from EPA. Non-biodegradable heavy metals in both soil and water present danger to both human health and the environment at large (Cannas et al. 2020). For example, it encompasses metallic deposition in the air through air pollution, the river sediments, town waste composts, agricultural chemicals such as fertilizers and pesticides, industrial wastes originating from factories, and so forth because of their human made origins. The issue of heavy metal contamination is becoming widespread with cadmium, copper, zinc, nickel, cobalt, lead, mercury and arsenic are most prevalent in agricultural soils in various countries around the globe. There are many other micro nutrients like iron, manganese, zinc, copper, magnesium, molybdenum, nickel, and cobalt which commonly occur in the soils and have significant impact on the growth. A number of metals such as cadmium, chromium, lead, cobalt, silver, selenium, and mercury may be in soil and water and thus capable of accumulation.

12.3 Environmental and Health Risks Associated with Heavy Metal Contamination

12.3.1 Ecological Impact on Soil and Aquatic Ecosystems

Anthropogenic and geological activities are the main medium of heavy-metal pollution in the environment. Activities such as mining, military activities, municipal waste, application of fertilizer, discharge of urban effluent, vehicle exhausts, wastewater, waste incineration, fuel production, and smelting cause the production of metal contaminants. Agricultural pesticides and the utilization of fertilizers on agricultural soil have raised the concentration of Cd, Zn, Cu, and As in soil. A constantly increasing need for agricultural produce has increased the application of pesticides, fertilizers, and herbicides (Ali et al. 2020). This excessive use of these agrochemicals may result in the accumulation of these pollutants in plants, soils, and water bodies as well as the use of phosphate fertilizer and inorganic fertilizers to control the diseases of crops, grain, and vegetables. An enormous quantity of fertilizers when applied to deliver the K, P, and N required to improve the growth of crops in turn increases the incidence of cadmium, lead, iron, and mercury in substantially high concentrations in the environmental media (Basavegowda and Baek 2021). Inputs of heavy metals to agricultural land through the excessive use of fertilizers are increasing apprehension about their probable hazard to the environment. Untreated wastewater irrigation leads to the buildup of various heavy metals such as cadmium, lead, nickel, copper, and zinc which are frequently present in the subsurface of the soil. Wastewater irrigation for long periods increase the concentration of heavy metal in the soil at toxic levels. The unregulated dumping of municipal solid waste is also another main source of raised soil and water contamination load. Open dumps and landfilling are the common practices used worldwide to dispose of municipal solid waste. Despite being a useful source of nutrients, these wastes are also a source of some harmful toxic metals as well. Maintenance and deicing operations on roads also generate groundwater/surface contaminants.

There has been an increased concentration of many toxic metals mainly attributed to human activities and which are considered to present certain serious hazards such as reduction in fish numbers and biodiverse and also destruction of the environment and decrease in number of microorganisms and quality of ground water, reduced fertile soil Anthropogenic heavy metals tend to be more mobile, thereby making them bioavailable as compared to pedogenic or lithogenic heavy metals (Saikat et al. 2022). The metal containing solids at contaminated sites may come from many different anthropogenic sources such as mine tailings, non-protected dumping of high metal waste products, fueled by leaded gasoline or using paint with leads, fertilizers, and animal manure.

The amount of soil has the limit it can absorb contaminated particles; if this occurs then those substances will influence also others environment sections like into the food chain. Because of this, soils are polluted as well, which lowers crop yields and their quality, leading to food insecurity. Air pollution is also a result of emissions generated through the release of volatile compounds produced by soil pollution. Additionally, the combustion of fossil fuels give rise to air pollution leading to acid rains which create a harsh condition on the soil surfaces. This destroys micro organisms which enhance soil construction by degrading organic materials and improving water movement. Some chemicals in soils are also leached into the ground water that finally ends up in the streams, lakes and oceans (Manisalidis et al. 2020). Highly nutritive soils can also lead to increased concentration of nitrogen and phosphorus in waterways resulting in algal blooms that make less oxygen for aquatic life. Similarly, soil erosion results in water polluting and sedimentation of rivers. The most widely prevalent environmental pollutant is the heavy metals which have a long term toxicity potential for both humans and animals. Several heavy metals that may create long term health hazard in humans, especially in children have been identified when the leachates contaminate ground water posing both environmental and health risks. For this reason and more some cleanup remedies have been developed for cleaning up of contaminated soil.

12.3.2 *Human Health Risks and Potential Exposure Pathways*

Currently, one of the major factors that are leading to environmental degradation is an environmental contaminate. This involves the presence of common industrial contaminants such as heavy metals. Water quality assessment is usually based on heavy metals that include Cd, Cr, Cu, Hg, Zn, As, Ni, and Pb. Some of these are the micronutrients necessary for plant growth while others like Pb, Co, and Cd are not biologically relevant. The major problem with these pollutants is that they are not bioavailable or chemically available for decomposition by an organism (Hama et al. 2023); therefore, they must be eliminated or changed physically into a less harmful chemical substance. As a result, this chemical substance fails to biodegrade and, therefore, leads to an increase in levels of heavy metals within organisms that feed on it. The amount of metal ions builds up over time until they reach human beings who take high-level contaminated food via the food chain. According to previous research, it was asserted that even though the human body penetration by heavy metals may occur via various means like direct swallows, dermal contacts or inhaling (Witkowska et al. 2021). It was revealed that the most common entry way for them is mainly into the digestive system by this becomes prominent especially for population groups residing in high anthropogenic areas where these contaminants are more frequent than in low loaded ones. It is dangerous even in elemental and dissolved form because heavy metals are usually lethal and extremely toxic. Despite being harmful at extremely small amounts, it is a threat to organisms. For example, the elements such as arsenic occur naturally yet are extremely poisonous for both man and any other living being (Ravenscroft et al. 2009). The other example is lead, poisonous for most of the flora, fauna and bacteria. It is toxic in humans; it causes many diseases associated with brain damage and mental retardation. Mercury is a persistent environmental pollutant due to its ability to accumulate in living organisms. Since mercury is capable of passing through the blood–brain barrier, it also affects the brain in mercury poisoning. The gills and gastrointestinal tracts of animals are also capable of adsorbing heavy metals, such as cadmium. There are a number of tissues that can accumulate this substance, most notably those around the heart and kidneys. Because cadmium forms semi-permanent metalation with proteins, it has a long residence time. By ingesting heavy metals (such as arsenic, chromium, and lead), humans experience decreased levels of vitamin C and iron in their bodies. When the human body is deficient in essential nutrients, it can suffer from weakened immune systems, functional disorders, disabilities, and malnutrition (Saikat et al. 2022) When ingested over an extended length of time, heavy metals can also cause harm to the immune system, gastrointestinal tract, central nervous system, and reproductive system. As previously said, it can result in blood poisoning, cancer, skin issues, lung ailments, cognitive retardation, and malfunctions in human DNA. Long-term exposure results in crippling conditions including Parkinson's and Alzheimer's, as well as collapses in the brain, optic nerves, and skeletal muscle. Heavy metals and other water contaminants have a detrimental effect on fish populations and biodiversity in the ecosystem (Nabi and Tabassum 2022). The World Health Organization's restriction on heavy metals can have detrimental effects on soil and aquatic systems by lowering groundwater quality, destroying microorganisms that are beneficial to agriculture, and stunting the growth of flora.

The majority of copper pollution in soil and groundwater occurs around mine sites. However, copper at super optimum levels (>100 mg/kg) can be extremely toxic to plants. In general, most Cu tolerant plants are excluders, and Cu accumulators have not been classified so far. Some hazardous heavy metals like copper exist, and can bind with albumen or albumin and some even free 'Cu' in humans' bodies and nerves (Grzegorz et al. 2021) The intestinal tract may suffer damage, such as nausea, vomiting, and diarrhea from eating too much copper.

However, the presence of high amount As as well as As being extremely toxic and carcinogen; therefore the remediation of As-polluted sites is essential (Saikat 2022). Copper and lead are common ores that can be found naturally in soil and rocks. Arsenate and arsenite are the most commonly encountered natural inorganic forms of As. Skin alterations are also associated with injury of blood vessels due to chronic exposure to As. These short-term effects are skin, liver, bladder, and lung cancer, sore throat, neurological problems among others. High amounts of inorganic arsenic exposure causes infertility and miscarriage in females, dermatitis, decreased immune function, encephalopathy, as well as cardiovascular effects, such as hypertension, coronary heart disease, vascular disease (Tchounwou et al. 2019). These mines may contain a lot of heavy metals in their tailings. One of the most prominent contaminates found on mine tailings is toxic metals. Studies too have shown that these heavy metals are capable of disturbing the functioning of the neural and mental system alongside the metabolism in the human body system. Impairs blood, cards vascular, endocrine, and immune, reproductive and urinary systems due to exposure to heavy metals.

Lead is not vital as a trace element to proper nutrition in human or animal systems. It can adversely affect the health of different organisms, including humans, as it accumulates and magnifies along the food chains (Assi et al. 2016). The human body absorbs lead through ingestion of contaminated food, water, air, soil, and consumer products. Lead concentration and duration of exposure play an important role in measuring lead toxicity. When one is exposed to a high concentration of lead for a short period of time, acute poisoning may happen with significant implications. In the acute condition, seizures, coma, and even death occur rapidly. On the other hand, most case studies involve chronic poisoning following long-term and low-level exposure. Toxic lead causes diseases of anemia, neurotoxicity, hemotoxicity, nephrotoxicity, and toxic metabolic encephalopathy. The organs and tissues it affects include the heart, bones, intestines, kidneys, and the reproductive system and this has devastating effects on life sustaining processes. On the other hand, lead toxicity leads to brain damages and mental retardation among the children.

The copper in normal human serum averages between 120 and 140 μg per gram bound through ceruloplasmin, albumin as well as other molecules. On the other hand, elevated amounts of copper could lead to illnesses like liver disease, kidney ailments, anemia or gastro-intestinal problems. Also in an indirect way, copper links with neurodegenerative disorders such as Alzheimer's Disease (Nabi and Tabassum 2022) Wilson's Disease, and Prion disease. Nevertheless, bio-accumulation and biomagnification of copper can lead to a dramatic increase in its concentration up to toxic levels that may appear as adverse effects for certain organisms and environmental conditions. The presence of copper in the soil may be equally troublesome, for it could also get into the body through food. Research shows that the grass absorbs copper that seeps through contaminated soils from the power lines, resulting in high level of copper found in the milk of grazing livestock for aquatic ecosystems, the non-point source of dissolved copper is the cation, and not only does the cation change the water chemistry over time, and bioaccumulate in fish, but bio-magnify through the food chain (Rehman et al. 2019). Nickel plays a vital role in growth and forms part of the co-enzymes of some living organisms and plants. However, nickel can become slightly toxic when present in high concentrations. When cultured in the laboratory, it is found to be necessary for the growth of Blue-green Algae *Oscillataria* spp. and Bacteria *Alcalegenes eutrophus*. Like copper, nickel also bio-accumulates and bio-masses along food chains, causing serious problems to living organisms. There is evidence of nickel bio-mass accumulation and magnification in fish (e.g., embryos, non-feeding larvae) phase to target organs such as gills, kidneys, liver, brain, and muscle. Short term overexposure to high relative concentrations of nickel is not known to cause health problems to humans, but long term exposure is a serious health risk (Alfano and Cavazza 2020). The primary health effects of nickel are: Roughness of the respiratory system Acute inflammation on the nasal membrane Hypersensitivity of the nasal mucosa Bronchial asthma Neoplastic transformation Kidney and liver damage Cardiovascular system poisoning and skin irritation.

12.4 Traditional Approaches for Heavy Metal Remediation in Soil

Conventional methods for heavy metal removal from contaminated soils use physical, chemical and biological methods that can be used to eliminate contaminants from contaminated soils. Although these methods are effective, most of them are expensive, time consuming, and harmful to the environment. The costs and challenges associated with technology make soil cleanup difficult. There are many limitations and risks associated with the use of conventional methods for soil restoration.

12.4.1 Physical Methods (e.g. Landfilling and Leaching, Excavation, Soil Washing, Calcination)

These processes are costly but allow for great removal efficiency and the cleanup of vast amounts of soil.

12.4.1.1 Physical Separation

- Large-scale removal of heavy metal/metalloid particles from polluted soils is possible through physical separation. This method takes advantage of the variations in size, density, magnetic, and hydrophobic surface qualities between metal particles and soil. The chemical characteristics of soil, such as pH, elemental content, and OMC, are unaffected by physical separation (Raffa et al. 2021). Physical separation may be accomplished through the following methods, depending on the particle characteristics:
  - Screening: This involves passing soil particles through standardized screen holes, the particle separation occurs on the basis of particle size.
- Hydrodynamic separation: This process takes advantage of the various particle settling velocities in a water flow; alternatively, the various effects of centrifugal force can be used (Chelgani and Asimi 2022).
- Froth flotation: The technique uses air bubbles injected into a soil slurry to separate metal-bearing particles from the soil matrix by taking advantage of differences in hydrophobic properties (Pawlik 2022).
- Magnetic separation: uses the differences in magnetic properties of particles to separate the soil particles.
- Soil washing: the contaminants are desorbed and extracted from soil with an extractant solution.

According to Liu et al. (2022), the concentrations of heavy metals like Pb, Cd, and Zn in soil rise as particle size decreases and when the soil contains a lot of organic matter or clays. When washing contaminated soil, mechanical mixing encourages direct physical contact between the contaminated soil particles and the washing solution. Using ultrasonic mechanical soil washing to remove heavy metals more effectively has also been researched. Physical methods have the advantages of high efficiency, simplicity, and speed of application, but they are expensive and can alter some soil properties, such as soil texture or particle size, which can reduce soil fertility.

12.4.1.2 Soil Replacement

Soil excavation is a method of completely or partially replacing contaminated soil. This method is recommended for small scale applications. The soil that is replaced is treated as waste. This method can also be used to reduce heavy metal and metallicoid concentrations. Soil spading requires digging deep and removing contaminated soil using a spade (USEPA 2017). The contaminated soil is then replaced by fresh soil. Clean soil is then added to the contaminated soil and combined with the new soil when it is imported. To prevent groundwater contamination before the soil is replaced, the contaminated area needs to be physically separated. Although laborious and costly, soil replacement can be used for small, very contaminated areas.

12.4.1.3 Thermal Treatment

Calcination, or thermal desorption, is the process of removing a volatile material from a medium by heating it to a specific temperature. The four main ways to heat soil are electrical resistive heating, radiofrequency heating, steam-based heating, and conductive heating. Once volatile pollutants have been desorbed from soil, they are collected using vacuum-negative pressure or a carrier gas. The process can occur at low-temperature desorption (90–320 °C) or high-temperature desorption (320–560 °C), depending on the pollutant boiling point. For example, this technique is widely used to treat soils contaminated with mercury. Although the process can harm the soil structure, It is only truly effective at high pollutant concentrations and comes with high capital costs and gas emission control. Because this metal can be boiled at 357 °C, thermal desorption is an effective way to extract it from soil (Karavolias et al. 2021).

12.4.1.4 Vitrification

At high temperatures (>1500 °C) heavy metals and metalloids are fixed into vitreous material, vitrification is a thermal technique that reduces the mobility of these substances. During the procedure, high-voltage electricity, an outside heat source, or electrical discharge-induced gas plasma are used to warm up the contaminated soil (Tawfik et al. 2022). The soil is transformed into molten lava by melting it, and the lava cools to become a vitrified structure. While the other contaminants are eliminated, the heavy metals/metalloids are encapsulated in this glassy matrix. Although it is an effective technique, vitrification is expensive, complicated, and harmful to the soil. Additionally, it does not apply to soil that contains a lot of organic matter, has a lot of moisture, or is contaminated with volatile or flammable organics.

12.4.2 *Chemical Methods (e.g., Precipitation, Ion Exchange)*

These techniques are very efficient but some could introduce new chemical pollutants into soils, such as soil washing.

12.4.2.1 Chemical Precipitation Technique Used in the Water Remediation Process

Chemical precipitation is a frequently employed technique for water purification, in removing metallic impurities from wastewater from industry due to its inexpensive nature and simplicity of use. In order to create insoluble particles, precipitating factors such as hydroxides, carbonates, sulphides, and dissolved metal ions in wastewater react chemically. Physical procedures like filtration, the coagulation process or sedimentation come after the process. Following the precipitation of heavy metals as carbonates and hydroxides, these materials can be extracted by gravity, magnetic, dissolved air flotation, or vortex separation techniques. Other chemicals are subsequently used to precipitate the remaining heavy metals.

Large-scale industries have long employed ion exchange as a method of softening water. It entails switching out an ion on insoluble and immobile surfaces with a charged ion associated with an exchange site.

12.4.2.2 Electrolysis and Flocculation/Coagulation Techniques Used in the Water Remediation Process

The process involves the use of an electrochemical reaction-based technique for separating heavy metal ions from both wastewater and water. This process entails, injecting a direct electric current into an Ionic material that induces an appropriate chemical reaction, at the electrodes Ma et al. (2017). It is also suitable in producing zinc and copper in bulk. The second stages of treatment are based on coagulation and flocculation process which leads to the improvement of clarity removal. Flocculation joins the destabilized colloidal particles together thereby bonding them together unlike coagulation that reduces the repelling forces between them. The use nano-flocculants has been developed for treating dye, microorganism and heavy metal water pollution Jumadi et al. (2020). Nonetheless, problems of toxicity and health risks linked to inorganic coagulants are important concerns.

12.4.2.3 Ultrafiltration and Adsorption Techniques Used in the Water Remediation Process

Filtration is done through a selectivity permeable membrane that separates the suspended particles and macromolecule from fluid or liquid feed. Large-size solute retention by ultrafiltration membranes occurs as opposed to RO membranes at relatively low pressures (Liang et al. 2021). These include procedures like sedimentation, filtration, coagulation, softening, dechlorination, ozonation, and disinfection which are applied on drinking water in order to get rid of sediments such as turbidity, colloidal particles, and dissolved organic material (Grzegorzek et al. 2020). Membrane applications are susceptible to fouling, like inorganic fouling, colloid and particle fouling, biological fouling, and organic fouling. The best technology to treat the oily waste water in the process of ultra filtration is enhanced morphology and mechanical strength. Surface phenomenon of removal of inorganic and organic contamination mainly depended upon the interaction between porous solid adsorbent and an adsorbable solute through physical or chemical bonds known as adsorption (Srivastava et al. 2015). This method of wastewater treatment is cheaper, it is more effective, it is possible to use it in various places for different purification tasks due to its simplicity, and it allows for the purification of all types of contaminants (Ali and Gupta 2006). Moreover, it is revealed that such nanoadsorbents based on nanoparticles show enhanced rate of adsorption, eco-friendliness, recyclability, removal efficiency, economical performance and even in wastewater solutions.

12.4.3 Limitations and Drawbacks of Traditional Approaches

Different strategies have been developed to remove heavy metals from water. Some of the disadvantages of conventional treatments include high energy consumption, partial removal, and the production of toxic sludge. Examples of these treatments include chemical precipitation, ion exchange, and electrochemical removal. The financial and technological prerequisites for installation, operation, and maintenance may present additional challenges. This could lead to an inadequate implementation of these technologies, especially in developing nations and decentralized situations. Development of efficient and eco-friendly solutions is therefore required; examples of such solutions include adsorption techniques utilizing biological, chemical, or nanotechnology processes (Kumar et al. 2021a; Lee et al. 2019; Tavker et al. 2021).

12.5 Sustainable and Greener Technologies for Heavy Metal Management

12.5.1 Phytoremediation

12.5.1.1 Overview of Phytoremediation and Its Mechanisms

Phytoremediation is an environmental cleanup technique that utilizes plants to eliminate or counteract contaminants from soil, water, and air. Its main aim is to reduce the detrimental impacts of pollutants and revive the ecosystems to a more favorable condition (Othmani et al. 2020). This procedure depends on different methods like phytoextraction, rhizofiltration, phytostabilization, phytodegradation, and rhizodegradation (Pedron et al. 2017, 2021). This utilizes the innate capabilities of plants to purify and immobilize harmful substances (Table 12.1). The effectiveness of phytoremediation is impacted by various factors such as the types and levels of pollutants, the choice of plant species, the properties of the soil, the surrounding environmental conditions, and the interactions with microorganisms (Fig. 12.1). This flexible technology is utilized in various areas, including the elimination of pollutants in industries, solving contamination issues in agriculture, restoring landfills, treating water and wastewater, and reducing urban pollution (Hao et al. 2014). Nevertheless, there are various obstacles and restrictions that phytoremediation encounters, including the need for significant time and financial resources, difficulties in obtaining and reaching contaminants, genetic and physiological limitations, the requirement for ongoing monitoring and upkeep, and the acceptance and perception of the general public. In spite of these difficulties, phytoremediation offers exciting possibilities for the future. This includes making progress in genetic engineering, combining with other technologies, discovering new ways plants can interact with microbes, creating guidelines and policies, and expanding its use for large-scale projects.

Table1 Different mechanisms by which plants can remediate contaminated medium

Mechanisms	Description
Phytoextraction	Plants absorb contaminants and store in above-ground shoots and the harvestable parts of roots
Phytostabilization	Roots and their exudates immobilize contaminants through adsorption, accumulation, precipitation within the root zone, and thus prevent the spreading of contaminants
Phytodegradation	Plant enzymatic breakdown of organic contaminants, both internally and through secreted enzymes
Rhizodegradatiom (phytostimulation)	Plant roots stimulate soil microbial communities in plant root zones to break down contaminants
Phytovolatilization	Contaminants taken up by the roots through the plants to the leaves and are volatized through stomata where gas exchange occurs

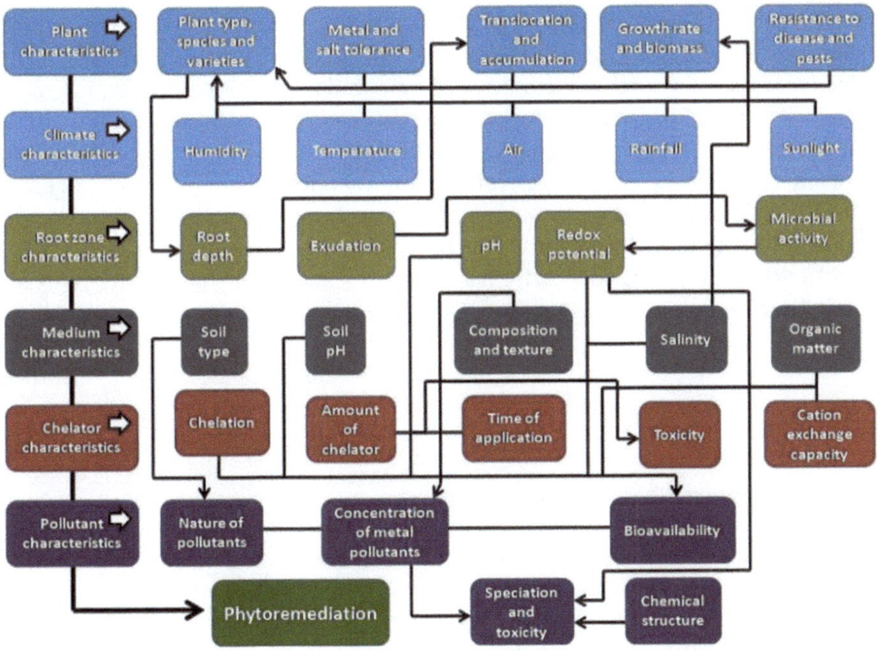

Fig. 12.1 Relationships among the factors affecting phytoremediation efficiency (*Source* Saxena et al. 2019)

12.5.1.2 Bioremediation

- Although bioremediation techniques like bioventing, biosparging, bioaugmentation, and biostimulation are economical and environmentally beneficial, their rate of degradation is slower than that of other treatments (Raffa et al. 2021).
- Microorganisms' function in the breakdown and immobilization of heavy metals

  Using soil amendments, heavy metal-contaminated soils can be remedied by adjusting their bioavailability. It is necessary to keep an eye on the immobilized heavy metals' long-term stability. The impact of organic and inorganic amendments on soil organic carbon sequestration and heavy metal immobilization was examined in a growth chamber study that lasted eighteen months. When it comes to the immobilization and remobilization of hazardous metals as well as their interactions with metals, yeasts are essential. Applications of biotechnology, like bioaugmentation that take advantage of yeasts' degradation capabilities, are highly pertinent in the modern world. Bacteria that reduce sulfate are important in lowering the mobility of heavy metals (Mosa et al. 2016). While both sediment adsorption and microbial removal are crucial for immobilizing heavy metals, microbial degradation is the primary factor in the breakdown of SMX. Lately, LDHs have shown exceptional long-term stability and have been employed as amendments to immobilize heavy metals (Mishra and Malik 2013). Application of bioaugmentation and biostimulation techniques
- Benefits and challenges of bioremediation

  Although the use of microorganisms living in sediment for bioremediation has gained acceptance as a feasible solution, its application is restricted by its low controllability and low efficiency. Compared to traditional MFCs, the setup and operation of BMFCs provide distinct problems due to their unique properties. There has been an attempt to pinpoint the difficulties, areas in need of investigation, and essential strategies for the effective use of bioremediation in the field. In order to bioremediate a variety of waste categories, vegetable grass may be essential. The generation of microalgae biomass in conjunction with the bioremediation of CO_2 and nutrients from livestock farms has been investigated, with possible advantages and disadvantages. Because microalgae are widely available, reasonably priced, and environmentally benign, they are also utilised in phycoremediation of heavy metals. An extensive analysis of MES for the bioremediation of soil-adsorbed PAHs and sediment is lacking.

12.5.1.3 Nanotechnology-Based Solutions

- Use of nanoparticles for heavy metal removal

  Various authors have reported that, nanotechnology can be referred to as one of the most promising and effective approaches used in bioremediation procedures, whereupon it possesses significant advantages that are capable of upgrading the environmental technologies implemented in the area of nanomaterials having enhanced The paper explores nanobioremediation as a new promising trend in environmental engineering (Singh et al. 2022). In addition, their nanosizing allows them to access other unreachable locations thereby supporting in-situ remediation as opposed to ex-situ remediation. Recently, Kumar et al. (2023), discussed about several processes to remove ECs involving different influencing parameters. Finally, Kumar et al. (2023), in brief, discussed the challenges and issues related to microalgae based EC's bioremeditation.

- Nanotechnology in remediation: advantages and limitations

  The use of nanotechnology in the agricultural sector can reduce the arbitrary use of agrochemicals and thus decrease the amount of chemical pollutants. A study by Das et al. (2015) examines the applications of nanotechnology in sustainable development, with a focus on renewable energy, air-, water-, and soil-remediation. Another study by Patil et al. (2016) highlights the successful use of nanotechnology for environmental clean-up, the potential ecological implications of environmental nanotechnology (E-Nano), and potential solutions to current E-Nano-related implications. Although there are concerns among the scientific community about nanotechnology risk management, the effectiveness of E-Nano is undeniable. Usman et al. (2020) reviewed recent developments in nanotechnology applications in agriculture, including crop production and protection with an emphasis on nanofertilizers, nanopesticides, nanobiosensors, and nano-enabled remediation strategies for contaminated soils. The efficiency and fate of nanomaterials are strongly influenced by their properties and interactions with soil components. Additionally, engineered nanomaterials (ENMs) and nanotechnology have shown great potential in addressing complex soil remediation problems due to their unique features of high reactivity, selectivity, and versatility. The development of cleaner and greener technologies with significant health and environmental benefits would be greatly influenced by nanotechnology, according to . The use of nanomaterials in the synthesis and production of water treatment solutions has been demonstrated to reduce energy consumption (Santhosh et al. 2016), facilitate recycling and manufacturing of new products, and utilize eco-friendly nanomaterials in water and wastewater remediation, as reported by Elgarahy et al. (2021).

12.5.2 Emerging Sustainable Technologies

By the early 2000s, new remediation techniques were required as alternative approaches to existing technology known as "Green Remediation". The new approach is intended to curb the secondary impact as well as clean up the soils. This is driven by the fact that all developed countries are turning their attention towards GR because it introduces new technologies, which also factor in the socio-economic impacts on remediations. This innovative method of remediation, based on NBS, also puts more focus on soil quality and limited use of renewables resources.

12.5.2.1 Phytoextraction and Hyperaccumulating Plants

Transgenic techniques can be used to improve phytoremediation of metal-contaminated locations by introducing genes from plants, microbes, and mammals that enable plants to resist, tolerate, and hyperaccumulate harmful metals. These genes produce the MTPs, MTs, and PCs that are necessary for phytoremediation and the detoxification of hazardous metals. Through chloroplast engineering, complete bacterial operons can be transferred into plants to increase the expression of the metal-binding enzymes (Mishra and Malik 2013). However, transgenic plants have not yet been used on a large scale in the field, and it is uncommon for transgenic genes to spread to wild relatives. The risks to the environment associated with using genetically engineered plants (GEPs) for increased metal phytoremediation in laboratory conditions have been examined, but they also include exposure to hazardous metals, unchecked spread, alteration of indigenous flora, and possibility of invasion by uncontrolled plants.

12.5.2.2 Biochar Application and Soil Amendment Techniques

Both natural and inorganic poisons can be bioremediated with the aid of biochar. According to Karppinen et al. (2017), the key mechanism is an increase in the variety of microorganisms that degrades petroleum (petrol) in soils that have been treated with biochar (Fig. 12.2). Heavy metals and metalloids can alter one structure before moving on to the next, from a higher fixation to a lower focus, but they cannot completely disappear from the environment. These heavy metals and metalloids can concentrate via different biochemical pathways and be accumulated into the tissues, cells or any part of an organism (Verma et al. 2021). Consequently, these heavy metals and non-metallic pollutants can be also treated in two ways (Li et al. 2019). The negative surface charge of biochar and basic solutions can be combined for removal of toxic metals. This encourages improvement of the more useful and nutrient-rich soils abundant in beneficial microbial populations, thereby providing favorable conditions for plant growth through repeated application of fertilizers as well as maintaining suitable soil properties and design characteristics.

12.5.3 Green Materials and Bio-based Sorbents

12.5.3.1 Bacteria

Within each group, there are also significant differences in the biosorptive capacities of various microbial biomasses. Nevertheless, the pre-history, pretreatment and test plan determines how much sorption capability every biosorbent possesses. Thus, it should be an inexpensive, effective, and fast growing and harvestable biosorbent (Gupta et al. 2015). In addition, the organisms' ability to adapt with alterations in

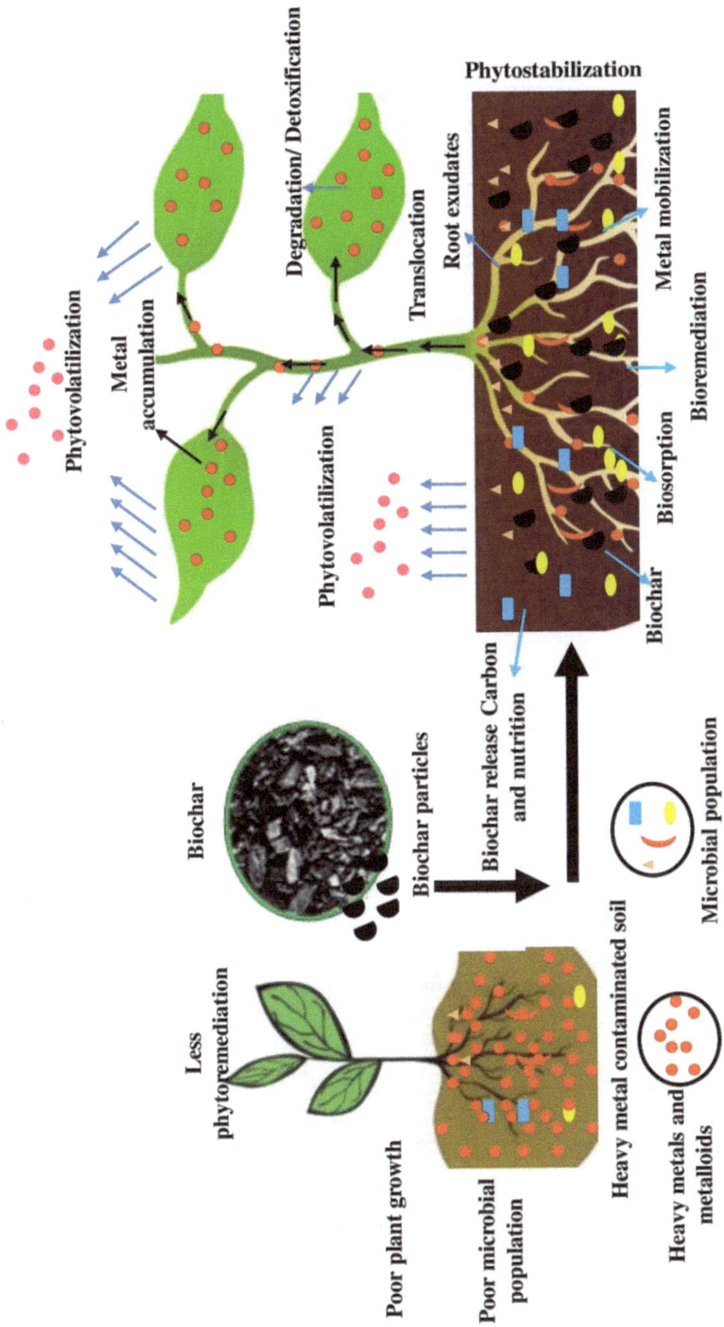

Fig. 12.2 Influence of biochar on phytoremediation process

the bioreactors' settings and physical-chemical circumstances could further improve the rate of biosorption (Fomina and Gadd 2014). Bacteria are good biosorbents due to their wide distribution, size and ability to grow in controlled environment with different environmental stresses. Some of the heavy metals which have been tested with bacteria include several strains such as Pseudomonas, Enterobacter, Bacillus and Micrococci. Their high surface-to-volume ratio explains their higher sorption capacity.

12.5.3.2 Fungi and Yeast

There are many filamentous fungus in the natural world. Due to their vegetative characteristics, filamentous fungus are some of the most affordable and environmentally acceptable biosorbents on the market right now (Shah and Daverey 2020; Akar et al. 2005). Additionally, by employing inexpensive growing media, it is simple to manufacture fungal biomasses. They also have a high biomass production and are simple to grow on a large scale. After fermentation, fungi are quite prevalent in industrial waste products and have a number of advantages over other microorganisms like bacteria (Dhankhar and Hooda 2011; Cárdenas González et al. 2019). *Aspergillus terreus* Thom (1918), according to Cerino-córdova et al. (2012), has the ability to biosorb 68.52% of copper from aqueous solutions. Although these findings suggest that *Aspergillus terreus* may be used as a biosorbent, with an efficiency of 104.1 mg/g, Sinha et al. (2012) devised a sequential bioreactor to remove mercury from synthetic effluent. Using *Micrococcus luteus*, which has an ideal removal capacity of 1965 mg/g (Puyen et al. 2012), a significant amount of lead was extracted from a synthetic medium. Kim et al. (2015) developed a batch method for the removal of Cu, Cr, and Ni from contaminated saltwater that uses zeolite-immobilized *Desulfovibrio desulfuricans*. Moulds and yeasts are frequently utilised as biosorbents to remove harmful metals from contaminated wastewaters since they are simple to grow and may be genetically and morphologically modified. They have created an intricate defensive mechanism to counteract the toxicity of heavy metals (Srivastava and Thakur 2006). In wasted chrome effluent, the fungus *Paecilomyces lilacinus* only accumulated 24% of the chromium, however in a synthetic media, 100% clearance was seen.

12.5.3.3 Algae

Because of their large sorption capacity, algae—especially brown algae—are useful biosorbents for the removal of heavy metals (Wang and Chen 2009). These autotrophic organisms generate a lot of biomass with little nutritional requirement. Six distinct types of algae have been found through studies to have strong sorption capabilities for a variety of metals, including Cd, Cu, Ni, Pb, and Zn. For instance, at pH 5, *Sargassum muticum* can adsorb up to 5.5 mg/g with only little pH-related effects.

12.6 Integration of Sustainable Technologies for Efficient Management

12.6.1 Combined Approaches and Synergistic Effects

Biomass valorization is an advantageous, environmentally friendly phyto-management approach. Phyto-biomass was originally treated as the trash, but now this byproduct has been recognized as an important resource for industrial purposes. This includes burning the plant matter directly for energy extraction or through further processing into biofuels (Lesmana et al. 2009). New integrated phytoremediation—bioenergy approaches are promising sustainable solutions from the points of views of the environment (decontamination of contaminated sites, waste reduction), the economy (bioenergy production) and society (green redevelopment) (European Commission 2009). The implementation of these methods fosters cost-efficient conservation in a context that endorses recycling, thus forming the basis for the circular economy model in Africa. Nevertheless, new phytoremediation methods need to be proposed in line with the SRR principles and adaptation to climatic changes.

An innovative approach to plant HM phytoremediation involves transfer of the entire operon from bacteria into plastics thus enabling plants to overexpress such enzymes (Dhankhar et al. 2011; Goel et al. 2009). Another approach involves introduction of new genes from natural metal hyperaccumulators or microbial species into fast growing metal hyperaccumulators that would be used additionally to enhance phytoremediation efficiency. Transgenic genes are certainly not going to cross over to wild relatives of transgenic plants because use of transgencics has not reached full blast in fields yet. Some researchers such as, Kotrba and co. (2009), Goel and co. (2009), Vangronsveld and others (2009), Chandra et al. (2015) have considered GEP's using them Some of the environmental risks that need to be considered on field-scale metal phytoremediation using GM plant species include. Examples of ecological threats include (a) exposing wildlife and people to toxic components such as genetic material in a more available structure (genetic contamination), (b) the uncensored spread of transgenic plants as a result of cross-pollination with wild relatives or increased fitness associated Risk assessment must be conducted prior to implementation of GEPs at the field scale [2009]. The engineering of symbiotic plant-associated microbes, such as PGPRs and endophytes (Pietrini et al. 2021) may also be a promising phytobacterial technology for enhancing tolerance to high metal concentrations and detoxification (Weyens et al. 2009, 2013; Huang et al. 2017). This involves the transferring, one or more genes that provide coded information on metal chelators, stress, uptake regulators, transporters, and enzymes promoting homeostasis (Ullah et al. 2015). However, engineering plant associated microbes is hardly applied as these bacteria are localized in rhizosphere mainly around the root zones and have little distribution outside it. Moreover, they depend on host plants (Wu et al. 2006).

The current "omics" technologies like proteomics, metabolomics, genomics, metagenomics, and transcriptomics have also increased the chance of recognizing these characteristics. As such, employing other strategies that may result to high affinity of metal binding, elevated metal and tolerance as well as enhanced metal accumulation in bacterial and plant system by introduction and over expression of the genes, and also metal-binding proteins may offer an alternative approach to phytoremediation.

12.6.2 Monitoring and Assessment of Remediation Effectiveness

This involves evaluating whether the objectives and criteria of the remediation have been achieved by examining the outcomes. This involves collecting and examining data on fluctuations in pollutant amounts, the state of the environment, and potential threats to human health and the ecosystem. Furthermore, assess the efficiency and influence of the remediation technology while identifying any insights gained or best practices that can be applied to future situations. In order to maintain sustainability, it is crucial to document and share evaluation findings and recommendations.

12.6.3 Suggested Ways to Measure and Evaluate Water Quality

- Physical Testing: This involves measuring the physical characteristics of water, such as temperature, pH, turbidity, color, and odor. These parameters can be measured using simple handheld instruments or more complex laboratory equipment.
- Chemical Testing: Chemical testing involves analyzing the concentration of various chemicals in the water, such as dissolved oxygen, nitrates, phosphates, metals, and pesticides. This can be done using various techniques, including titration, spectrophotometry, and chromatography.
- Biological Testing: Biological testing involves measuring the presence and abundance of various organisms in the water, such as bacteria, algae, and fish. This can be done through sampling, observation, or genetic techniques such as DNA sequencing.
- Microscopic Examination: This involves examining water samples under a microscope to look for the presence of bacteria, algae, and other microorganisms.
- Remote Sensing: This involves using satellite imagery and other remote sensing techniques to monitor water quality from a distance. This can provide valuable information on the distribution and extent of pollution and changes over time.

- Bioassays: This involves exposing living organisms, such as fish or bacteria, to water samples and measuring their response. This can provide information on the toxicity of the water and the potential harm to aquatic life.

12.7 Conclusion

The article underscores the significance of the reclamation of contaminated sites and the adverse health effects of heavy metals such as copper and lead. Additionally, it has discussed the association of high levels of some metals with adverse health effects, including liver and kidney damage, anemia, and gastrointestinal irritation, as well as its indirect association with neurological disorders. It highlights the resistance of heavy metals to biodegradation and their potential to cause toxic effects on soil, aquatic systems, and human health. Therefore the importance of efficient management of heavy-metal contaminated soil and water due to their negative impacts on biological systems and the environment cannot be overemphasized, as this paper suggests that it is only through the utilization of these techniques that we can maintain a sustainable environment.

References

Ahmad W, Alharthy RD, Zubair M et al (2021) Toxic and heavy metals contamination assessment in soil and water to evaluate human health risk. Sci Rep 11:17006. https://doi.org/10.1038/s41598-021-94616-4

Ajay Singh (2021) A review of wastewater irrigation: environmental implications. Resour Conserv Recycl 168:105454. https://doi.org/10.1016/j.resconrec.2021.105454. ISSN 0921-3449

Akar T, Tunali S, Kiran I (2005) Botrytis cinerea as a new fungal biosorbent for removal of Pb(II) from aqueous solutions. Biochem Eng J 25:227–235

Alfano M, Cavazza C (2020) Structure, function, and biosynthesis of nickel-dependent enzymes. Protein Sci 29(5):1071–1089. https://doi.org/10.1002/pro.3836. Epub 2020 Feb 18. PMID: 32022353; PMCID: PMC7184782

Ali I, Gupta VK (2006) Advances in water treatment by adsorption technology. Nat Protoc 1:2661–2667. https://doi.org/10.1038/nprot.2006.370

Ali S, Abbas Z, Rizwan M, Zaheer IE, Yavaş İ, Ünay A, Abdel-Daim MM, Bin-Jumah M, Hasanuzzaman M, Kalderis D (2020) Application of floating aquatic plants in phytoremediation of heavy metals polluted water: a review. Sustainability 12(5):1927. https://doi.org/10.3390/su12051927

Assi MA, Hezmee MNM, Sabri MYM, Rajion MA (2016) The detrimental effects of lead on human and animal health. Vet World 9(6):660

Bala S, Garg D, Thirumalesh BV, Sharma M, Sridhar K, Inbaraj BS, Tripathi M (2022) Recent strategies for bioremediation of emerging pollutants: a review for a green and sustainable environment. Toxics 10(8):484. https://doi.org/10.3390/toxics10080484. PMID: 36006163; PMCID: PMC9413587

Balali-Mood M, Naseri K, Tahergorabi Z, Khazdair MR, Sadeghi M (2021) Toxic mechanisms of five heavy metals: mercury, lead, chromium, cadmium, and arsenic. Front Pharmacol 12:643972. https://doi.org/10.3389/fphar.2021.643972. PMID: 33927623; PMCID: PMC8078867

Basavegowda N, Baek KH (2021) Current and future perspectives on the use of nanofertilizers for sustainable agriculture: the case of phosphorus nanofertilizer. 3 Biotech 11(7):357. https://doi.org/10.1007/s13205-021-02907-4. Epub 2021 Jun 28. PMID: 34268065; PMCID: PMC8239106

Beerling DJ, Leake JR, Long SP, Scholes JD, Ton J, Nelson PN, Bird M, Kantzas E, Taylor LL, Sarkar B, Kelland M, Hansen J (2018) Farming with crops and rocks to address global climate, food and soil security. Nat Plants 4(3):138–147

Briffa J, Sinagra E, Blundell R (2020) Heavy metal pollution in the environment and their toxicological effects on humans. Heliyon 6(9):e04691. https://doi.org/10.1016/j.heliyon.2020.e04691. ISSN 2405-8440

Cannas D, Loi E, Serra M, Firinu D, Valera P, Zavattari P (2020) Relevance of essential trace elements in nutrition and drinking water for human health and autoimmune disease risk. Nutrients 12(7):2074. https://doi.org/10.3390/nu12072074. PMID: 32668647; PMCID: PMC7400883

Cárdenas González JF, Rodríguez Pérez AS, Vargas Morales JM, Martínez Juárez VM, Rodríguez IA, Cuello CM, Muñoz Morales A (2019) Bioremoval of cobalt (II) from aqueous solution by three different and resistant fungal biomasses. Bioinorg Chem Appl 2019(1):8757149

Cerino-Córdova FD, García-León AM, Soto-Regalado E, Sánchez-González MN, Lozano-Ramírez T, García-Avalos BC, Loredo-Medrano JA (2012) Experimental design for the optimization of copper biosorption from aqueous solution by Aspergillus terreus. J Environ Manag 95:S77–S82

Chandra S, Singh S, Kumari D (2015) Evaluation of functional properties of composite flours and sensorial attributes of composite flour biscuits. J Food Sci Technol 52:3681–3688

Chang H, Hu R, Zou Y, Quan X, Zhong N, Zhao S, Sun Y (2020) Highly efficient reverse osmosis concentrate remediation by microalgae for biolipid production assisted with electrooxidation. Water Res 174. https://doi.org/10.1016/j.watres.2020.115642

Chelgani SC, Asimi Neisiani A (2022) Gravity separation. In: Dry mineral processing. Springer, Cham. https://doi.org/10.1007/978-3-030-93750-8_3

Conte A, Chiaberge S, Pedron F, Barbafieri M, Petruzzelli G, Vocciante M, Franchi E, Pietrini I (2021) Dealing with complex contamination: a novel approach with a combined bio-phytoremediation strategy and effective analytical techniques. J Environ Manag 288:112381

Das O, Sarmah AK, Bhattacharyya D (2015) Structure–mechanics property relationship of waste derived biochars. Sci Total Environ 538:611–620

Das S, Sultana KW, Ndhlala AR, Mondal M, Chandra (2023) Heavy metal pollution in the environment and its impact on health: exploring green technology for remediation. Environ Health Insights 5(17):11786302231201259. https://doi.org/10.1177/11786302231201259. PMID: 37808962; PMCID: PMC10559720

Deblina D, Arya S, Kumar S (2021) Industrial wastewater treatment: current trends, bottlenecks, and best practices. Chemosphere 285:131245. https://doi.org/10.1016/j.chemosphere.2021.131245. ISSN 0045-6535

Dhankhar R, Hooda A (2011) Fungal biosorption–an alternative to meet the challenges of heavy metal pollution in aqueous solutions. Environ Technol 32(5):467–491

Dhankhar S, Ruhil S, Balhara M, Dhankhar S, Chhillar AK (2011) Aegle marmelos (Linn.) Correa: a potential source of Phytomedicine. J Med Plant Res 5(9):1497–1507

EC—European Commission (2009) Directive 2009/28/EC of the European Parliament and of the Council of 23 April 2009 on the promotion of the use of energy from renewable sources and amending and subsequently repealing directives 2001/77/EC and 2003/30/EC. European Commission, Brussels

El Ati-Hellal M, Hellal F (2021) Heavy metals in the environment and health impact. Intech Open. https://doi.org/10.5772/intechopen.97204

Fasih UH, Liqun C, Coulter JA, Cheema SA, Wu J, Zhang R, Wenjun M, Farooq M (2021) Cadmium toxicity in plants: impacts and remediation strategies. Ecotoxicol Environ Saf 211:111887. https://doi.org/10.1016/j.ecoenv.2020.111887. ISSN 0147-6513

Fomina M, Gadd GM (2014) Biosorption: current perspectives on concept, definition and application. Biores Technol 160:3–14

Franson JC, Hollmén TE, Flint PL, Matz AC (2022) Trace elements in blood of sea ducks from Dutch Harbor and Izembek Lagoon, Alaska. J Fish Wildl Manag 14(1):41–50. https://doi.org/10.3996/JFWM-21-065

Gasparatos D (2022) Soil contamination by heavy metals and metalloids. Environments 9(3):32. https://doi.org/10.3390/environments9030032

Goel N, Rao H, Durmer JS, Dinges DF (2009) Neurocognitive consequences of sleep deprivation. In Seminars in Neurology, vol 29, No. 04, pp 320–339. © Thieme Medical Publishers.

Grillitsch M, Hansen T (2019) Green industry development in different types of regions. Eur Plan Stud 27(11):2163–2183. https://doi.org/10.1080/09654313.2019.1648385

Grzegorzek M, Majewska-Nowak K, Ahmed AE (2020) Removal of fluoride from multicomponent water solutions with the use of monovalent selective ion-exchange membranes. Sci Total Environ 722. https://doi.org/10.1016/j.scitotenv.2020.137681. https://doi.org/10.33263/LIANBS124.125. https://nanobioletters.com/

Grzegorz I, Mikula K, Skrzypczak D, Moustakas K, Witek-Krowiak A, Chojnacka K (2021) Potential environmental pollution from copper metallurgy and methods of management. Environ Res 197:111050. https://doi.org/10.1016/j.envres.2021.111050. ISSN 0013-9351

Gupta VK, Nayak A, Agarwal S (2015) Bioadsorbents for remediation of heavy metals: current status and their future prospects. Environ Eng Res 20:1–18

Hama KHA, Mustafa FS, Omer KM, Hama S, Hamarawf RF, Rahman KO (2023) Heavy metal pollution in the aquatic environment: efficient and low-cost removal approaches to eliminate their toxicity: a review. RSC Adv 13(26):17595–17610. https://doi.org/10.1039/d3ra00723e. PMID: 37312989; PMCID: PMC10258679

Hao Z, Yin H, Wang L, Meng Y (2014) Wear behavior of seven artificial resin teeth assessed with three-dimensional measurements. J Prosthet Dent 112:1507–1512. https://doi.org/10.1016/j.prosdent.2014.04.030

He C, Liu Z, Wu J et al (2021) Future global urban water scarcity and potential solutions. Nat Commun 12:4667. https://doi.org/10.1038/s41467-021-25026-3

Hou D, Bolan NS, Tsang DC, Kirkham MB, O'connor D (2020) Sustainable soil use and management: an interdisciplinary and systematic approach. Sci Total Environ 729:138961

Huang J, Qi F, Zeng G, Shi L, Li X, Gu Y, Shi Y (2017) Repeating recovery and reuse of SDS micelles from MEUF retentate containing Cd^{2+} by acidification UF. Colloids Surf A 520:361–368. https://doi.org/10.1016/j.colsurfa.2017.02.001

Joseph D (ed) (2023) Trace metals in the environment. IntechOpen. https://doi.org/10.5772/intechopen.1001530

Jumadi J, Kamari A, Hargreaves JSJ, Yusof N (2020) A review of nano-based materials used as flocculants for water treatment. Int J Environ Sci Technol 17:3571–3594. https://doi.org/10.1007/s13762-020-02723-y

Kandasamy S, De Silva S, Santosh SS, Sathiyaseelan A, Ganeshalingam A, Jamla M, Sankaranarayanan A, Veeraraghavan VP, MubarakAli D, Lee J, Thiripuranathar G, Wang M-H (2022) Impact of industrial effluents on the environment and human health and their remediation using MOFs-based hybrid membrane filtration techniques. Chemosphere 307(Part 1):135593. https://doi.org/10.1016/j.chemosphere.2022.135593. ISSN 0045-6535

Karavolias NG, Horner W, Abugu MN, Evanega SN (2021) Application of gene editing for climate change in agriculture. Front Sustain Food Syst 5:296

Karppinen EM, Siciliano SD, Stewart KJ (2017) Application method and biochar type affect petroleum hydrocarbon degradation in northern landfarms. J Environ Qual 46(4):751–759

Kim IH, Choi J-H, Joo JO, Kim Y-K, Choi J-W, Oh B-K (2015) Development of a microbe-zeolite carrier for the effective elimination of heavy metals from seawater. J Microbiol Biotechnol 25:1542–1546

Kotrba P, Najmanova J, Macek T, Ruml T, Mackova M (2009) Genetically modified plants in phytoremediation of heavy metal and metalloid soil and sediment pollution. Biotechnol Adv 27(6):799–810

Kshyanaprava R, Das AP (2023) Lead pollution: impact on environment and human health and approach for a sustainable solution. Environ Chem Ecotoxicol 5:79–85. https://doi.org/10.1016/j.enceco.2023.02.001. ISSN 2590-1826

Kumar D, Dhankher OP, Tripathi RD, Seth CS (2023) Titanium dioxide nanoparticles potentially regulate the mechanism (s) for photosynthetic attributes, genotoxicity, antioxidants defense machinery, and phytochelatins synthesis in relation to hexavalent chromium toxicity in Helianthus annuus L. J Hazard Mater 454:131418

Lee SY, Kim HU, Chae TU, Cho JS, Kim JW, Shin JH, Kim DI, Ko YS, Jang WD, Jang YS (2019). A comprehensive metabolic map for production of bio-based chemicals. Nat Catal 2(1):18–33

Lesmana SO, Febriana N, Soetaredjo FE, Sunarso J, Ismadji S (2009) Studies on potential applications of biomass for the separation of heavy metals from water and wastewater. Biochem Eng J 44:19–41

Liang D, He W, Li C, Wang F, Crittenden JC, Feng Y (2021) Remediation of nitrate contamination by membrane hydrogenotrophic denitrifying biofilm integrated in microbial electrolysis cell. Water Res 188. https://doi.org/10.1016/j.watres.2020.116498

Lianwen L, Li W, Song W, Guo M (2018) Remediation techniques for heavy metal-contaminated soils: principles and applicability. Sci Total Environ 633:206–219. https://doi.org/10.1016/j.scitotenv.2018.03.161. ISSN 0048-9697

Ma J, Fu K, Fu X, Guan Q, Ding L, Shi J, Zhu G, Zhang X, Zhang S, Jiang L (2017) Flocculation properties and kinetic investigation of polyacrylamide with different cationic monomer content for high turbid water purification. Sep Purif Technol 182:134–143. https://doi.org/10.1016/j.seppur.2017.03.048

Mallavarapu M, Ramakrishnan B, Venkateswarlu K, Sethunathan N, Naidu R (2011) Bioremediation approaches for organic pollutants: a critical perspective. Environ Int 37(8):1362–1375. https://doi.org/10.1016/j.envint.2011.06.003. ISSN 0160-4120

Manisalidis I, Stavropoulou E, Stavropoulos A, Bezirtzoglou E (2020) Environmental and health impacts of air pollution: a review. Front Public Health 8:14. https://doi.org/10.3389/fpubh.2020.00014. PMID: 32154200; PMCID: PMC7044178

Mishra A, Malik A (2013) Recent advances in microbial metal bioaccumulation. Crit Rev Environ Sci Technol 43:1162–1222

Mosa KA, Saadoun I, Kumar K, Helmy M, Dhankher OP (2016) Potential biotechnological strategies for the cleanup of heavy metals and metalloids. Front Plant Sci 7:1–14

Nabi M, Tabassum N (2022) Role of environmental toxicants on neurodegenerative disorders. Front Toxicol 11(4):837579. https://doi.org/10.3389/ftox.2022.837579. PMID: 35647576; PMCID: PMC9131020

Othmani B, Rasteiro MG, Khadhraoui M (2020) Toward green technology: a review on some efficient model plant-based coagulants/flocculants for freshwater and wastewater remediation. Clean Technol Environ Policy 22:1025–1040. https://doi.org/10.1007/s10098-020-01858-3

Patil SS, Shedbalkar UU, Truskewycz A, Chopade BA, Ball AS (2016) Nanoparticles for environmental clean-up: a review of potential risks and emerging solutions. Environ Technol Innov 5:10-21

Pawlik M (2022) Fundamentals of froth flotation. ChemTexts 8. https://doi.org/10.1007/s40828-022-00170-5

Pedron F, Grifoni M, Barbafieri M, Petruzzelli G, Rosellini I, Franchi E, Bagatin R, Vocciante M (2017) Applicability of a Freundlich-like model for plant uptake at an industrial contaminated site with a high variable arsenic concentration. Environments 4:67

Pedron F, Grifoni M, Barbafieri M, Petruzzelli G, Franchi E, Samà C, Gila L, Zanardi S, Palmery S, Proto A et al (2021) New light on phytoremediation: the use of luminescent solar concentrators. Appl Sci 11:1923

Pietrini I, Grifoni M, Franchi E, Cardaci A, Pedron F, Barbafieri M, Petruzzelli G, Vocciante M (2021) Enhanced lead phytoextraction by endophytes from indigenous plants. Soil Syst 5:55

Priya AK, Muruganandam M, Ali SS, Kornaros M (2023) Clean-up of heavy metals from contaminated soil by phytoremediation: a multidisciplinary and eco-friendly approach. Toxics 11(5):422. https://doi.org/10.3390/toxics11050422. PMID: 37235237; PMCID: PMC10221411

Puyen ZM, Villagrasa E, Maldonado J, Diestra E, Esteve I, Solé A (2012) Biosorption of lead and copper by heavy-metal tolerant Micrococcus luteus DE2008. Bioresour Technol 126:233–237

Raffa CM, Chiampo F, Shanthakumar S (2021) Remediation of metal/metalloid-polluted soils: a short review. Appl Sci 11(9):4134. https://doi.org/10.3390/app11094134

Ravenscroft P, Brammer H, Richards K (2009) Arsenic pollution: a global synthesis. Wiley-Blackwell

Raymond AW, Felix EO (2011) Heavy metals in contaminated soils: a review of sources, chemistry, risks and best available strategies for remediation. Int Sch Res Not Article ID 402647, 20 pp. https://doi.org/10.5402/2011/402647

Rehman M, Liu L, Wang Q et al (2019) Copper environmental toxicology, recent advances, and future outlook: a review. Environ Sci Pollut Res 26:18003–18016. https://doi.org/10.1007/s11356-019-05073-6

Robin M, Székács A, Zaller JG (2021) 1—Herbicides: brief history, agricultural use, and potential alternatives for weed control. In: Robin M, Zaller JG (eds) In emerging issues in analytical chemistry, herbicides. Elsevier, pp 1–20. https://doi.org/10.1016/B978-0-12-823674-1.00002-X. ISBN 9780128236741

Saikat M, Chakraborty AJ, Tareq AM, Emran TB, Nainu F, Khusro A, Idris AM, Khandaker MU, Osman H, Alhumaydhi FA, Simal-Gandara J (2022) Impact of heavy metals on the environment and human health: novel therapeutic insights to counter the toxicity. J King Saud Univ Sci 34(3)

Santhosh C, Velmurugan V, Jacob G, Jeong SK, Grace AN, Bhatnagar A (2016) Role of nanomaterials in water treatment applications: a review. Chem Eng J 306:1116–1137. https://doi.org/10.1016/j.cej.2016.08.053

Saxena G, Purchase D, Mulla S, Saratale G, Bharagava R (2019) Phytoremediation of heavy metal-contaminated sites: eco-environmental concerns, field studies, sustainability issues and future prospects. Rev Environ Contam Toxicol. https://doi.org/10.1007/398_2019_24

Shah V, Daverey A (2020) Phytoremediation: A multidisciplinary approach to clean up heavy metal contaminated soil. Environ Technol Innov 18:100774

Sinha A, Pant KK, Khare SK (2012) Studies on mercury bioremediation by alginate immobilized mercury tolerant Bacillus cereus cells. Int Biodeterior Biodegrad 71:1–8

Srivastava S, Thakur IS (2006) Isolation and process parameter optimization of Aspergillus sp. For removal of chromium from tannery effluent. Bioresour Technol 97:1167–1173

Srivastava S, Agrawal S, Mondal M (2015) A review on progress of heavy metal removal using adsorbents of microbial and plant origin. Environ Sci Pollut Res 22:15386–15415

Srivastava P, Bolan N, Casagrande V, Benjamin J, Adejumo SA, Sabir M, Farooqi ZUR, Saifullah, Sarkar A (2022) Chapter 14—Lead in soils: sources, bioavailability, plant uptake, and remediation. In: Kumar V, Sharma A, Setia R (eds) Appraisal of metal(loids) in the ecosystem. Elsevier, pp331–360. ISBN 9780323856218

Tawfik AS, Mustaqeem M, Khaled M (2022) Water treatment technologies in removing heavy metal ions from wastewater: a review. Environ Nanotechnol Monit Manag 17:100617. https://doi.org/10.1016/j.enmm.2021.100617. ISSN 2215-1532

Tchounwou PB, Yedjou CG, Udensi UK, Pacurari M, Stevens JJ, Patlolla AK, Noubissi F, Kumar S (2019) State of the science review of the health effects of inorganic arsenic: perspectives for future research. Environ Toxicol 34(2):188–202. https://doi.org/10.1002/tox.22673. Epub 2018 Dec 4. PMID: 30511785; PMCID: PMC6328315

Ullah A, Heng S, Munis MFH, Fahad S, Yang X (2015) Phytoremediation of heavy metals assisted by plant growth promoting (PGP) bacteria: a review. Environ Exp Bot 117:28–40

UNEP F (2024) Climate Risks in the Transportation Sector

Usman M, Farooq M, Wakeel A, Nawaz A, Cheema SA, ur Rehman H, Ashraf I, Sanaullah, M (2020) Nanotechnology in agriculture: current status, challenges and future opportunities. Sci Total Environ 721:137778

USEPA (2017) 2017 potable reuse compendium
Vijendra S, Daverey A (2020) Phytoremediation: a multidisciplinary approach to clean up heavy metal contaminated soil. Environ Technol Innov 18:100774. https://doi.org/10.1016/j.eti.2020.100774. ISSN 2352-1864
Wang C, Alpatova A, McPhedran KN, Gamal El-Din M (2015) Coagulation/flocculation process with polyaluminum chloride for the remediation of oil sands process-affected water: performance and mechanism study. J Environ Manag 160:254–262. https://doi.org/10.1016/j.jenvman.2015.06.025
Wang C, Wang Z, Wei X, Li X (2019) A numerical study and flotation experiments of bicyclone column flotation for treating of produced water from ASP flooding. J Water Process Eng 32. https://doi.org/10.1016/j.jwpe.2019.100972
Wang J, Chen C (2009) Biosorbents for heavy metals removal and their future. Biotechnol Adv 27:195–226
Weyens N, van der Lelie D, Taghavi S, Newman L, Vangronsveld J (2009) Exploiting plant–microbe partnerships to improve biomass production and remediation. Trends Biotechnol 27(10):591–598
Weyens N, Schellingen K, Beckers B, Janssen J, Ceulemans R, van der Lelie, D, Taghavi S, Carleer R, Vangronsveld J (2013) Potential of willow and its genetically engineered associated bacteria to remediate mixed Cd and toluene contamination. J Soils Sediments 13:176–188.
Wu SC, Cheung KC, Luo YM, Wong MH (2006) Effects of inoculation of plant growth-promoting rhizobacteria on metal uptake by Brassica juncea. Environ Pollut 140(1):124–135

Chapter 13
Role of Nanotechnology in the Remediation of Heavy Metals

Hridya Thrikkaikkal and P. S. Harikumar

Abstract When it comes to monitoring the quality of drinking water on a worldwide scale, heavy metal contamination is the most pressing worry. Water quality and human health are gravely threatened by the extensive pollution of heavy metals. Because it is poisonous, bioaccumulates, and does not biodegrade, drinking this polluted water can have devastating effects on communities and individuals. Decontamination procedures that are both quick and energy efficient are necessary to solve the problem. Optoelectronics, chemistry, device storage, physics, medicine, biology, and environmental restoration are just a few of the many fields that can benefit from nanotechnology, which is a fast expanding field. Among nanotechnology's numerous uses, the employment of nanoparticles to clean up polluted environments may stand out. An area of growing interest in the application of nanotechnology is the study of nanoparticles as a heavy metal adsorbent for use in water purification systems. The unique potential of nanoparticle-based remediation to decrease the need for treatment and disposal of polluted water by eliminating or converting organic or inorganic pollutants into harmless forms, as well as the high surface area that these materials offer in a very small volume, led to their selection. Because of their improved and synergistic activity, BNPs have recently been utilised as an additional layer of protection against heavy metal contamination in different kinds of nanoparticles. Because of their bigger surface area, bimetallic nanoparticles are more capable of adsorption than their monometallic equivalent.

Keywords Heavy metals · Bimetallic nanoparticle · Adsorption · Nanocomposite · Wastewater treatment

H. Thrikkaikkal · P. S. Harikumar (✉)
Ecology and Environment Research Group, Centre for Water Resources Development and Management, Kunnamangalam, Kozhikode 673571, Kerala, India
e-mail: drpshari@gmail.com

N. Kumar (ed.), *Heavy Metal Toxicity*, Environmental Science and Engineering,
https://doi.org/10.1007/978-3-031-56642-4_13

13.1 Introduction

13.1.1 Heavy Metal Contamination

In recent years, there has been a growing global concern over the presence of high amounts of heavy metals in aquatic resources. Heavy metals are poisonous, carcinogenic, and non-biodegradable, in contrast to organic pollutants. In addition, because they are not biodegradable, they have the potential to build up in living tissues, eventually concentrating in vital organs including the liver, kidneys, and brain (Pehlivan et al. 2008; Schwarzenbach et al. 2010). Some of the body's most important systems may become impaired due to exposure to these heavy metals. In addition, heavy metals are long-lived environmental contaminants because they are resistant to bacterial action and other processes that break them down.

The transfer of hazardous heavy metals from poorly or not treated effluents into the food chain has far-reaching effects on ecosystems. "A number of metal ions have worrying effects when they accumulate in industrial wastewater at high concentrations; among them are lead, mercury, cadmium, zinc, manganese, and chromium in their aqueous soluble forms, which are regarded as emergent water pollutants" (Simeonidis et al. 2019). According to Al-Othman et al. (2012) and Singh et al. (2011), even in small amounts, some heavy metals can harm the health of humans and animals. According to Singh and Prasad (2015), heavy metals can enter human tissues through many absorption routes, such as climbing the food chain, directly consuming, inhaling, swallowing, and skin contact. According to Wang and Chen (2009), the treatment of toxic metal-containing industrial effluents before to release is thus becoming an increasingly important concern.

The recent reports of hazardous metal releases into the environment in India, including lead (Pb), cadmium (Cd), nickel (Ni), cobalt (Co), chromium (Cr), barium (Ba), mercury (Hg), arsenic (As), selenium (Se), and others, have become a significant national concern (Marg 2011). Geochemical processes, soil erosion, urban runoff, aerosols, mineral weathering, and volcanic activity are all natural sources of heavy metals. The primary source of heavy metal contamination caused by humans is wastewater discharge from various industries, including but not limited to: electroplating, paints, pigment manufacturing, batteries, water cooling, leather tanning, paints, pigments, and mining (Tchounwou et al. 2012). Pandey and Pandey (2009) and Prasse et al. (2012) both cite emissions from cars and other urban activities as additional contributors. Also, surface and groundwater nearby might be contaminated with these heavy metals that are released from polluted places because of how easily they are transferred through the ecosystem. According to Nriagu and Pacyna (1988), this often leads to the dispersal of these metals over extensive regions, resulting in widespread pollution.

Due to their potential harmful effects on human health, the US Environmental Protection Agency (USEPA) determined the Maximum Contaminant Level (MCL) of heavy metals in drinking water (USEPA 2019; Babel and Kurniawan 2003). Chlorination, boiling, and sun disinfection are among the most popular methods for treating water for human consumption, however they are not successful in eliminating these heavy metals (De Kwaadsteniet et al. 2013). The massive expansion of both humans and their natural habitats makes the treatment of heavy metals an especially pressing issue. For this reason, developing in-situ treatment solutions for heavy metals that are both cost-effective and ecologically friendly is crucial. The US Environmental Protection Agency has established regulations regarding the release of heavy metals into wastewater and drinking water due to their non-biodegradable characteristics and extremely hazardous nature, even at extremely low quantities. In Table 13.1, we can see the main heavy metal sources together with the possible health impacts of these substances. According to the USEPA, WHO, and BIS, the maximum contaminant levels (MCCLs) of heavy metals in drinking water are displayed in Table 13.2.

Table 13.1 Characteristic of common heavy metals

Heavy metal	Health effects	Sources
Lead (Pb)	Carcinogenic, anemia, joint pains, high blood pressure, kidney problems, damage in circulatory system and nervous system	Lead acid batteries, paints, coal-based thermal power plants, ceramics, house hold plumbing system
Cadmium (Cd)	Kidney damage, renal disorder, carcinogenic, chronic lung disease	Zinc smelting, waste batteries, paint sludge, incinerations and fuel combustion
Manganese (Mn)	Toxic, causes growth retardation, muscles fatigue, eye blindness, sexual impotence	Coal mining, production of dry battery cells, matches, fireworks, glass-bonding materials
Zinc (Zn)	Stomach cramps, respiratory disorders, nausea, vomiting, skin irritations, anemia	Smelting, mining and metal production, dye, wood preservatives, ointments and electroplating industries
Arsenic (As)	Vascular disease, skin manifestations, visceral cancers	Pesticides and wood preservatives
Mercury (Hg)	Rheumatoid arthritis, diseases of the kidneys, circulatory system, and nervous system	Release from Au–Ag mining and coal combustion, electronics, medical waste
Nickel (Ni)	Dermatitis, nausea, chronic asthma, coughing, human carcinogen	Automobile batteries, industrial effluents, surgical instruments, steel alloys, kitchen appliances, etc.
Copper (Cu)	Wilson disease, liver damage, insomnia	Pesticides, fertilizers
Chromium (Cr)	Carcinogenic, diarrhea, nausea, headache, vomiting	Tanneries, electroplating, steel industries, fly ash

Table 13.2 Maximum contaminant level (MCL) of heavy metal according to USEPA and WHO

Heavy metal	Maximum contaminant level		
	USEPA[a] (mg/L)	WHO[b] (mg/L)	BIS IS:10500[c] (mg/L)
Pb	0.015	0.01	0.01
Cd	0.005	0.003	0.003
Mn	0.05	0.4	0.1
Zn	5	5	5
Fe	3	* *NGL*	3
Cu	1.3	2	0.05
Ni	0.1	0.07	0.02
Cr	0.1	0.05	0.05
Hg	0.002	0.006	0.001
As	0.01	0.01	0.01

[a] Value by United States Environment Protection Agency (USEPA 2008, 2019)
[b] Values prescribed by World Health Organization (WHO Chronicle 2011; WHO 2008)
[c] Values prescribed by Bureau of Indian Standards (BIS 2012)
* *NGL*: No Guideline, because it is not of health concern at amounts typically found in drinking water, but may affect the acceptability of water at concentration more than 300 μg/L

13.1.2 Methods for the Removal of Heavy Metal Ions

Many conventional methods have been used to remove heavy metals from water and industrial wastewater, including chemical precipitation (Meunier et al. 2006; Matlock et al. 2002), reverse osmosis (Weber and Holz 1991; Akin et al. 2011), ion exchange (Alyüz and Veli 2009; Xing et al. 2007), membrane filtration (Ersahin et al. 2012), coprecipitation (Kameda et al. 2014), electrochemical treatment (Feng et al. 2007; Walsh and Reade 1994) and adsorption (Sun et al. 2015; Alluri et al. 2007). Other than these, some methods include reduction and oxidation, screening, neutralization and remineralization, sedimentation, crystallization, fluidization, flotation, gravity separation, evaporation, solvent extraction, distillation and so on. Table 13.3 provides an overview of those methods along with their advantages and disadvantages.

13.1.2.1 Adsorption

Given its excellent performance, high efficiency, reduction of secondary waste, and economically viable nature, adsorption stands out among all these techniques (Ren et al. 2014; Wang et al. 2019). It is effective against both soluble and insoluble contaminants. Due to its simplicity and great efficacy in removing tiny amounts of heavy metal ions, adsorption has become one of the most popular strategies for treating wastewater that contains these ions (Zamboulis et al. 2011). The process of reversible adsorption entails transferring adsorbate ions from the adsorbent's surface

Table 13.3 Techniques for the removal of heavy metal ions along with their advantages and disadvantages

Methodology	Advantages	Disadvantages
Chemical precipitation	Simple operation, low capital cost, effective for removal of large quantities of metals	Removal of only a few metals, sludge generation, extra operational cost for sludge disposal
Ion exchange	High selectivity and efficiency, low sludge volume, high regeneration efficiency	High initial capital cost, high maintenance cost
Electrochemical treatment	Rapid process, capable of selective removal of metals	High maintenance and operation costs, large generation of by products, filtration process for flocs
Membrane filtration	Transport selectivity of the membrane, low chemical consumption, needs low energy, facile in operation small space requirement	Relatively expensive to install, ease of membrane blocking, prone to fouling, scaling or membrane degradation, limited flow rates
Coagulation and flocculation	Economically feasible, bacterial inactivation capability	High sludge formation and chemical consumption
Adsorption	Flexibility and simplicity of design, high surface area, accessible surface sites, high capacity, low cost, a wide verity of target pollutants, fast kinetics	Adsorbent requires regeneration, performance depends on type of adsorbent

to the solution; subsequent regeneration allows for reuse of the adsorbent (Mishra 2014).

When pollutant molecules (whether in a liquid or gaseous state) physically or chemically bind to a solid surface, adsorption takes place on the surface. Many chemicals can be extracted from liquids using this method. Adsorption takes place on a solid surface, which is known as the adsorbent, and the substance that is adsorbed is called the adsorbate. Figure 13.1 provides a schematic description of the adsorption process and the components involved in it. Figure 13.1 shows that adsorption is often accompanied by desorption, and that the regeneration efficiency of the adsorbent is proportional to the amount of adsorbate desorbed from the solid. Figure 13.1 shows that under certain conditions, an increase in desorption leads to an increase in regeneration efficiency of the adsorbent. The adsorption experiments are usually run in a column filled with porous adsorbents, either in a batch or continuous fashion.

The adsorbent's surface has an imbalanced force of attraction, which causes adsorption to happen. Therefore, the fields of forces on the surface of the adsorbent interact with the fields of forces on the liquid or gas when they come into contact. The surface of the adsorbent material tends to attract and retain gas or liquid molecules, atoms, or ions, thus satisfying these residual forces. The term "physical adsorption" can refer to either weak van der Waals forces, electrostatic attraction, or hydrogen bonding as the mechanisms by which the adsorbate adheres to the surface.

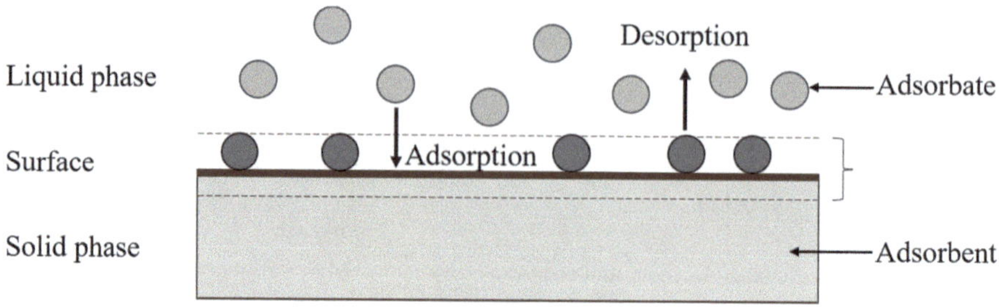

Fig. 13.1 The adsorption process is represented schematically

In chemical adsorption, also known as chemisorption, the adsorbate and surface sites form bonds through chemical processes, either covalently or ionicly (Fig. 13.2). The physical adsorption process is characterised by a low heat of adsorption of 15–30 kJ/mol, is reversible, and is not selective. By contrast, chemisorption is selective, permanent, and has an adsorption heat ranging from 10 to 100 kJ/mol (Singh and Gupta 2016; Tripathi et al. 2015; Tyagi et al. 2017).

The adsorption of pollutant on the surface of adsorbent takes place in three steps:

Step 1: Film diffusion-the transportation of the pollutant molecules from the bulk solution to the adsorbent's exterior surface.

Step 2: Pore diffusion-transport of material by diffusion from the external surface into the pores of the adsorbent.

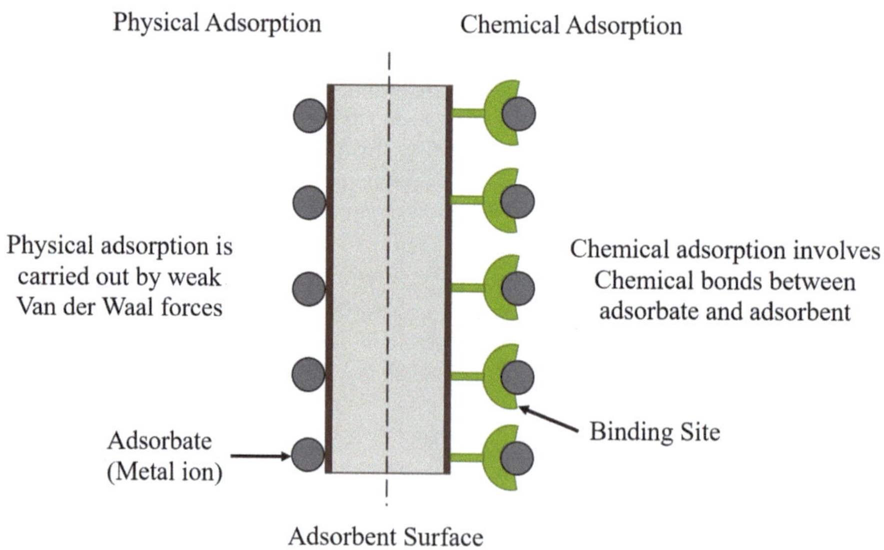

Fig. 13.2 Schematic representation of physical and chemical adsorption

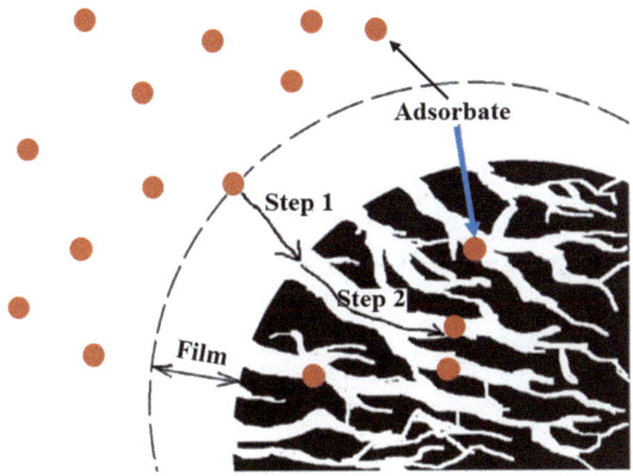

Fig. 13.3 Trajectory of an adsorbate molecule during adsorption

Step 3: Adsorption on the internal surface of the adsorbent via physisorption or chemisorption.

The first two steps in this case are transport steps, while the final step is a reaction step, and Fig. 13.3 represent the step 1 and 2 of an adsorbate molecule during adsorption.

Recent reports have indicated the use of a variety of adsorbents for the removal of heavy metals from wastewaters and aqueous solutions, including graphene oxide, activated carbon, zeolite, alumina, nanoparticles, nanocomposites, and metal oxides (Wang and Peng 2010; Vodyanitskii 2010; Wang et al. 2012). Size, shape, surface area, coating, charge, and other physicochemical properties of adsorbates and adsorbents are determined by their interactions with one another. Finding a new adsorbent material that effectively removes contaminants from water has thus been an ongoing research focus for a long time.

13.1.3 *Nanotechnology*

Nanotechnology, often referred as the science of the small, is a novel and emerging technology that dominating the various fields of research. It is regarded as a multidisciplinary field of study that combines engineering aspects with material science, physics, biology, chemistry, and medicine. Hence, it enhanced the area of scientific as well as industrial research.

13.1.3.1 Nanomaterials—At a Glance

In a speech he delivered at a December 1959 American Physical Society conference, Nobel laureate and "Father of nanotechnology" Richard Feyman first proposed the concept of nanotechnology with the words "There is a plenty of room at the bottom" (Feynman 1959). Additionally, Feynman outlined the behaviour of objects at the microscopic level and the various methods that can reduce particles to their microscopic forms (Feynman 1960).

A "nanometre" is one billionth of a metre, and anything with a size between 1 and 100 nm is called a nanomaterial (NM), meaning it is larger than molecular dimensions but smaller than macroscopic ones (Clifford et al. 2020). A material's physical and chemical properties fall somewhere in the middle between the bulk material's and the atomic or molecular structures when its diameter is less than 100 nm (Tarafdar et al. 2013). These features—a high surface-to-volume ratio, a large proportion of atoms at the surface, a high number of particles per mass unit, a high number of surface-reactive sites, a large surface free energy, and surface interactions—contribute to this phenomenon, which is associated with the quantum effects and surface effects—the main attraction of nanotechnology (Roduner 2006; Buzea et al. 2007).

While the physical properties of large materials do not change with size, the physicochemical properties of nanomaterials (NMs) can be influenced by variables such size, shape, surface characteristics, and interior structure. These variables encompass chemical, electrical, mechanical, and optical properties. Dehydrating a material into the nanoscale range increases its active surface area, a crucial parameter in many processes including catalysis and adsorption. Subtle changes in NMs' size, shape, and composition can lead to dramatic shifts in their characteristics. Li et al. (2011) and Pal et al. (2011) found that NMs have unexpected and intriguing applications in several fields, including medicine, optics, electronics, ceramics, catalysis, chemical industries, optics, and water treatment. Therefore, strategies are developed to leverage the regime of size-controlled attributes for the purpose of fabricating novel or enhanced physical, chemical, and biological capabilities that cannot be achieved with conventional bulk particles of comparable composition.

When discussing nanomaterials, a variety of words are frequently used, which might be confusing for less-experienced researchers or those who are new to the topic. In specifically, nanomaterials include hollow spheres, nanotubes, nanowires, nanospheres, nanofibres, nanofilms, nanofluids, nanoribbons, nanorods, and quantum dots in addition to NPs, which could be the most popular one (Kuhn et al. 2022).

13.1.3.2 Classification of Nanomaterials

There are several approaches for classifying nanomaterials. On the basis of dimensionality, morphology, uniformity, composition, and agglomeration, NMs are generally classified as follows (Saleh 2020).

Dimensionality

On the basis of dimensionality, nanomaterials can be categorised as zero dimensional materials (0D), one dimensional materials (1D), two dimensional materials (2D) and three-dimensional materials (3D) (Pokropivny and Skorokhod 2008).

(1) **Zero-dimensional nanomaterials (0D)**: This class of nanomaterials have all their three dimensions at the nanoscale level. e.g., nanoparticles, quantum dots, and fullerenes.
(2) **One-dimensional nanomaterials (1D)**: This class of nanomaterials have one dimension outside the nanometer scale while the other two dimensions are in nanoscale. e.g., nanofibers (like carbon nanofiber), nanotubes, nanorods, nanowires and nanocore shells.
(3) **Two-dimensional nanomaterials (2D)**: This class of nanomaterials have two dimensions outside the nanometer scale. e.g., nanosheets (graphene), nanofilms, and nanolayers (multilayered nanosheets).
(4) **Tree-dimensional nanomaterials (3D)**: Nanomaterials which are not confined to the nanoscale in any dimension. They create hybrid structures as a result of their unique confinement in a three-dimensional space. e.g., bulk powders, hybrid nanostructures, arrays of nanowires, carbon nanobuds, and nanotubes, etc.

Morphology

(1) **High aspect ratio**: The length exceeds the width by many folds. Nanotubes and nanowires fall under this category.
(2) **Low aspect ratio**: This covers spherical, oval, cubic, prism, helical, and pillar forms. e.g., nanospheres, nanocubes, nanopyramids, etc.

In addition to aspect ratio, NMs' morphological characteristics also include sphericity and flatness.

Uniformity

Depending on their uniformity, NMs can be categorised as dispersed and agglomerate or isometric and inhomogeneous.

Chemical Composition

NMs can be categorised into a number of groups according to their chemical composition, including organic, carbon-based, and inorganic, which in turn have further varieties (Joudeh and Linke 2022).

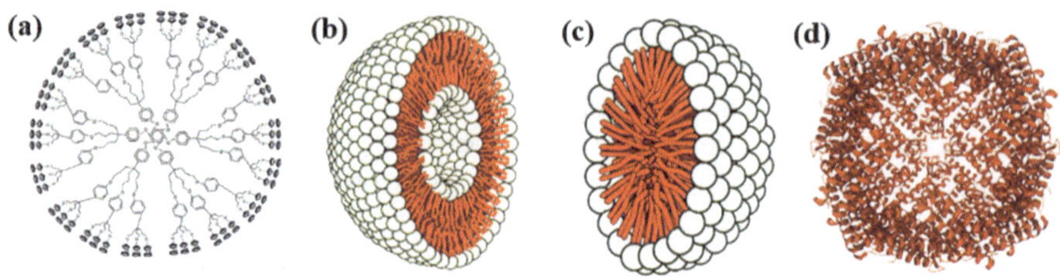

Fig. 13.4 Different types of organic NMs—**a** dendrimers; **b** liposomes; **c** micelles; and **d** ferritin

(1) **Organic nanomaterials**: NMs consisting of proteins, carbohydrates, lipids, polymers, or any other organic substances are generally known as organic NMs. They are most commonly employed in the biomedical industry for example drug delivery system and cancer therapy. e.g., dendrimers, micelles, liposomes and ferritin, etc. (Fig. 13.4).

(2) **Carbon-based nanomaterials**: NMs which mainly formed of carbon are knows as carbon based, e.g., fullerenes, CNTs, carbon nanofibers, carbon black and graphenes. Carbon-based NMs combine the distinct features of sp^2-hybridized carbon with their unique nanoscale physicochemical properties. Carbon-based NMs play a key role in several multidisciplinary fields and used in several applications, including as environmental sensing for the detection of microbiological infections, drug delivery, energy storage, bioimaging, adsorption and photovoltaic devices (Mauter and Elimelech 2008).

(3) **Inorganic nanomaterials**: Inorganic nanomaterials are those made of metals or metal oxides. e.g., metal, ceramic, and semiconductor NPs. Metal NPs can be monometallic, bimetallic, or polymetallic, and they are purely composed of metal precursors. It is possible to synthesize nanoparticles from almost all the metals. Aluminum (Al), iron (Fe), silver (Ag), cobalt (Co), copper (Cu), Nickel (Ni), lead (Pb), zinc (Zn) and gold (Au), are the most frequently utilized metals to synthesize nanoparticles. As compered to bulk materials, metal NPs have been shown to have peculiar chemical, optical, and electrical characteristics. To modify the characteristics of the corresponding metal nanoparticles, metal oxide-based nanoparticles are synthesized. e.g., Aluminium oxide (Al_2O_3), Magnetite (Fe_3O_4), Iron oxide (Fe_2O_3), Zinc oxide (ZnO), Cerium oxide (CeO_2), Silicon dioxide (SiO_2), Titanium oxide (TiO_2), etc. Due to the change in its surface properties, these nanoparticles exhibit a high reactivity and efficiency when compared to their metal counterparts.

13.1.3.3 Nanoparticles

To paraphrase Haydon (2015), "nano-objects having all external dimensions at the nanoscale, where the lengths of the longest and the shortest axes of the nano-object do not differ significantly" is how the International Organisation for Standardisation (ISO) defines nanoparticles (NPs). Nanofibers or nanoplates are alternatives to NPs when the size difference is very substantial, frequently more than three orders of magnitude. There is a wide range of sizes, forms, and architectures among nanoparticles, in addition to the compounds they contain. According to Ealia and Saravanakumar (2017), NPs can have several shapes and sizes, including spherical, cylindrical, tubular, conical, hollow core, spiral, flat, and so on. Their sizes can range from 1 to 100 nm. Nanoparticles (NPs) are well-suited for a variety of uses due to their unique physical and chemical properties that are absent in their larger-dimension equivalents.

Methods for the Synthesis of Nanoparticles

Nanoparticles with improved characteristics and lower manufacturing costs can be synthesised using a number of different traditional and modern processes. Three distinct methods exist for the synthesis of NPs: the top-down, bottom-up, and hybrid approaches. The top-down or destructive method is the first and involves physically or chemically drastically shrinking macroscopic material to nanometric scale. One of the most common top-down approaches to producing NPs is mechanical milling, followed by nanolithography, sputtering, laser ablation, and thermal breakdown. Two common applications of the top-down approach are the creation of structures with long-range order and the establishment of linkages at the macroscopic level (Kuhn et al. 2022). In the second case, NPs are prepared via the assembly of atoms, molecules or smaller particles to clusters or NPs and termed as bottom-up or constructive approach. The most widely utilised bottom-up techniques are chemical vapour deposition (CVD), sol–gel, chemical synthesis, pyrolysis, biosynthesis, and spinning. These methods are frequently employed to produce complex molecular devices, and also in the fabrication of structural composite nanomaterials. Figure 13.5 represents a simplified illustration of the above procedures. A hybrid technique that combines top-down and bottom-up techniques is another option for dealing with technical issues that arise with these two approaches (Kargozar and Mozafari 2018). Hybrid materials, or those that combine organic and inorganic components, are mostly produced via these hybrid process and have gained a lot of attention (Kuhn et al. 2022).

13.1.4 Engineered Nanoparticles for Heavy Metal Remediation

Engineered NPs presented numerous possibilities for heavy metal removal due to their singularity. Consequently, there has been a noticeable uptick in the use of nanomaterials for wastewater treatment over the last 20 years, and several types of nanomaterials have been found to effectively remove heavy metals from water (Zhang et al. 2016a, b; Abdelbasir and Shalan 2019). Due to their small size, large surface area, high adsorption capacity, regeneration efficiency, and quicker rates

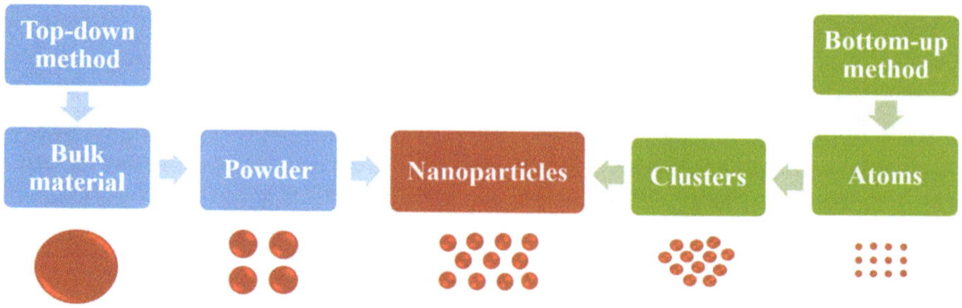

Fig. 13.5 Schematic representation of the top-down and bottom-up methods for nanoparticle preparation

in water treatment, as well as their chemical and thermal stability, adsorption by nanomaterial is considered the best available technology for the reduction of heavy metal contaminants.

Novel and cost-effective technologies for heavy metal remediation have been developed and used as a result of recent breakthroughs in nanotechnology. So, the most encouraging use of nanotechnology for water purification appears to be the creation of tailored nanoparticles that can filter out heavy metals from wastewater. Novel NPs have been the subject of a deluge of scientific and technical articles due to their increasing significance in water purification. Over the last 20 years, there has been a meteoric rise in the number of papers discussing the use of nanoparticles to remove heavy metals. This is indicative of the growing interest among scientists in this topic (Fig. 13.6). Heavy metal removal has been investigated using a variety of NPs, including metal oxides, carbon nanotubes, silica nanoparticles, chitosan, and a number of noble metals (Kaushal and Singh 2017; Wang et al. 2012).

Data on 25 sites that have directly used nanotechnology for site remediation has been made available by the US Environmental Protection Agency (Morris and Willis 2007). A sizable portion of the National Nanotechnology Initiative's (NNI) \$1.4 billion 2020 budget—1.04%—will be devoted to water sustainability initiatives that make use of nanotechnology (NNI 2020). The European Union's Horizon 2020 programme also includes water-related initiatives, such as Preserving Freshwater: Recycling Industrial Waters Industry (IA) (ID: CE-SPIRE-07-2020) and New Biotechnologies for Environmental Remediation (RIA) (ID: CE-BIOTEC-04-2018), as well as the integration of nanotechnology into each of these programmes (European Commission 2017).

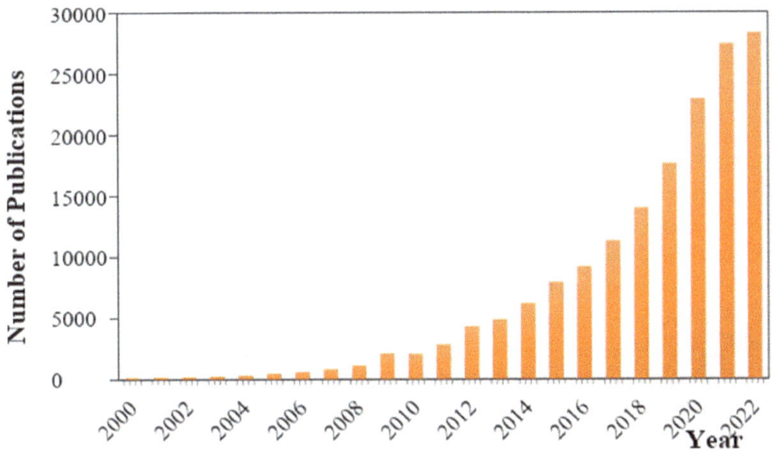

Fig. 13.6 Distribution of the number of publications per year from 2000 to 2022 with the keyword's "nanoparticles", "water" and "heavy metal" from Google Scholar

13.1.4.1 Metal Nanoparticles in Heavy Metal Removal

Over the last several decades, the concept of nanotechnology has rooted in water treatment research and have developed into the next-generation wastewater treatment technique. Nanoparticles especially metal NPs, purely made of metal precursors, are anticipated to play an important part in water purification. Metal nanoparticles (NPs) are used in water treatment for a number of reasons. Among these are their apparent high compatibility with aqueous chemistry, their multifaceted application in drinking water purification (removing heavy metals, organic compounds, and microorganisms), and their superior material properties with functional adaptability. Furthermore, Metal NPs may be created using relatively low-cost preparation processes and their great adsorption capacity and ease of separation and regeneration make them technologically and economically advantageous.

Further notable features include antibacterial capabilities, huge surface area to volume ratio, nanoscalar size, and modifiability of surface area (Musee et al. 2011). The surface atomic concentration increases relative to the accessible internal atomic concentration as the size of the metal NP approaches the nanometer range. The surface area and energy for chemical reactions are greatly enhanced because the unsaturated atoms on the surface can interact with other atoms more easily (Xu et al. 2012; Liu et al. 2005). Aggregation, chemical modification, and protection are all attempts by the system to lower their high surface energy. Consequently, NP surfaces are expected to host a diverse array of species. It may give more chemical reaction potential and highly active adsorption sites than the bulk material. As a result, surface-based sciences open up new avenues of inquiry.

According to several studies (Zhao et al. 2011; Zhang et al. 2012; Cheng et al. 2012; Montazer et al. 2012), metal NPs have great promise as antibacterial agents in many different fields, such as dentistry, synthetic fabrics, food packaging, and wastewater treatment. Due to their outstanding physiochemical characteristics, simplicity of synthesis, and modifiability, metal NPs are becoming more and more intriguing as potential adsorbents for wastewater purification. Consequently, heavy metals in water can be removed using these surface characteristics and their high reactivity.

Metal NPs can be monometallic, bimetallic, or polymetallic depending on how many component particles are present (Hassan and Ghadam 2020).

Monometallic Nanoparticles

As its name suggests, monometallic NPs (MNPs) consist entirely of a single metal. The metal atoms that compose these NPs determine their characteristics. Magnetic nanoparticles (MNPs) can be categorised as either metallic or transition metal nanoparticles (TMNPs), depending on the type of metal atoms present. According to Zhang et al. (2016a, b), Maruthupandy et al. (2017), Gawande et al. (2016), and Lima et al. (2020), these NPs have been widely used in many different disciplines, such as adsorption, optical, catalysis, and antibacterial agents. Similar to metal NPs, they have the unique ability to eliminate harmful aquatic species due to their increased chemical reactivity and surface area. That is why they are so useful for making water safe to drink. Over the years, several synthesis techniques have been studied for the preparation of MNPs, out of which the most often utilised techniques are chemical methods. MNPs have several drawbacks, including easy deactivation, slow absorption efficiency for impurities, their sensitivity to pH etc. (Huang et al. 2013, 2015; Lin et al. 2009; Wang et al. 2015).

Bimetallic Nanoparticles

To increase the reactivity and to overcome the limitations of MNPs, bimetallic NPs (BNPs) were created by the combination of two distinct metals where at least one of them has a nanoscale structure. Extensive investigations on BNPs were began barely a decade ago and have been researched for a number of applications including sensors, optoelectronics, catalysis, medicine, and UV protection gels, etc. Numerous BNPs with controlled size, morphology and compositions have been created because of their unusual magnetic (Singh 2003; Mattei et al. 2002), biological (Hong et al. 2005; Kim et al. 2005), electronic (Jo et al. 2005), and catalytic properties (Fortunelli and Velasco 2002).

According to Rao et al. (2003), the composition, nanometric size, and shape of BNPs are the defining features of these particles. The functioning and range of applications of BNPs are expanded by some surprising new properties compared to their monometallic equivalent (Luo et al. 2011). For instance, BNPs display remarkable properties as a result of the interaction between the two metals. These properties include improved antibacterial, electrical, magnetic, optical, and catalytic activity

(Yadav et al. 2018; Arora et al. 2020; González et al. 2011; Omori et al. 2011; Er et al. 2020). Additionally, they exhibit higher levels of selectivity and stability (Ali et al. 2010; Carrillo-Torres et al. 2016). The distinctive properties of BNPs are primarily shaped by the combined effects of the two metals, which work together to perform many functions (He et al. 2020; Sharma et al. 2019). The correct mix of metals with distinct features can be used to create BNPs with several functions (Srinoi et al. 2018). It is also possible to modify the chemical and physical properties of BNPs by changing their geometrical structures and individual components (Sobal et al. 2003; Kodama et al. 2006).

Additionally, BNPs can be synthesised in a variety of shapes and sizes because of the two metals (Belenov et al. 2017). There are other structural types that can be used to classify these materials, including core–shell, heterogeneous, intermetallic, and multi-shell (Zaleska-Medynska et al. 2016; Ferrando and Jellinek 2009). The combined metals' properties, including electrical and magnetic effects, surface energy, relative bond strength, atomic dimensions, charge transfer, and surface energy, largely dictate the variety of these nanostructures (Liu et al. 2014). Common methods for their synthesis include the sequential reduction of two metal ions or the simultaneous reduction of two metal ions and a coordinating radical; the former usually leads to the creation of alloys, while the latter usually yields core shell NP. Alloyed nanoparticles have better physical and structural properties than their bulk materials, including more solid solubility as particle size decreases.

13.1.5 Application of BNPs in Heavy Metal Removal

In recent years, there has been a meteoric rise in the amount of papers discussing BNPs, and there is a dearth of scientific data on the production and analysis of BNPs using different metal combinations. Recent years have seen a surge in scientific interest in BNPs as a potential solution for heavy metal removal from wastewater (Li et al. 2020; Hassan and Ghadam 2020; Fang et al. 2015; Wang et al. 2015). Many studies have addressed heavy metal removal from water and wastewater; Table 13.4 highlights a few of these. We are unaware of any research that have not focused on iron-based BNPs.

Nanotechnology has recently made strides in investigating BNPs as a potential new class of heavy metal adsorbents, which has piqued the curiosity of many other kinds of NPs. There have been numerous attempts to date to use different BNPs to extract or separate heavy metals from water (Table 13.4). According to several studies (Sharma et al. 2015; Toshima and Yonezawa 1998; Wadhawan et al. 2020), BNPs' enhanced adsorption power is due in large part to their larger surface area compared to monometallic counterparts. Additionally, the interaction between the two metals is crucial to the unique properties of BNPs. For a BNP to be engineered with desirable properties and activity—and hence to remove pollutants effectively—it is crucial to choose the right combination of metals and optimise their arrangement (Arora et al. 2020).

Table 13.4 Selected examples of BNPs for water treatment

S. no.	Type of BNPs	Method of synthesis	Application	References
1	Fe–Ni	Green synthesis	As(V) removal from aqueous solution	Lin et al. (2022)
2	Fe/Cu	Chemical reduction	Removal of hexavalent chromium (Cr(VI))	Mondal et al. (2022)
3	Cu/Fe	Liquid-phase chemical reduction	Removal of Cr(VI) from wastewater	Ye et al. (2021)
4	Fe/Cu	Co-reduction	Simultaneous removal of As(V) and Sb(V) from water	Wang et al. (2022)
5	Fe/Bi	Co-reduction	Removal of Cr(VI) from aqueous environment	Mahar et al. (2022)
6	Fe–Cu-chitosan	Coprecipitation	Antimony removal	Wang et al. (2022)
7	Fe/Al and Al/Fe	Chemical liquid-phase deposition and co-reduction	Remediation of Cr^{6+} contaminated groundwater	Ou et al. (2020)
8	Chitosan–Fe/Ag	Co-reduction	Co-removal of Cd, Ni, and Pb from aqueous solution	Van der Horsta et al. (2021)
9	Mg–Zn	–	Removal of Cr (VI) from industrial wastewater	Rahimi et al. (2021)
10	Fe/Cu	Co-reduction	Removal of As from aqueous solutions	Babaee et al. (2018)
11	Fe–Cu	Co-reduction	Removal of As(V) from synthetic contaminated water	Sepúlveda et al. (2018)
12	Montmorillonite-supported Fe/Ni	Co-reduction	Removal of Cr(VI) from wastewater	Lu et al. (2017)
14	Ni/Fe	Liquid-phase chemical reduction	Cr(VI) removal from synthetic contaminated water	Zhou et al. (2014)
15	Ni/Fe	Co-reduction	Remediation of Pb(II)	Saberi (2012)
16	Au–Fe	Successive reduction	Simultaneous removal of Cd and nitrate from aqueous solution	Su et al. (2014)

(continued)

Table 13.4 (continued)

S. no.	Type of BNPs	Method of synthesis	Application	References
17	Calcium alginate encapsulated Ni–Fe	Co-reduction	Simultaneous removal of Cu(II) and monochlorobenzene from aqueous media	Kuang et al. (2015)
18	Chitosan stabilized Fe/Ni	Liquid-phase reduction	Simultaneous removal of amoxicillin and Cd (II) from aqueous solutions	Weng et al. (2013)

13.1.6 Immobilization of Bimetallic Nanoparticle

Despite this, NPs' small size is a major hurdle to their effective incorporation into water treatment units due to factors such particle leaching, expensive replacement costs, and difficulties in separating them from wastewater. Conventional approaches for NP recovery from water have included nanofiltration, magnetic separation, and ultracentrifugation. However, if we look at typical approaches to water purification, we see that these procedures are tedious and could end up being the most expensive part of the process. Immobilisation or stabilisation of NPs is a common method for overcoming these restrictions. The immobilisation of NPs can be accomplished in several ways. Encapsulating nanoparticles (NPs) with the right materials allows for precise regulation of their release rate and good protection against NP oxidation and aggregation. Most of these methods, however, include binders and undergo high-temperature processing, which reduces their adsorption capability and makes them less cost-effective and less eco-friendly". Hence, in order to potentially remove heavy metals from water and wastewater, it is important to create immobilised NPs that are stable, efficient, easy to synthesise, economical, and reusable.

Silica, alumina, metal oxides, activated carbon, chitosan, alginate beads, and many more solid support materials have been utilised for the immobilisation of metal NPs (Li et al. 2014, 2016; Tabakci and Yilmaz 2008; Mallick and Scurrell 2003; Li and Tang 2014; You et al. 2014; Silva et al. 2002). According to Burguete et al. (2005) and King and Twyman (2002), polymers are among the most attractive substrate choices because they are non-volatile, simple, stable, flexible, and reusable. Because of their excellent mechanical properties, adaptable porosity structures, and chemically attached functional groups, polymers provide excellent support to the nanomaterials (Salem et al. 2022).

Polymer-based nanocomposites (PNCs) are being explored as attractive materials for water and wastewater treatment because they usually incorporate the inherent benefits of both nanoparticles and the polymeric matrix (Bustamante-Torres et al. 2021). One material that has recently attracted a lot of interest for immobilising metallic NPs is the porous polymer microsphere. This material's porosity determines how efficient it is at adsorption, and it also makes adsorbates more accessible by increasing the contact area between the two and having a larger specific surface area (Yang et al. 2015). Zha et al. demonstrated that compared to non-porous solid structured composite particles, multi-porous sulfonated polystyrene/chitosan/Ag microspheres had a greater catalytic degradation rate and were more efficient chromium ion adsorbents (Zha et al. 2015).

Tian et al. (2016) and Albuszis et al. (2016) found that styrene–divinylbenzene (St-DVB) microbeads, a type of porous polymer, have a large surface area, a controllable particle size, excellent physical and chemical durability, are easy to functionalize, and have tuneable pore architectures. Porous structure and surface area play a significant influence in determining the effectiveness of adsorption and release, particularly in water treatment (Wang et al. 2010; Lin et al. 2013; Ali and Gupta 2006).

13.1.7 Remediation of Heavy Metals Using Porous (Styrene–Divinylbenzene)/Cu-Ni Bimetallic Nanocomposite Microspheres

According to our earlier research (Harikumar and Hridya 2021), compared to other nanoadsorbents and individual Cu and Ni NPs, CuNi BNPs exhibit superior heavy metal adsorption effectiveness. In addition, we enhanced its adsorption capacity by immobilising CuNi BNPs in a porous (styrene–divinylbenzene) copolymer, which allowed us to control and reduce the NPs' release rate while effectively protecting them from oxidation and agglomeration (Referenced in Hridya et al. 2023 and 2021, respectively).

A number of fixed bed column tests were performed to assess the immobilised adsorbent's effectiveness. Results demonstrated the feasibility and efficacy of using a fixed bed column to remove various heavy metals from P(St-DVB)/CuNi BNC microbeads. Finding the optimal combination of system parameters—including starting metal concentration, bed height, flow rate, and removal rate—led to an increase in adsorption capacity when concentration and bed height rose, but flow rate fell. For all operating conditions studied, mathematical models like the Adams-Bohart, Yoon-Nelson, and Thomas models were able to successfully predict the column kinetic parameters and the obtained breakthrough. The experimental results were in good agreement with the Yoon-Nelson and Thomas models. Through four cycles of adsorption and desorption, heavy metals were successfully removed from P(St-DVB)/CuNi BNC microbeads using a NaOH solution, demonstrating excellent adsorbent reusability in the bed column. Reducing the active surface area of most NP immobilisation procedures leads to a partial loss of their activity. The porous poly(styrene–divinylbenzene) in P(St-DVB)/CuNi BNC, on the other hand, makes excellent interaction between the heavy metals and the CuNi BNPs. As a result, they may be able to show adsorption efficiency that is on par with CuNi BNPs.

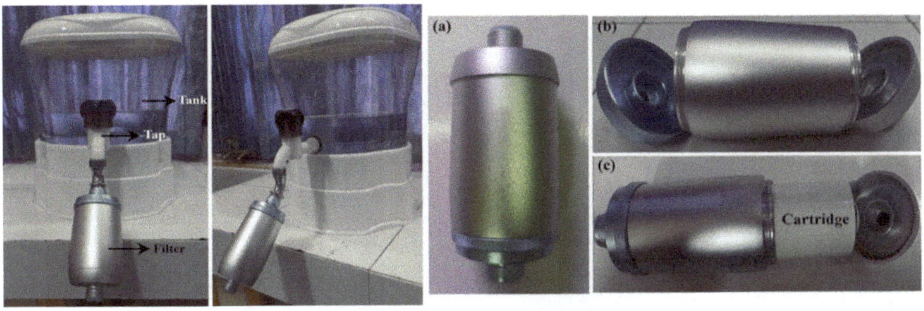

Fig. 13.7 Photographs of apparatus for the water purification device used for filtration containing lead, cadmium, manganese, and zinc contaminated water. **a** A side view of the filter, **b** a view of the apertures, and **c** the filter after the cartridge has been partially inserted

A portable, inexpensive, and easy-to-use system for heavy metal removal was developed after column studies evaluated the effectiveness of P(St-DVB)/CuNi BNC microspheres (Hridya 2023). A storage container, a tap, and a filtration unit make up the three main assemblies that make up the overall setup (Fig. 13.7). A cylindrical fibre container was utilised to store the water that was contaminated. To fill the container with raw water, simply remove the cover. A fiber tap was attached to the storage container, and it connects the storage container to the filtration unit. Filtration unit includes a cylindrical shaped hollow steel jar and a water purification cartridge (Fig. 13.7a). The cartridge is placed inside the steel jar. The steel jar had openings on both sides (Fig. 13.7b), and the cartridge was inserted into the jar through that opening (Fig. 13.7c). Both sides of the steel jar have a screw threaded socket adapted to fit on a water supply outlet tap.

The cartridge is made of fibre and has a cylindrical shape with a diameter of 50 mm and a height of 100 mm (Fig. 13.8). One side of the cartridge has an opening and through that opening, about 20 g of the composite (particle size: 110 m) were packed into the cartridge together with small pebbles, coarse sand, and fine sand (Fig. 13.8). Each layer is separated with medical grade cotton and Gauze.

After that, the cartridge was placed inside the steel jar, and the filtration unit was then assembled in an antigravity fashion. In a typical use, the polluted water is poured in the tank. When we turn on the tap, the pressure due to the water head in the tank drives the water into the water filter via the water tap and the water flow can be adjusted with the help of the tap. Then, the treated water was collected from the outlet of filtration unit (Fig. 13.8).

Using 20 g of the composite, the filled cartridge removed heavy metals (as per WHO guidelines) from approximately 66 L of water at an average flow rate of 25 mL/min, all without regenerating the material. Furthermore, the robustness of the BNCs prevents leaching of the CuNi BNPs and which makes it an excellent option for an industrially practicable material for supplying inexpensive water across the affected regions.

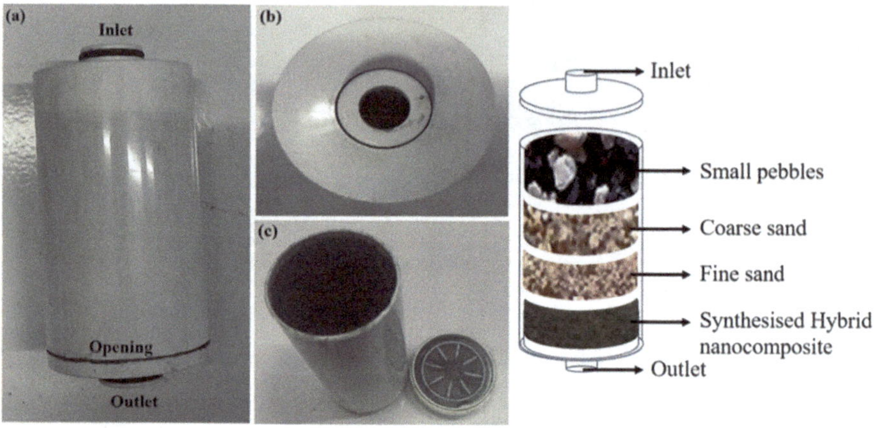

Fig. 13.8 Photographs of cartridge used in the filtration unit for the study

Acknowledgements The authors are gratefully acknowledging the Council of Scientific and Industrial Research (CSIR), India [Sanction no: 09/1195(0001)/2018-EMR-I], Centre for Water Resources Development and Management, Kunnamangalam, Kozhikode and University of Calicut for providing the research facilities and financial support.

References

Abdelbasir SM, Shalan AE (2019) An overview of nanomaterials for industrial wastewater treatment. Korean J Chem Eng 36(8):1209–1225

Akin I, Arslan G, Tor A, Cengeloglu Y, Ersoz M (2011) Removal of arsenate [As (V)] and arsenite [As (III)] from water by SWHR and BW-30 reverse osmosis. Desalination 281:88–92

Albuszis M, Roth PJ, Exnowitz F, Wong DL, Pauer W, Moritz HU (2016) Synthesis and in-depth characterization of reactive, uniform, crosslinked microparticles based on free radical copolymerization of 4-vinylbenzyl azide. Polym Chem 7(5):1168–1180

Ali I, Gupta VK (2006) Advances in water treatment by adsorption technology. Nat Protoc 1(6):2661–2667

Ali SMU, Nur O, Willander M, Danielsson B (2010) A fast and sensitive potentiometric glucose microsensor based on glucose oxidase coated ZnO nanowires grown on a thin silver wire. Sens Actuators B Chem 145(2):869–874

Alluri HK, Ronda SR, Settalluri VS, Bondili JS, Suryanarayana V, Venkateshwar P (2007) Biosorption: an eco-friendly alternative for heavy metal removal. African J Biotechnol 6(25)

Al-Othman ZA, Ali R, Naushad M (2012) Hexavalent chromium removal from aqueous medium by activated carbon prepared from peanut shell: adsorption kinetics, equilibrium and thermodynamic studies. Chem Eng J 184:238–247

Alyüz B, Veli S (2009) Kinetics and equilibrium studies for the removal of nickel and zinc from aqueous solutions by ion exchange resins. J Hazard Mater 167(1–3):482–488

Arora N, Thangavelu K, Karanikolos GN (2020) Bimetallic nanoparticles for antimicrobial applications. Front Chem 8:412

Babaee Y, Mulligan CN, Rahaman MS (2018) Removal of arsenic (III) and arsenic (V) from aqueous solutions through adsorption by Fe/Cu nanoparticles. J Chem Technol Biotechnol 93(1):63–71

Babel S, Kurniawan TA (2003) Low-cost adsorbents for heavy metals uptake from contaminated water: a review. J Hazard Mater 97(1–3):219–243
Belenov SV, Volochaev VA, Pryadchenko VV, Srabionyan VV, Shemet DB, Tabachkova NY, Guterman VE (2017) Phase behavior of Pt–Cu nanoparticles with different architecture upon their thermal treatment. Nanotechnol Russ 12(3):147–155
BIS (2012) Bureau of Indian Standards IS: 10500
Burguete MI, Fraile JM, García-Verdugo E, Luis SV, Martínez-Merino V, Mayoral JA (2005) Polymer-supported bis (oxazolines) and related systems: toward new heterogeneous enantioselective catalysts. Ind Eng Chem Res 44(23):8580–8587
Bustamante-Torres M, Romero-Fierro D, Arcentales-Vera B, Pardo S, Bucio E (2021) Interaction between filler and polymeric matrix in nanocomposites: magnetic approach and applications. Polymers 13(17):2998
Buzea C, Pacheco II, Robbie K (2007) Nanomaterials and nanoparticles: sources and toxicity. Biointerphases 2(4):MR17–MR71
Carrillo-Torres RC, García-Soto MJ, Morales-Chávez SD, Garibay-Escobar A, Hernández-Paredes J, Guzmán R, Barboza-Flores M, Álvarez-Ramos ME (2016) Hollow Au–Ag bimetallic nanoparticles with high photothermal stability. RSC Adv 6(47):41304–41312
Cheng L, Weir MD, Xu HH, Antonucci JM, Kraigsley AM, Lin NJ, Lin-Gibson S, Zhou X (2012) Antibacterial amorphous calcium phosphate nanocomposites with a quaternary ammonium dimethacrylate and silver nanoparticles. Dent Mater 28(5):561–572
Clifford CA, Stinz M, Hodoroaba VD, Unger WE, Fujimoto T (2020) International standards in nanotechnologies. In: Characterization of nanoparticles. Elsevier, pp 511–525
De Kwaadsteniet M, Dobrowsky PH, Van Deventer A, Khan W, Cloete TE (2013) Domestic rainwater harvesting: microbial and chemical water quality and point-of-use treatment systems. Water Air Soil Pollut 224(7):1–19
Ealia SAM, Saravanakumar MP (2017) A review on the classification, characterisation, synthesis of nanoparticles and their application. In: IOP conference series: materials science and engineering, vol 263, no 3. IOP Publishing, p 032019
Er OF, Caglar A, Ulas B, Kivrak H, Kivrak A (2020) Novel carbon nanotube supported Co@ Ag@ Pd formic acid electrooxidation catalysts prepared via sodium borohydride sequential reduction method. Mater Chem Phys 241:122422
Ersahin ME, Ozgun H, Dereli RK, Ozturk I, Roest K, van Lier JB (2012) A review on dynamic membrane filtration: materials, applications and future perspectives. Biores Technol 122:196–206
European Commission (2017) NanoData landscape compilation. Environment. https://doi.org/10.2777/031727
Fang W, Xu C, Zheng J, Chen G, Jiang K (2015) Fabrication of Cu–Ag bimetal nanotube-based copper silicates for enhancement of antibacterial activities. RSC Adv 5(49):39612–39619
Feng X, Wu Z, Chen X (2007) Removal of metal ions from electroplating effluent by EDI process and recycle of purified water. Sep Purif Technol 57(2):257–263
Ferrando R, Jellinek J (2009) Nano-particles as a catalysts incatalytic converters, RL Johnston. Chem Rev 108:845
Feynman RP (1959) Plenty of room at the bottom. In: APS annual meeting
Feynman RP (1960) There's plenty of room at the bottom. Eng Sci 23(5):22–36
Fortunelli A, Velasco AM (2002) A theoretical study of the catalytic properties of Pt/Fe nanoclusters. J Mol Struct: Thoechem 586(1–3):17–27
Gawande MB, Goswami A, Felpin FX, Asefa T, Huang X, Silva R, Zou X, Zboril R, Varma RS (2016) Cu and Cu-based nanoparticles: synthesis and applications in catalysis. Chem Rev 116(6):3722–3811
González E, Arbiol J, Puntes VF (2011) Carving at the nanoscale: sequential galvanic exchange and Kirkendall growth at room temperature. Science 334(6061):1377–1380
Harikumar PS, Hridya TK (2021) Application of CuNi bimetallic nanoparticle as an adsorbent for the removal of heavy metals from aqueous solution. Int J Environ Anal Chem 101(6):869–883

Hassan SA, Ghadam P (2020) Bimetallic nanoparticles with specific insight into nanoremediation
Haydon B (2015) Nanoengineering: a toolbox of standards for health and safety. In: Nanoengineering. Elsevier, pp 557–580
He Z, Zhang Z, Bi S (2020) Nanoparticles for organic electronics applications. Mater Res Express 7(1):012004
Hong R, Fischer NO, Emrick T, Rotello VM (2005) Surface PEGylation and ligand exchange chemistry of FePt nanoparticles for biological applications. Chem Mater 17(18):4617–4621
Hridya T, Varghese E, Harikumar PS (2021) Removal of heavy metals from aqueous solution using porous (Styrene-divinylbenzene)/CuNi bimetallic nanocomposite microspheres. Environ Nanotechnol Monit Manag 16:100606
Hridya T, Antu R, Harikumar PS (2023) Remediation of Pb (II), Cd (II), and Zn (II) from aqueous solutions using porous (styrene–divinylbenzene)/Cu–Ni bimetallic nanocomposite microspheres: continuous fixed-bed column study. Water Sci Technol 87(9):2277–2291
Hridya TK (2023) Synthesis and application of hybrid nanocomposites for the removal of heavy metals and bacteria from contaminated water. Doctoral dissertation, KSCSTE Centre for Water Resources Development and Management
Huang CC, Lo SL, Lien HL (2015) Vitamin B12-mediated hydrodechlorination of dichloromethane by bimetallic Cu/Al particles. Chem Eng J 273:413–420
Huang Q, Liu W, Huang W (2013) Reductive dechlorination of tetrachlorobisphenol A by Pd/Fe bimetallic catalysts. J Hazard Mater 262:634–641
Jo C, Lee JI, Jang Y (2005) Electronic and magnetic properties of ultrathin Fe−Co alloy nanowires. Chem Mater 17(10):2667–2671
Joudeh N, Linke D (2022) Nanoparticle classification, physicochemical properties, characterization, and applications: a comprehensive review for biologists. J Nanobiotechnol 20(1):1–29
Kameda T, Suzuki Y, Yoshioka T (2014) Removal of arsenic from an aqueous solution by coprecipitation with manganese oxide. J Environ Chem Eng 2(4):2045–2049
Kargozar S, Mozafari M (2018) Nanotechnology and nanomedicine: start small, think big. Mater Today: Proc 5(7):15492–15500
Kaushal A, Singh SK (2017) Adsorption phenomenon and its application in removal of lead from wastewater: a review. Int J Hydrol 1(2):38–47
Kim DK, Kan D, Veres T, Normadin F, Liao JK, Kim HH, Lee SH, Zahn M, Muhammed M (2005) Monodispersed Fe–Pt nanoparticles for biomedical applications. J Appl Phys 97(10):10Q918
King AS, Twyman LJ (2002) Heterogeneous and solid supported dendrimer catalysts. J Chem Soc Perkin Trans 1(20):2209–2218
Kodama D, Shinoda K, Sato K, Konno Y, Joseyphus RJ, Motomiya K, Takahashi H, Matsumoto T, Sato Y, Tohji K, Jeyadevan B (2006) Chemical synthesis of sub-micrometer-to nanometer-sized magnetic FeCo dice. Adv Mater 18(23):3154–3159
Kuang Y, Du J, Zhou R, Chen Z, Megharaj M, Naidu R (2015) Calcium alginate encapsulated Ni/Fe nanoparticles beads for simultaneous removal of Cu (II) and monochlorobenzene. J Colloid Interface Sci 447:85–91
Kuhn R, Bryant IM, Jensch R, Böllmann J (2022) Applications of environmental nanotechnologies in remediation, wastewater treatment, drinking water treatment, and agriculture. Appl Nano 3(1):54–90
Li D, Dong M, Besenbacher F, Huang Y, Chen M (2016) The preparation of a recyclable catalyst of silver nanoparticles dispersed in a mesoporous silica nanofiber matrix. RSC Adv 6(70):65613–65618
Li G, Tang Z (2014) Noble metal nanoparticle@ metal oxide core/yolk–shell nanostructures as catalysts: recent progress and perspective. Nanoscale 6(8):3995–4011
Li H, He Z, Ouyang Z, Palchoudhury S, Ingram CW, Harruna II, Li D (2020) Modifying electrical and magnetic properties of single-walled carbon nanotubes by decorating with iron oxide nanoparticles. J Nanosci Nanotechnol 20(4):2611–2616
Li S, Yan X, Yang Z, Yang Y, Liu X, Zou J (2014) Preparation and antibacterial property of silver decorated carbon microspheres. Appl Surf Sci 292:480–487

Li X, Xu H, Chen ZS, Chen G (2011) Biosynthesis of nanoparticles by microorganisms and their applications. J Nanomater

Lima RA, de Souza SLX, Lima LA, Batista ALX, de Araújo JTC, Sousa FFO, Rolim JPML, Bandeira TDJPG (2020) Antimicrobial effect of anacardic acid–loaded zein nanoparticles loaded on Streptococcus mutans biofilms. Braz J Microbiol 51(4):1623–1630

Lin CJ, Liou YH, Lo SL (2009) Supported Pd/Sn bimetallic nanoparticles for reductive dechlorination of aqueous trichloroethylene. Chemosphere 74(2):314–319

Lin Q, Pan J, Lin Q, Liu Q (2013) Microwave synthesis and adsorption performance of a novel crosslinked starch microsphere. J Hazard Mater 263:517–524

Lin Y, Jin X, Khan NI, Owens G, Chen Z (2022) Bimetallic Fe/Ni nanoparticles derived from green synthesis for the removal of arsenic (V) in mine wastewater. J Environ Manag 301:113838

Liu WJ, Qian TT, Jiang H (2014) Bimetallic Fe nanoparticles: recent advances in synthesis and application in catalytic elimination of environmental pollutants. Chem Eng J 236:448–463

Liu Y, Liang P, Guo L (2005) Nanometer titanium dioxide immobilized on silica gel as sorbent for preconcentration of metal ions prior to their determination by inductively coupled plasma atomic emission spectrometry. Talanta 68(1):25–30

Lu H, Tian B, Wang J, Hao H (2017) Montmorillonite-supported Fe/Ni bimetallic nanoparticles for removal of Cr (VI) from wastewater. Chem Eng Trans 60:169–174

Luo D, Zhang G, Liu J, Sun X (2011) Evaluation criteria for reduced graphene oxide. J Phys Chem C 115(23):11327–11335

Mahar AM, Alveroglu E, Balouch A, Talpur FN, Jagirani MS (2022) Fabrication of Fe/Bi bimetallic magnetic nano-oxides (IBBMNOs) as efficient remediator for hexavalent chromium from aqueous environment. Environ Sci Pollut Res 1–15

Mallick K, Scurrell MS (2003) CO oxidation over gold nanoparticles supported on TiO_2 and TiO_2-ZnO: catalytic activity effects due to surface modification of TiO_2 with ZnO. Appl Catal A 253(2):527–536

Marg BZ (2011) Hazardous metals and minerals pollution in India: sources, toxicity and management. A position paper. Indian National Science Academy, New Delhi

Maruthupandy M, Zuo Y, Chen JS, Song JM, Niu HL, Mao CJ, Zhang SY, Shen YH (2017) Synthesis of metal oxide nanoparticles (CuO and ZnO NPs) via biological template and their optical sensor applications. Appl Surf Sci 397:167–174

Matlock MM, Howerton BS, Atwood DA (2002) Chemical precipitation of heavy metals from acid mine drainage. Water Res 36(19):4757–4764

Mattei G, de Julián Fernández C, Mazzoldi P, Sada C, De G, Battaglin G, Sangregorio C, Gatteschi D (2002) Synthesis, structure, and magnetic properties of Co, Ni, and Co−Ni alloy nanocluster-doped SiO_2 films by sol−gel processing. Chem Mater 14(8):3440–3447

Mauter MS, Elimelech M (2008) Environmental applications of carbon-based nanomaterials. Environ Sci Technol 42(16):5843–5859

Meunier N, Drogui P, Montané C, Hausler R, Mercier G, Blais JF (2006) Comparison between electrocoagulation and chemical precipitation for metals removal from acidic soil leachate. J Hazard Mater 137(1):581–590

Mishra SP (2014) Adsorption–desorption of heavy metal ions. Curr Sci 601–612

Mondal A, Arora M, Dubey BK, Mumford K (2022) Comparative assessment of the characteristics and Cr (VI) removal activity of the bimetallic Fe/Cu nanoparticles pre-and post-coated with carboxymethyl cellulose. Chem Eng J 444:136343

Montazer M, Shamei A, Alimohammadi F (2012) Stabilized nanosilver loaded nylon knitted fabric using BTCA without yellowing. Prog Org Coat 74(1):270–276

Morris J, Willis J (2007) US Environmental Protection Agency nanotechnology white paper. US Environmental Protection Agency, Washington, DC

Musee N, Thwala M, Nota N (2011) The antibacterial effects of engineered nanomaterials: implications for wastewater treatment plants. J Environ Monit 13(5):1164–1183

NNI (2020) The national nanotechnology initiative supplement to the president's 2020 budget. https://www.whitehouse.gov/wp-content/uploads/2018/08/The-National-Nanotechnology-Initiative-Supplement-to-the-President%E2%80%99s-2019-Budget.pdf

Nriagu JO, Pacyna JM (1988) Quantitative assessment of worldwide contamination of air, water and soils by trace metals. Nature 333(6169): 134–139

Omori T, Ando K, Okano M, Xu X, Tanaka Y, Ohnuma I, Kainuma R, Ishida K (2011) Superelastic effect in polycrystalline ferrous alloys. Science 333(6038):68–71

Ou JH, Sheu YT, Tsang DC, Sun YJ, Kao CM (2020) Application of iron/aluminum bimetallic nanoparticle system for chromium-contaminated groundwater remediation. Chemosphere 256:127158

Pal SL, Jana U, Manna PK, Mohanta GP, Manavalan R (2011) Nanoparticle: an overview of preparation and characterization. J Appl Pharm Sci 228–234

Pandey J, Pandey U (2009) Atmospheric deposition and heavy metal contamination in an organic farming system in a seasonally dry tropical region of India. J Sustain Agric 33(4):361–378

Pehlivan E, Yanık BH, Ahmetli G, Pehlivan M (2008) Equilibrium isotherm studies for the uptake of cadmium and lead ions onto sugar beet pulp. Biores Technol 99(9):3520–3527

Pokropivny VV, Skorokhod VV (2008) New dimensionality classifications of nanostructures. Physica E 40(7):2521–2525

Prasse C, Zech W, Itanna F, Glaser B (2012) Contamination and source assessment of metals, polychlorinated biphenyls, and polycyclic aromatic hydrocarbons in urban soils from Addis Ababa, Ethiopia. Toxicol Environ Chem 94(10):1954–1979

Rahimi M, Pourmortazavi SM, Zandavar H, Mirsadeghi S (2021) Recyclable methodology over bimetallic zero-valent Mg: Zn composition for hexavalent chromium remediation via batch and flow systems in industrial wastewater: an experimental design. J Market Res 11:1–18

Rao CNR, Kulkarni GU, Thomas PJ, Edwards PP (2003) Size-dependent chemistry: properties of nanocrystals. In: Advances in chemistry: a selection of CNR Rao's Publications (1994–2003), pp 227–233

Ren X, Li J, Tan X, Shi W, Chen C, Shao D, Wen T, Wang L, Zhao G, Sheng G, Wang X (2014) Impact of Al_2O_3 on the aggregation and deposition of graphene oxide. Environ Sci Technol 48(10):5493–5500

Ribeiro MS, de Melo LS, Farooq S, Baptista A, Kato IT, Nunez SC, de Araujo RE (2018) Photodynamic inactivation assisted by localized surface plasmon resonance of silver nanoparticles: In vitro evaluation on Escherichia coli and Streptococcus mutans. Photodiagn Photodyn Ther 22:191–196

Roduner E (2006) Size matters: why nanomaterials are different. Chem Soc Rev 35(7):583–592

Saberi A (2012) Comparison of Pb^{2+} removal efficiency by zero valent iron nanoparticles and Ni/Fe bimetallic nanoparticles. Iranian (Iranica) J Energy Environ 3(2)

Saleh TA (2020) Nanomaterials: classification, properties, and environmental toxicities. Environ Technol Innov 20:101067

Salem SS, Hammad EN, Mohamed AA, El-Dougdoug W (2022) A comprehensive review of nanomaterials: types, synthesis, characterization, and applications. Biointerface Res Appl Chem 13(1):41

Schwarzenbach RP, Egli T, Hofstetter TB, Von Gunten U, Wehrli B (2010) Global water pollution and human health. Annu Rev Environ Resour 35(1):109–136

Sepúlveda P, Rubio MA, Baltazar SE, Rojas-Nunez J, Llamazares JS, Garcia AG, Arancibia-Miranda N (2018) As (V) removal capacity of FeCu bimetallic nanoparticles in aqueous solutions: the influence of Cu content and morphologic changes in bimetallic nanoparticles. J Colloid Interface Sci 524:177–187

Sharma G, Naushad M, Kumar A, Devi S, Khan MR (2015) Lanthanum/cadmium/polyaniline bimetallic nanocomposite for the photodegradation of organic pollutant. Iran Polym J 24(12):1003–1013

Sharma G, Kumar A, Sharma S, Naushad M, Dwivedi RP, ALOthman ZA, Mola GT (2019) Novel development of nanoparticles to bimetallic nanoparticles and their composites: a review. J King Saud Univ Sci 31(2):257–269

Silva AR, Freitas MMA, Freire C, de Castro B, Figueiredo JL (2002) Heterogenization of a functionalized copper (II) Schiff base complex by direct immobilization onto an oxidized activated carbon. Langmuir 18(21):8017–8024

Simeonidis K, Martinez-Boubeta C, Zamora-Pérez P, Rivera-Gil P, Kaprara E, Kokkinos E, Mitrakas M (2019) Implementing nanoparticles for competitive water purification. Environ Chem Lett 17(2):705–719

Singh A, Prasad SM (2015) Remediation of heavy metal contaminated ecosystem: an overview on technology advancement. Int J Environ Sci Technol 12(1):353–366

Singh N, Gupta SK (2016) Adsorption of heavy metals: a review. Int J Innov Res Sci Eng Technol 5(2):2267–2281

Singh PP (2003) Relativity and magnetism in Ni–Pd and Ni–Pt alloys. J Magn Magn Mater 261(3):347–352

Singh R, Gautam N, Mishra A, Gupta R (2011) Heavy metals and living systems: an overview. Indian J Pharmacol 43(3):246

Sobal NS, Ebels U, Möhwald H, Giersig M (2003) Synthesis of core−shell PtCo nanocrystals. J Phys Chem B 107(30):7351–7354

Srinoi P, Chen YT, Vittur V, Marquez MD, Lee TR (2018) Bimetallic nanoparticles: enhanced magnetic and optical properties for emerging biological applications. Appl Sci 8(7):1106

Su Y, Adeleye AS, Huang Y, Sun X, Dai C, Zhou X, Zhang Y, Keller AA (2014) Simultaneous removal of cadmium and nitrate in aqueous media by nanoscale zerovalent iron (nZVI) and Au doped nZVI particles. Water Res 63:102–111

Sun Y, Yang S, Chen Y, Ding C, Cheng W, Wang X (2015) Adsorption and desorption of U (VI) on functionalized graphene oxides: a combined experimental and theoretical study. Environ Sci Technol 49(7):4255–4262

Tabakci M, Yilmaz M (2008) Sorption characteristics of Cu (II) ions onto silica gel-immobilized calix [4] arene polymer in aqueous solutions: batch and column studies. J Hazard Mater 151(2–3):331–338

Tarafdar JC, Sharma S, Raliya R (2013) Nanotechnology: interdisciplinary science of applications. African J Biotechnol 12(3)

Tchounwou PB, Yedjou CG, Patlolla AK, Sutton DJ (2012) Heavy metal toxicity and the environment. Mol Clin Environ Toxicol 133–164

Tian Q, Yu D, Zhu K, Hu G, Zhang L, Liu Y (2016) Multi-hollow polymer microspheres with enclosed surfaces and compartmentalized voids prepared by seeded swelling polymerization method. J Colloid Interface Sci 473:44–51

Toshima N, Yonezawa T (1998) Bimetallic nanoparticles—novel materials for chemical and physical applications. New J Chem 22(11):1179–1201

Tripathi R, Sodaye S, Sudarshan K (2015) Non-compound nucleus fission in actinide and pre-actinide regions. Pramana 85(2):315–322

Tyagi I, Gupta VK, Sadegh H, Ghoshekandi RS, Makhlouf AH (2017) Nanoparticles as adsorbent; a positive approach for removal of noxious metal ions: a review. Sci Technol Dev 34(3):195–214

USEPA (U.S. Environmental Protection Agency) (2008) Draft nanomaterial research strategy. EPA/600/S-08/002. U.S. Environmental Protection Agency, Office of Research and Development, Washington, DC

USEPA (U.S. Environmental Protection Agency) (2019) Drinking water health advisories for PFOA and PFOS

Van der Horsta C, Silwanaa B, Makombea M, Iwuohab E, Somerseta V (2021) Application of a chitosan bimetallic nanocomposite for the simultaneous removal of cadmium, nickel, and lead from aqueous solution. Desalin Water Treat 220:168–181

Vodyanitskii YN (2010) The role of iron in the fixation of heavy metals and metalloids in soils: a review of publications. Eurasian Soil Sci 43(5):519–532

Wadhawan S, Jain A, Nayyar J, Mehta SK (2020) Role of nanomaterials as adsorbents in heavy metal ion removal from waste water: a review. J Water Process Eng 33:101038

Walsh FC, Reade GW (1994) Electrochemical techniques for the treatment of dilute metal-ion solutions. In: Studies in environmental science, vol 59. Elsevier, pp 3–44

Wang B, Wan Y, Zheng Y, Lee X, Liu T, Yu Z, Huang J, Ok YS, Chen J, Gao B (2019) Alginate-based composites for environmental applications: a critical review. Crit Rev Environ Sci Technol 49(4):318–356

Wang H, Sheng KC, Lan T, Adl M, Qian XQ, Zhu SM (2010) Role of surface treatment on water absorption of poly (vinyl chloride) composites reinforced by Phyllostachys pubescens particles. Compos Sci Technol 70(5):847–853

Wang J, Chen C (2009) Biosorbents for heavy metals removal and their future. Biotechnol Adv 27(2):195–226

Wang S, Peng Y (2010) Natural zeolites as effective adsorbents in water and wastewater treatment. Chem Eng J 156(1):11–24

Wang T, Jiao Y, He M, Ouyang W, Lin C, Liu X (2022) Facile co-removal of As (V) and Sb (V) from aqueous solution using Fe-Cu binary oxides: structural modification and self-driven force field of copper oxides. Sci Total Environ 803:150084

Wang X, Guo Y, Yang L, Han M, Zhao J, Cheng X (2012) Nanomaterials as sorbents to remove heavy metal ions in wastewater treatment. J Environ Anal Toxicol 2(7):154–158

Wang Z, Huang W, Fennell DE (2015) Rapid dechlorination of 1, 2, 3, 4-TCDD by Ag/Fe bimetallic particles. Chem Eng J 273:465–471

Weber B, Holz F (1991) Landfill leachate treatment by reverse osmosis. In: Effective industrial membrane processes: benefits and opportunities. Springer, Dordrecht, pp 143–154

Weng X, Lin S, Zhong Y, Chen Z (2013) Chitosan stabilized bimetallic Fe/Ni nanoparticles used to remove mixed contaminants-amoxicillin and Cd (II) from aqueous solutions. Chem Eng J 229:27–34

WHO (2008) Guidelines for drinking-water quality: second addendum. Vol. 1, Recommendations. World Health Organization

WHO, Edition F (2011) Guidelines for drinking-water quality. WHO Chronicle 38(4):104–108

Xing Y, Chen X, Wang D (2007) Electrically regenerated ion exchange for removal and recovery of Cr (VI) from wastewater. Environ Sci Technol 41(4):1439–1443

Xu P, Zeng GM, Huang DL, Feng CL, Hu S, Zhao MH, Lai C, Wei Z, Huang C, Xie GX, Liu ZF (2012) Use of iron oxide nanomaterials in wastewater treatment: a review. Sci Total Environ 424:1–10

Yadav N, Jaiswal AK, Dey KK, Yadav VB, Nath G, Srivastava AK, Yadav RR (2018) Trimetallic Au/Pt/Ag based nanofluid for enhanced antibacterial response. Mater Chem Phys 218:10–17

Yang Y, Liao H, Tong Z, Wang C (2015) Porous Ag/polymer composite microspheres for adsorption and catalytic degradation of organic dyes in aqueous solutions. Compos Sci Technol 107:137–144

Ye J, Wang Y, Xu Q, Wu H, Tong J, Shi J (2021) Removal of hexavalent chromium from wastewater by Cu/Fe bimetallic nanoparticles. Sci Rep 11(1):1–11

You J, Zhao C, Cao J, Zhou J, Zhang L (2014) Fabrication of high-density silver nanoparticles on the surface of alginate microspheres for application in catalytic reaction. J Mater Chem A 2(22):8491–8499

Zaleska-Medynska A, Marchelek M, Diak M, Grabowska E (2016) Noble metal-based bimetallic nanoparticles: the effect of the structure on the optical, catalytic and photocatalytic properties. Adv Coll Interface Sci 229:80–107

Zamboulis D, Peleka EN, Lazaridis NK, Matis KA (2011) Metal ion separation and recovery from environmental sources using various flotation and sorption techniques. J Chem Technol Biotechnol 86(3):335–344

Zha T, Song L, Chen P, Nie W, Zhou Y (2015) Nonsolvent/solvent-induced phase separation to multi-porous sulfonated polystyrene/chitosan/silver particles and their application in adsorbing chromium ion (III) and reduction of methylene blue. Colloids Surf A 481:423–430

Zhang F, Chen X, Wu F, Ji Y (2016a) High adsorption capability and selectivity of ZnO nanoparticles for dye removal. Colloids Surf A 509:474–483

Zhang M, Zhang K, De Gusseme B, Verstraete W (2012) Biogenic silver nanoparticles (bio-Ag0) decrease biofouling of bio-Ag0/PES nanocomposite membranes. Water Res 46(7):2077–2087

Zhang Y, Wu B, Xu H, Liu H, Wang M, He Y, Pan B (2016b) Nanomaterials-enabled water and wastewater treatment. Nanoimpact 3:22–39

Zhao L, Wang H, Huo K, Cui L, Zhang W, Ni H, Zhang Y, Wu Z, Chu PK (2011) Antibacterial nanostructured titania coating incorporated with silver nanoparticles. Biomaterials 32(24):5706–5716

Zhou S, Li Y, Chen J, Liu Z, Wang Z, Na P (2014) Enhanced Cr (VI) removal from aqueous solutions using Ni/Fe bimetallic nanoparticles: characterization, kinetics and mechanism. RSC Adv 4(92):50699–50707

SPRINGER NATURE

GPSR Compliance

The European Union's (EU) General Product Safety Regulation (GPSR) is a set of rules that requires consumer products to be safe and our obligations to ensure this.

If you have any concerns about our products, you can contact us on ProductSafety@springernature.com

In case Publisher is established outside the EU, the EU authorized representative is:

Springer Nature Customer Service Center GmbH
Europaplatz 3
69115 Heidelberg, Germany

FSC
www.fsc.org
MIX
Papier | Fördert gute Waldnutzung
FSC® C083411

Kolibri 360 GmbH
Ferdinand-Jühlke-Straße 7
99095 Erfurt, Deutschland
produktsicherheit@kolibri360.de